Evolutionary Biology: Applied Concepts and Mechanisms

Evolutionary Biology: Applied Concepts and Mechanisms

Editor: Maia Gustin

R CALLISTO REFERENCE

www.callistoreference.com

Callisto Reference,
118-35 Queens Blvd., Suite 400,
Forest Hills, NY 11375, USA

Visit us on the World Wide Web at:
www.callistoreference.com

ISBN: 978-1-64116-202-9 (Hardback)

Cataloging-in-Publication Data

Evolutionary biology : applied concepts and mechanisms / edited by Maia Gustin.
        p. cm.
Includes bibliographical references and index.
ISBN 978-1-64116-202-9
1. Evolution (Biology). 2. Biology. I. Gustin, Maia.
QH366.2 .E96 2019
576.8--dc23

Table of Contents

Preface

The purpose of the book is to provide a glimpse into the dynamics and to present opinions and studies of some of the scientists engaged in the development of new ideas in the field from very different standpoints. This book will prove useful to students and researchers owing to its high content quality.

Life on Earth has undergone massive transformations, from a single common ancestor to the emergence of diverse organisms that now inhabit the Earth. This is due to various evolutionary processes, such as common descent, natural selection and speciation. Besides these, sexual selection, biogeography and genetic drift are the mechanisms contributing to evolution. Modern evolutionary biology explores these factors along with the genetic architecture of adaptation and molecular evolution. This field integrates studies from other domains, including phylogenetics, biological systematics and paleobiology. This book covers in detail some existing theories and innovative concepts revolving around evolutionary biology. The topics included in this book on the concepts and mechanisms of evolutionary biology are of utmost significance and bound to provide incredible insights to readers. It is appropriate for students seeking detailed information in this area as well as for experts.

At the end, I would like to appreciate all the efforts made by the authors in completing their chapters professionally. I express my deepest gratitude to all of them for contributing to this book by sharing their valuable works. A special thanks to my family and friends for their constant support in this journey.

Editor

Climate drives adaptive genetic responses associated with survival in big sagebrush (*Artemisia tridentata*)

Lindsay Chaney[1] [ID] | Bryce A. Richardson[2] | Matthew J. Germino[3]

[1]Plant and Wildlife Sciences, Brigham Young University, Provo, UT, USA

[2]USDA Forest Service, Rocky Mountain Research Station, Provo, UT, USA

[3]U.S. Geological Survey, Forest and Rangeland Ecosystem Science Center, Boise, ID, USA

Correspondence
Lindsay Chaney, Department of Biology, Snow College, Ephraim, UT, USA.
Email: lindsay.chaney@snow.edu

Present address
Lindsay Chaney, Department of Biology, Snow College, Ephraim, UT, USA

Funding information
Great Basin Native Plant Program; USDA Forest Service National Fire Plan

Abstract

A genecological approach was used to explore genetic variation for survival in *Artemisia tridentata* (big sagebrush). *Artemisia tridentata* is a widespread and foundational shrub species in western North America. This species has become extremely fragmented, to the detriment of dependent wildlife, and efforts to restore it are now a land management priority. Common-garden experiments were established at three sites with seedlings from 55 source-populations. Populations included each of the three predominant subspecies, and cytotype variations. Survival was monitored for 5 years to assess differences in survival between gardens and populations. We found evidence of adaptive genetic variation for survival. Survival within gardens differed by source-population and a substantial proportion of this variation was explained by seed climate of origin. Plants from areas with the coldest winters had the highest levels of survival, while populations from warmer and drier sites had the lowest levels of survival. Survival was lowest, 36%, in the garden that was prone to the lowest minimum temperatures. These results suggest the importance of climatic driven genetic differences and their effect on survival. Understanding how genetic variation is arrayed across the landscape, and its association with climate can greatly enhance the success of restoration and conservation.

KEYWORDS
adaptation, atmospheric decoupling, cold adaptations, minimum temperatures, polyploidy, population differentiation, survival analysis

1 | INTRODUCTION

Adaptive differentiation among populations within a species range is well documented. Examples date back to the classical studies of Turesson (1925) and Clausen, Keck, & Hiesey (1940) and continue into recent literature (Anderson, Lee, Rushworth, Colautti, & Mitchell-Olds, 2013; Bischoff et al., 2006; Galen, Shore, & Deyoe, 1991; Galloway & Fenster, 2000; Hereford, 2009; Kawecki & Ebert, 2004; Leimu & Fischer, 2008; Pratt & Mooney, 2013; Vergeer & Kunin, 2013). In species with large geographic ranges, it is common to have populations that occupy different climatic and ecological conditions. Divergent selection can subsequently result in the evolution of phenotypic traits that are better adapted to their different habitats (Chevin & Lande, 2011; Kawecki & Ebert, 2004). Climate variation is a strong selective agent that leads to population differentiation (Jump & Peñuelas, 2005; McKay et al., 2008; R. C. Johnson, Cashman, & Vance-Borland, 2012). While climatic factors largely determine a species range (Sexton, McIntyre, Angert, & Rice, 2009), climatic differences can also stimulate local adaptation, population divergence, and, in some cases, speciation (Hua & Wiens, 2013).

Genecological studies (sensu Turesson, 1923) are a primary means for assessing how landscape and spatial variation causes genetic differences among plant populations. For example, cold temperatures have been found to be an important factor affecting adaptation in many temperate plant species, influencing how and when plants grow to maximize water availability or avoid freezing (Richardson et al., 2014; St. Clair, Mandel, & Vance-Borland, 2005). Responses to the environment can then be used to create geographically delineated seed zones to help ensure that seed can be transferred among areas of similar climate, increasing odds of appropriately adapted vegetation and successful restoration (Bower, St. Clair, & Erickson, 2014). Recent genecology studies have found that winter minimum temperatures are a primary agent of genetic variation in conifers (Rehfeldt, 2004; Rehfeldt & Jaquish, 2010; Rehfeldt, Jaquish, Sáenz-Romero, et al., 2014), native grasses (R. C. Johnson et al., 2012; St. Clair, Kilkenny, Johnson, Shaw, & Weaver, 2013), and shrubs (Horning, McGovern, Darris, Mandel, & Johnson, 2010; Richardson et al., 2014). However, sufficient information on climatic adaptations is lacking in the majority of native grasses, forbs, and shrubs used for restoration purposes.

Artemisia tridentata (big sagebrush) is a widespread desert shrub that dominates much of western North America. Three predominant subspecies of *A. tridentata* are defined by ecological niches affected principally by climate, soil type, and soil depth (McArthur, 1994). Common gardens or variation in subspecies habitats has revealed genetic divergence across local gradients, such as across hillslopes or topographic positions (Graham, Freeman, & McArthur, 1995; Wang, McArthur, & Freeman, 1999; Wang, McArthur, Sanderson, Graham, & Freeman, 1997) that can be at least partly explained by differences in climate (Kolb & Sperry, 1999). Local adaptation within (Brabec et al., 2015) and among (Brabec, Germino, & Richardson, 2016) subspecies of *A. tridentata* was evident from survival analysis of a small number of populations (four and 11, respectively). These considerations and previous findings suggest a high likelihood of local climate adaptation in big sagebrush, although a comprehensive, range-wide assessment is needed. As a foundational species in the Great Basin ecosystems (Prevéy, Germino, Huntly, & Inouye, 2010), it is host to sagebrush-obligate wildlife, such as the greater sage-grouse (*Centrocercus urophansianus*), and is of conservation concern (Knick & Connelly, 2011). Millions of hectares of sagebrush have been eliminated during the past century due to change in land use (e.g., grazing), exotic plant invasion, and increased frequency of wildfire (Miller et al., 2011). Restoration of *A. tridentata* ecosystems through seeding or planting has occurred on millions of hectares in the western United States in recent decades and continues to be a management priority. Restoration success of sagebrush is contingent on a number of factors including weather, soils, seed mix composition, and seeding and planting techniques (Hardegree et al., 2011). Genetic adaptation, however, is by far the most critical factor for long-term resiliency of these ecosystems, underscoring the need to better understand how climate affects genetic variation in this species.

In the current study, we assess adaptive genetic variation within and between 55 source-populations of *A. tridentata* using common-garden experiments at three sites. The use of common gardens allows researchers to test the genetic basis for phenotypic differences among a relatively large number of plant populations. In genecology studies, geospatial patterns of these genetic traits are then related to plant source climates using regression and bioclimatic modeling. In this study, we address two main questions: (i) How is survivorship influenced by genetics, environment and/or their interactions? (ii) What are the climate patterns that affect survivorship? These results will be fundamental in understanding adaptive genetic variation and fitness in *A. tridentata*, which are essential to restoration in a changing climate.

2 | METHODS

2.1 | Source seed collection

Artemisia tridentata includes three predominant and widely recognized subspecies and two cytotype (ploidy) levels. Subspecies *tridentata* (basin big sagebrush) is diploid or tetraploid, grows in lower elevations with deep soil, and exhibits rapid growth and tall stature. Subspecies *wyomingensis* (Wyoming big sagebrush) is tetraploid, occurs in relatively warm and dry areas with shallow soils, and exhibits slow growth and a shorter stature. Subspecies *vaseyana* (mountain big sagebrush) is diploid or tetraploid, occurs in cooler and relatively mesic conditions at higher elevations, and has a compact growth form (McArthur & Sanderson, 1999). Tetraploids are polyphyletic, deriving from multiple events within and between subspecies (Richardson, Page, Bajgain, Sanderson, & Udall, 2012). Seeds from each of the three subspecies of *A. tridentata* were collected from 55 sites, hereafter referred to as populations, in the fall of 2009 (Table S1). A population typically consisted of half-sib families collected from eight to ten parents. These seeds were collected at random over an area approximately 1 ha. Subspecies were identified by morphology (height and stature), ultraviolet iridescence (following the methods of McArthur, Welch, & Sanderson, 1988), cytotype, molecular genetic analyses (Richardson et al., 2012), and volatile organic compounds (Jaeger, Runyon, & Richardson, 2016). Cytotype was typically uniform within a source-population, but could vary among subspecies (Table S1), with the exception of *wyomingensis* that is only tetraploid.

2.2 | Experimental design

A five-year common-garden experiment was designed with three sites to understand how genetic and environmental factors contribute to *A. tridentata* survival (Table 1; Fig. S1). Garden sites were chosen based on the following criteria: (i) climates that were representative of each of the subspecies, (ii) adjacent to weather stations, and (iii) accessibility for data collection and maintenance. The garden in Ephraim, Utah, USA, is a relatively cold basin site with a dry climate typical of ssp. *tridentata*; the garden in Majors Flat, Utah, USA, is a higher-elevation mountainous site with a mesic climate typical of ssp. *vaseyana*; and the garden in Orchard, Idaho, USA, is a warm plain site with a dry climate typical of ssp. *wyomingensis*. Weather data at the common gardens were obtained from adjacent weather stations (USDA, Natural Resources Conservation Service and Forest Service).

TABLE 1　Geographic and climatic attributes of the three common gardens

Garden	Latitude	Longitude	Elevation (m)	MTCM (oC)	MTWM (oC)	TDIFF (oC)	MAP (mm)	SMRP (mm)
Ephraim, Utah	39.369	−111.578	1690	−9.7	21.2	30.9	276	54
Majors Flat, Utah	39.339	−111.520	2105	−4.7	20.8	25.5	442	72
Orchard, Idaho	43.322	−115.998	974	−2.9	25.0	27.9	257	4

MTCM, mean temperature of the coldest month; MTWM, mean temperature of the warmest month; TDIFF, temperature difference (MTWM-MTCM); MAP, mean annual precipitation; SMRP, summer precipitation (July and August).
Climate data are based on values from 2010 to 2013 water years.

Seeds were collected from 55 wild source-populations found throughout the species distribution (Table S1; Fig. S1). After collection, seeds were cleaned and placed into a freezer until sowing. Replicate seeds from each population were sown into six-inch cone-tainers with a custom soil mix for native plants, established in a greenhouse for approximately 3 months, and cold hardened outside for 2 weeks. Due to differences in climate (Table 1), the Ephraim and Orchard gardens were planted in late April, and the Majors Flat garden was planted in early June 2010. The germination and planting dates for Majors Flat seedlings were delayed to ensure plants were approximately the same age when transplanted. Seedlings were planted in a completely randomized design in a lightly tilled plot that had an above and below-ground fence to discourage ungulate and rabbit herbivores. Due to variation in seed yield and germination, populations in each garden ranged from 1 to 11 individuals with an average of 7.8 plants per population (Table S1). Plant spacing was 1.5 m between rows and 1 m within rows, and a border row was planted around the perimeter to minimize edge effects. To ensure establishment, plants were watered periodically during the first growing season in 2010.

2.3 | Data collection

Phenotypic measurements focused on A. tridentata survival in the three gardens. A plant was considered dead when all foliage was missing and no re-greening occurred during the growing season. Mortality for each experimental plant was recorded in each garden an average of 14.3 times during the experiment (approximately four census dates each year). A small proportion (<6%) of outplanted plants were excluded from the experiment because they did not survive the first growing season and thus could not address our main question of how climate affects sagebrush survival, leaving a total sample number of n = 1,299. The number of excluded and total number of plants at each garden are as follows: 10/459, Ephraim, 26/458, Majors Flat, 42/460, Orchard.

Climate for each source-population was utilized to determine how source climate affects survival. Geographic coordinates and elevation for each source-population were used to estimate climate variables from thin plate splines using ANUSPLIN v 4.1 (Hutchinson, 2000) based on normalized monthly means from years 1961 to 1990 with a 0.0083° resolution (~1 km^2) (Crookston & Rehfeldt, 2012; Rehfeldt, 2006). A total of 40 climate variables were examined for this study including 18 yearly and season temperature and precipitation indices

and 22 interactions among different variables that can be important to plant adaptation (Table S2). This follows the methods of other similar ecological genetic studies (e.g., Bansal, Harrington, Gould, & St.Clair, 2015; Joyce & Rehfeldt, 2013; Ledig, Rehfeldt, Sáenz-Romero, & Flores-López, 2010; Rehfeldt, Jaquish, López-Upton, et al., 2014; Richardson et al., 2014).

2.4 | Data analysis

Data were first analyzed to assess significance of genetics, environment, and their interaction for survival in A. tridentata across the three common gardens. This analysis was performed using a generalized linear mixed model with a quasi-binomial error distribution on survival proportion of each source-population across the three gardens. A multivariate response variable with number of survivors and the number of dead for each source-population was used to account for differences in sample sizes. Estimates of genetic effects were specified as fixed effects and included plant subspecies:cytotype group, and source-population climate variables. Estimates of environment and genetics by environment were specified as random effects and included garden and the interaction of garden and subspecies:cytotype group. Model was fit using the glmer function from the lme4 package (Bates, Mächler, Bolker, & Walker, 2015) with the objective to find the model with the highest predictive value, combined with the fewest number of climate variables. To do this, methods similar to R. C. Johnson et al. (2012) were used. More specifically, a forward selection procedure was used to select the source-population climate variables that best explain variation for survival in the model and variables were no longer added when not significant (p > .05). Significance for each of the effects was calculated based on likelihood-ratio chi-square tests. Conditional and marginal R^2 values were calculated with the rsquared.glmm function (Nakagawa & Schielzeth, 2013); conditional R^2 describes the variation explained by both fixed and random effects; and a marginal R^2 describes fixed effects alone. To provide a more accurate model, only populations with more than two samples were evaluated, resulting in n = 152 populations among the three gardens. This model is also referred to as the genecological model and is used in the mapping discussed below.

Next, a more detailed survival analysis was performed to determine the rate of survival within the Ephraim garden. To do this, the survival package in R (Therneau & Lumley, 2014) was utilized to fit a parametric survival model. This method accounts for the interval and right-censored nature of the data (i.e., the date of death is after some known date). Survival rate was determined to have a lognormal distribution, based on a lower AIC value and higher maximum likelihood of our statistical model. Kaplan–Meier estimators were used to generate survival curves for each subspecies:cytotype group. Differences in survival curves were tested using the log-rank test, and post hoc pairwise comparisons were performed with a Bonferroni-corrected confidence level for multiple comparisons. Median survival and probability of survival at different time intervals were calculated for both subspecies:cytotype group and source-population. All data analyses were performed in the R statistical environment (v 3.2.1; R Core Team 2015). Analysis code and data are available for public access (scripts: https://github.com/lchaney/Sagebrush_Mort; data: http://dx.doi.org/10.5061/dryad.32s2t).

2.5 | Climate mapping

A map was generated for the probability of survival utilizing the genetic effects of the generalized linear mixed model. Probability of survival was calculated for individual cell values, $0.00833°$ (approximately 1 km^2), using the *yalmpute* package in R (Crookston & Finley, 2007). The generalized linear mixed model was projected within the climatic niche boundaries of subspecies *wyomingensis* (Still & Richardson, 2015) using QGIS (QGIS Development Team 2015). This was completed using the log-link transformed coefficients of the intercept and slopes of the fixed effects in the generalized linear mixed model (i.e., the slope of the two climate variables and *wyomingensis*). We focused on subspecies *wyomingensis* because it is the subspecies of greatest conservation concern and restoration investment. Further, distributions of subspecies *tridentata* and *wyomingensis* can often be sympatric at this spatial scale, with the distinction that *tridentata* is usually controlled by local topographic features such as soil depth and higher moisture features (e.g., fence lines, roadways, washes) that provide additional moisture and are difficult to model (Still & Richardson, 2015).

3 | RESULTS

Survival of *A. tridentata* varied in the three common gardens as a function of environment, genetics, and their interaction. Survival was significantly different between gardens, indicating an effect of environment on *A. tridentata* survival (χ_1^2 = 15.417, *p* < .001; Table 2). Survival at Ephraim was much lower than the other two gardens; 36% of *A. tridentata* at Ephraim survived compared to 85% and 78% at Majors Flat and Orchard, respectively (Fig. S2). The genetic influence on survival was environmentally dependent, as indicated by a significant subspecies:cytotype group by garden effect (χ_1^2 = 14.107, *p* ≤ .001; Table 2) which denotes differences in phenotypic plasticity by subspecies:cytotype group. This was expected considering each garden represented the habitat type for a different subspecies. Subspecies:cytotype group was a significant predictor of survival in *A. tridentata*, indicating a genetic component of survival (χ_4^2 = 15.372,

TABLE 2 Generalized linear mixed genecological model that explains effects of genetics, population climate of origin and subspecies:cytotype group, and environment, garden, on survival of *A. tridentata* across the three common gardens

	df	Chi-square	p
Fixed effects			
TDIFF	1	32.764	<.001
SMRP	1	3.712	.054
Subspecies:cytotype	4	15.372	.004
Random effects			
Garden	1	15.417	<.001
Garden × subspecies:cytotype	1	14.107	<.001

TDIFF: temperature difference between mean temperature in the coldest month and mean temperature in the warmest month; SMRP: summer precipitation (July and August).

p = .004; Table 2). Evidence of genetic effects were further shown by population differentiation, specifically significant effects of climate of populations' origins on their survival. Two source-population climate variables, TDIFF (temperature difference between mean temperature in the coldest month and mean temperature in the warmest month) and marginally significant SMRP (summer precipitation, July and August), were found to best explain variation in survival (χ_1^2 = 32.764, *p* < .001 and χ_1^2 = 3.712, *p* = .054, TDIFF and SMRP, respectively; Table 2).

Mapping the generalized linear mixed genecological model revealed a relationship of survival in the gardens to continental-scale gradients in climate. Specifically, big sagebrush from regions with greater seasonal temperature differences and higher summer precipitation supported a greater probability of survival in the central range. These areas were predominantly found in interior regions of the continent with blue and green shades (Figure 1). In contrast, populations from climates that consisted of moderated winter temperatures and drier summers had lower probability of survival in the gardens, mapped in red and dark orange (Figure 1).

To further investigate the low rate of survival at Ephraim, we performed a detailed survival analysis for this garden. Survival increased at a log-normal rate with a scale of 0.649, that is, the rate of mortality decreased over time (Figure 2; Table S3). Rate of survival was significantly different for tetraploid *vaseyana* (V4x) than for the other subspecies:cytotype groups (χ_4^2 = 52.6, *p* < .001). Specifically, survival time of tetraploid *vaseyana* (V4x) was 0.445 times shorter than the baseline tetraploid *tridentata* (T4x) ($e^{-0.810}$; Table S3). The median survival of tetraploid *vaseyana* (V4x) was 23 months (*i.e.*, only a 50% chance of survival at 23 months), and there was less than 10% chance of survival at 48 months. In contrast, tetraploid *tridentata* (T4x) had a median survival of 47 months (Table S4). Survival varied considerably around the mean survival proportion of 0.65 among source-populations at Ephraim (Figure 3, Table S5). Some populations experienced 100% mortality after the first winter (i.e., population CAV4), while all families of other populations persisted throughout the experimental period (i.e., populations MTT1, MTW2).

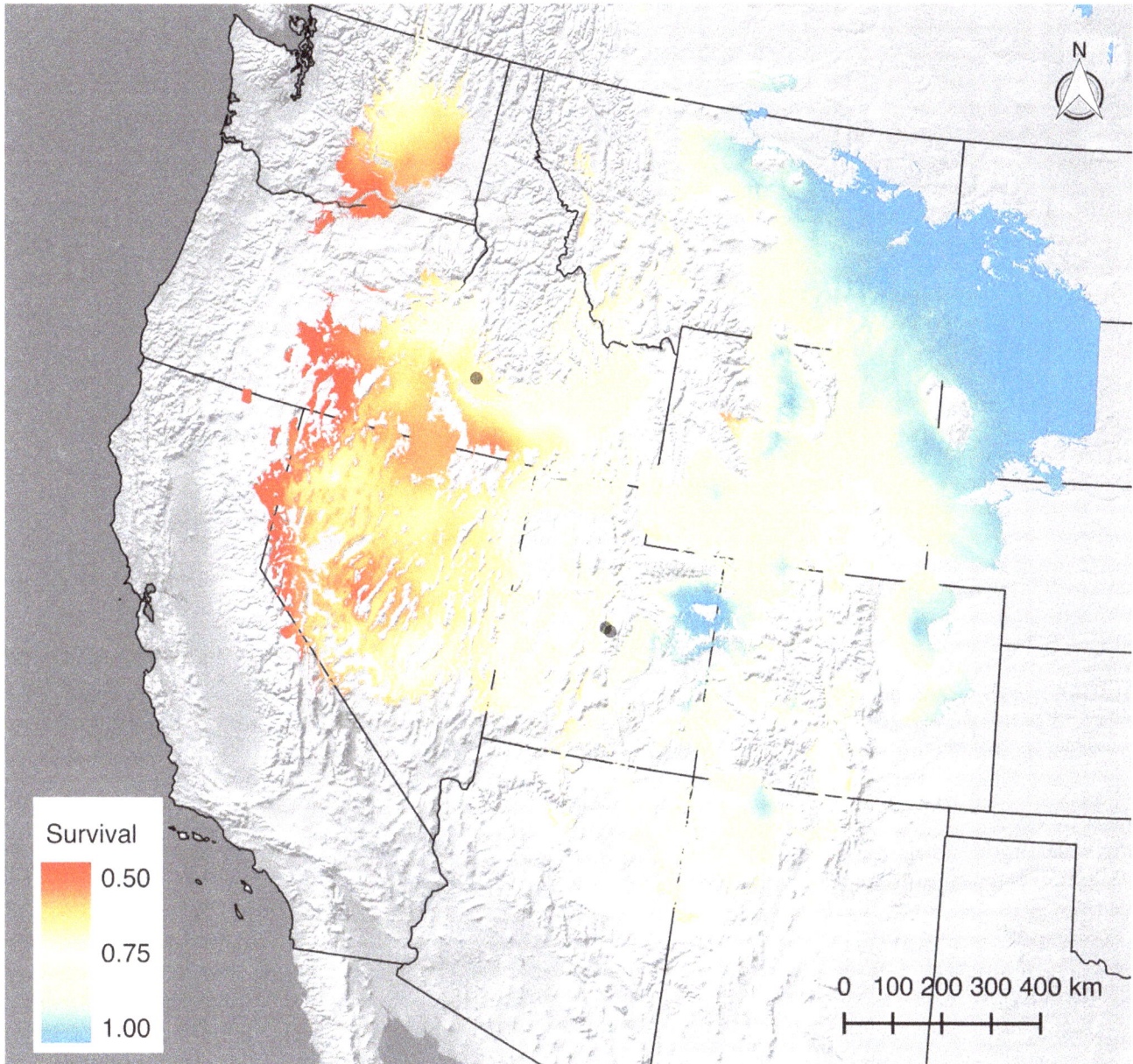

FIGURE 1 Genecological projection of survivorship in big sagebrush in three common gardens. Common-garden locations shown as black circles. Log-link transformed coefficients of slopes and intercepts of the generalized linear mixed model were used to project probability of survival into the niche boundaries of *wyomingnesis* big sagebrush (Still & Richardson, 2015). Climate predictors included in the model are TDIFF and SMRP. Areas that have greater temperature differences between summer and winter and wetter summers have higher probability of survival (blue), while areas with more moderated temperatures and drier summers have lower probability of survival (red). TDIFF: temperature difference between mean temperature in the coldest month and mean temperature in the warmest month; SMRP: summer precipitation (July and August). Hillshade background is based on the US Geological Survey Digital Elevation Model

4 | DISCUSSION

4.1 | Adaptive genetic responses to climate

Growth in *A. tridentata* occurs primarily in the spring, while soil moisture levels are high (Germino & Reinhardt, 2014; Schlaepfer, Lauenroth, & Bradford, 2014), yet minimum temperatures are often physiologically limiting and even lethal (Brabec et al., 2016). Strong differences in avoidance and tolerance were evidence among seedlings

of the subspecies:cytotype groups of *A. tridentata* related to their mortality and suggested the importance of freezing response as an axis for adaptive differentiation (Brabec et al., 2016). At the landscape level, selection pressures of different climates have potentially generated differences in phenology and growth strategies for *A. tridentata*. One hypothesis would be that interior regions, with a continental climate (greater summer–winter temperature extremes), support big sagebrush populations that have a growth strategy that confers resistance, tolerance, or avoidance of freezing temperatures through

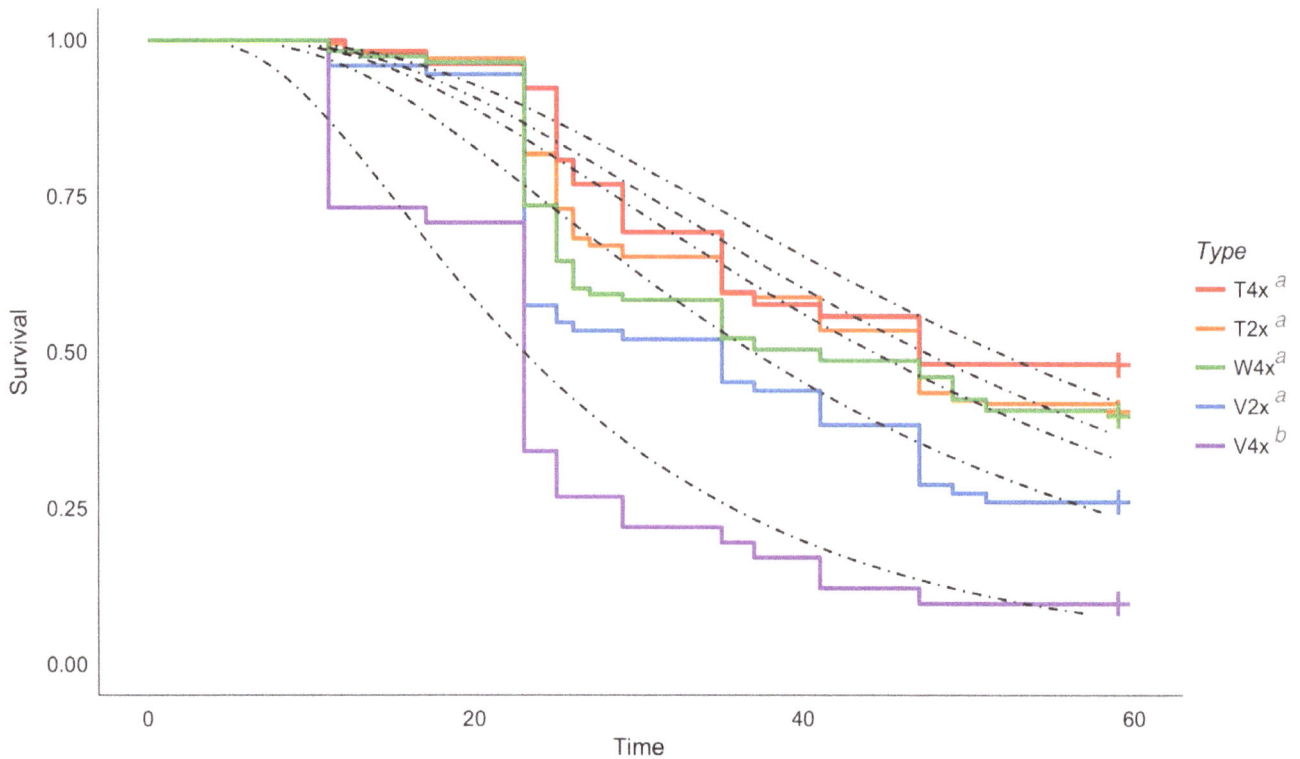

FIGURE 2 Kaplan–Meier survivorship curves by subspecies:cytotype group in big sagebrush at the Ephraim garden. Model estimates of the log-normal survivorship curve from the survivorship regression are overlaid to assess fit of model. Survivorship was significantly different by subspecies:cytotype group. Letters indicate significant difference from Bonferroni post hoc comparisons

physiological mechanisms or deferment/delay of growth until after the spring frost period. In more moderate climates to the south and west, we further hypothesize that *A. tridentata* initiates growth earlier in spring to capture spring soil moisture, resulting in greater risk of exposure to freezing in the gardens we evaluated. Below, we discuss how results from our analyses support this hypothesis.

Results from the generalized linear mixed model indicated that source-population climate is a strong predictor of survival in *A. tridentata*. The final generalized linear mixed genecological model explained 44% of variation (conditional R^2 = .439). The genetic effects of the final model, climate variables TDIFF and SMRP along with plant subspecies:cytotype group, explained 17% of the variation (marginal R^2 = .171). Survival patterns follow a southwest to northeast gradient where, in general, populations from higher latitudes and lower longitudes support higher survival (Figure 1). These continental regions, such as Montana and Colorado, have colder winters and have higher survival. These populations appear adapted to low minimum winter temperatures and later springs. Lower survival occurred in plants from areas to the southwest with more moderated climates. Growth in these regions likely begins earlier in the spring to maximize water availability from snowmelt and has low probability of freezing due to milder winters. However, early growth in Ephraim, where there was exposure to late winter extreme low minimum temperatures and more prevalent spring frosts (Fig. S3), puts the plant at risk for freezing and cold-related drought (Lambrecht, Shattuck, & Loik, 2007).

Our results support previous research that also indicated the adaptive importance of minimum temperatures (Bower et al., 2014; Erickson, Mandel, & Sorensen, 2004; Horning et al., 2010; Johnson, Sorensen, Bradley St Clair, & Cronn, 2004; Richardson et al., 2014; St. Clair et al., 2013). Although *A. tridentata* is generally thought of as being cold tolerant, winterkill has been documented in natural populations (Walser et al., 1990) and common gardens of seedlings at relatively warm sites (Brabec et al., 2016). Other studies of *A. tridentata* also found differences in freeze tolerance associated with source location, such as greater freezing acclimation for seedlings from higher-elevation seed sources (Loik & Redar, 2003).

While we found differences in survival rate among plant subspecies:cytotype groups, these differences were highly influenced by source-population. Post hoc comparisons of our survival analysis found that only tetraploid *vaseyana* (V4x) was significantly different from other subspecies:cytotype groups. This may be attributable to source-population origins rather than true taxonomical differences, considering how many tetraploid *vaseyana* (V4x) were from the southwest United States (e.g., CAV3, CAV4, NVV3). Previous work has suggested that *A. tridentata* subspecies are adapted to different habitats (Mahalovich & McArthur, 2004; McArthur, 1994; Wang et al., 1997), which can often be defined on local spatial scales. However, our findings suggest that within subspecies, populations are principally adapted to clines in continentality (Table 2).

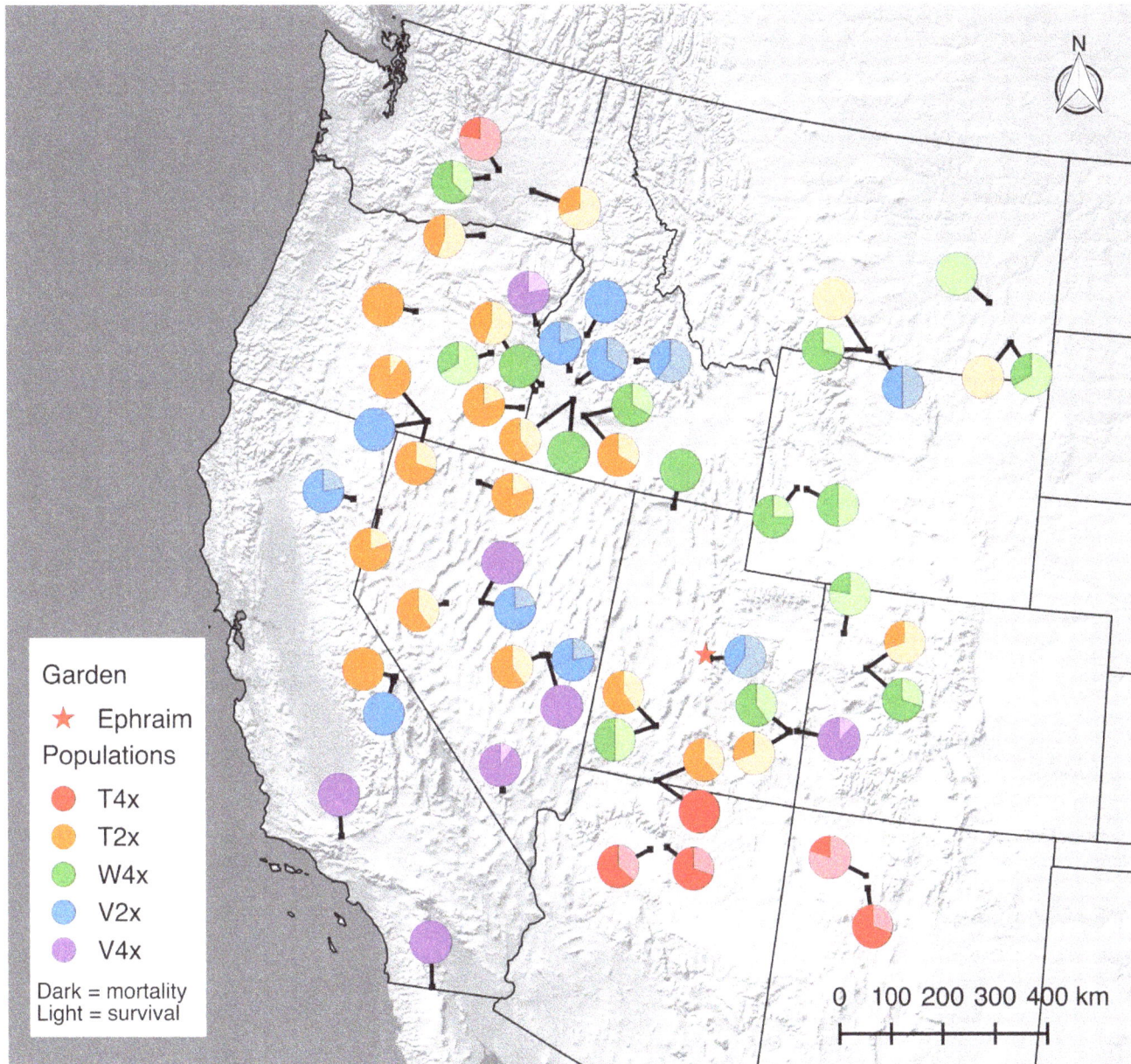

FIGURE 3 Proportion survival at the Ephraim garden for each source-population. The pie charts show the proportion of survival for each population, dark shade indicates mortality, while the light shade indicates survival. Color of each population is determined by subspecies:cytotype group. Hillshade background is based on the US Geological Survey Digital Elevation Model. T4x = tetraploid *tridentata*; T2x = diploid *tridentata*; W4x = tetraploid *wyomingensis*; V2x = diploid *vaseyana*; V4x = tetraploid *vaseyana*

4.2 | Microclimate variation influences survival

Survival of *A. tridentata* differed substantially between common-garden sites (Table 2, Fig. S2). These differences were most striking between Ephraim and Majors Flat, which are separated only by 8 km. Ephraim is located in a basin, whereas Majors Flat is located 415 m higher in the mountains. However, minimum temperatures were consistently lower at Ephraim (Table 1) as result of cold air drainage during the spring and summer and temperature inversions (i.e., "atmospheric decoupling"; Daly, Conklin, & Unsworth, 2010; Schuster, Kirchner, Jakobi, & Menzel, 2013) during the winter. Daily minimum temperatures during the winter in Ephraim were an average of 2.7°C

and 6.8°C less than Majors Flat and Orchard, respectively. These differences were most pronounced during subregional temperature inversion events (settling and trapping of cold air within meters to km above the earth surface) where minimum temperatures were at most 18.6°C and 28.5°C less at Ephraim than Majors Flat and Orchard, respectively (Fig. S3). Another climatic factor that may have impacted survival is low snow depth. A large winter precipitation gradient occurs between basins and ranges, with a larger accumulation of snow in the mountains rather than the basins. We hypothesize that low minimum temperatures and shallow snow depth in Ephraim impacted the low survival in the Ephraim garden, especially in subspecies *vaseyana* (20% survival). The current study design does not allow us to specifically test for the presence of microclimate adaptations; we suggest that future work examine this. Differences in microclimate, namely freezing temperatures, have been known to be an important determinant in other desert species distribution (Franco & Nobel, 1989; Loik & Nobel, 1993; Shreve, 1911). Further, research in *A. tridentata* has found low survival in *vaseyana* in areas where a typically high snow pack was absent, yet high survival in areas with a snow cover present (Hanson, Johnson, & Wight, 1982). Further, Loik and Redar (2003) suggested that *A. tridentata* seedlings at higher elevations may have less exposure to freezing due to greater prevalence of snow. Consistent with this, Brabec et al. (2016) found that subspecies *vaseyana* had the least and *wyomingensis* had the greatest physiological avoidance and resistance to freezing among subspecies, and they proposed the seemingly ironic differences were due to greater insulating snow cover at higher elevation during winter and spring. These results suggest that selection for cold tolerance may relate to local-scale microclimate variation, which we propose to be an important topic for further investigation.

Many traits can affect fitness, influencing growth and fecundity, along with survival. This study demonstrates moving warm-adapted plants into cold climates decreases their survival. While there is little evidence for the inverse, low survival among cold-adapted plants in a warm climate, trade-offs in other fitness traits may exist. For example, seed yield data suggest that cold-adapted plants had reduced fecundity in warmer climates compared to warm-adapted plants (B.A. Richardson, unpublished data). Another possibility, which was beyond the scope of this experiment, is that cold-adapted plants may have lower establishment success in warmer climates compared to warm-adapted plants.

4.3 | Applications

Future climate scenarios predict an increase in global mean temperatures and an increase in extreme weather patterns. As the climate that *A. tridentata* is adapted to is displaced, plants will need to migrate to new areas, a slow task for a sessile organism (Shaw & Etterson, 2012). The movement of plant species in response to rapid climatic warming will frequently be slower than phenotypic and adaptive genetic changes required to adjust to the novel climate. Due to population differentiation, the effects of climate change are likely to vary throughout a species range (Davis, Shaw, & Etterson, 2005). While locations subject to frequent cold air pooling are not likely to escape regionally increasing temperatures, they may act as microrefugia against the amplified temperature trends and variations (Dobrowski, 2011). Through genecological modeling of future climate scenarios, it is likely that as the climate changes, *A. tridentata* will not be genetically suited to the environment in which it currently grows, resulting in extirpation of some populations.

Movement of seed populations to areas outside of their adaptive breadth can have a negative impact on fitness (Hereford, 2009). Restoration of *A. tridentata* after fires or other disturbances has been a management priority over the past decade, yet success of these efforts has been varied (Arkle et al., 2014; Knutson et al., 2014). Planting *A. tridentata* seed outside their adaptive breadth could result

in unsuccessful establishment and/or low fitness and provide an opportunity for invasive species encroachment leading to a loss of species diversity and ecosystem degradation (reviewed in Dumroese, Luna, Richardson, Kilkenny, & Runyon, 2015). Previous research has shown that sagebrush species/subspecies are adapted to different ecological niches such as elevation and soil type (McArthur, 1994). Moving sagebrush populations to different climatic or edaphic conditions is therefore not recommended (Mahalovich & McArthur, 2004). Our work is the first step to creating climate-based guidelines for the transfer of seed for restoration purposes in *A. tridentata*. Being able to map traits relating to climate across the landscape is one practical advantage of genecology, which is particularly useful in highly heterogeneous environments such as western North America. This research will be the first step to delineate seed transfer zones, guidelines to ensure that seed used for restoration is adapted to the site.

ACKNOWLEDGEMENTS

We thank numerous volunteers, BLM, USFS, and Utah DNR staffs for assistance in collection of seed and garden maintenance. Funding was provided by Great Basin Native Plant Program and USDA Forest Service National Fire Plan. Any use of trade, firm, or product names is for descriptive purposes only and does not imply endorsement by the U.S. Government.

REFERENCES

Anderson, J. T., Lee, C-R, Rushworth, C. A., Colautti, R. I., & Mitchell-Olds, T. (2013). Genetic trade-offs and conditional neutrality contribute to local adaptation. *Molecular Ecology*, 22, 699–708.

Arkle, R. S., Pilliod, D. S., Hanser, S. E., Brooks, M. L., Chambers, J. C., Grace, J. B., ... Wirth, T. A. (2014). Quantifying restoration effectiveness using multi-scale habitat models: Implications for sage-grouse in the great basin. *Ecosphere*, 5, 1–32.

Bansal, S., Harrington, C. A., Gould, P. J., & St.Clair, J. B. (2015). Climate-related genetic variation in drought-resistance of douglas-fir (*Pseudotsuga menziesii*). *Global Change Biology*, 21, 947–958.

Bates, D., Mächler, M., Bolker, B., & Walker, S. (2015). Fitting linear mixed-effects models using lme4. *Journal of Statistical Software*, 67, 1–48.

Bischoff, A., Cremieux, L., Smilauerova, M., Lawson, C. S., Mortimer, S. R., Dolezal, J., Lanta, V., ... Müller-Schärer, H., (2006). Detecting local adaptation in widespread grassland species–the importance of scale and local plant community. *Journal of Ecology*, 94, 1130–1142.

Bower, A. D., St. Clair, J. B., & Erickson, V. (2014). Generalized provisional seed zones for native plants. *Ecological Applications*, 24, 913–919.

Brabec, M. M., Germino, M. J., & Richardson, B. A. (2016). Climate adaption and post-fire restoration of a foundational perennial in cold desert: Insights from intraspecific variation in response to weather. *Journal of Applied Ecology*, doi:10.1111/1365-2664.12679 [Epub ahead of print].

Brabec, M. M., Germino, M. J., Shinneman, D. J., Pilliod, D. S., McIlroy, S. K., & Arkle, R. S. (2015). Challenges of Establishing Big Sagebrush (*Artemisia tridentata*) in Rangeland Restoration: Effects of Herbicide, Mowing, Whole-Community Seeding, and Sagebrush Seed Sources. *Rangeland Ecology & Management*. Retrieved from http://www.sciencedirect.com/science/article/pii/S1550742415000950.

Chevin, L.-M., & Lande, R. (2011). Adaptation to marginal habitats by evolution of increased phenotypic plasticity. *Journal of Evolutionary Biology*, 24, 1462–1476.

Clausen, J., Keck, D. D., & Hiesey, W. M. (1940). Effect of Varied

Environments on Western North American Plants. Carnegie Institution.

Crookston, N. L., & Finley, A. O. (2007). "yaImpute: An R Package for kNN Imputation." Journal of Statistical Software 23 (10). Retrived from http://www.jstatsoft.org/v23/i10.

Crookston, N. L., & Rehfeldt, G. E. (2012). Climate Estimates and Plant-Climate Relationships. USDA Forest Service, Rocky Mountain Research Station, Moscow ID. Retrieved from http://forest.moscowfsl.wsu.edu/climate/ (assessed 2011-2012).

Daly, C., Conklin, D. R., & Unsworth, M. H. (2010). Local atmospheric decoupling in complex topography alters climate change impacts. International Journal of Climatology, 30, 1857-1864.

Davis, M. B., Shaw, R. G., & Etterson, J. R. (2005). Evolutionary responses to changing climate. Ecology, 86, 1704-1714.

Dobrowski, S. Z. (2011). A climatic basis for microrefugia: The influence of terrain on climate. Global Change Biology, 17, 1022-1035.

Dumroese, R. K., Luna, T., Richardson, B. A., Kilkenny, F. F., & Runyon, J. B. (2015). Conserving and restoring habitat for greater sage-grouse and other sagebrush-obligate wildlife: The crucial link of forbs and sagebrush diversity. Native Plants Journal, 16, 276-299.

Erickson, V. J., Mandel, N. L., & Sorensen, F. C. (2004). Landscape patterns of phenotypic variation and population structuring in a selfing grass, Elymus Glaucus (blue wildrye). Canadian Journal of Botany, 82, 1776-1789.

Franco, A. C., & Nobel, P. S. (1989). Effect of nurse plants on the microhabitat and growth of cacti. The Journal of Ecology, 77, 870-886.

Galen, C., Shore, J. S., & Deyoe, H. (1991). Ecotypic divergence in alpine polemonium viscosum: Genetic structure, quantitative variation, and local adaptation. Evolution, 45, 1218-1228.

Galloway, L. F., & Fenster, C. B. (2000). Population differentiation in an annual legume: Local adaptation. Evolution, 54, 1173-1181.

Germino, M. J., & Reinhardt, K. (2014). Desert shrub responses to experimental modification of precipitation seasonality and soil depth: Relationship to the two-layer hypothesis and ecohydrological niche. Journal of Ecology, 102, 989-997.

Graham, J. H., Freeman, D. C., & McArthur, E. D. (1995). Narrow hybrid zone between two subspecies of big sagebruh (artemisia tridentata: Asteraceae). II. selection gradients and hybrid fitness. American Journal of Botany, 82, 709-716.

Hanson, C. L., Johnson, C. W., & Wight, J. R. (1982). Foliage mortality of mountain big sagebrush [Artemisia tridentata subsp. vaseyana] in Southwestern Idaho during the Winter of 1976-77. Journal of Range Management, 35, 142-145.

Hardegree, S. P., Jones, T. A., Roundy, B. A., Shaw, N. L., Monaco, T. A., & Briske, D. D. (2011). Assessment of Range Planting as a Conservation Practice. Conservation Benefits of Rangeland Practices: Assessment, Recommendations, and Knowledge Gaps. Washington, DC, USA: USDA-NRCS, 171-212.

Hereford, J. (2009). A quantitative survey of local adaptation and fitness trade-offs. The American Naturalist, 173, 579-588.

Horning, M. E., McGovern, T. R., Darris, D. C., Mandel, N. L., & Johnson, R. (2010). Genecology of holodiscus discolor (Rosaceae) in the Pacific Northwest, U.S.A. Restoration Ecology, 18, 235-243.

Hua, X., & Wiens, J. J. (2013). How does climate influence speciation? The American Naturalist, 182, 1-12.

Hutchinson, M. F. (2000). ANUSPLIN User Guide Version 4.1. Centre for Resource and Environmental Studies: Australian National University, Canberra, Australia.

Jaeger, D. M., Runyon, J. B., & Richardson, B. A. (2016). Signals of speciation: Volatile organic compounds resolve closely related sagebrush taxa, suggesting their importance in evolution. New Phytologist, 211, 1393-1401.

Johnson, R. C., Cashman, M. J., & Vance-Borland, K. (2012). Genecology and seed zones for Indian Ricegrass Collected in the Southwestern United States. Rangeland Ecology & Management, 65, 523-532.

Johnson, G. R., Sorensen, F. C., Bradley St Clair, J., & Cronn, R. C. (2004).

Pacific Northwest Forest tree seed zones a template for native plants? Native Plants Journal, 5, 131-140.

Joyce, D. G., & Rehfeldt, G. E. (2013). Climatic niche, ecological genetics, and impact of climate change on eastern white pine (pinus strobus l.): Guidelines for land managers. Forest Ecology and Management, 295, 173-192.

Jump, A. S., & Peñuelas, J. (2005). Running to stand still: Adaptation and the response of plants to rapid climate change. Ecology Letters, 8, 1010-1020.

Kawecki, T. J., & Ebert, D. (2004). Conceptual issues in local adaptation. Ecology Letters, 7, 1225-1241.

Knick, S., & Connelly, J. W. (2011). Greater Sage-Grouse: Ecology and Conservation of a Landscape Species and Its Habitats. Vol. 38. Univ of California Press.

Knutson, K. C., Pyke, D. A., Wirth, T. A., Arkle, R. S., Pilliod, D. S., Brooks, M. L., ... Grace, J. B. (2014). Long-term effects of seeding after wildfire on vegetation in great basin shrubland ecosystems. Journal of Applied Ecology, 51, 1414-1424.

Kolb, K. J., & Sperry, J. S. (1999). Differences in drought adaptation between subspecies of Sagebrush (Artemisia Tridentata). Ecology, 80, 2373-2384.

Lambrecht, S. C., Shattuck, A. K., & Loik, M. E. (2007). Combined drought and episodic freezing effects on seedlings of low-and high-elevation subspecies of sagebrush (Artemisia tridentata). Physiologia Plantarum, 130, 207-217.

Ledig, F. T., Rehfeldt, G. E., Sáenz-Romero, C., & Flores-López, C. (2010). Projections of suitable habitat for rare species under global warming scenarios. American Journal of Botany, 97, 970-987.

Leimu, R., & Fischer, M. (2008). A meta-analysis of local adaptation in plants. PLoS ONE, 3, e4010.

Loik, M. E., & Nobel, P. S. (1993). Freezing tolerance and water relations of Opuntia fragilis from Canada and the United States. Ecology, 74, 1722-1732.

Loik, M. E., & Redar, S. P. (2003). Microclimate, freezing tolerance, and cold acclimation along an elevation gradient for seedlings of the great basin desert shrub, Artemisia tridentata. Journal of Arid Environments, 54, 769-782.

Mahalovich, M. F., & McArthur, E. D. (2004). Sagebrush (Artemisia spp.) seed and plant transfer guidelines. Native Plants Journal, 5, 141-148.

McArthur, E. D. (1994). Ecology, Distribution, and Values of Sagebrush Within the Intermountain Region. Retrieved from http://www.fs.fed.us/rm/pubs_int/int_gtr313/int_gtr313_347_351.pdf.

McArthur, E. D., & Sanderson, S. C. (1999). Cytogeography and chromosome evolution of subgenus tridentatae of artemisia (asteraceae). American Journal of Botany, 86, 1754-1775.

McArthur, E. D., Welch, B. L., & Sanderson, S. C. (1988). Natural and artificial hybridization between big sagebrush (Artemisia tridentata) subspecies. Journal of Heredity, 79, 268-276.

McKay, J. K., Richards, J. H., Nemali, K. S., Sen, S., Mitchell-Olds, T., Boles, S., ... Rausher, M. (2008). Genetics of drought adaptation in Arabidopsis thaliana II. Qtl Analysis of a new mapping population, Kas-1 × Tsu-1. Evolution, 62, 3014-3026.

Miller, R. F., Knick, S. T., Pyke, D. A., Meinke, C. W., Hanser, S. E., Wisdom, M. J., & Hild, A. L. (2011). "Characteristics of Sagebrush Habitats and Limitations to Long-Term Conservation". Greater Sage-Grouse: ecology and Conservation of a Landscape Species and Its Habitats. Studies in Avian Biology, 38, 145-184.

Nakagawa, S., & Schielzeth, H. (2013). A general and simple method for obtaining R2 from generalized linear mixed-effects models. Methods in Ecology and Evolution, 4, 133-142.

Pratt, J. D., & Mooney, K. A. (2013). Clinal adaptation and adaptive plasticity in Artemisia californica: Implications for the response of a foundation species to predicted climate change. Global Change Biology, 19, 2454-2466.

Prevéy, J. S., Germino, M. J., Huntly, N. J., & Inouye, R. S. (2010). Exotic plants increase and native plants decrease with loss of foundation species in sagebrush steppe. Plant Ecology, 207, 39-51.

QGIS Development Team. (2015). *QGIS Geographic Information System*. Open Source Geospatial Foundation. Retrieved from http://qgis.osgeo.org.

R Core Team. (2015). *R: A Language and Environment for Statistical Computing*. Vienna, Austria: R Foundation for Statistical Computing. Retrieved from http://www.R-project.org/.

Rehfeldt, G. E. (2004). Interspecific and Intraspecific Variation in *Picea engelmannii* and Its Congeneric Cohorts: Biosystematics, Genecology, and Climate Change.

Rehfeldt, G. E. (2006). A Spline Model of Climate for the Western United States.

Rehfeldt, G. E., & Jaquish, B. C. (2010). Ecological impacts and management strategies for Western Larch in the face of climate-change. *Mitigation and Adaptation Strategies for Global Change, 15*, 283–306.

Rehfeldt, G. E., Jaquish, B. C., López-Upton, J., Sáenz-Romero, J. C, St. Clair, B., Leites, L. P., & Joyce, D. G. (2014). Comparative genetic responses to climate for the varieties of *pinus ponderosa* and *Pseudotsuga menziesii*: Realized climate niches. *Forest Ecology and Management, 324*, 126–137.

Rehfeldt, G. E., Jaquish, B. C., Sáenz-Romero, C., Joyce, D. G., Leites, L. P., St Clair, J. B., & López-Upton, J. (2014). Comparative genetic responses to climate in the varieties of *Pinus ponderosa* and *Pseudotsuga menziesii*: Reforestation. *Forest Ecology and Management, 324*, 147–157.

Richardson, B. A., Kitchen, S. G., Pendleton, R. L., Pendleton, B. K., Germino, M. J., Rehfeldt, G. E., & Meyer, S. E. (2014). Adaptive responses reveal contemporary and future ecotypes in a desert shrub. *Ecological Applications, 24*, 413–427.

Richardson, B. A., Page, J. T., Bajgain, P., Sanderson, S. C., & Udall, J. A. (2012). Deep sequencing of amplicons reveals widespread intraspecific hybridization and multiple origins of polyploidy in big sagebrush (*Artemisia tridentata*; Asteraceae). *American Journal of Botany, 99*, 1962–1975.

Schlaepfer, D. R., Lauenroth, W. K., & Bradford, J. B. (2014). Natural regeneration processes in big sagebrush (*Artemisia tridentata*). *Rangeland Ecology & Management, 67*, 344–357.

Schuster, C., Kirchner, M., Jakobi, G., & Menzel, A. (2013). Frequency of inversions affects senescence phenology of *Acer pseudoplatanus* and *Fagus sylvatica*. *International Journal of Biometeorology, 58*, 485–498.

Sexton, J. P., McIntyre, P. J., Angert, A. L., & Rice, K. J. (2009). Evolution and ecology of species range limits. *Annual Review of Ecology, Evolution, and Systematics, 40*, 415–436.

Shaw, R. G., & Etterson, J. R. (2012). Rapid climate change and the rate of adaptation: Insight from experimental quantitative genetics. *New Phytologist, 195*, 752–765.

Shreve, F. (1911). The influence of low temperatures on the distribution of the giant cactus. *Plant World, 14*, 136–146.

St. Clair, J. B., Kilkenny, F. F., Johnson, R. C., Shaw, N. L., & Weaver, G. (2013). Genetic variation in adaptive traits and seed transfer zones for *Pseudoroegneria spicata* (Bluebunch Wheatgrass) in the Northwestern United States. *Evolutionary Applications., 6*, 933–948.

St Clair, J. B., Mandel, N. L., & Vance-Borland, K. W. (2005). Genecology of douglas fir in Western Oregon and Washington. *Annals of Botany, 96*, 1199–1214.

Still, S. M., & Richardson, B. A. (2015). Projections of contemporary and future climate niche for wyoming big sagebrush (*Artemisia tridentata* subsp. *wyomingensis*): A guide for restoration. *Natural Areas Journal, 35*, 30–43.

Therneau, T. M., & Lumley, T. (2014). *Package 'survival'*. Survival Analysis: Published on CRAN.

Turesson, G. (1923). The scope and import of genecology. *Hereditas, 4*, 171–176.

Turesson, G. (1925). The plant species in relation to habitat and climate. *Hereditas, 6*, 147–236.

Vergeer, P., & Kunin, W. E. (2013). Adaptation at range margins: Common garden trials and the performance of *Arabidopsis lyrata* across Its Northwestern European range. *New Phytologist, 197*, 989–1001.

Walser, R. H., Weber, D. J., Durant McArthur, E., & Sanderson, S. C. (1990). Winter Cold Hardiness of Seven Wildland Shrubs. In Proceedings-Symposium on Cheatgrass Invasion, Shrub Die-Off, and Other Aspects of Shrub Biology and Management; 5-7 April 1989; Las Vegas, NV, USA. Ogden, UT, USA: USDA Forest Service Intermountain Research Station, 115–18.

Wang, H., McArthur, E. D., & Freeman, D. C. (1999). Narrow hybrid zone between two subspecies of big sagebrush (*Artemisia tridentata*: Asteraceae). IX. elemental uptake and niche separation. *American Journal of Botany, 86*, 1099–1107.

Wang, H., McArthur, E. D., Sanderson, S. C., Graham, J. H., & Freeman, D. C. (1997). Narrow hybrid zone between two subspecies of big sagebrush (*Artemisia tridentata*: Asteraceae) IV. reciprocal transplant experiments. *Evolution, 51*, 95–102.

Benefits of gene flow are mediated by individual variability in self-compatibility in small isolated populations of an endemic plant species

Christopher T. Frye[1,2] | Maile C. Neel[3]

[1]Natural Heritage Program, Maryland Department of Natural Resources, Wildlife and Heritage Service, Wye Mills, MD, USA

[2]Department of Plant Science and Landscape Architecture, University of Maryland, College Park, MD, USA

[3]Department of Plant Science and Landscape Architecture and Department of Entomology, University of Maryland, College Park, MD, USA

Correspondence
Christopher T. Frye, Natural Heritage Program, Maryland Department of Natural Resources, Wildlife and Heritage Service, Wye Mills, MD, USA.
Email: chris.frye@maryland.gov

Abstract

Many rare and endemic species experience increased rates of self-fertilization and mating among close relatives as a consequence of existing in small populations within isolated habitat patches. Variability in self-compatibility among individuals within populations may reflect adaptation to local demography and genetic architecture, inbreeding, or drift. We use experimental hand-pollinations under natural field conditions to assess the effects of gene flow in 21 populations of the central Appalachian endemic *Trifolium virginicum* that varied in population size and degree of isolation. We quantified the effects of distance from pollen source on pollination success and fruit set. Rates of self-compatibility varied dramatically among maternal plants, ranging from 0% to 100%. This variation was unrelated to population size or degree of isolation. Nearly continuous variation in the success of selfing and near-cross-matings via hand pollination suggests that *T. virginicum* expresses pseudo-self-fertility, whereby plants carrying the same S-allele mate successfully by altering the self-incompatibility reaction. However, outcrossing among populations produced significantly higher fruit set than within populations, an indication of drift load. These results are consistent with strong selection acting to break down self-incompatibility in these small populations and/or early-acting inbreeding depression expressed upon selfing.

KEYWORDS
endemic, gene flow, index of self-incompatibility, mating system, pseudo-self-fertility, *Trifolium virginicum*

1 | INTRODUCTION

Plant mating systems mediate the frequencies of outcrossing and selfing, which, in turn, strongly affect the amount and distribution of genetic variation within and among populations (Charlesworth, 2006; Duminil, Hardy, & Petit, 2009; Loveless & Hamrick, 1984; Young, Broadhurst, & Thrall, 2012). Population size and connectivity also affect amounts and patterns of genetic variation in that small, isolated populations have lower levels of standing genetic variation and increased inbreeding (Eckert et al., 2010; Heschel & Paige, 1995;

Holsinger & Vitt, 1997; Jacquemyn, De Meester, Jongejans, & Honnay, 2012; Soulé, 1987; Young & Pickup, 2010). Specific effects of population size on inbreeding and fitness may have complex dependencies on life history characteristics of species (Angeloni, Ouborg, & Leimu, 2011), but inbreeding generally negatively affects fitness (Frankham, 2015).

Species that are of conservation concern due to recent reduction in population size through habitat loss may be at higher risk of fitness declines than chronically rare species (Holsinger & Vitt, 1997; Honnay & Jacquemyn, 2007). If population reduction is accompanied by increased isolation that eliminates gene flow among previously connected populations, increased selfing and mating among close relatives can reveal substantial genetic load that was masked in larger populations (Keller & Waller, 2002). The consequences of more frequent inbreeding within populations may be indicated by low fruit set or low-quality seed (early-acting inbreeding depression), as well as poor survival or reproduction of inbred progeny (late-acting inbreeding depression).

At the same time, selfing can provide reproductive assurance under conditions of low pollinator density, low mate availability, or otherwise marginal environmental conditions (Jacquemyn et al., 2012; Kalisz, Vogler, & Hanley, 2004; Karron et al., 2012; Lloyd, 1992). Benefits of reproductive assurance in small and isolated populations appear to outweigh the benefits of cross-pollination that are found in large populations (Delmas, Cheptou, Escarvage, & Pornon, 2014; Herlihy & Eckert, 2002; Holsinger, 2000; Igić, Bohs, & Kohn, 2006; Kalisz et al., 2004; Porcher & Lande, 2005). Shifts in the relative proportions of selfing and outcrossing within populations are the result of mating system evolution and the diversity of mating systems in plants is an indicator of the flexibility in responding to selection (Levin, 2012). Evidence is increasing that the mating system itself may respond adaptively to small population size and fragmentation through breakdown of self-incompatibility (Busch, Joly, & Schoen, 2010; Karron et al., 2012; Stephenson, Good, & Vogler, 2000; Willi, 2009). Levin (1996) suggested this breakdown is often due to the action of modifier genes that alter the effectiveness of self-incompatibility alleles (i.e., pseudo-self-fertility, PSF), a critical step in the evolution of self-fertility.

Gene flow among populations can alleviate inbreeding effects by introducing variation from relatively unrelated individuals that masks genetic load and restores compatible mating types (Cheptou & Donohue, 2011; Frankham, 2015; Spielman, Brook, & Frankham, 2004; Young & Pickup, 2010). The interaction between a species' dispersal ability and the distribution of habitat in a landscape determines patterns of gene flow under natural conditions. Long-term patterns of gene flow may be very different than current gene flow if habitat patches are smaller (fewer potential migrants) and are further apart (requiring longer dispersal distances) than they were under historical habitat distributions (Honnay & Jacquemyn, 2007).

Because mating system and gene flow are key to assessing the risks associated with small population size, the effects of crossing distance between individuals within and among populations have long been of interest to evolutionary biologists and conservation biologists (Edmands, 2007; Fenster & Sork, 1988; Frankham et al., 2011; Marsden, Engelhardt, & Neel, 2013; Weeks et al., 2011; Whitlock et al., 2013). Field studies linking the reproductive biology and mating system of species with an ecologically relevant scale of crossing distance can assist land managers in making informed decisions regarding the potential benefits associated with increased gene flow when conducting restoration activities or developing management plans (Marsden et al., 2013; Whitlock et al., 2013).

Thus, we sought to understand the effects of an artificial increase in gene flow in a species that exists in small and isolated populations, Kates Mountain Clover (*Trifolium virginicum* Small; Fabaceae), an

endemic to the central Appalachian shale barrens. We investigated relationships between crossing distance and reproductive success across populations of different sizes and degrees of isolation. *Trifolium virginicum* is a perennial herbaceous plant species that is restricted to small habitat patches within the dominant woodland habitat on shale substrate. Shale barrens and *T. virginicum* are globally rare (NatureServe 2014). Within the shale barren region, however, *T. virginicum* has a relatively broad distribution, occurring in discrete barren patches within the Ridge and Valley Physiographic Province from southwestern Virginia and adjacent West Virginia, north through western Maryland and south-central Pennsylvania (Figure 1).

As with many early successional habitats in eastern North America, shale barrens depend on periodic disturbance, such as wildfire, to retard succession to closed forest (Copeheaver, Fuhrman, Gellerstedt, & Gellerstedt, 2004; Foster et al., 2003; Norris & Sullivan, 2002; Tyndall, 2015). Lacking such disturbance, shale barren habitats have become restricted to small patches in which particularly harsh environmental conditions slow succession to woodland (Keener, 1983; Platt, 1951). Beyond forest succession, the shale barren region has experienced an increase in the number of potential barriers to gene flow via development and road construction (Copeheaver et al., 2004; Norris & Sullivan, 2002; Maryland Natural Heritage Program, Annapolis, MD). These changes in habitat structure leave *T. virginicum* in occupied areas within barrens that often cover only a few square meters. In Maryland, *T. virginicum* occurs on 95 barrens (Maryland Natural Heritage Program, Annapolis, MD). Seventy-six percent of these barrens have <50 plants and only 11% have ≥100 plants. The largest known population in Maryland had ~373 plants as of a 2016 census. Barrens are separated from their nearest neighboring barren by a minimum of 0.2 km to a maximum of 12.8 km (median = 1.0 km). Extirpation of four populations during construction of an interstate highway created a 300- to 400-m-wide gap that effectively divided a northern group of barrens from a southern group (Figure 1).

Trifolium virginicum individuals are long-lived, perhaps surviving many decades; marked plants at two sites have survived over 17 years. Plants have a deep taproot and produce multiple 2- to 3-cm-diameter spherical flower heads on 4- to 15-cm-long peduncles that lie prostrate on the ground and elongate with age. Pollination is likely affected by one or more native bee species. Fruits are slender legumes (pods) containing 1–3 seeds. The perianth and pods are long-persistent; the small seeds (~2.2 to 2.7 mm diameter) are released in late summer upon disintegration of the inflorescence and lack obvious means for long-distance dispersal.

The genus *Trifolium* is known to possess gametophytic self-incompatibility (GSI; Lawrence, 1996). GSI is a widespread genetic system that enables hermaphroditic plants to avoid self-fertilization and mating with close relatives by rejection in the pistil of pollen carrying the same S-allele. Species in the genus *Trifolium* are known to have a large number of S-alleles (Casey et al., 2010; Lawrence, 1996).

Given the broad distribution of small, isolated barrens within the extensive forested matrix and lack of adaptation for long-distance seed dispersal, they likely represent relics of a once more continuous distribution. If they are relics, the extant patches have highly

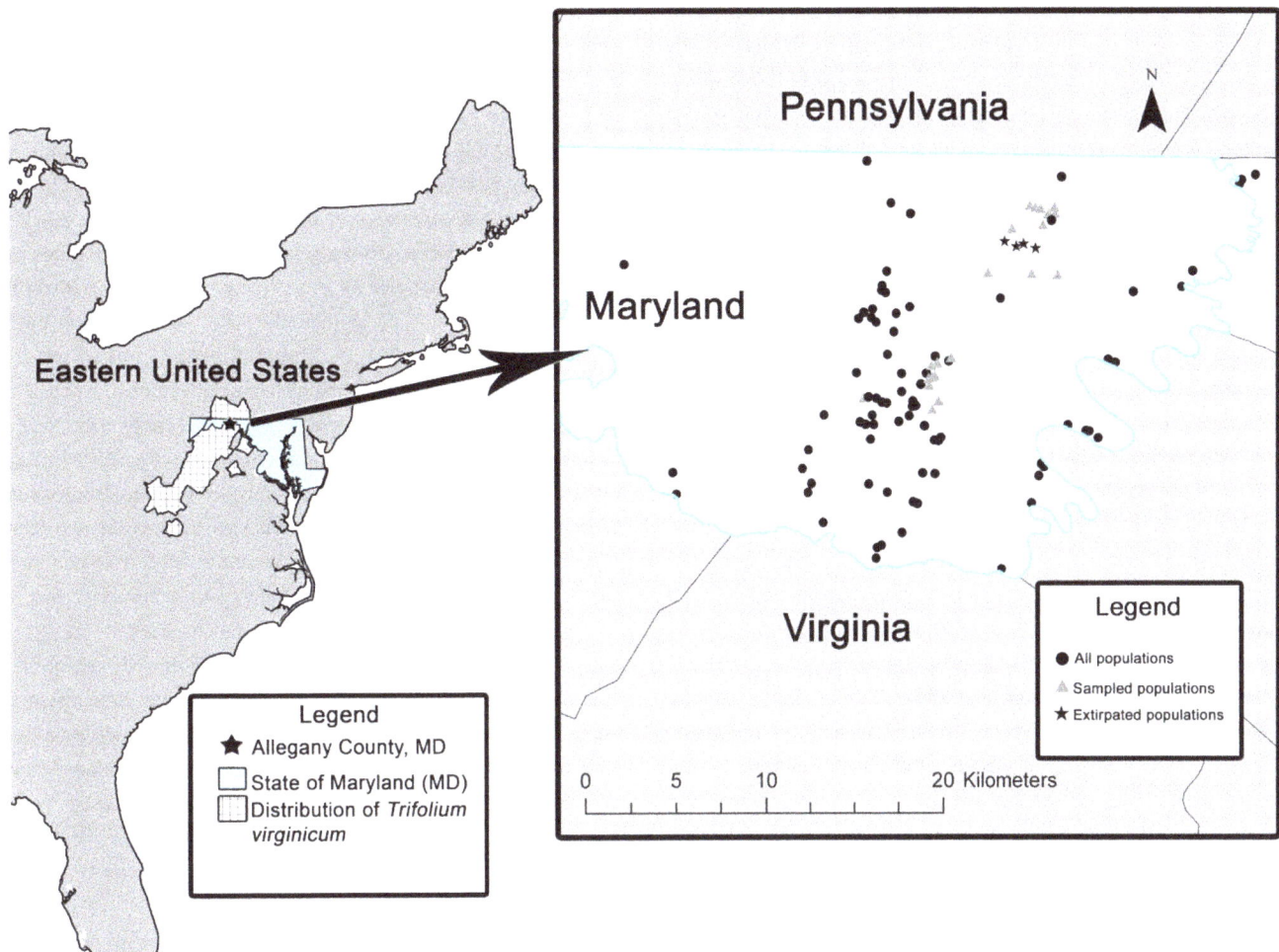

FIGURE 1 Geographic distribution of *Trifolium virginicum* in the eastern United States (hatched), location of the State of Maryland (blue), and the study area in Allegany County, Maryland (star). Inset: detail of the study area including sampled populations (triangles), unsampled but extant populations (filled circles), and extirpated populations (stars)

reduced population sizes and habitat areas that are far more isolated than they were prior to the last 100–160 years (Tyndall, 2015). This isolation of *T. virginicum* populations creates high risk of increased inbreeding, loss of genetic diversity, and loss of compatible mating types (S-alleles) via drift. These risks could be ameliorated by managing the habitat between barrens to establish larger openings and to enhance gene flow by linking currently isolated patches or by intentional supplementation of small populations that have experienced no increase in population size in >30 years. In this study, we seek to understand the risks and benefits of increasing gene flow in *T. virginicum* to inform management choices that range from maintaining the status quo (within population mating) to supplementing populations using pollen, plants, or seed from distant populations. We gain this understanding by performing experimental crosses using self-pollen (selfing), pollen from within sites (near-cross), and pollen from distant sites (far-cross). We quantify the effect of these cross-types on reproduction (fruit set and seed weight) as measures of fitness, and examine variability in self-compatibility (from hand self-pollinations) among maternal plants and sites across a range of population sizes and degrees of isolation. Because small plant populations are often pollen-limited (Knight et al.,

2005), we characterize fruit set in open-pollinated plants under natural conditions and examine relationships with population size and degree of isolation. We calculate the index of self-incompatibility (ISI) as a species average and examine variation between type of outcross pollen (near or far) and variability among individual maternal plants. Thus, we go beyond the traditional usage of ISI in which species mating systems are categorized as self-compatible (ISI < 0.2), mixed-mating (0.2 < ISI > 0.8), or self-incompatible (0.8 < ISI) (see Raduski, Haney, & Igić, 2011). Additionally, we test for reproductive assurance in the absence of pollinators (autogamy). Our analyses focus on answering the following questions:

1. Are there relationships between fruit set in self- and in open-pollinated flowers with population size, or metrics quantifying the degree of population isolation?
2. Does cross-type influence fitness (fruit set) and how much variability is due to maternal plants and sites?

We expected to find evidence of GSI in *T. virginicum* and concomitantly we predicted that the success of self-pollination would be extremely low. However, preliminary data from five *T. virginicum* plants at three sites indicated great variation among maternal plants in ability to form fruit from hand self-pollinations. Thus, we sought to investigate this variability in selfing success with a larger sample. We predicted that success of self-pollination would vary with population size or degree of isolation. Specifically, we predicted that selfing success would increase with decreasing population size and increasing isolation due to strong selection for breakdown of self-incompatibility in putatively mate-limited populations experiencing little gene flow. In contrast, we predicted that fruit set in open-pollinated plants would increase with increasing population size and decreasing isolation due to higher probabilities of compatible mates, larger floral displays to pollinators, and shorter pollinator flight distances. Finally, we predicted that far-cross success would be consistently greater due to higher probabilities of delivering a novel S-allele as well as the effects of heterosis.

The possibility that *T. virginicum* may be pseudo-self-fertile (PSF, sensu Levin, 1996) became evident during our field studies. Genes conferring PSF have been demonstrated in the genus by Atwood (1942) and Townsend (1965, 1966, 1969). *Trifolium repens* and *Trifolium pratense* are known to exhibit breakdown in self-incompatibility due to PSF loci whose function may vary over time (Riday & Krohn, 2010; Yamada, Fukuoka, & Wakamatsu, 1989) or under high temperatures (Townsend, 1965). If PSF is a natural component of the mating system, we expected some selfing success, but with great variation across maternal plants (Levin, 1996). Variation in selfing success in early versus late flowers is also expected but because this experiment was not designed to test for PSF, we did not test for temporal variation.

2 | METHODS

2.1 | Site selection

We used graph theoretic analysis as implemented by the computer program Conefor 2.6 (Saura & Torné, 2009, 2012) to select sites that varied independently in size and connectivity. Because of the large amount of information that can be gained with few data inputs, graph theory is being increasingly applied to conservation problems (Calabrese & Fagan, 2004; Neel, Tumas, & Marsden, 2014; Pascual-Hortal & Saura, 2006). Graphs provide a spatially explicit representation of landscapes based on the distances at which habitat patches (termed nodes) are connected to one another into networks. We used a layer of all known populations of *T. virginicum* in Maryland as nodes. We calculated pairwise geographic distance among patches in ArcMap 10.0 (ESRI 2011). Census size was used to represent habitat size and quality for each node. Census size of each population is based upon counts of individuals at peak flower when plants are most visible.

We selected two graph metrics ($IIC_{connector}$ and BC(IIC)) that provide meaningful and interpretable measures of connectivity (Bodin & Saura, 2010). $IIC_{connector}$ measures each patch's contribution to connectivity as a stepping stone through which other patches of the network are connected. BC(IIC) measures the degree to which a node sits among other nodes in a network based on the number of shortest paths for movement between all pairs of nodes that pass through that node

(Baranyi, Saura, Podani, & Jordan, 2011; Bodin & Saura, 2010). We assess the importance of each node using forms of $IIC_{connector}$ and BC(IIC) that are calculated as the change in values between the full network and a network from which each focal node has been sequentially removed (called $dIIC_{connector}$ and dBC(IIC)). $dIIC_{connector}$ and dBC(IIC) are independent measures of connectivity (Bodin & Saura, 2010). The former measures the importance of a node based on how much connectivity would be reduced if the patch was lost. The latter measures the importance of a patch based on its centrality in an existing landscape.

To assess connectivity at multiple scales, we calculated patch importance values for $dIIC_{connector}$ and dBC(IIC) for interpatch distances from 150 m (the minimum distance between two patches) to 36,000 m in 100-m increments (measured from approximate population centroids). We graphically assessed the behavior across this range of distances and noted the distance(s) at which local maximum values (thresholds) were achieved. At these distances, the network of patches is particularly susceptible to changes in connectivity. We noted three such distances: 500, 1,000, and 1,850 m that we examined further.

We examined the rank of node importance values of $dIIC_{connector}$ and dBC(IIC) at the three distances for use. Sites were categorized as connected if both metrics were >0 at two of the three distances and isolated if both metrics were zero at two of the three distances. Sites were subdivided into small (<50 plants), medium (≥50 and <100 plants), and large (≥100 plants) populations based on censuses by the Maryland Natural Heritage Program from 1984 to 2015. Sites were then evaluated for accessibility (e.g., not on private property) and appropriateness for study (e.g., supported >3 plants).

Twenty-one sites that covered the range of size and connectivity we sought were located in a 10 km × 15 km area that lies within the Green Ridge State Forest in Allegany County, Maryland (Figure 1). We updated the census for these sites in 2013 (Table 1). Sites occurred on the same geological formation and all experience the same general climate, thus limiting the possibility of local adaptation to different environments. The population size distribution of the study sites roughly mirrors the distribution of all *T. virginicum* populations in Maryland (Maryland Natural Heritage Program, Annapolis, MD): thirteen (62%) of the study sites are classified as small populations versus 76% of all sites, five (19%) are medium populations versus 13% of all sites, and three (14%) are large populations, versus 11% of the total (Table 1).

2.2 | Selection of maternal plants at sites

At each site, we selected maternal plants for pollination treatments (detailed in the next section) and deployed pollinator exclusion bags prior to flowering (Table 1). The choice of experimental plants was based on a combination of distance from other plants and accessibility. Even within larger patches of habitat, *T. virginicum* individuals are often clustered in small microsites. To increase chances of sampling less related individuals, we selected maternal plants at each site that were separated by at least 5 m as a general convention. We continued selection of maternal plants until all such isolated clusters of plants at a site were utilized. If all plants at a site were within 5 m, we chose only a single plant.

TABLE 1 Study sites with number of *Trifolium virginicum* censused in 2013, connectivity category (>0 = connected, 0 = isolated) using $dIIC_{connector}$ and $dBC(IIC)$ at two of three dispersal distances (500, 1,000, 1,850 m) and number of maternal plants receiving each of the three crossing treatments, tested for autogamous selfing, and open-pollination

Site	Census size (2013)	Population size category	Connectivity category	Number of mothers receiving crossing treatments	Autogamy	Open
1	64	Medium	Isolated	5	5	5
8	131	Large	Isolated	7	3	7
9	67	Medium	Isolated	4	0	4
15	8	Small	Isolated	1	1	1
32	130	Large	Isolated	7	3	7
33	46	Small	Connected	1	1	1
34	65	Medium	Connected	4	3	4
35	9	Small	Connected	1	0	2
38	158	Large	Connected	5	0	5
43	23	Small	Isolated	2	1	2
44	23	Small	Isolated	3	6	3
45	30	Small	Connected	2	3	2
46	9	Small	Isolated	1	1	1
51	9	Small	Connected	1	1	1
54	12	Small	Connected	1	0	1
55	54	Medium	Connected	5	2	5
57	46	Small	Isolated	3	1	3
60	6	Small	Isolated	1	1	1
74	19	Small	Connected	2	0	3
99	50	Medium	Isolated	2	1	2
100	38	Small	Isolated	2	2	2
$N = 21$	$\Sigma = 960$			$N = 60$	$N = 35$	$N = 62$

Σ maternal plants = 157

For each maternal plant treated at each site, we chose one additional maternal plant that received no treatment to represent open-pollination (Table 1). Finally, we chose 35 additional plants at 16 sites to test for autogamy (Table 1). These plants received no additional manipulation beyond bagging.

2.3 | Crossing design and procedures

On each experimental maternal plant, we established three treatments reflecting distance from pollen source. Implementing all treatments on the same plant controls for maternal genotype. The selfing treatment (S) used pollen from within the same inflorescence. The near-cross treatment (NX) used pollen from 20 to 130 anthers collected from 2 to 13 donors within the site that were ≥5 m from the maternal plant. The far-cross treatment (FX) used pollen from 20 to 130 anthers from 2 to 13 fathers from sites at least 1 km distant. We pooled pollen from multiple fathers to increase the probability of delivering at least one compatible S-allele. Variation in number of donors resulted from differences in the number of plants in populations and the number of flowers with pollen available. In some of the smallest populations, additional plants other than the focal maternal plant were needed to serve as pollen donors for the near-cross; these plants were by necessity sometimes within 5 m. Pollen for the NX treatment was collected

while at the site and was applied within 30–60 min of collection. Pollen for the FX treatment was collected 1–4 hr prior to application. Anthers were mixed in a vial and kept on ice until they were applied to the mothers. In some cases, entire heads (with attached peduncle) were kept overnight, and anthers were harvested from freshly open flowers.

All crosses were performed under field conditions and sites were visited daily (weather permitting) between May 5 and May 15, 2013 to perform crosses. Pollination treatments were performed on flowers when the banner was expanded and reflexed exposing magenta-colored nectar guides to pollinators. We had previously determined that the stigma was receptive in this period as assessed by testing for peroxidase activity using a 3% solution of hydrogen peroxide (Dafni, Kevan, & Husband, 2005). All flowers were emasculated using forceps. Anthers were applied directly to the stigma in the self-treatment and outcross treatments used stamens with dehiscing anthers from the NX or FX anther pools as appropriate. One to two anthers were haphazardly selected from the appropriate tube for application to the stigma using forceps. Forceps were dipped in ethanol and flamed before moving to the next treatment.

We attempted to complete all treatments on one maternal plant (ranging from 35 to 60 min per plant) on a single day. If few flowers were receptive during a single visit, we returned the following day in an attempt to pollinate at least five flowers per head per treatment. We succeeded in pollinating 897 flowers with an average of 5.5 flowers per treatment. The number of flowers per head in each pollination treatment varied because the number of flowers available for pollination during any time interval was unpredictable. In total, 180 heads on 60 individual maternal plants received manipulative treatments (S, NX, FX).

Flower heads from all manipulative treatments, flower heads testing for autogamy and heads from open-pollinated plants were collected and brought to the laboratory for dissection after 4–6 weeks in the field. There was occasional loss of individual treatments on some maternal plants due to destruction of pollinator exclusion bags by wildlife. Percent fruit set $\left(\frac{\text{number of flowers producing fruit}}{\text{number of flowers treated}} \times 100 \right)$ was calculated for each treatment on each maternal plant. We determined seed weight (nearest 0.01 mg) from experimental crosses and open-pollination using a balance.

2.4 | Fruit set and number of seeds per pod in open-pollinated flowers under natural conditions

We quantified natural fruit set from 62 open-pollinated heads from 62 maternal plants at 21 sites. We also quantified the average number of flowers per head and the number of seed per pod in 1,199 legumes containing at least one seed. All statistics reported as mean ± standard error (SE) or median and range of values.

2.5 | Data analyses

We performed all statistical analyses (with the exception of a generalized linear mixed model detailed below) using Systat 13 (Systat Software, Inc., San Jose, CA, USA). We calculated mean (±SE) seed weight for each crossing treatment in each site (S, 17 sites, $N = 167$; NX, 19 sites, $N = 157$; FX, 16 sites, $N = 150$) and for open-pollination (O, 17 sites, $N = 1,001$). We tested for differences in seed weight among crossing treatments using a general linear mixed model with cross-type as fixed effect and site as a random effect. We assessed variability in mating system using percent fruit set in each of the three manipulative pollination treatments remaining on maternal plants (S, NX, FX; $N = 55, 54, 50$, respectively), open-pollinated flowers (O, $N = 62$), and heads testing for autogamy (A, $N = 35$). Percent fruit set violated assumptions of normality (Shapiro–Wilk tests, $p < 0.05$) so we report both mean and median values. We tested the significance of differences in the variances for manipulative treatments and open-pollination using Levene's test, confirming that our data also violated assumptions of homogeneity of variances ($F = 3.314_{2, 135}$, $p = 0.026$).

We calculated the index of self-incompatibility (ISI) following Raduski et al. (2011) as

$$\text{ISI} = 1 - (\text{selfed success/outcrossed success})$$

We computed Spearman rank correlations between percent fruit set in self- and open-pollination and population size at each site ($N = 21$) using 1,000 bootstraps to assess the significance of the correlation. We also tested for relationship between self- and open-pollination averaged within sites using Spearman rank.

We explored relationships between the calculated values for connectivity metrics ($d\text{IIC}_{\text{connector}}$, $d\text{BC(IIC)}$) at 500, 1,000, and 1,850 m, and percent fruit set in self- and open-pollination at each site ($N = 21$) using Spearman rank correlation analyses.

We examined the effect of cross-type on probabilities of fruit set using a hierarchical generalized linear mixed model (GLMM; as implemented by PROC GLIMMIX, SAS v. 9.4). GLMMs are the appropriate tool for analyzing non-normal data with random effects (Bolker et al., 2008). We modeled fruit set as a binary outcome (failure = 0, success = 1) for each cross-type (self, near, far) resulting from individual hand-pollinated flowers ($N = 766$), nested within maternal plants ($N = 46$), which were nested within site ($N = 15$). We restricted our analysis to the 46 maternal plants with no missing data (all cross-types present) to control for maternal genotype. We specified a binomial distribution and a logarithmic link (logit) to transform the dichotomous outcome into a continuous variable (the log-odds). The logit transformation allows us to establish a linear relationship between our binary outcome variable (fruit set) and the predictor variables (cross-type as fixed effect and maternal plant and site as random effects). The log-odds of fruit set and errors for cross-type, maternal plant and site was estimated using residual pseudo-likelihood (modeled in PROC GLIMMIX as the probability of fruit-set failure). To account for the hierarchical nature of the data, log-odds is calculated using different intercepts for random subjects (site, maternal plant nested within site) to estimate variance parameters of subject and subject × cross-type interactions. The intercepts are the average log-odds of fruit set for the near-cross because this represents the most probable outcross event among maternal plants within sites. We estimate log-odds solutions for the fixed effects (cross-type = self, near, and far) predicting the probabilities of fruit set failure and test for significant differences in the probability of fruit set failure between cross-types.

For each random subject and subject × cross-type interaction, we report the variance parameter estimates, standard errors, and the percent of the total variance attributable to subjects by dividing the estimate by the sum of variances. We did not calculate log-likelihood ratio tests for significance of random effects because these methods have not been resolved for binary data (Bolker et al., 2008).

Because log-odds range from zero to positive infinity, we converted the log-odds of fruit set to predicted probabilities (expressed as a percent) of fruit set success and failure for fixed effects using the equation

$$\phi_{ij} = \frac{e^{\eta ij}}{1 + e^{\eta ij}}$$

where e takes a value of approximately 2.72, ηij is the log-odds of failure, and ϕ is the probability of failure.

3 | RESULTS

The sample of 62 open-pollinated *T. virginicum* heads yielded a total of 2,410 flowers with a mean of 38.3 (±8.7) flowers per head (median = 37, range = 22–61). Median fruit set per head was 48.7% (Table 2). Flowers matured centripetally within flower heads and each

Benefits of gene flow are mediated by individual variability in self-compatibility in small isolated populations...

17

TABLE 2 Summary statistics for percent fruit set within heads in each treatment. Columns present statistics for distribution of the values, and rows represent treatments (S, self; NX, near-cross; FX, far-cross; A, autogamy; O, open-pollination). Variances with the same superscript letter are not significantly different (p < 0.05)

Cross-type	Number of heads	Mean fruit set	Median fruit set	Range	Variance
S	55	49.7	50.0	0, 100	979.35[a]
NX	54	55.8	60.0	0, 100	1174.04[a]
FX	50	69.6	63.5	17, 100	581.62[b]
A	35	0.7	0	0, 9.8	N/A
O	62	49.3	48.7	0, 87	542.42[b]

one was available to pollinators for a single day. These open-pollinated heads yielded 1,199 well-formed pods that each most often contained a single seed (mean = 1.06 ± 0.02, median = 1, range 1–1.81).

Trifolium virginicum set fruit upon selfing (median = 50%) and outcrossing (FX median = 63.5%; NX median = 60.0%) (Table 2; Figure 2). There was essentially no autonomous selfing based on fruit set in the 35 heads testing for autogamy (median = 0) (Table 2; Figure 2). There was no main effect or interaction of cross-type and site on seed weight ($F_{3, 45}$ = 1.08, p = 0.22) so seed weight was eliminated as a variable of interest.

The mean ISI value of 0.05 (N = 53) would place *T. virginicum* in the category of self-compatible species according to a survey of ISI values for >1,200 angiosperm taxa (Raduski et al., 2011). Observed values of ISI across maternal plants ranged from 1 to −4. This variation is obscured if only species-level averages are reported (as in Raduski et al., 2011; Schoen & Lloyd, 1992), and if negative ISI values are set to zero (as in Raduski et al., 2011). These negative values indicate when selfing outperforms outcrossed matings. The success of self-pollination varied dramatically among maternal plants finding instances of both full self-compatibility (100% fruit set) and apparent self-incompatibility (0% fruit set) (Table 2; Figure 3).

The mean ISI value indicated little difference in the overall success of self- versus outcross pollen in terms of fruit set when FX and NX, as outcross pollen, were combined. However, this result masks the contrasting ISI values for the two types of outcross. For the S/NX ratio, ISI of −0.12 (N = 43) indicates higher fruit set on average with self-pollen.

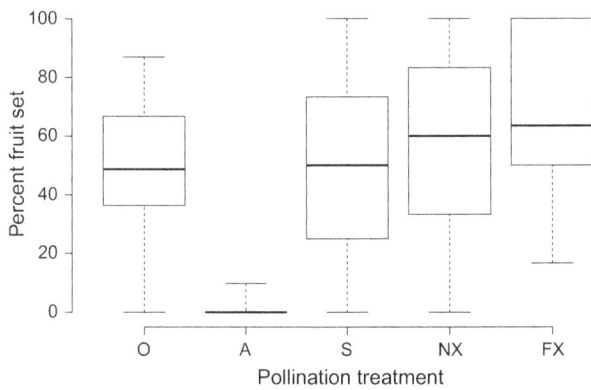

FIGURE 2 Percent fruit set within treatments for all maternal plants: Open-pollination (O), bagged heads testing for autogamy (A), self-pollination (S), near-cross (NX), far-cross (FX). Median marked by the central line within the box. The area of the box limits 25th and 75th percentiles (the interquartile range). Vertical lines extending from the box extend to minimum and maximum values

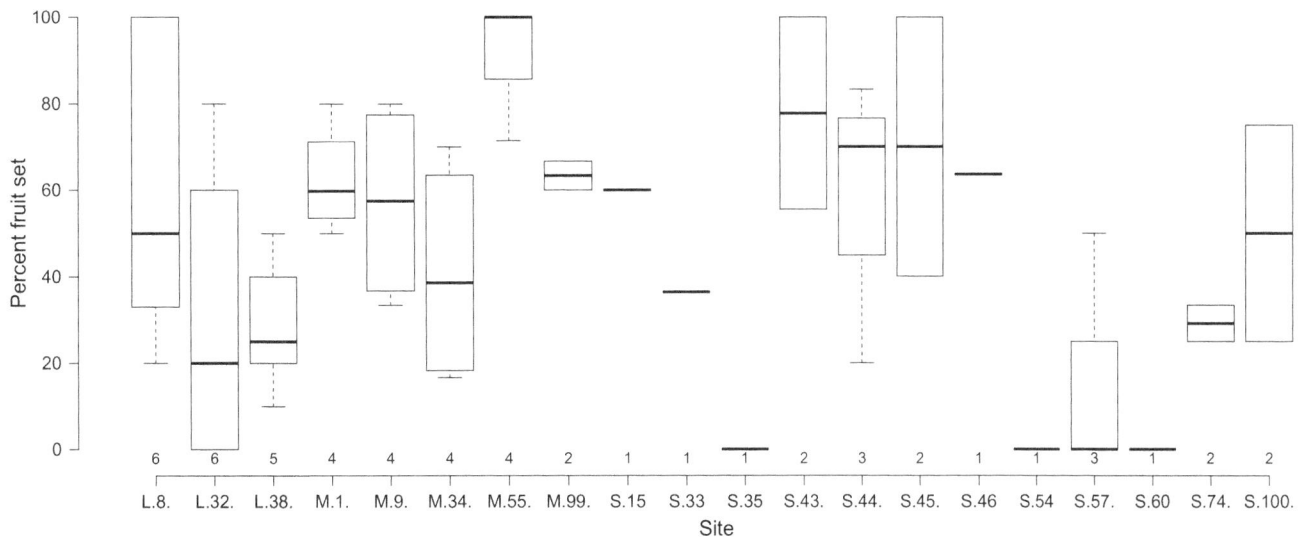

FIGURE 3 Variability in selfing success (percent fruit set) of maternal plants within 20 sites. On the horizontal axis, site census size (L, M, S) followed by site number is shown below the axis and sample size above the axis (no data for site 51). Median marked by the central line within the box. The area of the box (when N > 1) limits 25th and 75th percentiles (the interquartile range). Vertical lines extending from the box extend to minimum and maximum values

TABLE 3 Log-odds estimates and standard errors from a generalized linear mixed-model (GLMM) analysis predicting the probability of fruit set for fixed effects (cross-type). Significant ($p < 0.05$) contrasts with near-cross (intercept) are indicated by asterisks. Log-odds estimates were converted to predicted probabilities of fruit set failure (PP_0) in the last column. Probability of fruit set success (PP_1) = 1 − PP_0

Effect	Cross-type	Estimate	Error	DF	t value	Pr > t	PP_0
Intercept		−0.2065	0.2371	14	−0.87	0.40	55.2
Cross-type	Self	0.2258	0.2909	28	0.78	0.44	55.6
Cross-type	Far	−0.6480	0.3002	28	−2.16	0.04*	34.3
Cross-type	Near	0					

Covariance parameter	Subject	Estimate	Error	Estimate/sum of variances
Intercept	Site	0.0300	0.1597	2.39%
Cross-type	Site	0.1044	0.1686	8.21%
Intercept	Maternal plant(site)	0.4466	0.2621	35.13%
Cross-type	Maternal plant(site)	0.6899	0.2684	54.27%

TABLE 4 Logistic covariance parameter estimates, standard errors, and percent variance accounted for by random subjects and subject × cross-type interactions. Variance in the log-odds of fruit set modeled as a binary outcome (0 = failure, 1 = success)

By contrast, for the S/FX ratio, ISI of 0.21 ($N = 48$) suggested mixed-mating and a strong advantage to far-cross pollen (see Appendix S1).

We found no significant correlations between population size and percent fruit set resulting from self- or open-pollination. Self- and open-pollination success within sites was positively but not significantly correlated ($R = 0.32$, $p = 0.16$). As expected, open-pollination success was positively correlated with population size ($R = 0.32$), but we found no clear evidence of pollen limitation as low fruit set was observed in some small, isolated populations (e.g., site 60, 0%; site 46, 9%) but not in others (e.g., site 15, 62%; site 100, 58%). Self-success shows little relationship with population size ($R = −0.02$). We also found no significant correlations between fruit set in self- and open-pollination with connectivity with neither the contribution of habitat patches as a stepping stone ($dIIC_{connector}$) nor contribution of patch centrality ($dBC(IIC)$) at any of the three threshold distances ($N = 21$ sites, maximum $R = 0.22$, $p = 0.34$ between open-pollination success and $dIIC_{connector}$ at 1,850 m).

The GLMM revealed a significant effect of cross-type while controlling for maternal plant and site characteristics ($F = 4.66_{2, 28}$, Pr > F = 0.02). The probability of fruit set failure for self- and near-cross flowers was >0.55, whereas the probability of fruit set failure with far-cross flowers is significantly reduced (<0.35) (Table 3). The probability of fruit set varied considerably across maternal plants within sites, but not among sites, demonstrating individual variability in mating system among maternal plants at all sites (Table 4).

4 | DISCUSSION

We have shown that plants receiving pollen from another population (far-cross) have significantly greater reproductive success than self- or near-cross-pollinations. The beneficial effects of gene flow are demonstrated by both a significantly higher probability of individual flowers setting fruit when treated with far-cross pollen and positive ISI values for the S/FX ratio. At the same time, we found evidence of

reproductive assurance resulting from successful hand self-pollination in this putatively self-incompatible plant. Most interestingly, we found great variation among selfed mothers that was independent of census population size and two measures of connectivity.

Census sizes of all sampled populations are small enough to generate population bottlenecks in S-allele diversity (Thrall, Encinas-Viso, Hoebee, & Young, 2014; Young et al., 2012) and to yield high inbreeding (Leimu, Mutikainen, Koricheva, & Fischer, 2006). The benefits of gene flow as mediated by artificial pollination could act by adding S-allele diversity and by alleviating inbreeding depression (e.g., Frankham, 2015). The lack of relationship between self-success and population size would appear to contradict evidence of the importance of population size in mediating the effects of inbreeding (Angeloni et al., 2011; Leimu et al., 2006). However, given that all of our populations have <150 individuals and most have <50 (Table 1), there is little room for variation in inbreeding effects. The lack of relationship of fruit set in self- and open-pollination with connectivity measures could likewise indicate that gene flow does not vary over the distances that separate the remnant shale barrens. It is possible that the median separation of 1.0 km from the nearest neighbor is sufficient to functionally isolate barrens from one another and to preclude any gradient in gene flow. Additionally, variability in self-compatibility in response to smaller population size and increase in isolation is expected as long-lived species like *T. virginicum* may maintain levels of genetic variation that existed prior to the population declines for an extended period and may not exhibit expected effects of small, isolated populations. For example, Schleuning, Niggemann, Becker, and Matthies (2009) concluded that the longevity of *Trifolium montanum* has likely delayed extirpation of some populations. A lag in fitness declines in perennial species may be particularly long if there is strong selection against selfed progeny (Davies et al., 2015; Delmas et al., 2014; Young, Boyle, & Brown, 1996). Strong selection against inbred progeny in *T. virginicum* is suspected as census size estimates for some populations in 1984 were exactly the same in 2013; however, we cannot eliminate other factors such as poor pollinator visitation rates or habitat degradation.

Although we did find evidence for reproductive assurance, selfing success was significantly more variable than far-cross or open-pollination, ranging from instances of full self-compatibility to complete failure indicating apparent rejection of self-pollen (Figure 3). Although error is a potential explanation, it is unlikely because the selfing treatment required little manipulation. This variability has two potential explanations: (1) random drift of alleles modifying the self-incompatibility reaction, that is, pseudo-self-fertility and (2) early-acting inbreeding depression due to expression of deleterious/lethal recessives. Variation in self-incompatibility among maternal plants has been observed in a wide range of species (Good-Avila, Mena-Alí, & Stephenson, 2008) including naturally rare or endemic species (Schleuning et al., 2009 *Trifolium montanum*; Busch et al., 2010 *Leavenworthia alabamica*; Alonso & Garcia-Sevilla, 2013, *Erodium cazorlanum*).

We confirmed one characteristic of PSF identified by Levin (1996): higher fruit production with outcross pollen than self-pollen but a nearly continuous distribution of fruit set following self-pollination (Figure 3). Additionally, two components of PSF identified by Levin (1996), the age of flowers and the temperature at the time of treatment, may represent gene by environment interactions that may have played a role in the observed variability. Although we completed three pollination treatments on a single plant within 1 hr on a single day, treatments were applied to different plants throughout a day and replicate treatments occurred over a 2-week period. Thus, there is the possibility that time (flower age) and temperature contributed to the observed variability in self- and near-cross success. These results strongly suggest the presence of alleles (with perhaps environmental triggers) that modify the self-incompatibility reaction. Our finding that variability in the probability of fruit set was explained by individual variability in self-compatibility among maternal plants across all sites supports this hypothesis (Table 4).

Alternatively, variability in the success of self- and near-cross-pollination could be due to early-acting inbreeding depression because the former is the most extreme form of inbreeding and the latter likely due to mating by individuals that were already closely related. According to theory, maintenance of a self-incompatible genetic system is linked with strong inbreeding depression (Gervais, Awad, Roze, Castric, & Billiard, 2014). A self-compatible mutant that arises within a self-incompatible population should spread rapidly as long as the costs of selfing do not exceed the 50% transmission advantage and the benefit of reproductive assurance. The severity of inbreeding depression will be affected by genetic relatedness among individuals and the efficacy of purging in these small populations, both primarily functions of population size (Young & Pickup, 2010). Purging of genetic load has been invoked to explain why small populations persist despite increased inbreeding (Garcia-Dorado 2015; Husband and Schemske 1996; Winn et al. 2011). However, the effectiveness of purging is dependent upon the distribution of deleterious alleles and severity of their effects. Recessive mutations with major fitness effects are more easily purged in an inbreeding population, whereas mutations of small effect may be fixed by drift (Carr & Dudash, 2003; Gervais et al., 2014) or sheltered by linkage to loci under frequency dependent selection

such as the self-incompatibility locus (Glémin, Bataillon, Ronfort, Mignot, & Olivieri, 2001). The fitness consequences of increased inbreeding in our populations could theoretically be ameliorated by purging of genetic load and strong selection against inbred progeny (Crnokrak & Barrett, 2002; Fox, Scheibly, & Reed, 2010). However, the fraction of mutation load fixed by drift (drift load) is expressed even under random mating and may form a feedback loop where drift load lowers population size, which in turn enhances drift load (Carr & Dudash, 2003; Willi, Griffin, & Van Buskirk, 2013). A highlight of our study was finding a significant heterotic benefit of far-cross relative to near-cross, which is the best indicator of drift load in populations. Fitness declines caused by drift load that is fixed within small and isolated populations cannot recover without increased gene flow (Willi et al., 2013).

We cannot distinguish between PSF and inbreeding depression with our data and doing so will be important for this species (and many other endemics) because of the implications for conservation. If *T. virginicum* exhibits PSF, then this may be an evolutionary adaptive mechanism that allows persistence of small and isolated populations by ensuring some reproduction and thus the demographic integrity of these small populations. However, the benefits of reproductive assurance gained by breakdown in self-incompatibility may occur at a cost of reduced fitness in resulting progenies, and lifetime inbreeding depression may be particularly acute for long-lived perennial species (Alonso & Garcia-Sevilla, 2013; Delmas et al., 2014; Morgan, Schoen, & Bataillon, 1997). Given the current landscape of isolated patches, increasing the ratio of self- relative to outcross progeny may have long-term population-level-fitness consequences as genetic relatedness increases among remaining individuals and inbreeding reduces fruit and seed set.

Importantly, we found that reproductive assurance can occur in *T. virginicum* through pollinator-mediated selfing as determined analogously by hand-pollination in our study. However, we found no evidence that autogamous self-pollination is a reliable mechanism for reproductive assurance in absence of pollinators (Table 2; Figure 2); thus, pollinator abundance and floral attractiveness to pollinators in small populations will impact population persistence (Levin, 2012).

Current shale barren restoration and management strategies of selective tree removal and prescribed fire may theoretically restore available habitat for *T. virginicum* (see Tyndall, 2015). Although there are no data on *T. virginicum* dispersal distances, Matter, Kettle, Ghazoul, Hahn, and Pluess (2013) found that most dispersal events for the congener *T. montanum* in calcareous grassland fragments were <1 m and the maximum distance was 324 m. Thus, there is a high probability that *T. virginicum* seeds are dispersed near the maternal plant or into adjacent nonhabitat forest. Potential for colonizing newly created habitat patches will be limited unless they are extremely close to existing patches. Limitation in the colonizing ability of plant species is recognized as a major obstacle to habitat-based restoration (Donohue, Foster, & Motzkin, 2000). Even if historical disturbance regimes are reinstituted, many previously occupied patches may remain unoccupied and patches remain fragmented. An understanding of long-term gene flow patterns in *T. virginicum* represents an unmet need but is critical to understanding historical connectivity among populations. If historical patterns of gene flow are precluded by the current levels of isolation, then restoration goals should focus on re-establishing gene flow between sites rather than on only single-site management.

A reasonable hypothesis is that the historical landscape configuration facilitated gene flow, at least among proximal populations (e.g., those barrens occurring along the same ridgeline) and that changes in the landscape configuration and alteration in natural processes have resulted in severe fragmentation. Fire exclusion likely plays a role in restricting populations of T. virginicum to open-canopy sites over shallow soils. Prescribed burns are potential management treatments that can increase populations. For example, a prescribed burn at one site in 1999 resulted in apparent population growth of T. virginicum according to census performed at dates preburn (May 1984, N = 74) and postburn (May 2013, N = 131). The question for long-term conservation of T. virginicum and other species restricted or endemic to shale barrens is whether management of a few focal sites having relatively large populations (e.g., >100) is enough to conserve the evolutionary potential of these species. This management focus presumes that the larger barrens capture both general levels of genetic diversity and sufficient S-allele diversity. Given the small sizes of even the largest populations and habitat areas, this assumption may not be warranted. Additionally, the success of single-site management is measured by population increase, but the plants are long-lived and there may be a substantial lag time in discerning demographic trends, a characteristic of extinction debt (Hanski & Ovaskainen, 2002; Tilman, May, Lehman, & Nowak, 1994). Future work on T. virginicum focuses on assessing whether the heterosis observed after far-cross extends to lifetime fitness (germination to reproduction) of F1 progeny. Additionally, we need to assess genetic relatedness of individuals within sites to determine the extent to which inbreeding versus lack of S-allele diversity explains the differential success between near- and far-cross in this study. Finally, and critical for management, we need to elucidate long-term (historical) patterns of gene flow to develop appropriately scaled reserves.

ACKNOWLEDGEMENTS

This project was made possible by funding from the Chesapeake Bay and Endangered Species Fund administered by the Maryland Department of Natural Resources, Wildlife and Heritage Service. Field assistance was provided by Jennifer Selfridge and Michael Baranski. Edward Thompson, Wayne Tyndall, and Mark Beals assisted with logistics and site access. Yvonne Willi provided expertise with statistical modeling and interpretation.

REFERENCES

Alonso, C., & Garcia-Sevilla, M. (2013). Strong inbreeding depression and individually variable mating system in the narrow endemic Erodium cazorlanum (Geraniaceae). Anales del Jardin Botánico de Madrid, 70, 72–80.

Angeloni, F., Ouborg, N. J., & Leimu, R. (2011). Meta-analysis on the association of population size and life-history with inbreeding depression in plants. Biological Conservation 144, 35–43.

Atwood, S. S. (1942). Genetics of pseudo-self-incompatibility and its relation to cross-incompatibility in Trifolium repens. Journal of Agricultural Research, 64, 699–709.

Baranyi, G., Saura, S., Podani, J., & Jordan, F. (2011). Contribution of habitat patches to network connectivity: redundancy and uniqueness of topological indices. Ecological Indicators, 11, 1301–1310.

Bodin, O., & Saura, S. (2010). Ranking individual habitat patches as connectivity providers: Integrating network analysis and patch removal experiments. Ecological Modeling, 221, 2393–2405.

Bolker, B. M., Brooks, M. E., Clark, C. J., Geange, S. W., Poulsen, J. R., Stevens, M. H. H., & White, J.-S. S. (2008). Generalized linear mixed models: A practical guide for ecology and evolution. Trends in Ecology and Evolution, 24, 127–135.

Busch, J. W., Joly, S., & Schoen, D. J. (2010). Does mate limitation in self-incompatible species promote the evolution of selfing? The case of Leavenworthia alabamica. Evolution, 64, 1657–1670.

Calabrese, J. M., & Fagan, W. F. (2004). A comparison shopper's guide to connectivity metrics: Trading off between data requirements and information content. Frontiers in Ecology and the Environment, 2, 529–536.

Carr, D. E., & Dudash, M. R. (2003). Recent approaches into the genetic basis of inbreeding depression in plants. Philosophical Transactions of the Royal Society of London, B, Biological Sciences, 358, 1071–1084.

Casey, N. M., Milbourne, D., Barth, S., Febrer, M., Jenkins, G., Abberton, M. T., ... Thorogood, D. (2010). The genetic location of the self-incompatibility locus in white clover (Trifolium repens L.). Theoretical and Applied Genetics, 121, 567–576.

Charlesworth, D. (2006). Evolution of plant breeding systems. Current Biology, 16, 726–735.

Cheptou, P.-O., & Donohue, K. (2011). Environment dependent inbreeding depression: Its ecological and evolutionary significance. New Phytologist, 189, 395–407.

Copeheaver, C. A., Fuhrman, N. E., Gellerstedt, L. S., & Gellerstedt, P. A. (2004). Tree encroachment in forest openings: A case study from Buffalo Mountain, Virginia. Castanea, 69, 297–308.

Crnokrak, P., & Barrett, S. C. (2002). Perspective: Purging the genetic load: A review of the experimental evidence. Evolution, 56, 2347–2358.

Dafni, A., Kevan, P. G., & Husband, B. C. (Eds.) (2005). Practical pollination biology. Ontario: Enviroquest Ltd.

Davies, S. J., Cavers, S., Finegan, B., White, A., Breed, M. F., & Lowe, A. J. (2015). Pollen flow in fragmented landscapes maintains genetic diversity following stand-replacing disturbance in a neotropical pioneer tree, Vochysia ferruginea. Heredity, 115, 125–129.

Delmas, C., Cheptou, P.-O., Escarvage, N., & Pornon, A. (2014). High lifetime inbreeding depression counteracts reproductive assurance benefit of selfing in a mass-flowering shrub. BMC Evolutionary Biology, 14, 243. doi:10.1186/s12862-014-0243-7.

Donohue, K., Foster, D. R., & Motzkin, G. (2000). Effects of the past and the present in species distribution: Land-use history and demography of wintergreen. Journal of Ecology, 88, 303–316.

Duminil, J., Hardy, O. J., & Petit, R. J. (2009). Plant traits correlated with generation time directly affect inbreeding depression and mating system and indirectly genetic structure. BMC Evolutionary Biology, 9, 177–191.

Eckert, C. G., Kalisz, S., Geber, M. A., Sargent, R., Elle, E., Cheptou, P.-O., ... Johnston, M. O. (2010). Plant mating systems in a changing world. Trends in Ecology and Evolution, 25, 35–43.

Edmands, S. (2007). Between and rock and a hard place: Evaluating the relative risks of inbreeding and outbreeding for conservation and management. Molecular Ecology, 16, 463–475.

ESRI (2011). ArcGIS desktop: Release 10. Redlands, CA: Environmental Systems Research Institute.

Fenster, C. B., & Sork, V. L. (1988). Effect of crossing distance and male parent on in vivo pollen tube growth in Chamaecrista fasciculata. American Journal of Botany, 75, 1898–1903.

Foster, D., Swanson, F., Aber, J., Burke, I., Brokaw, N., Tilman, D., & Knapp, A. (2003). The importance of land-use legacies to ecology and conservation. *BioScience*, *53*, 77–88.

Fox, C. W., Scheibly, K. L., & Reed, D. H. (2010). Experimental evolution of the genetic load and its implications for the genetic basis of inbreeding depression. *Evolution*, *62*, 2236–2249.

Frankham, R. (2015). Genetic rescue of small inbred populations: Meta-analysis reveals large and consistent benefits of gene flow. *Molecular Ecology*, *24*, 2610–2618.

Frankham, R., Ballou, J. D., Eldrige, M. D., Lacy, R. C., Ralls, K., Dudash, M. R., & Fenster, C. B. (2011). Predicting the probability of outbreeding depression. *Conservation Biology*, *25*, 465–475.

Garcia-Dorado, A. (2015). On the consequences of ignoring purging on genetic recommendations for minimum viable population rules. *Heredity*, *115*, 185–187.

Gervais, C., Awad, D., Roze, D., Castric, V., & Billiard, S. (2014). Genetic architecture of inbreeding depression and the maintenance of gametophytic self-incompatibility. *Evolution*, *68*, 3317–3324.

Glémin, S., Bataillon, T., Ronfort, J., Mignot, A., & Olivieri, I. (2001). Inbreeding depression in small populations of self-incompatible plants. *Genetics*, *159*, 1217–1229.

Good-Avila, S. V., Mena-Alí, J. I., & Stephenson, A. G. (2008). Genetic and environmental causes and evolutionary consequences of variations in self-fertility in self-incompatible species. In V. E. Franklin-Tong (Ed.), *Self-incompatibility in flower plants. Evolution, diversity, and mechanisms* (pp. 33–52). Berlin: Springer-Verlag.

Hanski, I., & Ovaskainen, O. (2002). Extinction debt at extinction threshold. *Conservation Biology*, *16*, 666–673.

Herlihy, C. R., & Eckert, C. G. (2002). Genetic cost of reproductive assurance in a self-fertilizing plant. *Nature*, *416*, 320–323.

Heschel, M. S., & Paige, K. N. (1995). Inbreeding depression, environmental stress, and population size variation in scarlet gilia (*Ipomopsis aggregata*). *Conservation Biology*, *9*, 126–133.

Holsinger, K. E. (2000). Demography and extinction in small populations. In A. G. Young (Ed.), *Genetics, demography and viability of fragmented populations* (pp. 55–74). Cambridge: Cambridge University Press.

Holsinger, K. E., & Vitt, P. (1997). The future of conservation biology: What's a geneticist to do? *The ecological basis of conservation* (pp. 202–216). New York, NY: Springer.

Honnay, O., & Jacquemyn, H. (2007). Susceptibility of common and rare plant species to the genetic consequences of habitat fragmentation. *Conservation Biology*, *21*, 823–831.

Husband, B. C., & Schemske, D. W. (1996). Evolution of the magnitude and timing of inbreeding depression in plants. *Evolution*, *50*, 54–70.

Igić, B., Bohs, L., & Kohn, J. R. (2006). Ancient polymorphism reveals unidirectional breeding system transitions. *Proceedings of the National Academy of Sciences, USA*, *103*, 1359–1363.

Jacquemyn, H., De Meester, L., Jongejans, E., & Honnay, O. (2012). Evolutionary changes in plant reproductive traits following habitat fragmentation and their consequences for population fitness. *Journal of Ecology*, *100*, 76–87.

Kalisz, S., Vogler, D. W., & Hanley, K. M. (2004). Context-dependent autonomous self-fertilization yields reproductive assurance and mixed mating. *Nature*, *430*, 884–887.

Karron, J. D., Ivey, C. T., Mitchell, R. J., Whitehead, M. R., Peakall, R., & Case, A. L. (2012). New perspectives on the evolution of plant mating systems. *Annals of Botany*, *109*, 1–11.

Keener, C. S. (1983). Distribution and biohistory of the endemic flora of the mid-Appalachian shale barrens. *The Botanical Review*, *49*, 65–115.

Keller, L. F., & Waller, D. M. (2002). Inbreeding effects in wild populations. *Trends in Ecology and Evolution*, *17*, 230–241.

Knight, T. M., Steets, J. A., Vamosi, J. C., Mazer, S. J., Burd, M., Campbell, D. R., ... Ashman, T. (2005). Pollen limitation of plant reproduction: Pattern and process. *Annual Review of Ecology, Evolution, and Systematics*, *31*, 467–497.

Lawrence, M. J. (1996). Number of incompatibility alleles in clover and other species. *Heredity*, *76*, 610–615.

Leimu, R., Mutikainen, P., Koricheva, J., & Fischer, M. (2006). How general are positive relationships between plant population size, fitness and genetic variation. *Journal of Ecology*, *94*, 942–952.

Levin, D. A. (1996). The evolutionary significance of Pseudo-Self-Fertility. *The American Naturalist*, *148*, 321–332.

Levin, D. A. (2012). Mating system shifts on the trailing edge. *Annals of Botany*, *109*, 613–620.

Lloyd, D. G. (1992). Self- and cross-fertilization in plants, II. The selection of self-fertilization. *International Journal of Plant Sciences*, *153*, 370–380.

Loveless, M. D., & Hamrick, J. L. (1984). Ecological determinants of genetic structure in plant populations. *Annual Review of Ecology and Systematics*, *15*, 60–90.

Marsden, B. W., Engelhardt, K. A. M., & Neel, M. C. (2013). Genetic rescue versus outbreeding depression in *Vallisneria americana*: Implications for mixing seed sources for restoration. *Biological Conservation*, *167*, 203–214.

Matter, P., Kettle, C. J., Ghazoul, T., Hahn, T., & Pluess, A. R. (2013). Evaluating contemporary pollen dispersal in two common grassland species *Ranunculus bulbosus* L. (Ranunuculaceae) and *Trifolium montanum* L. (Fabaceae) using an experimental approach. *Plant Biology*, *15*, 583–592.

Morgan, M. T., Schoen, D. J., & Bataillon, T. M. (1997). The evolution of self-fertilization in perennials. *American Naturalist*, *150*, 618–638.

NatureServe (2014).NatureServe explorer: An online encyclopedia of life [web application]. Version 7.1. NatureServe, Arlington, Virginia. Available http://www.natureserve.org/explorer (Accessed: November 12, 2014).

Neel, M. C., Tumas, H. R., & Marsden, B. W. (2014). Representing connectivity: Quantifying effective habitat availability based on area and connectivity for conservation status and recovery. *PeerJ*, *2*(2014), e622.

Norris, S. J., & Sullivan, R. E. (2002). Conservation assessment for the mid-Appalachian shale barrens. Unpublished report, West Virginia Division of Natural Resources, Wildlife Diversity Program, PO Box 67, Elkins WV 26241.

Pascual-Hortal, L., & Saura, S. (2006). Comparison and development of new graph-based landdcape connectivity indices: Towards the prioritization of habitat patches corridors for conservation. *Landscape Ecology*, *21*, 959–967.

Platt, R. B. (1951). An ecological study of the mid-Appalachian shale barrens and of the plants endemic to them. *Ecological Monographs*, *21*, 269–300.

Porcher, E., & Lande, R. (2005). Loss of gametophytic self-incompatibility with evolution of inbreeding depression. *Evolution*, *59*, 46–60.

Raduski, A. R., Haney, E. B., & Igić, B. (2011). The expression of self-incompatibility in angiosperms is bimodal. *Evolution*, *66*, 1275–1283.

Riday, H., & Krohn, A. L. (2010). Increasing population hybridity by restricting self-incompatibility in red clover populations. *Crop Science*, *50*, 853–860.

Saura, S., & Torné, J. (2009). Conefor Sensinode 2.2: A software package for quantifying the importance of habitat patches for landscape connectivity. *Environmental Modeling & Software*, *24*, 135–139.

Saura, S., & Torné, J. (2012). Conefor 2.6 user manual (April 2012). Universidad Politécnica de Madrid. Available at http://www.conefor.org.

Schleuning, M., Niggemann, M., Becker, U., & Matthies, D. (2009). Negative effects of habitat degradation on the declining grassland plant *Trifolium montanum*. *Basic and Applied Ecology*, *10*, 61–69.

Schoen, D. J., & Lloyd, D. G. (1992). Self-and cross-fertilization in plants. III. Methods for studying modes and functional aspects of self-fertilization. *International Journal of Plant Sciences*, *153*, 381–393.

Soulé, M. E. (1987). *Viable populations for conservation*. Cambridge: Cambridge University Press.

Spielman, D., Brook, B. W., & Frankham, R. (2004). Most species are not driven to extinction before genetic factors impact them. *Proceedings of the National Academy of Sciences, USA, 101,* 15261–15264.

Stephenson, A. G., Good, S. V., & Vogler, D. W. (2000). Interrelationships among inbreeding depression, plasticity in self-incompatibility system, and the breeding system of *Campanula rapunculoides* L. (Campanulaceae). *Annals of Botany, 85,* 211–219.

Thrall, P. H., Encinas-Viso, F., Hoebee, S. E., & Young, A. S. (2014). Life history mediates mate limitation and population viability in self-incompatible plant species. *Ecology and Evolution, 4,* 673–687.

Tilman, D., May, R. M., Lehman, C. L., & Nowak, M. A. (1994). Habitat destruction and the extinction debt. *Nature, 371,* 65–66.

Townsend, C. E. (1965). Seasonal and temperature effects on self-incompatibility in tetraploid *Trifolium hybridum* L. *Crop Science, 5,* 329–332.

Townsend, C. E. (1966). Self-incompatibility studies with diploid alsike clover, *Trifolium hybridum* L. II. Inheritance of a self-incompatibility factor with gametophytic and sporophytic characteristics. *Crop Science, 6,* 415–419.

Townsend, C. E. (1969). Self-compatibility studies with diploid alsike clover, *Trifolium hybridum* L. IV. Inheritance of type II self-compatibility in different genetic backgrounds. *Crop Science, 9,* 443–446.

Tyndall, R. W. (2015). Restoration results for a Maryland Shale Barren after pignut hickory management and a prescribed burn. *Castanea, 80,* 77–94.

Weeks, A. R., Sgro, C. M., Young, A. G., Frankham, R., Mitchell, N. J., Miller, K. A., Byrne, M., et al. (2011). Assessing the benefits and risks of translocations in changing environments: A genetic perspective. *Evolutionary Applications, 4,* 709–725.

Whitlock, R., Stewart, G. B., Goodman, S. J., Piertney, S. B., Butlin, R. K., Pullin, A. S., & Burke, T. (2013). A systematic review of phenotypic responses to between-population outbreeding. *Environmental Evidence, 2,* 13. doi:10.1186/2047-2382-2-13.

Willi, Y. (2009). Evolution towards self-compatibility when mates are limited. *Journal of Evolutionary Biology, 22,* 1967–1973.

Willi, Y., Griffin, P., & Van Buskirk, J. (2013). Drift load in populations of small size and low density. *Heredity, 110,* 296–302.

Winn, A. A., Elle, E., Kalisz, S., Cheptou, P., Eckert, C. G., Goodwillie, C., … Vallejo-Marin, M. (2011). Analysis of inbreeding depression in mixed-mating plants provides evidence for selective interference and stable mixed mating. *Evolution, 65,* 3339–3359.

Yamada, T., Fukuoka, H., & Wakamatsu, T. (1989). Recurrent selection programs for white clover (*Trifolium repens* L.) using self-compatible plants. I. Selection of self-compatible plants and inheritance of a self-compatibility factor. *Euphytica, 44,* 167–172.

Young, A. G., Boyle, T., & Brown, T. (1996). The population genetic consequences of habitat fragmentation for plants. *Trends in Ecology and Evolution, 11,* 413–418.

Young, A. G., Broadhurst, L. M., & Thrall, P. H. (2012). Non-additive effects of pollen limitation and self-incompatibility reduce plant reproductive success and population viability. *Annals of Botany, 109,* 643–653.

Young, A. G., & Pickup, M. (2010). Low S-allele numbers limit mate availability, reduce seed set and skew fitness in small populations of a self-incompatible plant. *Journal of Applied Ecology, 47,* 541–548.

3

Genetic admixture and heterosis may enhance the invasiveness of common ragweed

Min A. Hahn[1] | Loren H. Rieseberg[1,2]

[1]Department of Botany and Biodiversity Research Centre, University of British Columbia, Vancouver, BC, Canada

[2]Department of Biology, Indiana University, Bloomington, IN, USA

Correspondence
Min A. Hahn, Department of Botany and Biodiversity Research Centre, University of British Columbia, Vancouver, BC, Canada.
Email: min.hahn.a@gmail.com

Funding information
Swiss National Science Foundation (SNSF), Grant/Award Number: P2FRP3_151666; Natural Sciences and Engineering Research Council of Canada, Grant/Award Number: 03760

Abstract

Biological invasions are often associated with multiple introductions and genetic admixture of previously isolated populations. In addition to enhanced evolutionary potential through increased genetic variation, admixed genotypes may benefit from heterosis, which could contribute to their increased performance and invasiveness. To deepen our understanding of the mechanisms and management strategies for biological invasions, we experimentally studied whether intraspecific admixture causes heterosis in common ragweed (*Ambrosia artemisiifolia*) by comparing the performance of crosses (F1) between populations relative to crosses within these populations for each range (native, introduced) under different ecologically relevant conditions (control, drought, competition, simulated herbivory). Performance of admixed genotypes was highly variable, ranging from strong heterotic effects to weak outbreeding depression. Moreover, heterosis was not uniformly observed among between-population crosses, but certain native population crosses showed considerable heterosis, especially under simulated herbivory. In contrast, heterosis was largely absent in crosses from the introduced range, possibly implying that these populations were already admixed and benefit little from further mixing. In conclusion, these results support the hypothesis that heterosis may contribute to biological invasions, and indicate the need to minimize new introductions of exotic species, even if they are already present in the introduced range.

KEYWORDS

Ambrosia artemisiifolia, biological invasion, common ragweed, genetic admixture, heterosis, intraspecific hybridization, outbreeding depression

1 | INTRODUCTION

With enhanced global trade and transport over the past several centuries, the number of species that have either intentionally or accidentally become introduced into new regions has dramatically increased. Some of these exotics thrive in their new ranges and can cause serious problems for the environment, agriculture or human health (Pimentel, Lach, Zuniga, & Morrison, 2000; Sakai et al., 2001; Vitousek, DAntonio, Loope, Rejmanek, & Westbrooks, 1997). Strategies for the management and control of current invasions, as well as the prevention of future biological invasions, may be aided by a deeper understanding of the processes and mechanisms that underlie invasion success. While early research on this question focused mainly on ecological aspects of invasions (Keane & Crawley, 2002; Levine, Adler, & Yelenik, 2004), the role of evolutionary changes in invasions has increasingly gained attention (Blossey & Nötzold, 1995; Lee, 2002; Müller-Schärer, Schaffner, &

Steinger, 2004; Prentis, Wilson, Dormontt, Richardson, & Lowe, 2008). But, despite significant advances in this field in recent years, a number of unresolved questions concerning the genetic processes associated with invasions remain (Bock et al., 2015).

Due to the strong connection between species introductions and global trade and transport, it is not surprising that many invaders have become introduced to new regions multiple times (Bossdorf et al., 2005; Dlugosch & Parker, 2008). In addition to enhanced propagule pressure, which increases the likelihood that an introduced species will persist (Simberloff, 2009), the introduction of individuals from genetically differentiated source populations and subsequent genetic admixture may have important consequences for invasion success (Rius & Darling, 2014; Verhoeven, Macel, Wolfe, & Biere, 2011). In the long term, invaders may benefit from increased genetic variation, which can reduce negative effects of genetic bottlenecks and drift, and facilitate rapid adaptation of introduced populations to novel conditions (Lavergne & Molofsky, 2007). On a shorter timescale, genetically admixed individuals may benefit from heterosis (hybrid vigor), that is, the phenotypic superiority of hybrid genotypes compared to their parents (Lippman & Zamir, 2007), which may contribute to the often observed increased performance of introduced genotypes relative to genotypes from the native range when compared in common gardens (Blossey & Nötzold, 1995; Bossdorf et al., 2005).

Three main genetic models have been proposed to explain the increased performance of newly formed hybrids: 1) the dominance hypothesis, which attributes heterosis to the masking (complementation) of undesirable recessive alleles from one parent by desirable dominant alleles from the other parent, 2) overdominance, which refers to the enhanced performance of heterozygous genotypes compared to homozygotes at a given locus, and 3) epistasis, which ascribes heterosis to complex interactions between genes (Hochholdinger & Hoecker, 2007; Lippman & Zamir, 2007). According to these models, heterosis is expected to be maximal in crosses between strongly differentiated and presumably inbred populations. However, analogous mechanisms could also reduce the fitness of hybrids, leading to outbreeding depression. This may be the case, for example, if genetic admixture results in genetic incompatibilities, underdominance or the loss of local adaptation through the introduction of maladapted alleles, or the breakup of adapted gene complexes (Lynch, 1991).

There is increasing empirical evidence that genetic admixture and heterosis may play important roles during biological invasions (Rius & Darling, 2014), with important implications for management and control strategies. Several population genetic studies have revealed mixed ancestries of invasive populations reflecting genetic admixture of multiple divergent native source populations (e.g., Kolbe et al., 2004; Rosenthal, Ramakrishnan, & Cruzan, 2008; Chun, Fumanal, Laitung, & Bretagnolle, 2010; Stephen R. Keller, Gilbert, Fields, & Taylor, 2012). Moreover, some studies also provide evidence for phenotypic changes and effects on fitness associated with admixture (Kolbe, Larson, & Losos, 2007; Facon, Pointier, Jarne, Sarda, & David, 2008; S. R. Keller & Taylor, 2010). However, although these observational studies suggest a strong link of genetic admixture and invasion success, it remains difficult to disentangle the direct effects of genetic admixture (e.g., heterosis) from long-term effects of increases in genetic variation (evolutionary potential), as well as other confounding effects such as propagule pressure that may be associated with multiple introductions. So far, only a few experimental studies have addressed these questions. For example, Turgeon et al. (2011) showed that experimental crosses of the invasive harlequin ladybird *Harmonia axyridis* benefited from admixture of different source strains, which may have contributed to the invasiveness of this species. Likewise, in a recent study, Van Kleunen, Roeckle, and Stift (2015) found increased fitness in crosses between different populations of the invasive plant *Mimulus guttatus* compared to within-population crosses, in particular in crosses between native and invasive populations. While these studies provide important mechanistic insights into the role of genetic admixture and heterosis for invasions, more experimental work is needed to assess the generality of these findings. Theoretical questions of particular interest include the relative importance of positive (i.e., heterosis) vs. negative (i.e., outbreeding depression) outcomes of admixture in invasions, the expression of heterosis under different environmental conditions, and the contributions of selection toward preserving heterotic effects in subsequent generations. From a practical perspective, a better understanding of the role of heterosis in invasions is critical for guiding management efforts (i.e., prevention of new introductions and gene flow, monitoring admixture in introduced species) and to improve control options (i.e., evaluating the efficacy of control measures).

In this study, we used an experimental approach to address these questions in *Ambrosia artemisiifolia* L. (common ragweed, Asteraceae), which is one of the most problematic plant invaders in Europe and in several other regions of the world. It is difficult to control because of its large seed production, resulting in yield losses in crop fields (Cowbrough, Brown, & Tardif, 2003), as well as major human health problems because of its highly allergenic pollen (Laaidi, Laaidi, Besancenot, & Thibaudon, 2003). From its native range in North America, *A. artemisiifolia* has been inadvertently introduced to Europe as a seed contaminant. Previous population genetic studies provide clear evidence for multiple introductions and genetic admixture in European populations of *A. artemisiifolia* (Chun et al., 2010; Gaudeul, Giraud, Kiss, & Shykoff, 2011; Genton, Shykoff, & Giraud, 2005). In addition, invasive genotypes have been found to show increased performance compared to native genotypes, which suggests that evolutionary changes may underlie the invasion success of this species (Hodgins & Rieseberg, 2011). Given the high levels of genetic admixture reported in this species, as well as the large effects of heterosis commonly observed in many crop species (Schnable & Springer, 2013), we hypothesized that heterosis may have contributed to the increased performance of invasive genotypes of *A. artemisiifolia* and could play a key role for the invasion success of this species.

To deepen our understanding on the potential mechanistic role of heterosis in the invasion of *A. artemisiifolia* as well as its implications for management and control, we studied the outcomes of admixture in experimentally reconstructed admixed genotypes (between-population crosses) from putative native source populations by estimating heterosis based on their relative performance to nonadmixed genotypes (within-population crosses) from the same populations. In addition, we created admixed genotypes from the introduced range, which allowed us to evaluate differences in heterosis between crosses from presumably highly differentiated native populations as compared to crosses from less differentiated, admixed populations from the introduced range. Furthermore, we studied the expression of heterosis under a range of environmental conditions (control, drought, competition, simulated herbivory), which may be relevant for the invasion under current and future conditions, as well as to inform specific management and control options. Specifically, we asked (1) whether plants of *A. artemisiifolia* show evidence for heterosis or outbreeding depression, (2) whether heterosis is more pronounced in the crosses between native populations than those from the invaded range, (3) whether the level of heterosis varies among the particular population cross-combinations, and (4) whether the expression of heterosis or outbreeding depression differs among environmental conditions.

2 | MATERIALS AND METHODS

2.1 | Population sampling

We used seed material collected from North American populations in the fall 2008 and 2013 and from European populations in the fall 2008. We selected four populations from each range (Table 1, Figure 1). The selected European populations are located across France covering a region in Europe where *A. artemisiifolia* is especially invasive, and the populations from North America are from regions that have previously been identified as putative source regions of western European populations (Gaudeul et al., 2011).

2.2 | Production of crosses

We produced crosses between individuals from different populations (between-population crosses) as well as from the same populations

TABLE 1 Description of the populations of *Ambrosia artemisiifolia* used in the experiment (population code, location, geographical coordinates, and year of collection)

Population	Range	Country	State/Province	N	W	Year
AA5	Native	USA	MN	46.217083	−96.050194	2008
MO	Native	USA	MO	37.00644	−94.35011	2013
MN2	Native	Canada	ON	44.44716	−79.80385	2013
QC3	Native	Canada	QC	47.67876	−69.022	2013
FR7	Introduced	France	–	47.175819	3.014628	2008
FR6	Introduced	France	–	46.800028	4.972428	2008
FR1	Introduced	France	–	45.080225	4.757443	2008
FR8	Introduced	France	–	44.216656	4.264008	2008

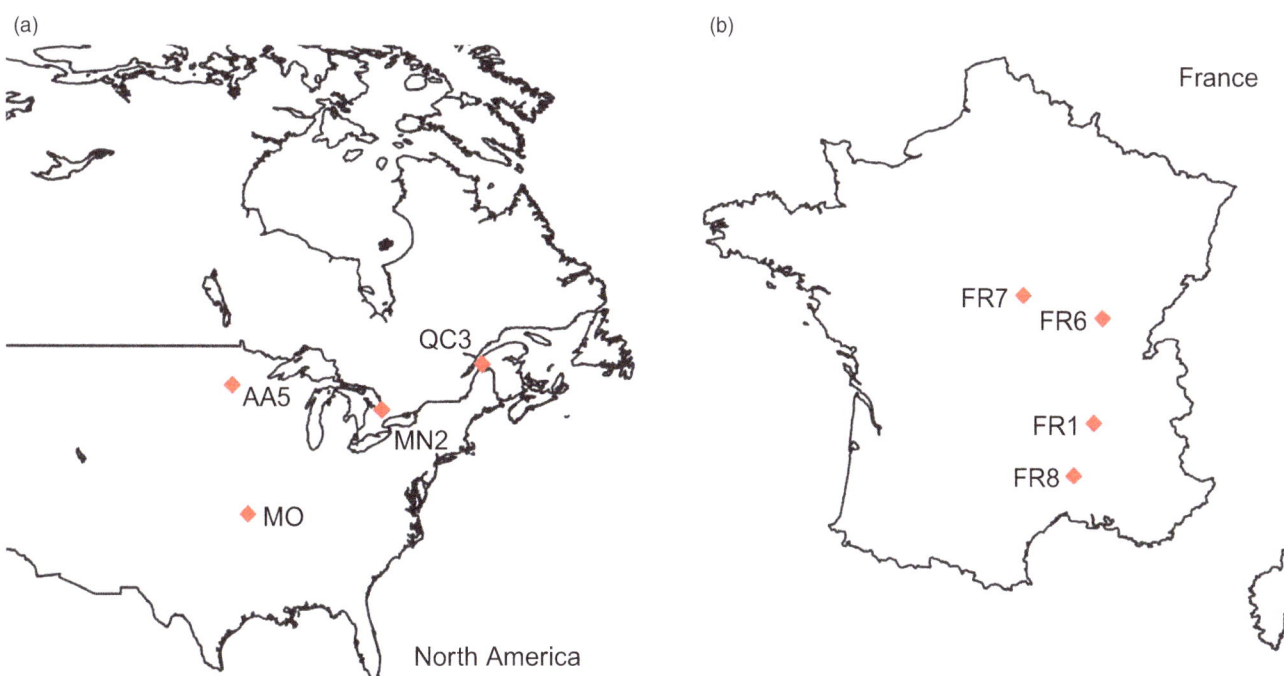

FIGURE 1 Sampling locations of the populations of *Ambrosia artemisiifolia* used in the experiment from (a) the native range (North America) and (b) the introduced range (France)

(within-population crosses) with four different populations or combinations of populations ("population combination" in the following) used in the within-population and between-population crosses, respectively. Notably, not all crosses from the possible population combinations have been produced, but only the four crosses between the geographically most distant populations in each range. From each population, seeds from three different seed families were used to obtain three independent biological replicates for each cross. The seeds were stratified for a period of 8 weeks following the procedures suggested by Willemsen (1975) starting in April 2014. In June 2014, the seeds were germinated on damp filter paper in Petri dishes with 1% plant preservative mixture in a growth chamber with a 24°C day and 18°C night and a 14:10-hr light/dark cycle. After 2 weeks, on June 23, the emerged seedlings were transplanted into seedling trays (2.5 × 2.5 cm compartments) with a 1:1 mixture of potting soil and sand under ambient conditions in the horticultural greenhouse at the University of British Columbia. On July 15, pairs of plants of similar size for each planned cross were planted into single pots (15 cm diameter, 18 cm height) with the same soil–sand mixture. On August 5, the plant pairs were covered with pollen-proof pollination bags (PBS International, UK) to prevent cross-contamination of pollen from non-target individuals. The bags were shaken every few days to assure cross-pollination of the plants within the bags. As previous studies reported strong self-incompatibility mechanisms in *A. artemisiifolia* (Friedman & Barrett, 2008), selfing was expected to be negligible. The mature seeds were collected between December 1, 2014 and March 27, 2015. After excluding crosses that did not set seed because of nonoverlapping flowering times, we ended up with a total of 68 crosses (including 28 reciprocal crosses to account for maternal effects), with roughly equal numbers of crosses for the different cross types (15 native within-population crosses, 16 native between-population crosses, 20 invasive within-population crosses, 17 invasive between-population crosses).

2.3 | F1 experiment

To study heterosis under different ecologically relevant conditions, we grew plants from the F1 generation in a common garden greenhouse experiment under four different treatments (control, drought, competition, and simulated herbivory). Four individuals from each cross were used, each of which was subjected to a different experimental treatment, for a total of 272 plants. The seeds were stratified and germinated as described above. Germination was initiated on June 2, 2015, and the emerged seedlings were transplanted into seedling trays in the greenhouse on June 11, 2015. On June 22, 2015, the plants were transplanted into bigger pots (10 cm diameter) with a 2:1:1 mixture of potting soil : coarse forestry sand : fine industrial sand. After 3 days of acclimatization and adequate watering, the experimental treatments were initiated. Plants in the control treatment were automatically watered by flooding the greenhouse bench every other day for the first 4 weeks and daily afterward. Plants in the drought treatment were placed into trays to prevent automatic watering and received equal amounts of water when the first plants (typically the majority of plants) started

wilting (~every 3 days on average). In the competition treatment, plants were under light and nutrient limitation due to the addition of the grass *Poa pratensis*, but received similar watering as the controls. To ensure a reasonably dense and high grass cover in the pots, we twice sowed 0.3 g of *Poa pratensis* seeds into the pots prior to the start of the experiment (on May 13 and June 12, 2015). For the simulated herbivory treatment, 50% of the total leaf area of all newly emerged leaves longer than 2 cm was cut every week. Moreover, after removal from the bench, the plants were sprayed with 5 mM methyl jasmonate until all leaves were soaked and left to dry before they were moved back. Methyl jasmonate is commonly used to elicit defense responses against herbivores in many plant species (McConn, Creelman, Bell, Mullet, & Browse, 1997). The watering regime was similar to the control plants. Three times during the experiment, plants of all treatments were fertilized with a minimal amount of 0.2 ml all-purpose fertilizer (20-20-20N-P-K; Plant-Prod Ultimate; Premier Tech Home & Garden Inc.). All plants were grown in a completely randomized design 25 cm apart, and the positions were rerandomized weekly throughout the experiment.

2.4 | Measurements

Throughout the experiment, we measured several traits on the experimental plants related to growth and reproduction in regular intervals (weekly for most traits). But, as preliminary time series analyses (results not shown) did not reveal any significant interactions of time with cross types or treatments, we present only the results based on the final measurements (or single measurements for traits that were measured only once). The traits included the final plant height, stem diameter at the base of the stem, the number of leaves of 3-week-old plants, the number of branches, and the number of flower heads at the time of flowering onset. At the time of flowering onset for each plant (first opening of male flowers and release of pollen, between July 21 and October 16, 2015), the plants were harvested and the aboveground biomass was determined after drying to a constant weight at 65°C for 10 days.

2.5 | Statistical analyses

We explicitly estimated the level of heterosis (and outbreeding depression) for each individual between-population cross based on its performance relative to the mean of the two corresponding within-population crosses (midparent heterosis). Heterosis was calculated as ((F1 − MP)/MP) × 100, in which F1 is the trait value in a given between-population cross and MP is the mean of the trait values of the two corresponding within-population crosses. The trait values of the within-population crosses were based on averages of all crosses derived from the same parental lineage as the between-population crosses to obtain most accurate estimations accounting for different genetic backgrounds. Differences in heterosis for each trait among crosses from different ranges (native, introduced) and treatments were analyzed using linear mixed-effects models using the "lme" function from the "nlme" package in R (Pinheiro, Bates, DebRoy, & Sarkar, 2015) with range, treatment, and their interaction as fixed factors and population combination as a random factor. Differences in heterosis among crosses from specific population combinations for each range were analyzed using linear models with population combination, treatment, and their interaction as explanatory variables. Significance of model terms was evaluated by stepwise removal of the least significant term (interactions first) using likelihood ratio tests for mixed-effects models and F-tests for linear models. Differences among experimental groups were tested using least square means with the R package "lsmeans" (Lenth, 2015), and p-values of pairwise Tukey contrasts were adjusted using Bonferroni corrections for multiple comparisons. All analyses were conducted using the statistical software R version 3.2.2 (R Core Team 2015).

3 | RESULTS

3.1 | Heterosis in crosses from different ranges and treatments

The outcomes of admixture (heterosis and outbreeding depression) varied significantly among the experimental crosses from the different ranges and treatments for plant height, biomass, and the number of flower heads (Table 2). Significant heterosis (estimate bigger than zero) was only observed for crosses between native populations (Table 3, Figure 2a), which showed an average increase of 217% in the production of flower heads compared to the crosses within the corresponding native populations (Table 3). However, there was also

considerable variation in levels of heterosis and outbreeding depression among individual crosses, ranging from strong heterotic effects to weak outbreeding depression (Table 3, Figure 2a). In contrast to the crosses from the native range, no significant heterosis was found for crosses between populations from the introduced range (Table 3).

We also detected significant interactions between ranges and treatments for the estimate of heterosis in plant height and biomass (Table 2). In native crosses, heterosis for plant height was significantly higher under simulated herbivory compared to plants in the control (df = 118, t = −4.19, $p < 0.001$) and drought treatment (df = 118, t = −4.72, $p < 0.0001$; Table 3, Figure 2b). Also for biomass, heterosis was higher in crosses from the native range under simulated herbivory compared to plants in the control (df = 118, t = −3.70, $p < 0.002$), drought (df = 118, t = −3.25, $p < 0.009$), and competition treatment (df = 118, t = −3.38, $p < 0.006$; Table 3). Again, no significant differences in heterosis were found in the crosses from the introduced range even when tested under different environmental conditions.

3.2 | Heterosis in crosses from different population combinations and treatments

The outcomes of admixture also differed significantly among population combinations and treatments (Table 4). Significant heterosis for certain population combinations was detected for most traits,

TABLE 2 Effects of range (native, introduced), treatments (control, drought, competition, simulated herbivory), and their interactions on the level of heterosis in different traits of between-population crosses of *Ambrosia artemisiifolia*. Table shows results of likelihood ratio (LR) tests of models with and without a given term, following stepwise removal of nonsignificant terms starting with interactions. Significant p-Values are shown in bold

Trait	Range			Treatment			Range × Treatment		
	df	LR	p-Value	df	LR	p-Value	df	LR	p-Value
Plant height	1	0.02	.8952	3	12.27	.0065	3	15.10	.0017
Stem diameter	1	0.51	.4747	3	7.47	.0585	3	5.33	.1493
Leaves	1	1.28	.2572	3	0.52	.9138	3	3.00	.3923
Branches	1	2.10	.1474	3	3.27	.3515	3	2.66	.4476
Biomass	1	0.00	.9874	3	10.20	.0170	3	8.50	.0367
Flower heads	1	18.47	<.0001	3	1.90	.5941	3	0.77	.8570

TABLE 3 Estimated means and standard errors of heterosis estimates (%) for different traits of between-population crosses of *Ambrosia artemisiifolia* from different ranges (native, introduced) and treatments (control, drought, competition, simulated herbivory) from minimal adequate models. Estimates that significantly differ from zero after Bonferroni correction for multiple comparisons are highlighted in bold

Trait	Range	Control		Drought		Herbivory		Competition		Overall	
		Mean	SE	Mean	SE	Mean	SE	Mean	SE	Mean	SE
Plant height	Native	−13.00	11.54	−17.85	11.54	25.28	11.65	5.26	11.54	–	–
	Introduced	−7.28	11.41	0.80	11.41	−3.90	11.41	2.87	11.41	–	–
Biomass	Native	−16.75	19.44	−8.48	19.44	50.95	19.71	−10.86	19.44	–	–
	Introduced	−8.68	19.15	12.78	19.15	6.74	19.15	4.61	19.15	–	–
Flower heads	Native	–	–	–	–	–	–	–	–	**217.16**	**32.51**
	Introduced	–	–	–	–	–	–	–	–	19.19	30.79

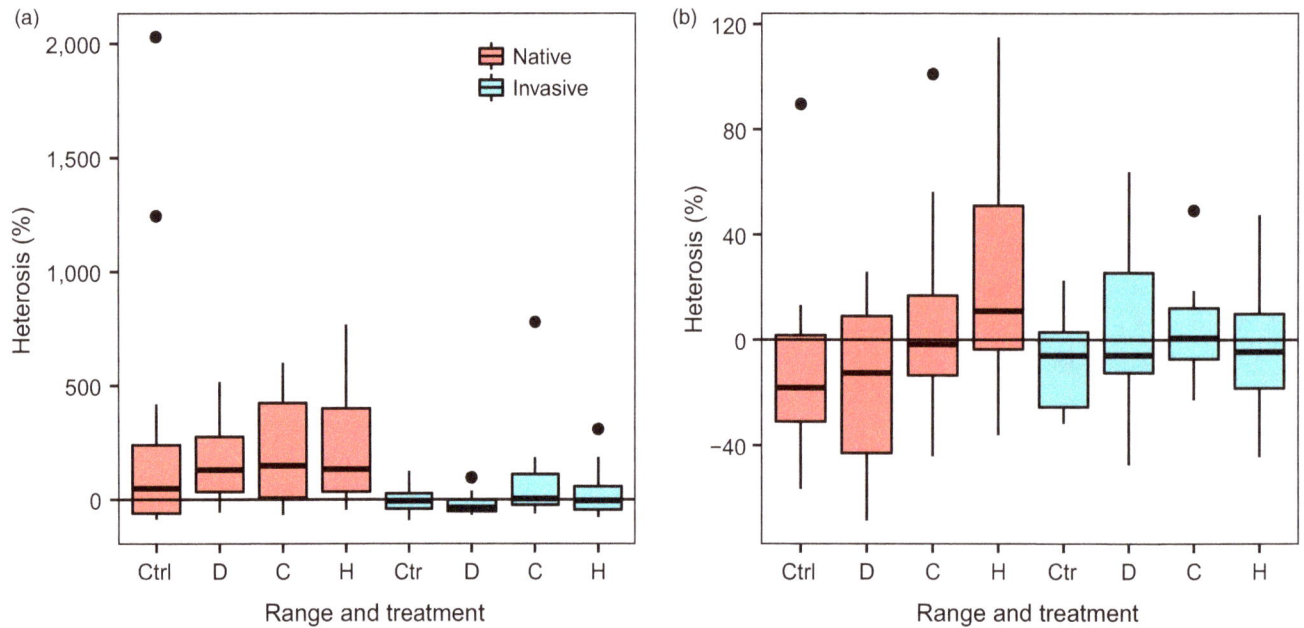

FIGURE 2 Boxplots of heterosis [%] in (a) the number of flower heads and (b) final plant height for each individual between-population cross of *Ambrosia artemisiifolia* from the native and introduced range (native and invasive crosses, respectively) in the different experimental treatments (Ctrl: control, D: drought, C: competition, H: simulated herbivory). Heterosis estimates are based on the performance of between-population crosses relative to the average performance of the corresponding within-population crosses derived from the same parental lineages

TABLE 4 Effects of population combinations, treatments (control, drought, competition, simulated herbivory), and their interactions on the level of heterosis in different traits of between-population crosses of *Ambrosia artemisiifolia*. Significance of given terms was determined by *F*-tests, following stepwise removal of nonsignificant terms starting with interactions. Separate models were fitted for crosses from the native and introduced ranges. Significant *p*-Values are shown in bold

		Pop. combination			Treatment			Pop comb. × Treat.		
Range	Trait	df	F	p-Value	df	F	p-Value	df	F	p-Value
Native	Plant height	3	13.42	**<.0001**	3	7.01	**<.001**	9	0.69	.7100
	Stem diameter	3	3.65	**.0175**	3	2.71	.0539	9	0.47	.8898
	Leaves	3	3.47	**.0216**	3	0.60	.6164	9	0.75	.6654
	Branches	3	6.29	**<.001**	3	0.79	.5032	9	0.53	.8484
	Biomass	3	5.50	**.0022**	3	4.41	**.0075**	9	0.79	.6266
	Flower heads	3	0.61	.6084	3	0.22	.8804	9	1.15	.3505
Introduced	Plant height	3	3.52	**.0199**	3	0.79	.5066	9	0.34	.9569
	Stem diameter	3	1.15	.3352	3	0.75	.5294	9	0.72	.6890
	Leaves	3	4.00	**.0113**	3	0.40	.7515	9	1.74	.1041
	Branches	3	0.47	.7038	3	1.48	.2283	9	1.35	.2372
	Biomass	3	7.41	**.0002**	3	0.79	.5030	9	1.47	.1825
	Flower heads	3	1.64	.1888	3	1.94	.1313	9	1.08	.3948

but almost exclusively for crosses from the native range (Table 5). In addition, there was considerable variation in levels of heterosis and apparent outbreeding depression, which was not consistent across population combinations or traits. While crosses from all native population combinations showed heterosis for the number of flower heads (Table 5, Figure 3a), there were variable levels of heterosis and even contrasting patterns (heterosis and outbreeding depression) for plant height, stem diameter, the number of leaves, the number of branches, and biomass depending on the particular population combination (Table 5, Figure 3b,c). In contrast, the outcomes of admixture in crosses from the introduced range were less variable (Table 5, Figure 3). Moreover, the only significant case of heterosis in invasive between-population crosses was found for biomass in the population combination FR7/FR8 (Table 5, Figure 3c),

TABLE 5 Estimated means and standard errors of heterosis estimates (%) for different traits of between-population crosses of *Ambrosia artemisiifolia* from different population combinations and treatments (control, drought, competition, simulated herbivory) from minimal adequate models. Estimates that significantly differ from zero after Bonferroni correction for multiple comparisons are highlighted in bold

		Population combination									
		AA5/QC3		AA5/MN2		MO/QC3		MN2/MO		Overall	
Range	Trait	Mean	SE	Mean	SE	Mean	SE	Mean	SE	Mean	SE
Native	Plant height	**23.04**	**8.42**	**−22.23**	**6.52**	**−24.79**	**8.42**	**23.69**	**6.70**	–	–
	Stem diameter	**24.72**	**8.10**	3.75	6.27	−13.05	8.10	3.74	6.44	–	–
	Leaves	−6.14	14.40	22.59	11.16	−24.63	14.40	26.66	11.45	–	–
	Branches	18.68	9.84	17.94	7.62	−24.28	9.84	**28.15**	**7.82**	–	–
	Biomass	31.03	17.07	−15.80	13.22	−37.87	17.07	**36.59**	**13.58**	–	–
	Flower heads	–	–	–	–	–	–	–	–	**217.16**	**44.35**
		FR1/FR7		FR1/FR6		FR7/FR8		FR6/FR8			
		Mean	SE	Mean	SE	Mean	SE	Mean	SE		
Introduced	Plant height	−7.72	5.30	−9.26	4.74	12.09	5.30	−2.47	5.30	–	–
	Leaves	**−18.13**	**6.12**	2.73	5.47	**−17.66**	**6.12**	2.66	6.12	–	–
	Biomass	−15.62	10.43	−20.32	9.33	**38.90**	**10.43**	13.29	10.43	–	–

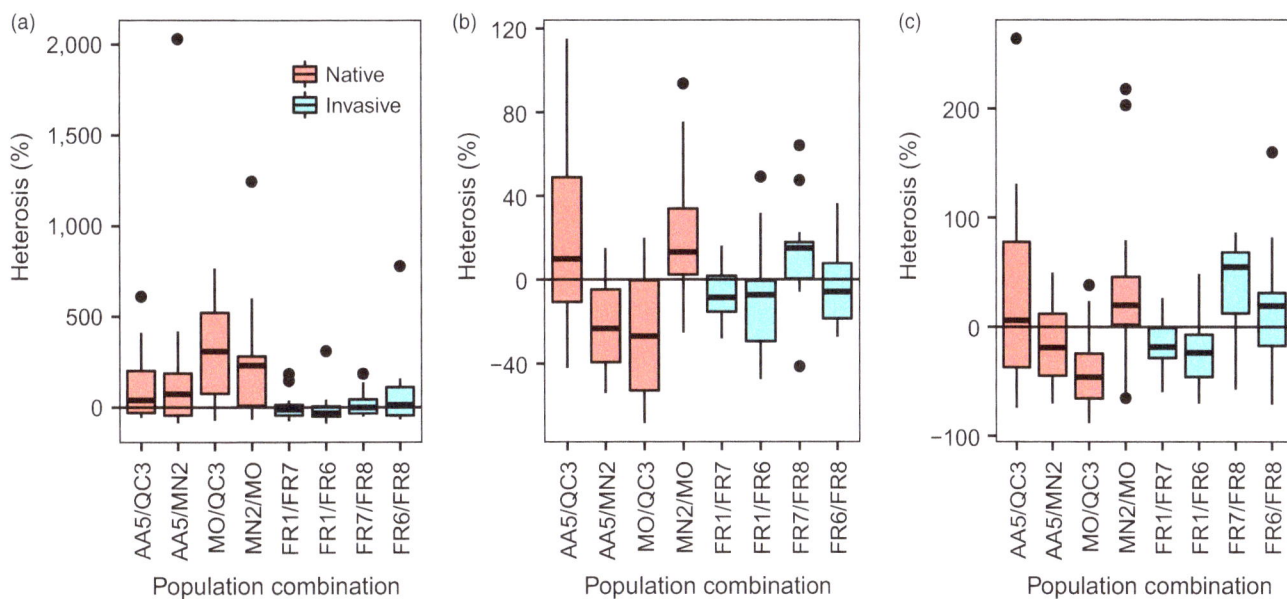

FIGURE 3 Boxplots of heterosis [%] in (a) the number of flower heads, (b) final plant height, and (c) aboveground biomass for each individual between-population cross of *Ambrosia artemisiifolia* from different population combinations in the native and introduced range (native and invasive crosses, respectively). Heterosis estimates are based on the performance of between-population crosses relative to the average performance of the corresponding within-population crosses derived from the same parental lineages

which represents the cross between the two geographically most distant populations studied from the introduced range (Table 1, Figure 1).

4 | DISCUSSION

Our analyses of the effects of genetic admixture in *A. artemisiifolia* revealed large variation in the level of heterosis across different traits,

crosses from the different ranges, population combinations, and environmental conditions. Interestingly, heterosis was found almost exclusively in crosses from the native range, which suggests that genetic admixture resulting from multiple introductions of differentiated native populations could have contributed to the enhanced performance of *A. artemisiifolia* genotypes from the invaded range (Hodgins & Rieseberg, 2011). The lack of heterosis in crosses from the introduced range, in contrast, may indicate that these populations are already admixed and thus no longer show significant short-term benefits of genetic admixture.

4.1 | Large variation in heterosis in native crosses

The immediate effects of genetic admixture between different native populations of *A. artemisiifolia* ranged from highly beneficial heterosis to weak outbreeding depression, as well as variable effects among particular population combinations, traits, and environmental conditions. Given this variation, few general patterns emerged from the data. The only consistent evidence for heterosis across all population combinations and treatments was observed in flower head production in crosses from the native range. However, certain population combinations also showed evidence of heterosis in a number of other traits, indicating that the outcomes of genetic admixture depend on context, that is, the populations, genotypes, environmental conditions, and trait of interest. This is not surprising, given the numerous possible genetic interactions during genetic admixture (Hochholdinger & Hoecker, 2007; Lippman & Zamir, 2007). While dominance, overdominance, or epistasis could result in heterosis, a similar set of mechanisms (epistasis or underdominance) can lead to outbreeding depression depending on the genotypes involved.

Studies of intraspecific crosses in other species indicate that heterosis is likely if populations have previously experienced inbreeding depression (Verhoeven et al., 2011) or are genetically differentiated (Rius & Darling, 2014). The latter prediction is in line with the large variation in performance of native populations in our study as well as previous studies showing high differentiation and significant isolation by distance among native populations of *A. artemisiifolia* (Genton et al., 2005). On the other hand, admixed genotypes can be expected to exhibit outbreeding depression due to genetic incompatibilities arising from hybridization between differentiated populations, or through the loss of local adaptation following the introduction of unfavorable alleles or the breakup of adapted gene complexes (in particular in the F2 and later generations after recombination) (Johansen-Morris & Latta, 2006). Although in our experiment, there was some evidence for outbreeding depression, the overall pattern in our study indicates that the benefits outweighed potential negative effects of admixture in the majority of crosses.

Interestingly, the level of heterosis in the crosses from the native range in our experiment also varied among the environmental conditions with the largest effects found under presumably stressful conditions, in particular under simulated herbivory. While some effects of genetic admixture are expected to be mainly independent of the environment (e.g., reduction in inbreeding depression), others are likely to be strongly condition dependent (Rius & Darling, 2014). For example, novel gene and allele combinations resulting from hybridization between distinct parental genotypes may be better adapted to environments outside the range of conditions previously experienced by parental populations. This is seen in crop hybrids, which often show increased tolerance to stress compared to their parental lineages (Schnable & Springer, 2013). Whether these benefits eventually may contribute to the invasion success depends in part on the extent to which the environmental conditions differ between the native and introduced range (Rius & Darling, 2014). In *A. artemisiifolia*, increased growth and reproduction in introduced populations suggest that they have adapted to more competitive environments in Europe (Hodgins & Rieseberg, 2011), which is in line with the findings of our study. The high levels of heterosis observed under simulated herbivory in particular further imply that admixed genotypes may have experienced lower biotic resistance against their invasion due to increased tolerance of herbivory. This may have important implications for management strategies, as admixed genotypes may become less susceptible to control measures, such as cutting or biological control using herbivores, and therefore more difficult to control.

The large variation in outcomes of admixture observed further suggests that selection may play an important role in the subsequent evolution of admixed populations. Despite some evidence for outbreeding depression, these negative effects of intraspecific hybridization appeared to be relatively weak. This is in line with the hypothesis that the loss of local adaptation through admixture may be less crucial for introduced populations than for native populations, because they presumably lack local adaptation initially and therefore can more freely benefit from the positive effects of admixture (Verhoeven et al., 2011). Heterotic genotypes, in contrast, may be advantageous under the novel conditions and—depending on the underlying genetic mechanisms—could be fixed by selection. In the case of overdominance, heterosis results from heterozygosity, and hence, the effects are expected to be maximal in the F1 generation and will be reduced in subsequent generations due to decreasing heterozygosity. Nevertheless, even such transient effects could be beneficial for invasions, as they may assist populations in overcoming demographic challenges of small population sizes at initial stages of invasions (Drake, 2006). In contrast, if dominance or epistasis (which does not require heterozygous allele combinations) is the main mechanism of heterosis, selection may favor individuals with combinations of favorable alleles at multiple loci that increase fitness by complementing deleterious alleles (the dominance model) or through favorable epistatic interactions. Subsequently, the heterotic effects may be preserved in successive generations and potentially become fixed (Bock et al., 2015). Such heterotic gene combinations may contribute to the generation of transgressive phenotypes, which most frequently arise through the complementary action of additive loci (Rieseberg, Archer, & Wayne, 1999).

There appears to be some confusion in the invasion biology about selection on loci with dominance. While selection does not act on dominance variance, as long as a recessive allele exists at some frequency, then some of the genetic variance is additive and can be acted on by natural selection. Such additive variance can be increased in invasive populations due to founding events and population bottlenecks, potentially increasing the efficacy of selection for heterotic gene combinations (Robertson, 1952).

4.2 | No effects of admixture in crosses from the introduced range

Strikingly, in contrast to the large variation in admixture effects in native crosses, little evidence of either heterosis or outbreeding depression was observed in crosses between populations from the introduced range. This pattern may indicate that these populations are already admixed and therefore may not obtain additional short-term benefits from further admixture. This finding is consistent with previous studies that revealed high levels of genetic admixture and gene flow in European populations of *A. artemisiifolia*, and as a result high genetic variation within populations, but low genetic differentiation among populations (Chun et al., 2010; Genton et al., 2005).

4.3 | Heterosis may contribute to the invasion of *Ambrosia artemisiifolia*

In summary, the findings of our study support the hypothesis that heterosis may have contributed to the invasion of *A. artemisiifolia*. Although the limited sample size in our experiment does not allow firm general conclusions, building on previous studies that reported evidence for genetic admixture (Genton et al., 2005), our work suggests that heterosis may be a frequent result of native between-population crosses. Nevertheless, the variation we observed in admixture effects indicates that heterosis is not an invariable outcome of admixture and that selection may play an important role in maintaining the expected benefits in a population. In addition, given the few studied populations, which may not be representative for all populations and possible cross-combinations, as well as the partial nonindependence of the cross-combinations within each range due to our crossing design, it is likely that we have underestimated the range of outcomes as well as the potential benefits from heterosis (Kolbe et al., 2007). Moreover, admixture between populations from the native and invasive ranges, which may occur following recurrent introductions, might result in additional heterosis (Van Kleunen et al., 2015).

4.4 | Implications for biological invasions

Our findings add to a growing body of literature suggesting that heterosis may play an important role in the success of invasions through the resulting increases in performance of admixed genotypes. Heterosis may not only be relevant in the short term, but it may also have longer-lasting effects if heterotic gene combinations (favorable alleles at multiple loci) are fixed by selection, although evidence for the latter claim is admittedly weak. An important future goal, and beyond the scope of the present study, is to identify such heterotic genotypes and assess whether they contribute importantly to the often observed increased performance of invasive genotypes (Bossdorf et al., 2005).

In addition to these theoretical implications, our results have important practical implications, as knowledge of the potential consequences of genetic admixture is important for the prediction and prevention of biological invasions, as well as for the development of management and control options (Hulme, 2009). Most importantly, our results indicate the need to concentrate efforts on minimizing new introductions of exotic species, even if they already are present in the introduced range. In contrast, as heterosis was less apparent in crosses from the introduced range, the prevention of gene flow and admixture among populations in the introduced range may be less critical than the prevention of introductions of new genotypes from the native range. Furthermore, the elevated levels of heterosis under simulated herbivory and other stress conditions in *A. artemisiifolia* in our experiment suggest that heterosis may contribute to an increased tolerance to these conditions. Hence, admixed genotypes may be less susceptible to management and control efforts, which in turn need to be optimized. Our results further imply that also admixture among populations in a species' native range could be problematic, especially given rapidly changing environmental conditions, potentially giving rise to problematic native species (Chunco, 2014). And finally, because not all native species can be practically monitored, it probably makes the most sense to focus attention on native species such as common ragweed that are problematic and have become invasive elsewhere.

ACKNOWLEDGEMENTS

We would like to thank Kathryn Hodgins for providing the seed collections, Mike Whitlock for helpful discussions about selection on heterotic gene combinations, the Swiss National Science Foundation (SNSF) for funding to MAH, and the Natural Sciences and Engineering Research Council of Canada (NSERC, Grant ID 03760) for funding to LHR.

REFERENCES

Blossey, B., & Nötzold, R. (1995). Evolution of increased competitive ability in invasive nonindigenous plants – a hypothesis. *Journal of Ecology*, *83*(5), 887–889.

Bock, D. G., Caseys, C., Cousens, R. D., Hahn, M. A., Heredia, S. M., Hübner, S., ... Rieseberg, L. H. (2015). What we still don't know about invasion genetics. *Molecular Ecology*, *24*(9), 2277–2297.

Bossdorf, O., Auge, H., Lafuma, L., Rogers, W. E., Siemann, E., & Prati, D. (2005). Phenotypic and genetic differentiation between native and introduced plant populations. *Oecologia*, *144*(1), 1–11.

Chun, Y. J., Fumanal, B., Laitung, B., & Bretagnolle, F. (2010). Gene flow and population admixture as the primary post-invasion processes in common ragweed (*Ambrosia Artemisiifolia*) populations in France. *New Phytologist*, *185*(4), 1100–1107.

Chunco, A. J. (2014). Hybridization in a warmer world. *Ecology and Evolution*, *4*(10), 2019–2031.

Cowbrough, M. J., Brown, R. B., & Tardif, F. J. (2003). Impact of common ragweed (*Ambrosia Artemisiifolia*) aggregation on economic thresholds in soybean. *Weed Science*, *51*(6), 947–954.

Dlugosch, K. M., & Parker, I. M. (2008). Founding events in species invasions: Genetic variation, adaptive evolution, and the role of multiple introductions. *Molecular Ecology*, *17*(1), 431–449.

Drake, J. M. (2006). Heterosis, the catapult effect and establishment success of a colonizing bird. *Biology Letters*, *2*(2), 304–307.

Facon, B., Pointier, J.-P., Jarne, P., Sarda, V., & David, P. (2008). High genetic variance in life-history strategies within invasive populations by way of multiple introductions. *Current Biology*, *18*(5), 363–367.

Friedman, J., & Barrett, S. C. H. (2008). High outcrossing in the annual colonizing species *Ambrosia Artemisiifolia* (Asteraceae). *Annals of Botany*, *101*(9), 1303–1309.

Gaudeul, M., Giraud, T., Kiss, L., & Shykoff, J. A. (2011). Nuclear and chloroplast microsatellites show multiple introductions in the worldwide invasion history of common ragweed, *Ambrosia Artemisiifolia*. Edited by Hans Henrik Bruun. *PLoS ONE*, *6*(3), e17658.

Genton, B. J., Shykoff, J. A., & Giraud, T. (2005). High genetic diversity in french invasive populations of common ragweed, *Ambrosia*

Artemisiifolia, as a result of multiple sources of introduction. *Molecular Ecology, 14*(14), 4275–4285.

Hochholdinger, F., & Hoecker, N. (2007). Towards the molecular basis of heterosis. *Trends in Plant Science, 12*(9), 427–432.

Hodgins, K. A., & Rieseberg, L. H. (2011). Genetic differentiation in life-history traits of introduced and native common ragweed (*Ambrosia Artemisiifolia*) populations: Evolution and invasion. *Journal of Evolutionary Biology, 24*(12), 2731–2749.

Hulme, P. E. (2009). Trade, transport and trouble: Managing invasive species pathways in an era of globalization. *Journal of Applied Ecology, 46*(1), 10–18.

Johansen-Morris, A. D., & Latta, R. G. (2006). Fitness consequences of hybridization between ecotypes of *Avena Barbata*: Hybrid breakdown, hybrid vigor, and transgressive segregation. *Evolution, 60*(8), 1585–1595.

Keane, R. M., & Crawley, M. J. (2002). Exotic plant invasions and the enemy release hypothesis. *Trends in Ecology & Evolution, 17*(4), 164–170.

Keller, S. R., Gilbert, K. J., Fields, P. D., & Taylor, D. R. (2012). Bayesian inference of a complex invasion history revealed by nuclear and chloroplast genetic diversity in the colonizing plant, *Silene Latifolia*. *Molecular Ecology, 21*(19), 4721–4734.

Keller, S. R., & Taylor, D. R. (2010). Genomic admixture increases fitness during a biological invasion. *Journal of Evolutionary Biology, 23*(8), 1720–1731.

Kolbe, J. J., Glor, R. E., Schettino, L. R., Lara, A. C., Larson, A., & Losos, J. B. (2004). Genetic variation increases during biological invasion by a Cuban Lizard. *Nature, 431*(7005), 177–181.

Kolbe, J. J., Larson, A., & Losos, J. B. (2007). Differential admixture shapes morphological variation among invasive populations of the lizard *Anolis Sagrei*. *Molecular Ecology, 16*(8), 1579–1591.

Laaidi, M., Laaidi, K., Besancenot, J. P., & Thibaudon, M. (2003). Ragweed in France: An invasive plant and its allergenic pollen. *Annals of Allergy Asthma & Immunology, 91*(2), 195–201.

Lavergne, S., & Molofsky, J. (2007). Increased genetic variation and evolutionary potential drive the success of an invasive grass. *Proceedings of the National Academy of Sciences of the United States of America, 104*(10), 3883–3888.

Lee, C. E. (2002). Evolutionary genetics of invasive species. *Trends in Ecology & Evolution, 17*(8), 386–391.

Lenth, R. (2015). Lsmeans: Least-Squares Means. R Package Version 2.20-2.

Levine, J. M., Adler, P. B., & Yelenik, S. G. (2004). A meta-analysis of biotic resistance to exotic plant invasions. *Ecology Letters, 7*(10), 975–989.

Lippman, Z. B., & Zamir, D. (2007). Heterosis: Revisiting the magic. *Trends in Genetics: TIG, 23*(2), 60–66.

Lynch, M. (1991). The genetic interpretation of inbreeding depression and outbreeding depression. *Evolution, 45*(3), 622–629.

McConn, M., Creelman, R. A., Bell, E., Mullet, J. E., & Browse, J. (1997). Jasmonate is essential for insect defense in *Arabidopsis*. *Proceedings of the National Academy of Sciences of the United States of America, 94*(10), 5473–5477.

Müller-Schärer, H., Schaffner, U., & Steinger, T. (2004). Evolution in invasive plants: Implications for biological control. *Trends in Ecology & Evolution, 19*(8), 417–422.

Pimentel, D., Lach, L., Zuniga, R., & Morrison, D. (2000). Environmental and economic costs of *Nonindigenous* species in the United States. *BioScience, 50*(1), 53.

Pinheiro, J., Bates, D., DebRoy, S., Sarkar, D., & R Core Team. (2015). Nlme: Linear and Nonlinear Mixed Effects Models. R Package Version 3.1-122.

Prentis, P. J., Wilson, J. R. U., Dormontt, E. E., Richardson, D. M., & Lowe, A. J. (2008). Adaptive evolution in invasive species. *Trends in Plant Science, 13*(6), 288–294.

R Core Team. (2015). *R: A language and environment for statistical computing*. Vienna, Austria: R Foundation for Statistical Computing.

Rieseberg, L. H., Archer, M. A., & Wayne, R. K. (1999). Transgressive segregation, adaptation, and speciation. *Heredity, 83*, 363–372.

Rius, M., & Darling, J. A. (2014). How important is intraspecific genetic admixture to the success of colonising populations? *Trends in Ecology & Evolution, 29*(4), 233–242.

Robertson, A. (1952). The effect of inbreeding on the variation due to recessive genes. *Genetics, 37*, 188–207.

Rosenthal, D. M., Ramakrishnan, A. P., & Cruzan, M. B. (2008). Evidence for multiple sources of invasion and intraspecific hybridization in *Brachypodium Sylvaticum* (Hudson) Beauv. in North America. *Molecular Ecology, 17*(21), 4657–4669.

Sakai, A. K., Allendorf, F. W., Holt, J. S., Lodge, D. M., Molofsky, J., With, K. A., … Weller, S. G. (2001). The population biology of invasive species. *Annual Review of Ecology and Systematics, 32*(1), 305–332.

Schnable, P. S., & Springer, N. M.(2013). Progress toward understanding heterosis in crop plants. In S. S. Merchant (Ed.), *Annual review of plant biology* (Vol. 64, pp. 71–88). Palo Alto: Annual Reviews.

Simberloff, D. (2009). The role of propagule pressure in biological invasions. *Annual review of ecology evolution and systematics* (pp. 40:81–102). Palo Alto: Annual Reviews.

Turgeon, J., Tayeh, A., Facon, B., Lombaert, E., De Clercq, P., Berkvens, N., Lundgren, J. G., & Estoup, A. (2011). Experimental evidence for the phenotypic impact of admixture between wild and biocontrol Asian ladybird (*Harmonia Axyridis*) Involved in the European invasion. *Journal of Evolutionary Biology, 24*(5), 1044–1052.

Van Kleunen, M., Roeckle, M., & Stift, M. (2015). Admixture between native and invasive populations may increase invasiveness of *Mimulus guttatus*. *Proceedings of the Royal Society B-Biological Sciences, 282*(1815), 20151487.

Verhoeven, K. J. F., Macel, M., Wolfe, L. M., & Biere, A. (2011). Population admixture, biological invasions and the balance between local adaptation and inbreeding depression. *Proceedings of the Royal Society B: Biological Sciences, 278*(1702), 2–8.

Vitousek, P. M., DAntonio, C. M., Loope, L. L., Rejmanek, M., & Westbrooks, R. (1997). Introduced species: A significant component of human-caused global change. *New Zealand Journal of Ecology, 21*(1), 1–16.

Willemsen, R. W. (1975). Dormancy and germination of common ragweed seeds in the field. *American Journal of Botany, 62*(6), 639–643.

4

Varying selection differential throughout the climatic range of Norway spruce in Central Europe

Stefan Kapeller[1,2] | Ulf Dieckmann[1] | Silvio Schueler[2]

[1]Department of Forest Genetics, Federal Research and Training Centre for Forests, Natural Hazards and Landscape, Vienna, Austria

[2]Evolution and Ecology Program, International Institute for Applied Systems Analysis, Laxenburg, Austria

Correspondence
Silvio Schueler, Department of Forest Genetics, Federal Research and Training Centre for Forests, Natural Hazards and Landscape Seckendorff-Gudent-Weg 8, 1131 Vienna, Austria
Email: silvio.schueler@bfw.gv.at

Abstract

Predicting species distribution changes in global warming requires an understanding of how climatic constraints shape the genetic variation of adaptive traits and force local adaptations. To understand the genetic capacity of Norway spruce populations in Central Europe, we analyzed the variation in tree heights at the juvenile stage in common garden experiments established from the species' warm-dry to cold-moist distribution limits. We report the following findings: First, 47% of the total tree height variation at trial sites is attributable to the tree populations irrespective of site climate. Second, tree height variation within populations is higher at cold-moist trial sites than at warm-dry sites and higher within populations originating from cold-moist habitats than from warm-dry habitats. Third, for tree ages of 7–15 years, the variation within populations increases at cold-moist trial sites, whereas it remains constant at warm-dry sites. Fourth, tree height distributions are right-skewed at cold-moist trial sites, whereas they are nonskewed, but platykurtic at warm-dry sites. Our results suggest that in cold environments, climatic conditions impose stronger selection and probably restrict the distribution of spruce, whereas at the warm distribution limit, the species' realized niche might rather be controlled by external drivers, for example, forest insects.

KEYWORDS
adaptive capacity, among-population variation, climate change, conifers, gene flow, intraspecific variation, phenotypic variation, *Picea abies*, provenance trials, within-population variation

1 | INTRODUCTION

Understanding the constraints and drivers of species' distribution ranges is a prerequisite for predicting the consequences of climate change on natural ecosystems and for managing endangered species and populations. Within the last decade, ecologists have developed a wide variety of species distribution models to understand species' climatic and migrational limitations and to analyze the impact of climate change on biodiversity, ecosystem functions, and conservation activities (Araújo, Alagador, Cabeza, Nogués-Bravo, & Thuiller, 2011; Hanewinkel, Cullmann, Schelhaas, Nabuurs, & Zimmermann, 2012; Randin et al., 2013; Summers, Bryan, Crossman, & Meyer, 2012; Svenning & Skov, 2004; Sykes, Prentice, & Cramer, 1996; Thomas et al., 2004; Thuiller et al., 2011). Beyond the scope of immediate abiotic and biotic interactions, evolutionary biologists aim to understand which traits determine the species' genetic capacity to adapt and expand their present ranges across certain limits (Bridle & Vines, 2007; Polechova & Barton, 2015). Strong gene flow toward marginal habitats and across heterogeneous environments was found to be the major cause of restricted ranges because it results in higher genetic load and prevents local adaptation (Haldane, 1956; Kirkpatrick & Barton, 1997; Ronce & Kirkpatrick, 2001). A key determinant of the adaptive capacity of a population in a peripheral habitat and under changing environmental conditions is the genetic variation of traits related to survival, growth, and reproduction within such populations. Populations at range limits are expected to harbor lower genetic variance within populations, because they experience stronger selection than populations under optimal conditions (Kopp & Matuszewski, 2014). On the other hand, models that allow for evolving genetic variance show that a gradually changing environment with moving optima tends to increase the genetic variance as a result of an increase in rare alleles, particularly if the population size is large (Burger & Lynch, 1995; Burger, 1999; Kopp & Matuszewski, 2014). Thus, understanding whether genetic variance varies among populations throughout a species' range and how environmental constraints affect genetic variance is required for evaluating the long-term prospects of species in times of global change.

For trees, climate conditions are among the most important determinants of species' distributions (e.g., Araújo & Pearson, 2005; Prentice et al., 1992), meaning that the patterns of population phenotypic traits and patterns of climate conditions are related (e.g., Hannerz, Sonesson, & Ekberg, 1999; Hurme, Repo, Savolainen, & Pääkkönen, 1997; Rehfeldt et al., 2002). Provenance trials, where tree populations from a wide range of the natural distribution are planted in one or more climates, have revealed that these phenotypic responses are often based on both phenotypic plasticity and local adaptation (Morgenstern, 1996). The plastic response of populations can be used to model mean trait values both with uni- and multivariate climate response and genecological functions (Rehfeldt, Wykoff, & Ying, 2001; Wang, Hamann, Yanchuk, O'Neill, & Aitken, 2006) and with combined universal response functions (Wang, O'Neill, & Aitken, 2010). Such models have been found to be valuable tools for estimating the effect of climate change on growth traits and tree productivity, and for improved provenance selections (Kapeller, Lexer, Geburek, & Schueler, 2012; Mátyás, 1994; Rehfeldt et al., 2002). However, the variance of the trait means and the contribution of phenotypic plasticity to trait variance have rarely been analyzed in relation to the environmental conditions.

For Norway spruce (Picea abies [L.] Karst.), the most widespread conifer in Central Europe, we recently analyzed the intraspecific variation in climate response on the basis of an extensive provenance test (Nather & Holzer, 1979) where populations from almost the complete climatic distribution in Central Europe were tested across an equally wide range of test environments (Kapeller et al., 2012). This provenance test provides a unique opportunity to analyze the trait plasticity and variation throughout the species' climatic range, as Norway spruce occurs naturally from approximately 300 m up to 2,000 m above the sea level. Although a significant part of populations at low elevations are considered as secondary spruce forests, there is a long history of spruce populations in Austria, dating back to a refugial population in the alpine forelands (Ravazzi, 2002; Terhürne-Berson, 2005). Our previous analysis (Kapeller et al., 2012) focused on the relationship between trait means and climate parameters and thus on the immediate phenotypic response to climate. Based on the observed phenotypic plasticity and the genetic variation among provenance groups, we found that populations from warm and drought-prone areas may be appropriate candidates for extended silvicultural utilization under future climate conditions. In the present study, we aim to complement the previous analysis by investigating the phenotypic variance within and among populations across the main climate factors. The objective of this study was to quantify the phenotypic variation of height

growth within and among populations of Norway spruce. To account for environmental and genetic sources of phenotypic variation, we test for the relationships between height variation and the climate of both trial sites and population origin. Moreover, we study the temporal development of phenotypic variation at the juvenile stage. Finally, we explore the distribution of the potential selection differential in populations across the species climatic niche. For this purpose, we analyze the density distributions for climatically similar groups of populations and trial sites.

2 | MATERIALS AND METHODS

2.1 | Phenotypic data: Norway spruce provenance test 1978

We used tree height measurements from 29 trial sites of a Norway spruce provenance test series established in 1978 in the eastern Alpine region by Nather and Holzer (1979). The original trial series comprised 44 test sites, but measurement data are available for only 29 sites. These span a wide range of altitudes from 250 to 1,750 m above the sea level; they thus comprise a large part of the climatic niche of the Norway spruce, where sites at low altitudes mark the warm and dry distribution limit and sites at high altitudes close to the tree line indicate the cold distribution limit (Fig. 1).

The seed material for the trial series was collected from 480 Austrian Norway spruce populations during commercial seed harvests in 1971. Sixty populations from other countries were also included. The Austrian harvest comprised presumably autochthonous stands and included several trees as a representative sample of the stand (Kapeller et al., 2012; Nather & Holzer, 1979). Seeds were sown over six repetitions at the central forest nursery of the Austrian Federal Forest in Arndorf (Austria) and one repetition at the experimental nursery Mariabrunn of the Austrian

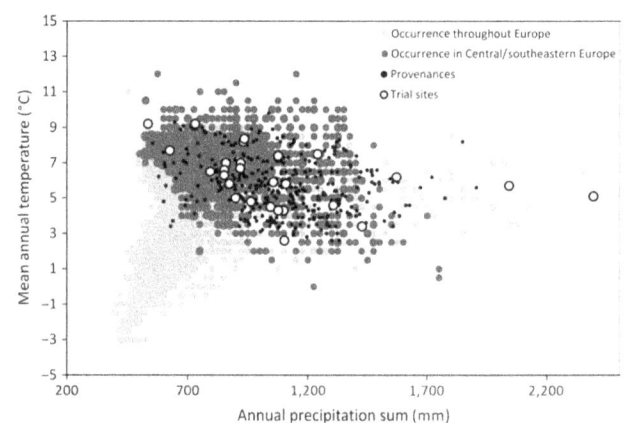

FIGURE 1 Distribution of 29 test sites (white circles) and tested populations (black dots) within the climatic range of Norway spruce. Light gray circles indicate mean annual temperature and annual precipitation of the complete Norway spruce distribution in Europe according to ICP Forests Level I monitoring plots (ICP Forests 2010). As the natural distribution in Europe can be divided into two nonoverlapping, genetically distinct ranges with different population history (Tollefsrud et al., 2008), dark gray circles indicate the central and southeastern distribution, which represent the majority of provenances

Federal Research and Training Centre for Forests, Natural Hazards and Landscape (BFW) in Vienna (Schulze, 1985). After 2 years, the seedlings were transferred into rows of tree nursery fields with 15 cm distance between seedlings (Schulze, 1985). In 1978, 5-year-old trees were transferred to the trial sites (Nather & Holzer, 1979). Each trial site was set up in a randomized complete block design with three blocks, except for sites 1 and 20, where there was only one block, and site 24, where only two blocks could be established. Because of the large number of populations sampled, not all populations could be planted at all sites. Instead, the number of tested populations per site ranged from 19 to 53 populations with an average of 28 (Table S1). The initial number of seedlings per single population per block averaged to 46.9. The seedlings of each population were planted in rectangular tree plots at a spacing of 1.5 m × 1.5 m.

The 29 trial sites were measured in 1983 and 1988, at 5 and 10 years after their establishment. This provided height data for the trees at the age of 10 and 15 years. During the 1983 measurement, the shoot length for the preceding 3 years was also measured, which provided heights for the trees at the age of 7, 8, and 9 years. Of the 109,101 trees initially planted in 1978, 83,304 could be measured in the year 1988 (on average 38.8 trees per plot). This reduction was caused by mortality, because there had been no forest management activities for the duration of the trial (1978–1988).

2.2 | Climate data

Climate data from all the trial sites and population origins were compiled by the Austrian Central Institute for Meteorology and Geodynamics for a previous analysis (Kapeller et al., 2012). The mean climate data of trial sites strictly refer to the growing period from 1978 to 1988, whereas the climate data of population origin are based on long-term means (1971–2008). These data include mean temperatures, mean monthly minima and maxima, and precipitation sums for both the complete year and the approximate growing season from April to September. In addition, the length of vegetation period (given as the number of days with an average temperature above 5°C), growing degree days (i.e., a thermal index accumulating degree days above a threshold of 5°C), the average day of the first frost in fall, and an annual heat moisture index according to Wang et al. (2006) were used as climate parameters.

To reduce the number of climate predictors and to obtain uncorrelated variables for the subsequent analysis, we performed a principal component analysis (PCA) using the statistical environment R (R Core Team 2014), with PCA functions from the R package "FactoMineR" (Husson, Josse, Le, & Mazet, 2015). The contribution of climate variables to each dimension of the PCA is given in Table S2. The first principal component explains 59.7% of observed variance and aggregates mainly temperature-related climate parameters. The second principal component adds another 17.3% variance and aggregates mainly precipitation- and drought-related parameters (Fig. 2). Therefore, we refer to the first dimension as "temperature-related principal component" (TempPC; large values indicate warm conditions) and to the second dimension as "precipitation-related principal component" (PrecPC; large values indicate moist conditions).

2.3 | Statistical analyses

Our analyses were performed in four steps. First, we applied a linear model at each test site to partition among-population variance from the total phenotypic variance. Second, we calculated the standardized within-population variation for each population and test site and related it to the climate of the trial site and the climate of the population origin. The standardized within-population variation of tree height was calculated for the same trees in different years at the ages of 7, 8, 9, 10, and 15 years; this was performed to assess the temporal changes in within-population variation throughout the juvenile stage. Third, we analyzed the density distributions of tree heights with mixture-model analysis to understand the environmental and genetic effects on phenotypic variation. In a fourth step, we validated the robustness of our results with respect to the potential biases introduced by unequal climatic distribution of tested populations across sites and varying survival rates between trial sites.

2.3.1 | Among-population variation

To estimate the variance among populations from the total variance at the individual trial sites, we used a linear model in the R package "lme4" (Bates, Maechler, Bolker, & Walker, 2014). Populations and repetitions (blocks) were treated as random effects, which allowed us to extract the estimates of the among-population variance σ^2_{ap}, the among-block

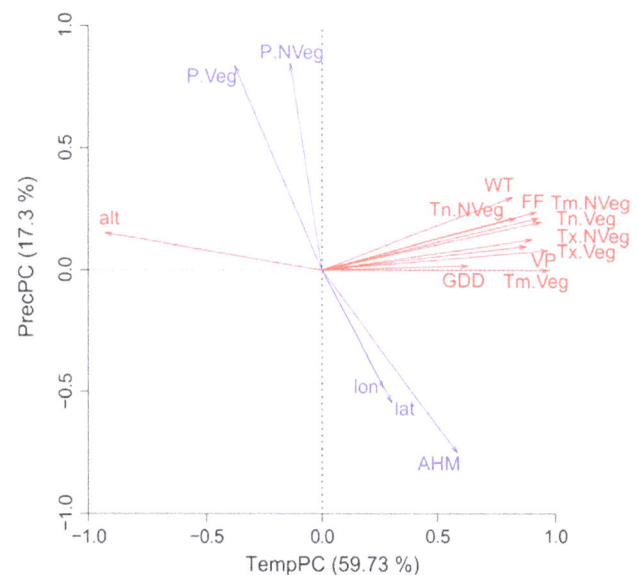

FIGURE 2 Principal component analysis of climatic parameters. The biplot shows that temperature-related factors and altitude (red) are oriented along dimension TempPC (first principal component). Precipitation and drought index (blue) are oriented along PrecPC (second principal component). Parameters included in this analysis were temperature mean (Tm), temperature maximum (Tx), temperature minimum (Tn) and mean precipitation sums (P) during the vegetation season (.Veg) and outside the vegetation season (.NVeg), growing degree days (GDD), length of vegetation period (VP), annual heat moisture index (AHM; see Wang et al., 2006), mean winter temperature (WT), average day of first frost in fall (FF), as well as longitude (lon), latitude (lat), and altitude (alt)

variance σ^2_b, and the total site variance σ^2_s of tree heights from the fitted models. To remove the effects of different growth rates among sites and to allow for comparisons between the individual test sites, these variance components (σ^2_s, σ^2_{ap}, σ^2_b) were standardized as coefficients of variation using the formulas $CV_s = \sigma_s/\mu_s$, $CV_{ap} = \sigma_{ap}/\mu_s$, and $CV_b = \sigma_b/\mu_s$, where μ_s is the mean tree height at the site. After calculating these variation measures for each of the 29 sites separately, the relationship of the standardized total variation CV_s and the standardized among-population variation CV_{ap} to the first two principal components of site climate TempPC and PrecPC was analyzed by linear regression analysis. Additionally, the ratio of among-population variation to the total site variation was calculated as CV_{ap}/CV_s (which equals σ_{ap}/σ_s) to test whether the portion of explained variance by population relative to the total site variance relates to site climate. Therefore, we applied linear regression analyses using the first two PCA components of the climate parameters (TempPC and PrecPC) from the test sites as explanatory variables.

2.3.2 | Within-population variation

To analyze the within-population variation along a climatic gradient of test sites and populations, we calculated the coefficient of variation of tree heights at age 15 for each population on each site separately as $CV_{wp} = \sigma_{wp}/\mu_{wp}$, where σ_{wp} and μ_{wp} refer to all individual tree heights measured for a specific population at a specific site. The coefficient of variation was calculated to remove the effect of different growth rates among sites. In total, we obtained 819 values of CV_{wp} for all population–site combinations. We used multiple linear regression analysis to investigate the potential relations between within-population variation CV_{wp} and the first two PCA components of the climate parameters (TempPC and PrecPC) from the test sites as well as from the population origins. This analysis of the within-population variation CV_{wp} was repeated with tree heights measured in earlier years, when trees aged 7, 8, 9, and 10 years, in order to identify the temporal changes in tree height variation within the juvenile stage. In addition, interactions between site and population climate were assessed in bivariate plots.

2.3.3 | Mixture-model analysis

To understand the effect of site and population climate on phenotypic variation, test sites and populations were both categorized into three climatic groups. The climatic ranges of the first two PCA components (TempPC and PrecPC) were subdivided into equal intervals and populations and sites were then assigned to the corresponding climatic subgroup. The groups were labeled S1, S2, and S3 for sites and P1, P2, and P3 for populations, referring to low, medium, and high levels of TempPC or PrecPC, respectively. For low TempPC, group S1 and P1 pooled "cold" sites and populations originating from a "cold" environment, respectively, while for low PrecPC, S1 and P1 represented "dry" sites and populations originating from dry locations. The density distributions of all nine subset combinations (three population subsets X three site subsets) were plotted separately and analyzed with mixture model analysis (R package "mixtools"; Benaglia & Chauveau, 2009). Mixture model analysis provide the density probabilities of hypothetical normal-distributed subgroups

within each subset. Such probabilistic models have been used to identify subpopulations within an overall population (Benaglia & Chauveau, 2009; Ni, Baiketuerhan, Zhang, Zhao, & Von Gadow, 2014) and to test for admixture within populations using quantitative trait data. Here, we used mixture model analysis not to identify truly distinct subpopulations in a strict sense of population genetics, but to visualize the subtle patterns in the density distributions of tree heights. With our data, the mixture-model analysis allowed us to differentiate two phenotypic subgroups of "tall" and "small" trees within each of the climatic subsets. The ratios of these subgroups provided by the mixture model analysis can be used as additional statistics to describe the shape of a density distribution.

2.3.4 | Robustness to stratification

The distribution of provenances to the trial sites did not follow a fully randomized procedure, but was specified in some cases according to the altitude of the test sites and populations. Thus, some populations from higher altitudes (with colder and wetter climate) were preferentially tested on (colder and wetter) sites at higher altitudes, and vice versa (see Fig. 3). As this provenance distribution had the potential to affect our analysis, we aimed to reduce the data imbalance and tested for the effects of a slightly unequal population distribution by applying two different data stratification approaches. Stratification was performed by weighting each test unit (=population × site combination) according to its relative frequency in the bivariate climate spectrum of populations and sites (Fig. 3). We divided the bivariate climate spectrum into 25 climate strata according to regular subdivisions of TempPC at sites and population origins. We then calculated weights for each unit using the relative frequency of each unit in relation to

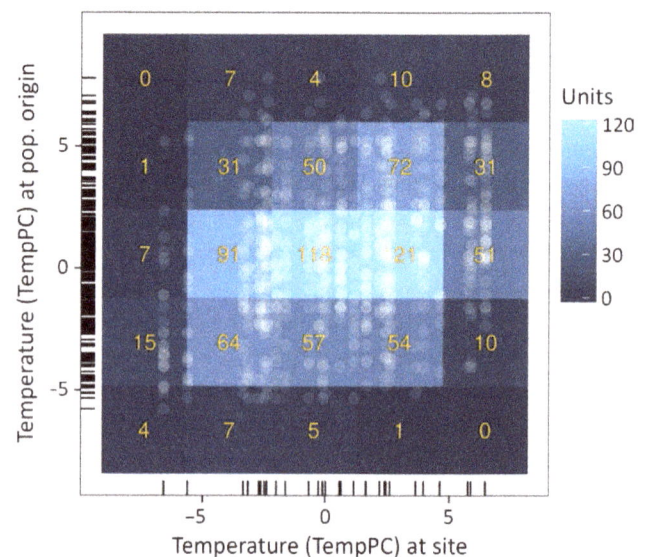

FIGURE 3 Climatic distribution of sites and populations. The bivariate climate spectrum of sites and population origins was subdivided into 25 climate strata, each representing a class of climatic site–population combinations. Relative frequency of test units, that is, site–population combinations (transparent white dots), was utilized to weight each unit and to test whether the stratification of the trial design has an effect on the outcomes of the study

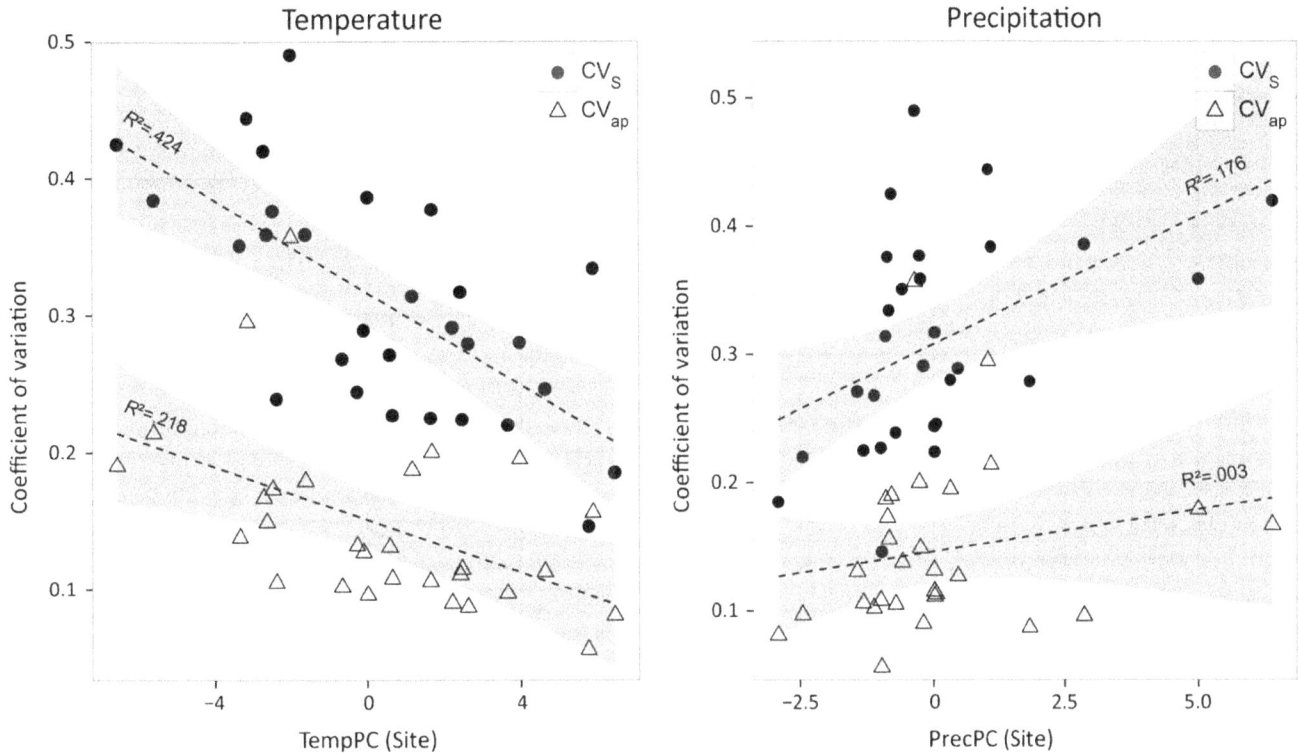

FIGURE 4 Total and among-population variation at trial sites. Coefficients of variation at each site (CV_S, full circles) and among-population variation (CV_{ap}, open triangles) are shown for each site along the climatic gradients of TempPC (left) and PrecPC (right). CV_S are high at cold-moist sites (low TempPC, high PrecPC) and low at warm-dry sites (high TempPC, low PrecPC)

the total number of units (weight#1) or its deviation from the joint frequency probability of the populations and sites of each respective stratum (weight#2). Taking into account these weights for all population x site combinations, we reanalyzed our multiple linear regression analyses of CV_{wp}.

2.3.5 | Effects of mortality

Of the 109,101 trees planted in 1978, 76% survived and could be measured in the year 1988 (i.e. 83,304 trees). The remaining 24% of trees died off for unknown reasons. As this mortality might also bias our estimates of height variation, we tested for the differences in the survival rate of individual populations across both the climate gradients of the trial sites and the gradient of the population origin using multiple regression analysis.

3 | RESULTS

3.1 | Total tree height variation and among-population variation are negatively related to temperature at trial sites

The mixed-effect model yielded variance components for each site, with one variance component explained by the differences among tested populations (on average 21.8% of the total variance σ^2_s), one component explained by the differences among repetition blocks (on average 4.2% of σ^2_s), and a residual variance (on average 74% of σ^2_s). Results for each site are given in supplementary Table S1.

Standardized coefficients of variation allow for comparisons among trial sites. The standardized total variation in tree heights CV_S ranged from 0.15 to 0.49 between the trial sites (Fig. 4, Table S1). When we partitioned the among-population variation from the total tree height variation at each site, the standardized among-population variation CV_{ap} ranged from 0.06 to 0.36 (Fig. 4, Table S1). The CV_{ap}/CV_S ratio ranged across all sites from 0.25 to 0.73 and averaged 0.47.

The overall variation in tree heights CV_S was found to be significantly related to the climate of test sites for both TempPC (adjusted $R^2 = 0.44$, $p < 0.001$) and PrecPC (adjusted $R^2 = 0.19$, $p = 0.011$). Here, CV_S was found to be high at cold-moist sites (low TempPC, high PrecPC) and low at warm-dry sites (high TempPC, low PrecPC). The ratio of among-population variation (CV_{ap}) to the overall variation (CV_{ap} / CV_S) was not significantly related to the climate at test sites for either TempPC ($p = 0.48$) or PrecPC ($p = 0.58$).

3.2 | Within-population variation is negatively related to temperatures at trial site and population origin

The variation within populations CV_{wp} ranged from 0.12 to 0.58 for individual population–site combinations with an average of 0.28. Across single trial sites, the average CV_{wp} ranged from 0.14 to 0.41 (Table S1).

Multiple regression analysis revealed a significant relationship between the within-population variation (CV_{wp}) and the climate at the test sites and the population origin (see summary of the multiple linear model fit shown in Table 1). Regression analyses were performed separately for

TABLE 1 Parameter estimates of multiple regression analyses, predicting CV_{wp} by climate parameters TempPC and PrecPC at test sites (*_S) and population origins (*_P) (adjusted R^2 = 0.424, F-statistic = 151.3 on 4 and 814 degrees of freedom)

	Estimate	SE	t value	Pr (>|t|)	
(Intercept)	0.2896	0.0024	123.1	<0.001	***
TempPC_S	−0.0116	0.0007	−15.5	<0.001	***
TempPC_P	−0.0017	0.0008	−2.1	0.0333	*
PrecPC_S	0.0134	0.0013	10.6	<0.001	***
PrecPC_P	0.0024	0.0014	1.8	0.0746	.

TempPC and PrecPC and for site and population-origin climate (Fig. 5). Regressions of CV_{wp} to the site climate revealed negative relationships with TempPC and positive relationships with PrecPC. Therefore, at warm and dry sites, within-population variation is low. In contrast, at cold and moist sites, within-population variation is high (Fig. 5, left). Moreover, within-population variation (CV_{wp}) is negatively related to the climate of the population origin for TempPC (Table 1; Fig. 5, center).

The multiple regression analysis for the 15-year-old trees demonstrated clear effects of trial site climate on CV_{wp} after the trees had been growing for 10 years in the field. When the trees were planted at the age of 5 years, they had a similar variation within populations CV_{wp}

at all test sites. To reveal the temporal course of CV_{wp} development throughout the growing period in the field, we also analyzed CV_{wp} for trees aged 7, 8, 9, and 10 years and compared CV_{wp} for populations at the coldest trial site (i.e., the lowest TempPC at site 11) and the warmest site (i.e., the highest TempPC at site 42) (Fig. 6). In 7-year-old trees, the variation within populations CV_{wp} after two growing periods was already higher at the coldest trial site than at the warmest site. At the coldest site, the within-population variation increased from age of 10 to 15. A contrasting pattern was found at the warmest trial site on the upper temperature limit. Here, within-population variation increased only slightly from tree ages of 7–10, but then remained constant at a relatively low level until the tree age of 15 years (Fig. 6).

These differences in CV_{wp} temporal development between trials at cold and warm sites can be observed at all trial sites (Fig. 7; the results of multiple regression analyses in Table S3): During the period from tree age of 7 to 15 years, CV_{wp} stayed at a more or less constant level at sites with warm (high TempPC) and dry (low PrecPC) climates, while CV_{wp} increased with tree age at sites with cold (low TempPC) and moist (high PrecPC) climates (Fig. 7, row 1, row 3). Thus, the significant relationships found between trial climate and CV_{wp} when a tree is 15 years old are due to increasing CV_{wp} at cold sites, but not to decreasing CV_{wp} at warm sites. We found the most pronounced response of CV_{wp} to site climate in 15-year-old trees. However, these

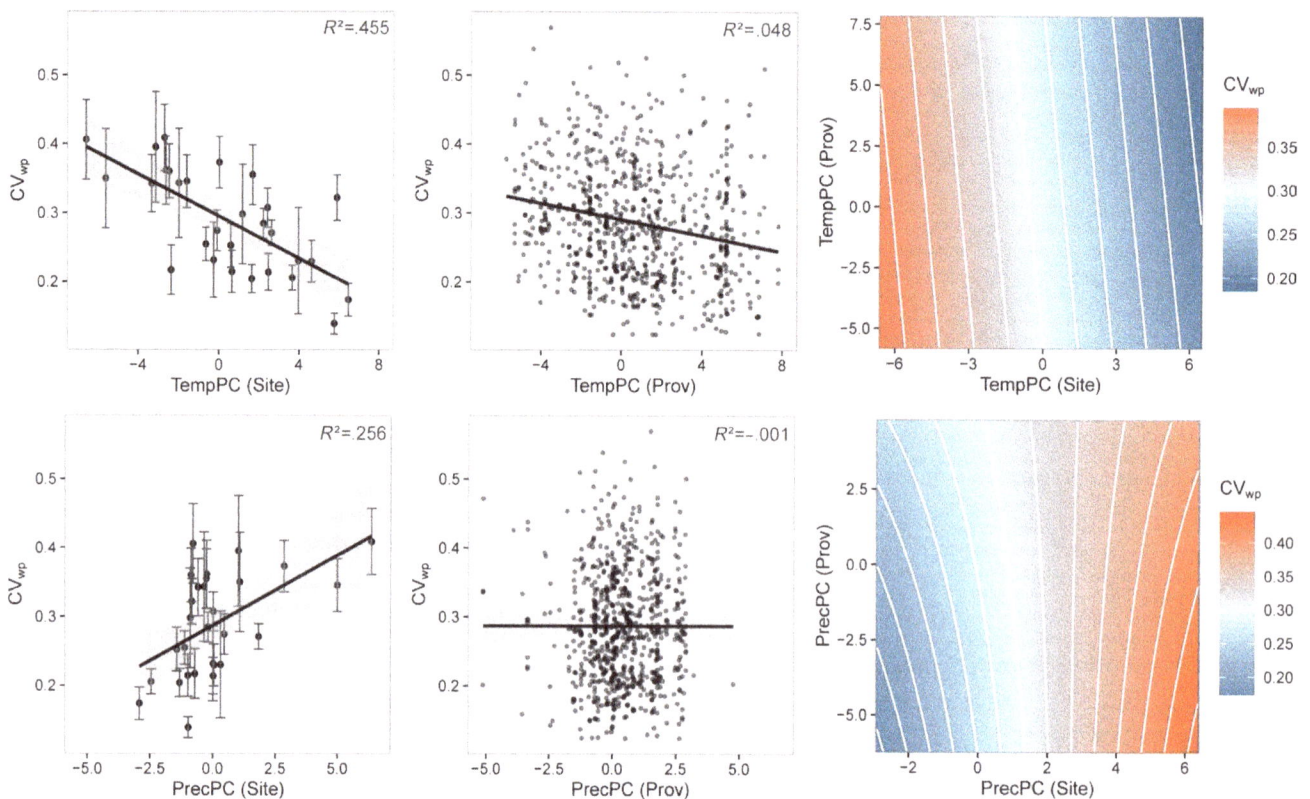

FIGURE 5 Within-population variation for each population–site combination. Coefficients of variation for within-population variation (CV_{wp}) are shown along the climate gradients TempPC (top) and PrecPC (bottom), where large values of TempPC and PrecPC represent warm and wet locations, respectively. Left: response along climate at the test sites. Error bars represent CV_{wp} values from all populations tested at a respective site; Center: response along population climate. Points represent CV_{wp} values of one population tested at one specific site; Right: bivariate plot along both site and population climates. Color indicates the level of within-population variation (CV_{wp}) (see legend)

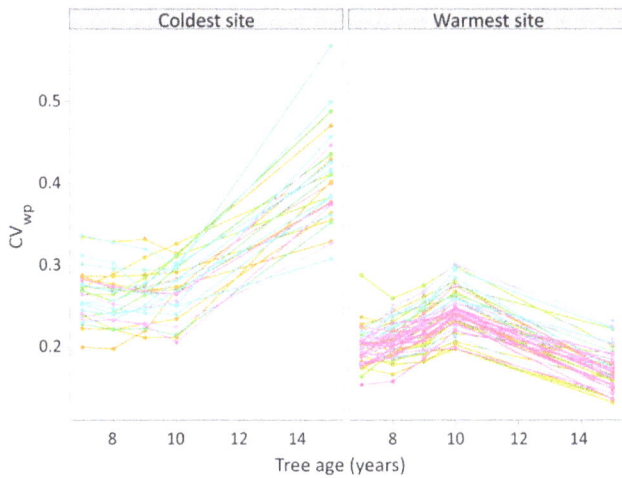

FIGURE 6 Within-population variation and tree age. CV_{wp} of populations at climatically most extreme sites 11 (coldest) and 42 (warmest) at tree ages of 7, 8, 9, 10, and 15 years

relationships (including a significant linear regression) between sites climate and CV_{wp} were already found for trees at the age of 7, 8, 9, and 10 (Table S3), but the slope of the relationship increased with tree age.

Beyond a significant relation of CV_{wp} to site climate, we found a significant relation of CV_{wp} to the temperature-related climate predictor (TempPC) of population origins for trees at the age of 7, 8, 9, 10, and 15 (Fig. 7, row 2; Table S3).

3.3 | Shapes of density distributions of tree height

The frequency density distributions of tree heights at the age of 15 reveal contrasting patterns for the categorized climatic subsets of the site and population climate. Figure 8 displays histograms of different site and population subsets (S1, S2, S3, P1, P2, P3) where the effects of the site climate are shown on the vertical axis and the effects of the climate at population's origins are shown on the horizontal axis. With respect to site climate, we found increasing mean tree heights with increasing

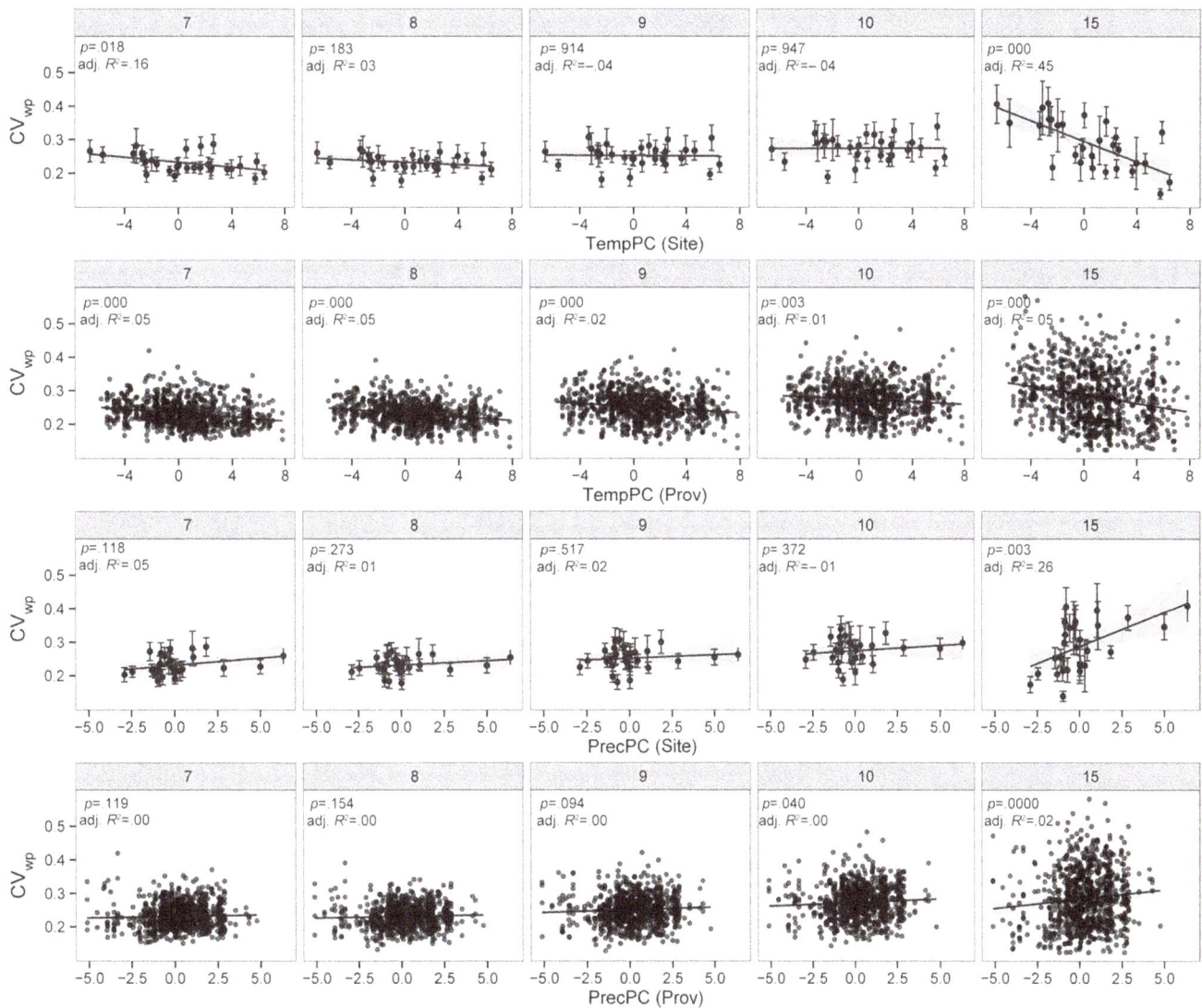

FIGURE 7 Within-population variation, tree age, and climatic gradients. Values of CV_{wp} are shown along climatic gradients of TempPC (1st and 2nd row) and PrecPC (rows 3 and 4) for tree ages of 7, 8, 9, 10, and 15 years. Error bars in site climate responses (rows 1 and 3) represent within-population variations of all populations on a site (mean ± SD of CV_{wp}). Dots in population climate responses (rows 2 and 4) represent within-population variation of a specific population at a specific site

TempPC and decreasing PrecPC. The shape of the density distributions suggests that at warm and dry sites, the variation in tree heights is also larger than at cold and moist sites (in contrast to our previous analyses, where we used the coefficient of variation). We found tree height distributions at cold and moist sites to be left-skewed, whereas at warm and dry sites they were not skewed or slightly right-skewed (see Table 2 for mean, variance, skewness, and kurtosis of the distributions). Kurtosis refers to the presence or absence of pronounced peaks in a density distribution. At cold sites, we found "peaked" (leptokurtic) distribution shapes (positive kurtosis) and "flat" (platykurtic) distribution shapes (negative kurtosis) at warm and dry sites.

Population subsets (P1, P2, P3) reveal only small differences in histogram patterns (Fig. 8). Overall, the mean tree height and the variation increased from (cold) P1 populations to (warm) P3 populations. Moreover, skewness increases from cold/moist populations to warm/dry populations, whereas kurtosis decreases. Thus, the populations from cold/moist sites are rather left-skewed and peaked, whereas the populations from warm/dry sites are nonskewed or slightly right-skewed and flat.

Density probability functions from mixture analyses provide additional insights into these patterns. In mixture analysis, each tree in a subset combination is attributed to one of two hypothetical components of the distribution (curves in Fig. 8), representing "relatively small" and "relatively tall" trees. Means and standard deviations of each component, as well as probabilities of data being classified as one of these components, are shown in Table S4. At warm S3 sites (high TempPC), 67% of trees from cold populations are classified as small (blue curve), while 33% of trees are classified as tall (red curve). Trees from warm populations at the same S3 sites are classified 46% small and 54% tall. At intermediate S2 sites, we again have a larger number of "small trees" in P1 populations, but conversely a larger number of "tall trees" in P3 populations. At cold S1 sites, most trees are classified as small, regardless of population climate.

3.4 | Robustness to stratification

To avoid the biases in our analyses due to the unbalanced distribution of populations to trial sites, we corrected our dataset by stratifying population x site combinations according to their relative frequency in the bivariate spectrum of the population and site climate (see Fig. 3). The reanalysis of the multiple linear regression analyses of the response variable CV_{wp} and the predictors TempPC and PrecPC of site and population origins (Table 1) with the two stratification weights (weight#1 and weight#2) did not alter the significance of the regression analysis compared to the unweighted model (Table S5). Using weight#1 resulted in a higher significance level for TempPC of population origin, and using weight#2 resulted in a higher significance level even for PrecPC of population origin. Therefore, stratification

TABLE 2 Statistics describing forms of distribution of each of the nine combinations of climate subset (see Fig. 8)

S	P	Trees	Sites	Pop	Mean	Var	Skew	Kurt
TempPC								
1	1	8,019	8	60	141.4	4,084.1	0.87	0.92
1	2	9191	8	82	147.7	4,356.0	0.96	0.93
1	3	2346	8	20	153.1	3,929.0	0.95	0.94
2	1	7457	11	45	282.8	11,796.4	0.25	−0.39
2	2	15436	11	109	300.4	13,894.7	0.17	−0.57
2	3	4811	11	29	309.1	14,496.8	0.00	−0.69
3	1	6109	10	35	341.8	15,792.7	0.30	−0.35
3	2	21266	10	122	388.8	20,029.1	0.03	−0.70
3	3	9413	10	38	398.4	22,514.9	−0.04	−0.87
PrecPC								
1	1	3,635	18	17	353.5	24,419.5	−0.10	−1.04
1	2	39,289	21	184	310.5	22,335.8	0.28	−0.89
1	3	15,619	21	74	294.8	21,815.6	0.31	−0.91
2	1	1,225	4	9	361.3	34,765.4	0.35	−0.81
2	2	11,230	6	90	293.9	24,381.4	0.60	−0.02
2	3	7,081	6	74	291.8	23,994.6	0.62	−0.07
3	1	129	1	1	295.4	13,147.1	0.15	−0.91
3	2	3,436	2	29	214.4	11,627.4	0.60	−0.28
3	3	2,404	2	19	214.3	11,596.1	0.75	−0.01

S, P = IDs of the climate subsets of sites and populations indicating the levels of the respective climate variable TempPC or PrecPC (1 = low, 2 = medium, 3 = high); trees = the number of individual trees pooled in the combination of climate subsets; sites = the number of pooled sites; pop = the number of pooled populations; mean = mean tree height at the age of 15; var = variance of tree heights; skew = skewness of tree height distribution; kurt = kurtosis of tree height distribution

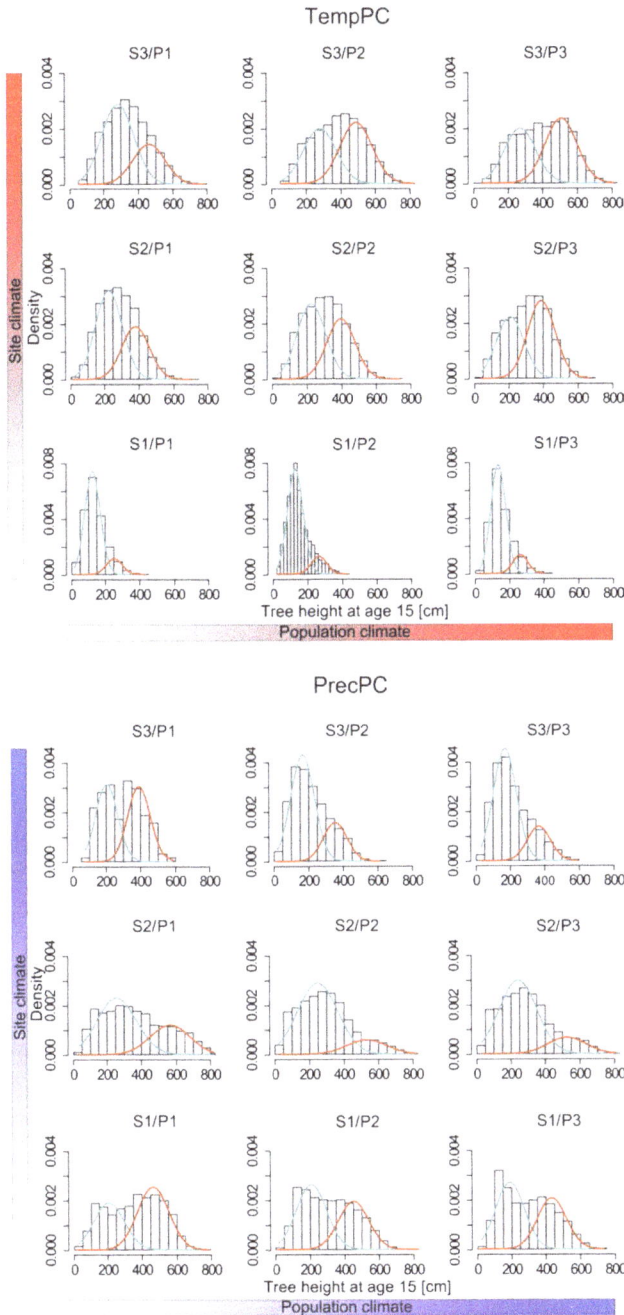

FIGURE 8 Density distributions of absolute tree heights at the age of 15. All records were divided into nine climate subsets by TempPC (top) and PrecPC (bottom). Each subset aggregates records of one of three test site groups (S1, S3, or S3) and one of three population groups (P1, P2, or P3). Red and blue curves are the results from a mixture-model analysis and represent density probabilities of two hypothetical subcomponents for "small" and "high" trees within each subset. Parameter estimates from the mixture-model analysis (portions of attributed data to each component, means of components and standard deviations) are presented in supplementary Table S4

by using two different weighting methods did not reveal any false significant model parameters in the linear regression analysis of the unweighted data. Instead, both weighted models resulted in a slightly stronger relevance of TempPC of population origins (compare Table 1 and Table S5).

3.5 | Effect of mortality

The analysis of survival rates across our trial sites did not indicate a bias of tree height variation measures. A significant relationship was found between the survival rate (calculated as the proportion of surviving trees after 10 years in the field) and TempPC of the trial sites, but not to TempPC of provenance origin. However, the total effect of site climate on mortality seems to be negligible, as the coefficient of determination of this relationship is very low (adjusted R^2 = 0.023) and the regression slopes close to zero (Table 3). Also, the relation of survival to PrecPC was not found to be significant for trial sites and provenance origin.

4 | DISCUSSION

The persistence of tree populations in future climates depends crucially on their adaptive capacity to adjust to new environmental conditions. Phenotypic variation in fitness-related traits is both a result of environmental selection pressures and a prerequisite for the adaptation to changing environments. Here, we used one of the largest common garden trial series of Norway spruce established more or less throughout its complete climate distribution in Central Europe and tested for the effects of climate conditions on the phenotypic variation within and among populations. Both the climate at trial sites and the climate at the geographic origin of populations had a significant impact on the phenotypic variation within populations (Fig. 5). The effect of site climate on phenotypic variation was much larger than the effect of climate of population origins, as previously shown also for other conifers (Chakraborty et al., 2015; Wang et al., 2010). The observed effect of site climate suggests stronger climatic selection pressures at the colder end of the species distribution resulting in increasing tree height variation in colder and moister environments. Survival rates were neither affected by site climate nor by provenance origin climate (see the numbers of planted and measured trees in Table S1).

4.1 | Environmental sensitivity of selection

Absolute mean heights were found to be strongly associated with site climate as a result of phenotypic plasticity. At the age of 15, trees

TABLE 3 Parameter estimates of multiple regression analyses predicting the survival rate by climate parameters TempPC and PrecPC at test sites (*_S) and population origins (*_P)

	Estimate	SE	t value	Pr (>|t|)	
(Intercept)	0.9713	0.0093	104.03	<0.001	***
TempPC_S	0.0131	0.0029	4.41	<0.001	***
TempPC_P	−0.0058	0.0032	−1.78	0.0757	.
PrecPC_S	0.0125	0.0050	2.48	0.0135	*
PrecPC_P	0.0090	0.0054	1.69	0.0923	.

Survival rate was calculated for each population–site combination as the percentage of living trees after 10 years in the field. We used arcsine transformation to meet normality assumptions. Adjusted R^2 = 0.023, F-statistic = 5.73 on 4 and 814 df.

reached on average 5 m at the warmest site and 1.2 m at the coldest site. Trees at warm sites also showed a higher absolute variance of tree heights. However, the overall coefficient of variation per trial site decreased significantly with increasing temperatures and ranged from 0.49 at the coldest site to 0.15 at the warmest (Fig. 4). This higher variation at colder sites was found to result mainly from the variation within populations (CV_{wp}), because the ratio of variation among populations to the total variation was not related to the site climate.

The strongest relationship between CV_{wp} and climate was found at tree age of 15 years (i.e., growing for 10 years in the field). The analysis of the temporal course of CV_{wp} by including data from earlier height measurements showed that CV_{wp} increased from age 7 to age 15 at the coldest trial sites, whereas CV_{wp} remained approximately constant at trial sites at the warm end of the species distribution (Figs. 6 and 7). This suggests that the climate conditions at the colder trial sites increase the differentiation among trees within populations, but climate conditions at warm sites do not affect CV_{wp}. Thus, at cold sites, relatively few well-performing trees have an advantage in height growth and will likely dominate and outcompete the slower-growing trees in the future stand. Given the huge intraspecific competition among juvenile trees (Vieilledent, Courbaud, Kunstler, & Dhôte, 2010), this will ultimately result in a higher potential selection differential at the cold edge of the spruce distribution. This is also supported by the mixture-model analysis across different environments (Fig. 8 and Table S4). At cold sites (labeled S1 in Fig. 8), we found right-skewed, leptokurtic distributions with a pronounced peak (negative kurtosis) resulting in a prevailing component of relatively small trees and a second much smaller component of taller trees. If climate was of similar importance at the upper temperature limit of the species' niche, we would expect a left-skewed distribution at the warmer trial sites (labeled S3 in Fig. 8), where physiological limits set a threshold for height growth. Instead, we did not find left-skewed distributions at the warmest trial sites, but nonskewed, platykurtic (positive kurtosis) distributions without pronounced peaks (Fig. 8, top). From such shapes of tree height density distributions, we conclude that Norway spruce is not under strong selection by climate constraints at any of our warm-dry test sites—some of which are located at the upper temperature limit of the species range (Fig. 1).

Potential reasons for the increasing height variation within populations at cold sites are manifold. One possible explanation for the increasing height variation at colder sites could be the strong relation between height growth and phenology (e.g., Kleinschmit, Sauer-Stegmann, Lunderstadt, & Svolba, 1981). In particular, tree height is strongly correlated with flushing and bud set within and across populations. Both traits, flushing and bud set, possess high additive genetic variation (Hannerz et al., 1999) and are driven by the annual temperature course in spring and late summer (Søgaard, Oystein, Jarle, & Olavi, 2008). Thus, at cold sites, small differences in temperature sum requirements for flushing among individual trees may accumulate into large differences in the onset of growth and resulting height growth pattern. As temperature accumulates much faster at warm sites, the same genetic differences in temperature sum requirements result in smaller differences of bud burst and height growth in warmer environments. Such behavior was also observed for other species: Davi et al.

(2011), for example, analyzed flushing of various tree species from 960 to 1530 m a.s.l. and found a significant altitude effect for *Pinus sylvestris*, which decreased with faster spring development. Generally, our results on climate constraints at the cold end of the species distribution are in good agreement with the manifold traits that were found to be related to cold adaptation in conifers (Aitken & Hannerz, 2001; Howe et al., 2003; Morgenstern, 1996). Besides bud burst and bud set, a significant variation within and among families of Norway spruce has been found for frost hardiness (Skrøppa, 1991) and populations from southern Finland were found to be more sensitive to frost events than northern populations (Pulkkinen, 1993). Narrow-sense heritability of frost resistance ranged from 0.04 to 0.28 in a Swedish progeny trials (Hannerz et al., 1999). Howe et al. (2003) concluded that cold adaptation traits appear to be under strong natural selection. In contrast, evidence for genetic adaptations of Norway spruce to warm temperatures, namely drought resistance or stomatal conductance, is limited. In part, this may be due to the fact that fewer studies have addressed the intraspecific variation in adaptation to the warm temperature edge of a species' range (but see Mátyás, Nagy, & Jármay, 2009 or Lamy et al., 2011); but it may also be that Norway spruce rarely reaches its physiological limits and has thus developed fewer local adaptations. Another potential cause for the unequal climate selection could be the structure of the trial sites itself: Warm trials are rather located at low elevations with relatively flat and homogeneous site conditions. In contrast, cold trial sites are located on mountain slopes and likely provide higher on-site heterogeneity that may result into increasing differentiation among individual trees. A larger heterogeneity of both land surface and soil structure might provide advantageous microclimatic or edaphic conditions under which a few young trees are able to outperform their competitors. Also, snow cover in early spring is likely to increase the variation among trees, as it may delay the onset of growth for small trees completely covered with snow in particular. Larger trees that extend already beyond the snow cover are able to receive environmental signals for the start of the growing period much earlier than their smaller counterparts.

4.2 | Phenotypic evolution and environmental heterogeneity

Besides the climate at the trial sites, the climate of population origin was also related to CV_{wp} with higher variations within populations from colder regions (Fig. 7, row 2; Table S3). This finding seems counterintuitive, considering that with stronger selection at colder sites one might expect lower phenotypic variation. One explanation could be the ongoing phenotypic evolution at a marginal habitat and within changing environments where, for large and recombining populations, the genetic variance was found to increase under directional selection as a result of an increasing frequency of rare alleles (e.g., Burger, 1999; Burger & Lynch, 1995). This type of situation has been observed in several theoretical studies and may fit for Norway spruce at its cold edge, namely at higher elevations in its alpine distribution. Here, permanent selection pressures might cause maladaptations and thus a permanent lag of the population mean behind the environmental optimum (see

Kopp & Matuszewski, 2014). Another explanation for the increased variation could be the immediate effect of phenotypic plasticity, which was found to increase the genetic variation within one generation because of variations in the slope of reaction norms (Chevin & Lande, 2011). To differentiate between the effects of phenotypic plasticity or directional selection as described by Burger (1999), the immediate effects of mortality during changes in the genetic variance need to be considered. Our analyses of survival rates at trial sites indicated only a negligible effect of climate on mortality, as the coefficient of determination of this analysis is very low (adjusted R^2 = 0.023) and regression slopes are close to zero (Table 3). This indicates that between the age of 5–15 years, tree height variation is rather shaped by phenotypic plasticity than natural selection. Theoretical analysis of plasticity and phenotypic variation in populations within marginal habitats along environmental gradients support this conclusion. Chevin and Lande (2011), for example, found an increase in genetic variance within one generation. However, the effects of changing environments on variance and the contributions of plasticity are still not fully understood, because genetic variances are often assumed to be constant in quantitative models (Kopp & Matuszewski, 2014). Another explanation for higher variation within populations from colder origins could be intensive gene flow and environmental heterogeneity in such environments. In our experiment, cold populations originate mainly from alpine locations at higher elevations. Within its distribution in the Alpine region, Norway spruce occurs from the valley floors up to subalpine habitats near the tree line. This high environmental heterogeneity likely results in manifold local adaptations. The intense gene flow caused by pollen flow across spatially close but climatically distant populations could increase the genetic variation by introducing maladapted genotypes, as shown, for example, by Yeaman and Jarvis (2006). The gene flow and environmental heterogeneity explanation is supported by a recent meta-analysis of progeny tests of Norway spruce in Sweden (Kroon, Ericsson, Jansson, & Andersson, 2011). This study compared the genetic variation across a latitudinal gradient from 56°N to 65°N covering a similar climatic range. In contrast to our study, Kroon et al. (2011) found a significant decrease in genetic variation with increasing latitude and thus decreasing temperatures. Across the large spatial distance in Sweden, gene flow is much less likely to connect the same environmental gradients as in the Alpine landscape.

4.3 | Among-population variation

The ratio of among-population variation to the total phenotypic variation ranged from 25% to 73% (with a mean of 47%, Table S1) and did not change along the climatic gradient of TempPC or PrecPC. This is in agreement with our initial hypothesis that the distribution of populations across trial locations (although not completely balanced) results in similar differences between populations at each site. However, the higher selection differential acting on within-population variation at colder trial sites would imply that selection also increases the differences among populations. Instead, the ratio of among-population variation was found to be independent from site climate, indicating that besides local adaptations to climate, other factors might contribute

to differentiation among populations. Such a factor is most likely the phylogeographic pattern of *P. abies*. The majority of tested populations originated from the eastern Alpine range and from surrounding countries. Indeed, the geographic origins of the populations cover the three main refugial lineages of Norway spruce in Central and Western Europe. The long population history of these lineages can still be observed with various molecular markers (e.g., Maghuly, Pinsker, Praznik, & Fluch, 2006; Mengl, Geburek, & Schueler, 2009), and provenance trials throughout Europe have recognized a strong variation among regional groups (Giertych, 1992; Krutzsch, 1992). In comparison with other provenance experiments with Norway spruce, the provenance effects on tree height variation in our study (25%–73%) seems higher than observed elsewhere (e.g., Weisgerber, Dimpflmeier, Ruetz, Kleinschmitt, & Widmaier, 1984; : 44%; Liesebach, Rau, & König, 2010: 0%–26%; Ujvári-Jármay, Nagy, & Mátyás, 2016: 12.6%). This is, however, because we calculated the coefficient of variation in order to correct for the strong variation in height growth among trial sites. The untransformed ratio of the among-population variation to total phenotypic variation ranged from 6% to 53% and is comparable to the values reported by Weisgerber et al. (1984), Liesebach et al. (2010), and Ujvári-Jármay et al. (2016).

4.4 | Understanding Norway spruce range limits

The focus of our empirical study is on tree height data, but not on direct measures of a tree's fitness in terms of survival and reproduction. Such direct measures of fitness for tree species are difficult to obtain, because many forest trees do not flower below the age of 20 years and unbiased estimates of reproductive performance would require reproductive success to be measured throughout an individual tree's life cycle. Nevertheless, measures of growth performance at the juvenile age are considered to be closely connected to fitness, as the intraspecific competition among trees is strongest among seedlings and juvenile trees (Vieilledent et al., 2010); only the very few trees that are able to dominate others will win the race for light and survive. During periods of extreme environmental conditions (e.g., drought or frost) as well as at later life stages, other physiological or phenological traits may have a higher impact and obscure height growth performance, but then the selection for tree height is expected to have already shaped the genetic structure of populations.

Under the assumption that tree height is strongly correlated with fitness, the varying selection differential at opposite ends of the spruce distribution can be discussed in terms of the species' fundamental and realized niche. At the cold end of the species distribution, we see high selection differential that very likely translates into a sharp range margin. Polechova and Barton (2015) recently demonstrated such an intrinsic limit to adaptation by modeling the joint evolution of trait mean and population size along environmental gradients within a genetic model. The existence of such range margins depends on the relation between fitness costs and the efficacy of selection relative to genetic drift (Polechova & Barton, 2015). Even small environmental gradients are able to generate intrinsic genetically determined range limits in the case of interspecific competition (Case & Taper, 2000), which tends to be the reality in a subalpine forest ecosystem. Our data neither allow us to estimate the model parameters of Polechova and Barton (2015) nor to prove the effects of interactions with other conifers, but they clearly demonstrate that climate selection in Norway spruce acts at the cold, but not the warm species distribution limit. Thus, the colder species distribution limit can be considered close to the limits of its fundamental niche, whereas the warm limit rather mirrors the species realized niche. Here, the species distribution might instead be defined by drivers other than climate, such as bark beetle attacks or competition with other tree species, which were found to be important drivers of species niches on ecological timescales (Hellmann, Prior, & Pelini, 2012; Meier, Edwards, Kienast, Dobbertin, & Zimmermann, 2011). However, given the short observation time in relation to a trees life span, it is also possible that we are not able to identify direct causes of mortality, or reduced reproduction at the warm temperature limit, which might also be related to intrinsic limits of the species range. To test whether the observed differences in selection differential are being carried into adult tree populations, we could design a simple genetic experiment: as our analysis predicts that no selection on tree height occurs on populations growing on the warm temperature limit, we would expect low heritability for height growth performance in the offspring of such trees. In contrast, offspring from populations growing at cold sites should have a higher degree of genetic determination.

Under climate change, the cold distribution limit is the leading edge that might spread to higher altitudes and latitudes (Sykes et al., 1996). The warm distribution limit is considered to be the trailing edge, where Norway spruce is expected to experience strong reductions in its present range with significant consequences for forest ecosystems, wood production (Hanewinkel et al., 2012), but also with losses of genetic diversity (Schueler et al., 2014). As our analysis indicates that natural selection at this trailing edge is limited, local adaptation at such sites seems to be impossible, in particular if we consider the high velocity of change. Thus, management actions to conserve the existing genetic diversity, for example, the establishment of gene conservation forests, are urgently needed at the warm limit of the species distribution range.

4.5 | Implications for forest tree breeding and distribution modeling

Our analysis aims to improve tree genetic conservation and to guide assisted migration measures. Assisted migration, that is, the translocation of forest reproductive material to areas with expectably favorable climates in the future, is widely discussed as a key forest management strategy to reduce climate maladaptation (e.g., Lu et al., 2014; McLachlan, Hellmann, & Schwartz, 2007; Wang et al., 2010). Based on the same Norway spruce dataset, we have already shown that populations from currently warm and drought-prone areas are appropriate candidates for continued silvicultural use in the future (Kapeller et al., 2012). In the mountainous area of the eastern Alps, this mainly means a shift of seed material upward in order to keep pace with global warming. In the present analysis, we found significantly stronger selection differential at the species colder distribution limit, although this has not resulted in reduced genetic variation in populations originating from such sites, likely because of intensive

gene flow and environmental heterogeneity in alpine environments. This suggests that reasonable seed transfers upward bears only a small risk of maladaptations, as the variation within the populations is only weakly correlated with the temperature gradient, and thus, also populations from warmer seed origin display a broad adaptive capacity to grow and survive on colder sites. In general, however, genetic conservation and seed transfer activities should not be based on the variation in single traits alone. A valid risk-benefit analysis might also consider further physiological or phenological traits related to the adaptive potential of populations (e.g., frost hardiness, drought resistance, and pest insect tolerance).

Similarly, attempts to model future tree species distributions need to take into account the variable selection differential along climatic gradients and the relative importance of adaptive traits at specific areas within the species range. For Norway spruce, it is widely believed and shown with various species distribution models that the species will undergo strong reductions in its present range mainly on the species' warm and dry distribution limit (e.g., Hanewinkel et al., 2012; Sykes et al., 1996; Zimmermann et al., 2013). This is in contrast to our data as they do not show signs of climate selection on tree height on warm trial sites, whereas cold conditions impose a stronger selection with highly skewed density functions. Thus, our analysis helps to decipher individual mechanisms that may trigger range contractions or expansions that should be included in future mechanistic distribution models. This will help to improve models of the species' fundamental niche because under climate change it is more important to understand where a species *could* occur than where it *currently* occurs (Wiens, Stralberg, Jongsomjit, Howell, & Snyder, 2009).

5 | CONCLUSIONS

Under climate change, populations throughout the entire climatic range will experience shifts of mean temperatures, related climate parameters, or both. In Norway spruce, populations at the currently cold sites harbor higher phenotypic variation and will likely be able to adapt to the prospective conditions. At the warm edge of its distribution, populations are not necessarily maladapted, as we have not observed climatic constraints on phenotypic variation even in our warmest trial sites at the border of the climatic range. In our analysis, temperature shapes the phenotypic variation much more strongly than precipitation-related parameters. As climatic predictions for temperature are more reliable than for precipitation, our results could be integrated into mechanistic models of population persistence and species distributions. Datasets similar to those used in the present study are available for many tree species and have been used to select appropriate populations for future reforestations based on the population's mean climate response (e.g., Leites, Robinson, Rehfeldt, Marshall, & Crookston, 2012; Lu et al., 2014; Rehfeldt, Tchebakova, & Barnhardt, 1999; Rehfeldt et al., 2001, 2002). Our analysis suggests that an in-depth analysis of the phenotypic variation within such datasets can provide additional knowledge on the population's adaptive capacities.

ACKNOWLEDGEMENTS

We would like to particularly thank K. Holzer—head of the Department of Forest Genetics at the BFW until 1989—and former staff at the BFW for establishing and measuring this extensive provenance experiment. We thank Kathryn Platzer and Daisy Brickhill at IIASA for English editing the article and two anonymous reviewers who provided constructive comments on an earlier version of this manuscript. This research was supported by the INTERREG Alpine Space programme (Project MANFRED: Management strategies to adapt Alpine Space forests to climate change risk) and the Austrian research program StartClim.

LITERATURE CITED

Aitken, S. N., & Hannerz, M. (2001). Genecology and gene resource management strategies for conifer cold hardiness. In F. J. Bigras, & S. J. Columbo (Eds.), *Conifer cold hardiness* (pp. 23–53). Dordrecht: Kluwer Academic Publishers.

Araújo, M. B., Alagador, D., Cabeza, M., Nogués-Bravo, D., & Thuiller, W. (2011). Climate change threatens European conservation areas. *Ecology Letters, 14*, 484–492.

Araújo, M., & Pearson, R. (2005). Equilibrium of species' distributions with climate. *Ecography, 28*, 693–695.

Bates, D., Maechler, M., Bolker, B., & Walker, S. (2014). *Lme4: Linear mixed-effects models using eigen and S4.* doi:http://lme4.r-forge.r-project.org/.

Benaglia, T., & Chauveau, D. (2009). Mixtools: an R package for analyzing finite mixture models. *Journal of Statistical Software, 32*, 1–29.

Bridle, J. R., & Vines, T. H. (2007). Limits to evolution at range margins: When and why does adaptation fail? *Trends in Ecology and Evolution, 22*, 140–147.

Burger, R. (1999). Evolution of genetic variability and the advantage of sex and recombination in changing environments. *Genetics, 153*, 1055–1069.

Burger, R., & Lynch, M. (1995). Evolution and extinction in a changing environment: A quantitative-genetic analysis. *Evolution, 49*, 151–163.

Case, T. J., & Taper, M. L. (2000). Interspecific competition, environmental gradients, gene flow, and the coevolution of species' borders. *The American Naturalist, 155*, 583–605.

Chakraborty, D., Wang, T., Andre, K., Konnert, M., Lexer, M. J., Matulla, C., & Schueler, S. (2015). Selecting populations for non-analogous climate conditions using universal response functions: The case of Douglas-fir in Central Europe. *PLoS ONE, 10*, e0136357.

Chevin, L. M., & Lande, R. (2011). Adaptation to marginal habitats by evolution of increased phenotypic plasticity. *Journal of Evolutionary Biology, 24*, 1462–1476.

Davi, H., Gillmann, M., Ibanez, T., Cailleret, M., Bontemps, A., Fady, B., & Lefèvre, F. (2011). Diversity of leaf unfolding dynamics among tree species: New insights from a study along an altitudinal gradient. *Agricultural and Forest Meteorology, 151*, 1504–1513.

Giertych, M. (1992). Summary results of the IUFRO 1938 Norway spruce (*Picea abies* (L.) Karst.) provenance experiment. Height growth. *Silvae Genetica, 25*, 154–164.

Haldane, J. B. S. (1956). The relation between density regulation and natural selection. *Proceedings of the Royal Society B: Biological Sciences, 145*, 306–308.

Hanewinkel, M., Cullmann, D. A., Schelhaas, M.-J., Nabuurs, G.-J., & Zimmermann, N. E. (2012). Climate change may cause severe loss in the economic value of European forest land. *Nature Climate Change, 3*, 203–207.

Hannerz, M., Sonesson, J., & Ekberg, I. (1999). Genetic correlations between growth and growth rhythm observed in a short-term test and performance in long-term field trials of Norway spruce. *Canadian Journal of Forest Research, 29*, 768–778.

Hellmann, J. J., Prior, K. M., & Pelini, S. L. (2012). The influence of species interactions on geographic range change under climate change. *Annals of the New York Academy of Sciences, 1249*, 18–28.

Howe, G. T., Aitken, S. N., Neale, D. B., Jermstad, K. D., Wheeler, N. C., & Chen, T. H. H. (2003). From genotype to phenotype: Unraveling the complexities of cold adaptation in forest trees. *Canadian Journal of Botany, 81*, 1247–1266.

Hurme, P., Repo, T., Savolainen, O., & Pääkkönen, T. (1997). Climatic adaptation of bud set and frost hardiness in scots pine (*Pinus sylvestris*). *Canadian Journal of Forest Research, 27*, 716–723.

Husson, F., Josse, J., Le, S., & Mazet, J. (2015). *Factominer: Multivariate exploratory data analysis and data mining. R package version 1.29.* http://cran.r-project.org/package=FactoMineR.

Kapeller, S., Lexer, M. J., Geburek, T., & Schueler, S. (2012). Intraspecific variation in climate response of Norway spruce in the eastern alpine range: Selecting appropriate provenances for future climate. *Forest Ecology and Management, 271*, 46–57.

Kirkpatrick, M., & Barton, N. H. (1997). Evolution of a species' range. *American Naturalist, 150*, 1–23.

Kleinschmit, J., Sauer-Stegmann, A., Lunderstadt, J., & Svolba, J. (1981). Charakterisierung von Fichtenklonen (*Picea abies* Karst.) II. Korrelation der Merkmale. *Silvae Genetica, 30*, 74–82.

Kopp, M., & Matuszewski, S. (2014). Rapid evolution of quantitative traits: Theoretical perspectives. *Evolutionary Applications, 7*, 169–191.

Kroon, J., Ericsson, T., Jansson, G., & Andersson, B. (2011). Patterns of genetic parameters for height in field genetic tests of *Picea abies* and *Pinus sylvestris* in Sweden. *Tree Genetics & Genomes, 7*, 1099–1111.

Krutzsch, P. (1992). IUFRO's role in coniferous tree improvement: Norway spruce (*Picea abies* (L.) Karst.). *Silvae Genetica, 41*, 134–150.

Lamy, J.-B., Bouffier, L., Burlett, R., Plomion, C., Cochard, H., & Delzon, S. (2011). Uniform selection as a primary force reducing population genetic differentiation of cavitation resistance across a species range. *PLoS ONE, 6*, e23476.

Leites, L. P., Robinson, A. P., Rehfeldt, G. E., Marshall, J. D., & Crookston, N. L. (2012). Height-growth response to climatic changes differs among populations of Douglas-fir: A novel analysis of historic data. *Ecological Applications, 22*, 154–165.

Liesebach, M., Rau, H.-M., & König, A.O. 2010. *Fichtenherkunftsversuch von 1962 und IUFRO-Fichtenherkunftsversuch von 1972. Ergebnisse von mehr als 30-jähriger Beobachtung in Deutschland. Beiträge aus der Nordwestdeutschen Forstlichen Versuchsanstalt Band 5.* Universitätsverlag Göttingen, Göttingen, Germany 467p.

Lu, P., Parker, W. H., Cherry, M., Colombo, S., Parker, W. C., Man, R., & Roubal, N. (2014). Survival and growth patterns of white spruce (*Picea glauca* [Moench] Voss) rangewide provenances and their implications for climate change adaptation. *Ecology and Evolution, 4*, 2360–2374.

Maghuly, F., Pinsker, W., Praznik, W., & Fluch, S. (2006). Genetic diversity in managed subpopulations of Norway spruce [*Picea abies* (L.) Karst.]. *Forest Ecology and Management, 222*, 266–271.

Mátyás, C. S. (1994). Modeling climate change effects with provenance test data. *Tree Physiology, 14*, 797–804.

Mátyás, C. S., Nagy, L., & Ujvári-Jármai, É. (2009). Genetic background of response of trees to aridification at the xeric forest limit and consequences for bioclimatic modelling. In: K. Strelcová, C. S. Mátyás, A. Kleidon, M. Lapin, F. Matejka, M. Blazenec, J. Skvarenina, J. Holecy, (Eds.) *Bioclimatology and natural hazards* (pp. 179–196). Springer: Berlin, Germany .

McLachlan, J. S., Hellmann, J. J., & Schwartz, M. W. (2007). A framework for debate of assisted migration in an era of climate change. *Conservation Biology: The Journal of the Society for Conservation Biology, 21*, 297–302.

Meier, E. S., Edwards, T. C. Jr, Kienast, F., Dobbertin, M., & Zimmermann, N. E. (2011). Co-occurrence patterns of trees along macro-climatic gradients and their potential influence on the present and future distribu-tion of *Fagus sylvatica* L. *Journal of Biogeography, 38*, 371–382.

Mengl, M., Geburek, T., & Schueler, S. (2009). Geographical pattern of hap-lotypic variation in Austrian native stands of *Picea abies. Dendrobiology, 61*, 117–118.

Morgenstern, E. (1996). *Geographic variation in forest trees: Genetic basis and application of knowledge in silviculture*. Vancouver, Canada: UBC press.

Nather, J., & Holzer, K. (1979). *Über die Bedeutung und die Anlage von Kontrollflächen zur Prüfung von anerkanntem Fichtenpflanzgut* (p. 181). Vienna, Austria: Informationsdienst Forstliche Bundesversuchsanstalt Wien.

Ni, R., Baiketuerhan, Y., Zhang, C., Zhao, X., & Von Gadow, K. (2014). Analysing structural diversity in two temperate forests in northeastern China. *Forest Ecology and Management, 316*, 139–147.

Polechova, J., & Barton, N. H. (2015). Limits to adaptation along environ-mental gradients. *PNAS, 112*, 6401–6406.

Prentice, I. C., Cramer, W., Harrison, S. P., Leemans, R., Monserud, R. A., & Solomon, A. M. (1992). A global biome model based on plant physiology and dominance, soil properties and climate. *Journal of Biogeography, 19*, 117–134.

Pulkkinen, P. (1993). Frost hardiness development and lignification of young Norway spruce seedlings of southern and northern Finnish ori-gin. *Silva Fennica, 27*, 47–54.

R Core Team(2014). *R: A language and environment for statistical computing.* Vienna, Austria: R Foundation for statistical computing. http://www.r-project.org/.

Randin, C. F., Paulsen, J., Vitasse, Y., Kollas, C., Wohlgemuth, T., Zimmermann, N. E., & Körner, C. (2013). Do the elevational limits of deciduous tree species match their thermal latitudinal limits? *Global Ecology and Biogeography, 22*, 913–923.

Ravazzi, C. (2002). Late Quaternary history of spruce in southern Europe. *Review of Palaeobotany and Palynology, 120*, 131–177.

Rehfeldt, G. E., Tchebakova, N. M., & Barnhardt, L. K. (1999). Efficacy of climate transfer functions: introduction of Eurasian populations of *Larix* into Alberta. *Canadian Journal of Forest Research, 29*, 1660–1668.

Rehfeldt, G. E., Tchebakova, N. M., Parfenova, Y. I., Wykoff, W. R., Kuzmina, N. A., & Milyutin, L. I. (2002). Intraspecific responses to climate in *Pinus sylvestris. Global Change Biology, 8*, 912–929.

Rehfeldt, G. E., Wykoff, W. R., & Ying, C. C. (2001). Physiologic plasticity, evolution, and impacts of a changing climate on *Pinus contorta. Climatic Change, 50*, 355–376.

Ronce, O., & Kirkpatrick, M. (2001). When sources become sinks: Migrational meltdown in heterogeneous habitats. *Evolution; International Journal of Organic Evolution, 55*, 1520–1531.

Schueler, S., Falk, W., Koskela, J., Lefevre, F., Bozzano, M., Hubert, J., … Olrik, D. C. (2014). Vulnerability of dynamic genetic conservation units of trees in Europe to climate change. *Global Change Biology, 20*, 1498–1511.

Schulze, U. (1985). Fichtenherkunftsprüfung 1978. Internationaler Verband Forstlicher Versuchsanstalten IUFRO, Arbeitsgruppe S 2.02-11 Fichtenherkünfte. 1-12.

Skrøppa, T. (1991). Within-population variation in autumn frost hardi-ness and its relationship to bud-set and height growth in *Picea abies. Scandinavian Journal of Forest Research, 6*, 353–363.

Søgaard, G., Oystein, J., Jarle, N., & Olavi, J. (2008). Climatic control of bud burst in young seedlings of nine provenances of Norway Spruce. *Tree Physiology, 28*, 311–320.

Summers, D. M., Bryan, B. A., Crossman, N. D., & Meyer, W. S. (2012). Species vulnerability to climate change: impacts on spatial conservation priorities and species representation. *Global Change Biology, 18*, 2335–2348.

Svenning, J., & Skov, F. (2004). Limited filling of the potential range in European tree species. *Ecology Letters, 7*, 565–573.

Sykes, M., Prentice, I., & Cramer, W. (1996). A bioclimatic model for the potential distributions of north European tree species under present and future climates. *Journal of Biogeography, 23*, 203–233.

Terhürne-Berson, R. (2005). *Changing distribution patterns of selected coni-fers in the Quaternary of Europe caused by climatic variations*. PhD Thesis. Bonn, Germany: Rheinischen Friedrich-Wilhelms-Universität Bonn.

Thomas, C. D., Cameron, A., Green, R. E., Bakkenes, M., Beaumont, L. J., Collingham, Y. C., Erasmus, B. F. N., de Siqueira, M. F., Grainger, A., Hannah, L., Hughes, L., Huntley, B., van Jaarsveld, A. S., Midgley, G. F., Miles, L. M., Ortega-Huerta, A., Peterson, A. T., Phillips, O. L. & Williams, S. E. (2004). Extinction risk from climate change. *Nature, 427*, 145–148.

Thuiller, W., Lavergne, S., Roquet, C., Boulangeat, I., Lafourcade, B., & Araujo, M. B. (2011). Consequences of climate change on the tree of life in Europe. *Nature, 470*, 531–534.

Tollefsrud, M. M., Kissling, R., Gugerli, F., Johnsen, Ø., Skrøppa, T., Cheddadi, R., Van Der Knaap, W. O., Latałowa, M., Terhürne-Berson, R., Litt, T., Geburek, T., Brochmann, C., Sperisen, C. (2008). Genetic conse-quences of glacial survival and postglacial colonization in Norway spruce: Combined analysis of mitochondrial DNA and fossil pollen. *Molecular Ecology, 17*, 4134–4150.

Ujvári-Jármay, É., Nagy, L., & Mátyás, C. S. (2016). The IUFRO 1964/68 in-ventory provenance trial of Norway spruce in Nyírjes, Hungary—results and conclusions of five decades Documentary study. *Acta Silvatica & Lignaria Hungarica, 12*(Special Edition), 178p.

Vieilledent, G., Courbaud, B., Kunstler, G., & Dhôte, J.-F. (2010). Mortality of silver fir and Norway spruce in the Western Alps—A semi-parametric approach combining size-dependent and growth-dependent mortality. *Annals of Forest Science, 67*, 305–305.

Wang, T., Hamann, A., Yanchuk, A. D., O'Neill, G. A., & Aitken, S. N. (2006). Use of response functions in selecting lodgepole pine populations for future climates. *Global Change Biology, 12*, 2404–2416.

Wang, T., O'Neill, G. A., & Aitken, S. N. (2010). Integrating environmental and genetic effects to predict responses of tree populations to climate. *Ecological Applications, 20*, 153–163.

Weisgerber, H., Dimpflmeier, R., Ruetz, W., Kleinschmitt, J., & Widmaier, T.(1984). Ergebnisse des internationalen Fichten-Provenienzversuches 1962. Entwicklung bis zum Alter 18. *Allgemeine Forst- und Jagdzeitung, 155*, 110–121.

Wiens, J. A., Stralberg, D., Jongsomjit, D., Howell, C. A., & Snyder, M. A. (2009). Niches, models, and climate change: Assessing the assumptions and uncertainties. *Proceedings of the National Academy of Sciences, 106*, 19729–19736.

Yeaman, S., & Jarvis, A. (2006). Regional heterogeneity and gene flow main-tain variance in a quantitative trait within populations of lodgepole pine. *Proceedings of the Royal Society of London B: Biological Sciences, 273*, 1587–1593.

Zimmermann, N. E., Jandl, R., Hanewinkel, M., Kunstler, G., Kölling, C., Gasparini, P., Breznikar, A., Meier, E. S., Normand, S., Ulmer, U., Gschwandtner, T., Veit, H., Naumann, M., Falk, W., Mellert, K., Rizzo, M., Skudnik, M., Psomas, A. (2013). Potential Future Ranges of Tree Species in the Alps, Management Strategies to Adapt Alpine Space Forests to Climate Change Risks, Dr. Gillian Cerbu (Ed.), InTech. Available from: http://www.intechopen.com/books/management-strategies-to-adapt-alpine-space-forests-to-climate-change-risks/potential-future-ranges-of-tree-species-in-the-alps.

5

Profiling the immunome of little brown myotis provides a yardstick for measuring the genetic response to white-nose syndrome

Michael E. Donaldson[1],[*] ID | Christina M. Davy[1,2,*] ID | Craig K. R. Willis[3] |
Scott McBurney[4] | Allysia Park[4] | Christopher J. Kyle[5]

[1]Environmental and Life Sciences Graduate
Program, Trent University, Peterborough, ON,
Canada

[2]Wildlife Research and Monitoring
Section, Ontario Ministry of Natural Resources
and Forestry, Peterborough, ON, Canada

[3]Department of Biology and Centre for Forest
Interdisciplinary Research (C-FIR), University
of Winnipeg, Winnipeg, MB, Canada

[4]Canadian Wildlife Health Cooperative,
Atlantic Region, Atlantic Veterinary College,
University of Prince Edward Island,
Charlottetown, PEI, Canada

[5]Forensic Science Department, Trent
University, Peterborough, ON, Canada

Correspondence
Christopher J. Kyle, Forensic Science
Department, Trent University, Peterborough,
ON, Canada.
Email: christopherkyle@trentu.ca
and
Michael Donaldson, Environmental and Life
Sciences Graduate Program, Trent University,
Peterborough, ON,
Canada.
Email: michaeldonaldson@trentu.ca

Funding information
Species at Risk Research Fund for Ontario;
Natural Sciences and Engineering Research
Council of Canada, The Canadian Wildlife
Health Cooperative; Liber Ero Fellowship
Program

Abstract

White-nose syndrome (WNS) has devastated populations of hibernating bats in eastern North America, leading to emergency conservation listings for several species including the previously ubiquitous little brown myotis (*Myotis lucifugus*). However, some bat populations near the epicenter of the WNS panzootic appear to be stabilizing after initial precipitous declines, which could reflect a selective immunogenetic sweep. To investigate the hypothesis that WNS exerts significant selection on the immunome of affected bat populations, we developed a novel, high-throughput sequence capture assay targeting 138 adaptive, intrinsic, and innate immunity genes of putative adaptive significance, as well as their respective regulatory regions (~370 kbp of genomic sequence/individual). We used the assay to explore baseline immunogenetic variation in *M. lucifugus* and to investigate whether particular immune genes/variants are associated with WNS susceptibility. We also used our assay to detect 1,038 putatively neutral single nucleotide polymorphisms and characterize contemporary population structure, providing context for the identification of local immunogenetic adaptation. Sequence capture provided a cost-effective, "all-in-one" assay to test for neutral genetic and immunogenetic structure and revealed fine-scale, baseline immunogenetic differentiation between sampling sites <600 km apart. We identified functional immunogenetic variants in *M. lucifugus* associated with WNS susceptibility. This study lays the foundations for future investigations of rangewide immunogenetic adaptation to WNS in *M. lucifugus* and provides a blueprint for studies of evolutionary rescue in other host–pathogen systems.

KEYWORDS
genotype-by-sequencing, immunogenetics, *Myotis lucifugus*, white-nose syndrome

1 | INTRODUCTION

Host–pathogen dynamics are changing at an unprecedented rate as climate change and human-mediated transport expand the range of pathogens into previously inhospitable/inaccessible environments (Fisher et al., 2012). As pathogen ranges shift, disease-related population declines in naïve wildlife populations often threaten population persistence, as evidenced by several emerging wildlife diseases

---

[*]These authors contributed equally to this work.

(Gallana, Ryser-Degiorgis, Wahli, & Segner, 2013; Smith et al., 2012). Selective forces exerted by infectious diseases can rapidly influence the distribution of adaptive genetic variants associated with disease susceptibility over short timescales (Gallana et al., 2013). For infectious diseases of conservation significance, this process of local adaptation can result in evolutionary rescue of a population, where disease-resistant animals survive a strong selective sweep from disease and pass their resistance to their offspring (Carlson, Cunningham, & Westley, 2016; Maslo & Fefferman, 2015). Spatial patterns of local adaptation to strong selective sweeps may be linked to particular gene variants favored in local interactions (Hansen, Olivieri, Waller, & Nielsen, 2012; Kyle et al., 2014; Rico, Morris-Pocock, Zigouris, Nocera, & Kyle, 2015; Schoville et al., 2012). Determining how these variants are spread or localized among populations is essential to understanding and managing the emergence of new selective pressures, such as emerging infectious diseases (Eizaguirre, Lenz, Kalbe, & Milinski, 2012; Kyle et al., 2014).

White-nose syndrome (WNS) is a recently emerged disease in hibernating bats caused by the fungal pathogen *Pseudogymnoascus destructans*. The fungus was introduced from Eurasia to North America, where it was first documented in Schoharie County, New York, in 2006 (Blehert et al., 2009; Leopardi, Blake, & Puechmaille, 2015). WNS has spread rapidly across North America, causing >80% declines in some eastern bat populations (Frick et al., 2010, 2015; Langwig et al., 2012; Lorch et al., 2016). While several North America bats are highly susceptible to WNS, European bats do not experience mortality from infection with *P. destructans* (Puechmaille, Fuller, & Teeling, 2011; Puechmaille, Wibbelt, et al., 2011). Controlled experiments with captive bats show that identical strains of *P. destructans* cause mortality in North American little brown myotis (*Myotis lucifugus*) but not in European greater mouse-eared bats (*Myotis myotis*; Davy et al., 2017), suggesting a genetic basis for immunotolerance or immunoprotection.

There are several promising leads for the development of treatments for WNS (e.g., Cheng et al., 2016; Cornelison, Gabriel, Barlament, & Crow, 2014; Wilcox & Willis, 2016), but no effective mitigation or treatment protocols are currently available. However, some populations near the epicenter of WNS may be stabilizing following their initial, precipitous declines (Dobony et al., 2011; Langwig et al., 2012, 2017). Persistence of these populations does not seem to be associated with immigration (Maslo, Valent, Gumbs, & Frick, 2015), but may indicate evolution of resistance or tolerance to the disease (Langwig et al., 2017). Thus, adaptation and evolutionary rescue may be the best hope for recovery of bat populations affected by WNS (Maslo & Fefferman, 2015). Understanding patterns of immunogenetic adaptation to WNS is therefore critical to determining disease management strategies and recovery programs for the affected populations.

Immune genes mediate the initial response of individuals to pathogens and in many cases, the acquisition of immunity. At the population level, genetic diversity of immune genes influences resistance or tolerance to disease via pathogen-mediated balancing selection (Eizaguirre et al., 2012; Rico et al., 2015). Studies of wildlife populations generally focus on adaptive immunity, which has often been assessed by using genetic diversity in the major histocompatibility complex (MHC). Diversity at the MHC provides a proxy for potential to adapt to shifting pathogen pressures, due to the role of MHC in pathogen recognition and pathogen susceptibility (Acevedo-Whitehouse & Cunningham, 2006; Eizaguirre et al., 2012; Kyle et al., 2014). Some studies of immunogenetic diversity also include receptor genes associated with innate immunity (e.g., Toll-like receptors and interleukins) and these markers have revealed spatial patterns of resistance to emerging infectious diseases such as chytridiomycosis and mycoplasmosis (Bonneaud, Balenger, Zhang, Edwards, & Hill, 2012; Savage & Zamudio, 2011).

The MHC Drb1 locus in *M. lucifugus* is among the most polymorphic recorded in mammals to date (Palmer et al., 2016). Pyrosequencing of 160 individuals sampled across Canada suggests that balancing selection has maintained similar MHC diversity among genetically differentiated subpopulations, which may be disrupted by WNS-mediated immunogenetic selection (Davy et al., 2017). However, the extreme observed polymorphism of the Drb1 locus in *M. lucifugus* is due partly to multiple gene duplications, which limits the use of these data. Furthermore, susceptible bats infected with *P. destructans* upregulate multiple, complementary immune responses (Field et al., 2015; Lilley et al., 2017; Moore et al., 2013; Rapin et al., 2014), so immunogenetic selection by WNS cannot be fully captured by experimental designs that target single, candidate genes. No other population-level immunogenetic analyses exist for *M. lucifugus*, or for any other North American species of bats threatened by WNS. Fortunately, new molecular tools allow more comprehensive investigation of immunogenetic adaptation (Harrisson, Pavlova, Telonis-Scott, & Sunnucks, 2014).

Genotype-by-sequencing (GBS) assays have emerged as a cost-effective method for obtaining population-level assessments of neutral and functional genetic variation, and identifying local adaptation (Tiffin & Ross-Ibarra, 2014). GBS assays involve enriching for genomic subsets of DNA via restriction enzyme-, amplicon-, or hybridization-based methods (Jones & Good, 2016), conducting high-throughput sequencing and identifying single nucleotide polymorphisms (SNPs). Targeted approaches, including amplicon- and hybridization-based GBS, have been used in wildlife studies to identify SNPs in specific coding and regulatory regions of immune genes, collectively called the "immunome." Targeted GBS can identify population-level immunogenetic shifts in response to pathogens, and has been applied to a range of species, including the Tasmanian devil (*Sarcophilus harrisii*; Morris, Wright, Grueber, Hogg, & Belov, 2015), turkey (*Meleagris gallopavo*; Reed, Mendoza, & Settlage, 2016), gray wolf (*Canis lupus*; Schweizer et al., 2016), thinhorn sheep (*Ovis dalli*; Roffler et al., 2016), and red fox (*Vulpes vulpes*; Donaldson et al., unpublished). GBS is an attractive option for understanding the impacts of WNS on immunogenetic diversity in bat populations, because it allows accurate characterization of diversity at duplicated loci, and cost-effective targeting of multiple, relevant genes. Regardless of the genomic coverage of high-throughput sequencing methods, population genetic analyses still rely on adequate sample sizes to detect the genetic signature of selection by pathogens or other selective pressures, reinforcing the importance of a cost-effective approach.

We developed a novel hybridization-based GBS assay to characterize the genetic diversity of the *M. lucifugus* immunome. Our assay includes 170 loci, including 120 immune genes and their regulatory areas, 18 Drb1-like exon 2 regions, and 32 neutral loci to allow characterization of neutral population structure, against which hypotheses of local adaptation can be tested. We applied this assay to test the hypothesis that WNS exerts significant selection on the immunome of affected bat populations. Controlling for neutral genetic population structure, we predicted immunogenetic divergence would be detectable between WNS-naïve populations and populations affected by WNS. This study provides a foundation for future investigations of rangewide immunogenetic adaptation to WNS in *M. lucifugus* and other affected species of bats.

2 | MATERIALS AND METHODS

2.1 | Microsatellite markers, immune genes and probe development

We developed our assay for primary application to *M. lucifugus* because this species' genome is publicly available (Myoluc2.0 genome assembly, Ensembl release version 81; Cunningham et al., 2015), and recent research has identified putative "WNS-response" genes for this species (Rapin et al., 2014), which informed our selection of target genes for sequence capture.

To assess functional immunogenetic variation, we assembled a list of 120 candidate genes related to immune system processes based on (i) the Human Innate & Adaptive Immune Responses RT[2] Profiler PCR Array (Qiagen); (ii) a review of innate and adaptive immunity, development, and signaling (Knight, 2013); (iii) a study of gene expression in *M. lucifugus* following infection with *P. destructans* (Rapin et al., 2014); and (iv) a gene ontology (GO) term search in the *M. lucifugus* Ensembl database for GO records related to fungi (including cellular response to molecule of fungal origin, defense response to fungus, and neutrophil-mediated killing of fungus). We used this candidate gene list to query the *M. lucifugus* Myoluc2.0 genome assembly and created a BED-formatted file containing coordinates for all exons. Additionally, we targeted potential regulatory regions by including coordinates for the 1,500-bp region upstream from the first exon for each gene. Finally, we added exon 2 coordinates for 18 Drb1-like genes identified in Ensembl that putatively encode functional full-length proteins.

To target putative neutral markers for the detection of genetic population structure, we selected 32 microsatellite markers for *M. lucifugus* from the published literature (Burns, Broders, & Frasier, 2012; Castella & Ruedi, 2000; Johnson et al., 2014; Oyler-McCance & Fike, 2011; Piaggio, Figueroa, & Perkins, 2009; Trujillo & Amelon, 2009). Using these primer sets, we added coordinates for these markers to the BED-formatted file. In total, the final BED-formatted file contained coordinates for 170 loci. Descriptions for protein-coding and microsatellite regions are provided in the Supporting Information (Tables S1–S2).

Custom NimbleGen SeqCap EZ probes (Roche) were produced for "primary targets" using the BED-formatted file and the *M. lucifugus*

Myoluc2.0 genome assembly as a reference. We added 100-bp "padding" to each target to increase the efficiency of the sequence capture, and we used a "relaxed" probe design that allowed up to 20 close matches to the *M. lucifugus* reference genome. We compared our probes to the *M. lucifugus* reference genome to ensure that our assay had a low likelihood for "off-target" sequence capture: 91% of the probes matched only their target sequence, and 99% had five or fewer matches to the *M. lucifugus* reference genome.

2.2 | Sample collection, DNA extraction, and quantification

All work was conducted under approved animal care protocols from the University of Winnipeg and the Ontario Ministry of Natural Resources and Forestry. To test the relative impacts of geographic location and exposure to *P. destructans* on neutral and immunogenetic population structure in *M. lucifugus*, we assigned bat samples collected in eastern Canada to three post hoc groups (Table S3). The first group included bats that were nonharmfully sampled at a hibernaculum in Manitoba, Canada (MB, *n* = 28), that did not contain *P. destructans* at the time of sampling. The second group contained bats from two hibernacula near Thunder Bay, Ontario, which were also sampled before the arrival of *P. destructans* (ON, *n* = 36). Wing biopsies from these bats were immediately stored in RNA*later* (Qiagen) following sampling. These two groups represent our "pre-WNS treatment." The third group came from populations of bats in Atlantic Canada that had been exposed to WNS for at least one winter, but were found moribund or dead in the winter of 2014 in the Atlantic provinces of Nova Scotia and Prince Edward Island (ATL, "post-WNS treatment," *n* = 28). These bats were submitted to the Canadian Wildlife Health Cooperative (CWHC), Atlantic Region for necropsy. Of these post-WNS bats, 15 were diagnosed as positive for WNS, 12 were diagnosed as suspect D for WNS, and 1 was negative for WNS using the approved diagnostic categories for WNS found in the Canadian Bat WNS Necropsy Protocol (CWHC, 2014), and they were assumed to not be immunotolerant nor immunocompetent to WNS. Wing tissue was collected from the left dactylopatagium major during these necropsies and stored in lysis buffer (4 M urea, 0.2 M NaCl, 0.5% n-lauroyl sarcosine, 10 mM 1,2-cyclohexanediaminetetraacetic acid, 0.1 mM Tris–HCl pH 8.0) until analysis. We dissolved all tissue samples in lysis buffer containing 600 U/ml proteinase *K* at 56°C for 2 hr. We extracted DNA using either the automated 96-well MagneSil Blood Genomic Max Yield System (Promega) or the DNeasy Blood and Tissue Kit (Qiagen). We then quantified all DNA extractions using the Quant-iTPicoGreen dsDNA Assay Kit (ThermoFisher Scientific).

To investigate the possibility that our assay could also be used to investigate immunogenetic variation and adaptation in other species affected by WNS, we also isolated DNA from "post-WNS" *Eptesicus fuscus* (*n* = 2) from New Brunswick, both suspect B for WNS (CWHC, 2014) and *M. septentrionalis* (*n* = 2) from Nova Scotia and Prince Edward Island, positive and suspect D for WNS, respectively (CWHC, 2014), and included these samples in the assay.

2.3 | DNA library preparation, sequence capture, and high-throughput sequencing

We prepared DNA libraries using the KAPA HTP Lib Prep Kit (Roche) and performed the sequence capture using the NimbleGen SeqCap EZ Developer Library kit v5.1 (Roche) with the following modifications to the manufacturer's protocol. Each DNA library preparation used 150 ng total DNA. TruSeq HT Dual-Index Adapters (Integrated DNA Technologies) resuspended in Nuclease Free Duplex Buffer (Integrated DNA technologies) were used at a final concentration of 600 nM instead of the SeqCap Adapter Kits A and B (Roche) during adapter ligation. We performed 11 cycles during the LM-PCR, and initial DNA library quality was assessed by ethidium bromide-stained gel electrophoresis using a 2% E-Gel (ThermoFisher Scientific). We used 1 µl of the xGen Universal Blocking Oligo TS HT-i5 (Integrated DNA Technologies) and 1 µl xGen Universal Blocking Oligo TS HT-i7 (Integrated DNA Technologies) instead of the NimbleGen Multiplex Hybridization Enhancing Oligo Pool (Roche), and we used NimbleGen SeqCap EZ Developer Reagent (Roche) instead of NimbleGen COT Human DNA (Roche) during hybridization sample preparation. The hybridization was carried out at 47°C for 72 hr. We assessed the pooled target-enriched DNA quality using a bioanalyzer (Agilent Technologies) and performed high-throughput sequencing on a HiSeq 2500 rapid run using 2 × 100-bp reads on a single flow cell (Illumina).

2.4 | Sequence alignment, variant annotation, and SNP/INDEL analysis

We used the bwa-mem command in the BURROWS-WHEELER ALIGNER v0.7.12 (BWA; Li, 2013) to align paired-end reads to the Myoluc2.0 genome sequence and compiled sequence alignment metrics using SAMTOOLS v1.2 (Li et al., 2009). We used the GENOME ANALYSIS TOOLKIT v3.5 (GATK; McKenna et al., 2010) for base quality score recalibration, realignment of insertions/deletions (INDELs), duplicate removal, depth of coverage calculations, SNP/INDEL discovery, and genotyping across all samples, using standard hard filtering parameters or variant quality score recalibration according to GATK best practices recommendations (DePristo et al., 2011; Van der Auwera et al., 2013).

2.5 | Analysis of targeted microsatellites

We used two different approaches to assign microsatellite genotypes. The first method (GATK) relied on sequence alignment to the *M. lucifugus* genome. We identified a single INDEL to represent each microsatellite by selecting the short tandem repeat that yielded the highest: (i) percentage of heterozygotes; (ii) GATK "quality" score; or (iii) number of alleles. For each of these three scenarios, we used the GATK to calculate the number of heterozygotes for each marker using a subset of our data that included only the 36 ON samples and the 28 MB samples. These 64 samples were previously genotyped based on traditional PCR amplification and sequencing of 11 microsatellite markers (Davy et al., 2017). We calculated the number of heterozygotes for each of these markers, to assess whether our sequence capture

assay could be used to build on previous microsatellite-based studies. Our second genotyping method used the Galaxy platform (Afgan et al., 2016) to run STR-FM (Galaxy Version 1.0.0; Fungtammasan et al., 2015) and identify di- and tetra-nucleotide STRs from the raw Illumina.fastq data, without genome alignment.

2.6 | Analysis of functional loci and identification of novel, putatively neutral SNPs

We used GATK to assemble a master variant call format file (.vcf) that included SNPs with a maximum missing genotype frequency of 5% and a minimum minor allele frequency of 2%. We then used GATK to generate subdatasets of SNPs from specific categories (exon, intron, regulatory region, and Drb1-like exon 2). For the "off-target" SNPs, we used the Ensembl variant effect predictor tool to determine the bp distance between a SNP and the closest gene and generated a list of putatively neutral SNPs that were at least 100,000 bp from a gene (e.g., Kawakami et al., 2014), which we considered to be in linkage disequilibrium. We "binned" these SNPs based on the minor allele frequency, and tested for genetic structure (see below) using the SNPs with minor allele frequency values of 2% and 25%. All.vcf files were reformatted using PGDSPIDER v2.0.9.2 (Lischer & Excoffier, 2012) for downstream analyses.

To explore variation in functional regions, we ran two LOSITAN analyses (Antao, Lopes, Lopes, Beja-Pereira, & Luikart, 2008; Beaumont & Nichols, 1996) to identify F_{ST} outliers that are putatively under selection. LOSITAN parameters included 1,000,000 iterations, a 99.5% confidence interval, a false discovery rate (FDR) threshold of 0.05, and a stepwise mutation model. We enabled the "Neutral mean F_{ST}" and the "Force mean F_{ST}" options. The first analysis used population priors based on geography (MB, ON, and ATL) and the second considered exposure to WNS (pre-WNS, post-WNS). We extracted the subset of directional F_{ST} outliers identified in each analysis with VCFTOOLS v0.1.14 (Danecek et al., 2011) and used them to explore immunogenetic population structure (see below).

SNPs that alter amino acids or affect splicing regions can have major effects on the function of encoded proteins. We considered F_{ST} outliers that had these particular consequences as the most likely signals of either local adaptation to pre-occurring pathogens (in the geographic comparison) or alleles disproportionately selected against by WNS (in the pre- and post-WNS comparison). Mutations in regulatory regions can also influence gene expression and ultimately affect disease outcome (Fraser, 2013) so we also identified SNPs within regulatory regions, although the functional results of these mutations cannot be inferred from sequence capture data alone.

2.7 | Characterization of neutral and immunogenetic population structure

We used two a priori groupings to test for neutral genetic and immunogenetic population structure: (i) geographic grouping (MN, ON, and ATL), or (ii) grouping by exposure to WNS (pre- and post-WNS). We explored genetic structure based on the different SNP datasets using

STRUCTURE v2.3.4 (Pritchard, Stephens, & Donnelly, 2000) and parallelized the runs using the STRAUTO v0.3.1 script (Chhatre & Emerson, 2017). We ran STRUCTURE with a burn-in length of 50,000 followed by 200,000 iterations for K = 1 through 4, and each run was performed 20 times. We used STRUCTURE HARVESTER WEB v0.6.94 (Earl & VonHoldt, 2012) to calculate the ΔK statistic (Evanno, Regnaut, & Goudet, 2005). Multiple structure runs were combined with CLUMPP v1.1.2 (Jakobsson & Rosenberg, 2007) using the Greedy option (10,000 repeats), and we visualized the results using DISTRUCT v1.1 (Rosenberg, 2004). We also performed principal component analysis (PCA) using ADEGENET v2.0.0 (Jombart & Ahmed, 2011). We obtained the required "genlight" objects for the ADEGENET analysis using a combination of VCFTOOLS and PLINK v1.07 (Purcell et al., 2007) to reformat the.vcf files to PLINK-formatted files (.raw).

3 | RESULTS

3.1 | High-throughput sequencing, sequence alignment, and depth of coverage

NimbleGen sequence capture and high-throughput sequencing yielded 717 million paired-end reads for 96 libraries. We mapped 712 million of these reads to the *M. lucifugus* genome (Table S4). Sequencing alignment and depth of coverage metrics (Tables 1 and 2) indicated that the *M. lucifugus* probes were successful in capturing the targeted loci in *E. fuscus* and *M. septentrionalis*. Primary target enrichment was 42.1%, 41.4%, and 37.3% for *M. lucifugus*, *E. fuscus*, and *M. septentrionalis*, respectively (Table 1) and coverage for sequenced samples from *E. fuscus* (59.5X, 78.2X) and *M. septentrionalis* (66.5X, 171.0X) fell within the observed range for *M. lucifugus* (26.2X–463.3X; Table 2). To visualize the variation in depth of coverage across samples and the primary targets, we plotted the mean depth of coverage for primary targets across all samples (Figure 1) and the depth of coverage obtained from each sample for the primary targets (Figure 2). Overall, we determined average coverage was high for microsatellite markers (135X), Drb1-like exon 2 targets (121X), and targeted immune genes (145X).

3.2 | Microsatellite genotyping via INDEL detection

When processing the 32 microsatellite loci included in our assay, GATK analysis identified 400 INDELs for the 32 loci, demonstrating that unique microsatellite regions contained multiple INDEL calls. However, the relatively short read length obtained with our sequencing method failed to reliably capture entire short tandem repeat (STR) regions. Thus, microsatellite genotypes could not be recovered for all samples. As a result, heterozygous genotypes scored from our high-throughput sequencing differed from the previous results obtained using traditional PCR methods (Davy et al., 2017) by −25% to −42%. The STR-FM analysis, which does not rely on aligning reads to the genome, was also unable to generate genotypes for more than two microsatellite markers using a subset of our samples (data not shown). Therefore, we did not conduct further analyses with the microsatellite data.

TABLE 1 High-throughput sequencing and read alignment summary statistics

Sample category	Total mapped reads	Total mapped reads filtered (%)	Duplicates (%)	Mapping quality (%)	Multimapped reads (%)	Total mapped reads [pass filter]	Reads mapped to primary targets [pass filter]	Primary target enrichment [pass filter] (%)
Myotis lucifugus (n = 92)								
Mean	7,459,167	80.39	73.70	6.60	0.09	1,411,920	656,230	42.1
Minimum	4,203,530	54.45	47.70	5.95	0.06	524,146	118,002	15.5
Maximum	11,483,159	91.95	85.37	9.43	0.14	3,061,461	2,023,112	66.3
Eptesicus fuscus (n = 2)								
Minimum	3,935,930	83.44	67.23	15.58	0.22	651,945	277,998	40.2
Maximum	6,080,356	84.98	69.17	15.98	0.23	913,531	367,312	42.6
M. septentrionalis (n = 2)								
Minimum	6,548,133	80.67	70.80	9.62	0.19	913,750	300,166	32.8
Maximum	9,467,611	86.05	76.22	9.68	0.21	1,830,053	765,227	41.8

TABLE 2 Primary target depth of coverage summary

Sample category	N	Mean	Minimum	Maximum
Myotis lucifugus (total)	92	148.1	26.2	463.3
Ontario (Hibernaculum 1)	12	164.9	108.2	220.9
Ontario (Hibernaculum 2)	24	44.4	26.2	61.1
Manitoba	28	159.8	78.1	294.6
Atlantic Canada	28	218.2	62.9	463.3
Eptesicus fuscus	2	–	59.5	78.2
M. septentrionalis	2	–	66.5	171.0

3.3 | Analysis of neutral genetic structure

The GATK analysis identified 16,115 "off-target" SNPs. The Ensembl variant effect predictor tool found that 1,038 of these SNPs were located >100,000 bp from a neighboring gene (Table 3; Table S5). The putatively neutral SNPs map to 111 different clusters (>100,000 bp from the next cluster) on 88 different scaffolds of the *M. lucifugus* genome sequence assembly. We found no evidence for genetic structure based on these putatively neutral SNPs in the ADEGENET- or STRUCTURE-derived plots based on geography (Figure 3a), or based on the presence of WNS in those areas (Figure 3b), regardless of the minor allele frequency cutoff used in the analysis (data not shown).

3.4 | Immunome SNP detection and analyses

We identified 17,495 SNPs within the primary target loci, located in exons (3,536 SNPs), introns (5,482 SNPs), and regulatory regions (5,482 SNPs). LOSITAN identified 328 and 299 directional outlier SNPs in the geography- and WNS-based analyses, respectively, 32 of which were detected in both analyses. We acknowledge that false positives for SNPs under selection are common in outliers detected using F_{ST}-based methods (Narum & Hess 2011); therefore, the candidate SNPs identified in this experiment will require further validation in future studies. The predicted impacts of each of those 595 directional outlier SNPs are summarized in Tables 5 and S6. Focusing on SNPs most likely to cause major functional changes, we found that 23 outlier SNPs in 19 genes in the geographic comparison resulted in an amino acid change, as did 28 SNPs in 21 genes in the WNS comparison (Table 6). In the WNS comparison, an outlier SNP in the intron region of HLA-DPB1 resulted in likely modification of the splice donor sequence. We also detected 194 SNPS in the regulatory regions of 78 genes (Table S6), of which 11 were identified in both comparisons (Table 4).

Analyses of genetic structure in ADEGENET and STRUCTURE did not identify geography- or WNS-associated genetic structure using the entire primary target locus, exon, intron, or regulatory region SNP datasets (data not shown). However, using the LOSITAN-predicted outlier SNPs, we observed subtle immunogenetic structure based

on geography or the presence of a brief period of co-occurrence with *P. destructans* (Figure 4). The analysis based on geography identified immunogenetic differentiation between *M. lucifugus* in MB and conspecifics in ON and ATL (Figure 4a), while analysis based on co-occurrence with *P. destructans* grouped ON and MB together (pre-WNS), differentiated from the post-WNS samples from ATL (Figure 4b).

4 | DISCUSSION

4.1 | Sequence capture and high-throughput sequencing

Reduced representation genomic profiling strategies have emerged as valuable alternatives to whole-genome sequencing (Narum, Buerkle, Davey, Miller, & Hohenlohe, 2013) where population-level assessments are not yet feasible for nonmodel organisms with larger genomes. Reduced representation approaches can include both transcriptome studies (all expressed genes) or GBS that can include restriction site association DNA (RAD) marker, target capture, and amplicon sequencing. While RAD sequencing has many advantages when genomic resources for the species are sparse, it has many limitations in identifying patterns of local adaptation (Andrews, Good, Miller, Luikart, & Hohenlohe, 2016). Amplicon sequencing of a large number of loci has many advantages in elucidating the genetic variation from known targets; however, in this instance, we chose a sequence capture approach to also pull down large segments of the immunome that included upstream regulatory regions of the genes of interest. As such, the candidate gene GBS approach employed in this study provided several advantages over other means in obtaining immunogenetic information that is likely to be influenced by the selective pressures from disease such as that caused by *P. destructans* (Table 6; Figure 4).

Here, we found target enrichment led to even sequencing uniformity/coverage, which has been noted by other research groups (Powell, Amish, Haynes, Luikart, & Latch, 2016; Samorodnitsky et al., 2015; Schweizer et al., 2016). The assay we developed provided a high on target means to obtain moderate to high coverage of each target (26–463X; Table 1) that was relatively even across samples and loci (Figures 1 and 2). The assay also worked across other species (*M. septentrionalis* and *E. fuscus*) that are known to also be impacted by WNS to varying degrees (Frank et al., 2014; Frick et al., 2015; Langwig et al., 2012). As such, this assay sets the stage for cross-species analyses to further our understanding of the variable immune responses to this disease. The assay, however, yielded far too high a percentage of duplicates (54%–92%; Table 1) that compromised the level of coverage. The duplicates were likely a matter of too many PCR cycles at the adapter ligation stage during DNA library preparations. In the future, we would decrease from 11 to 6–8 cycles in the LM-PCR step. One aspect of the assay that did not meet expectations was the amplification of microsatellite loci, largely as a matter of the sequencing technology used (HiSeq 2500 rapid run using 2 × 100-bp reads on a single flow cell). The

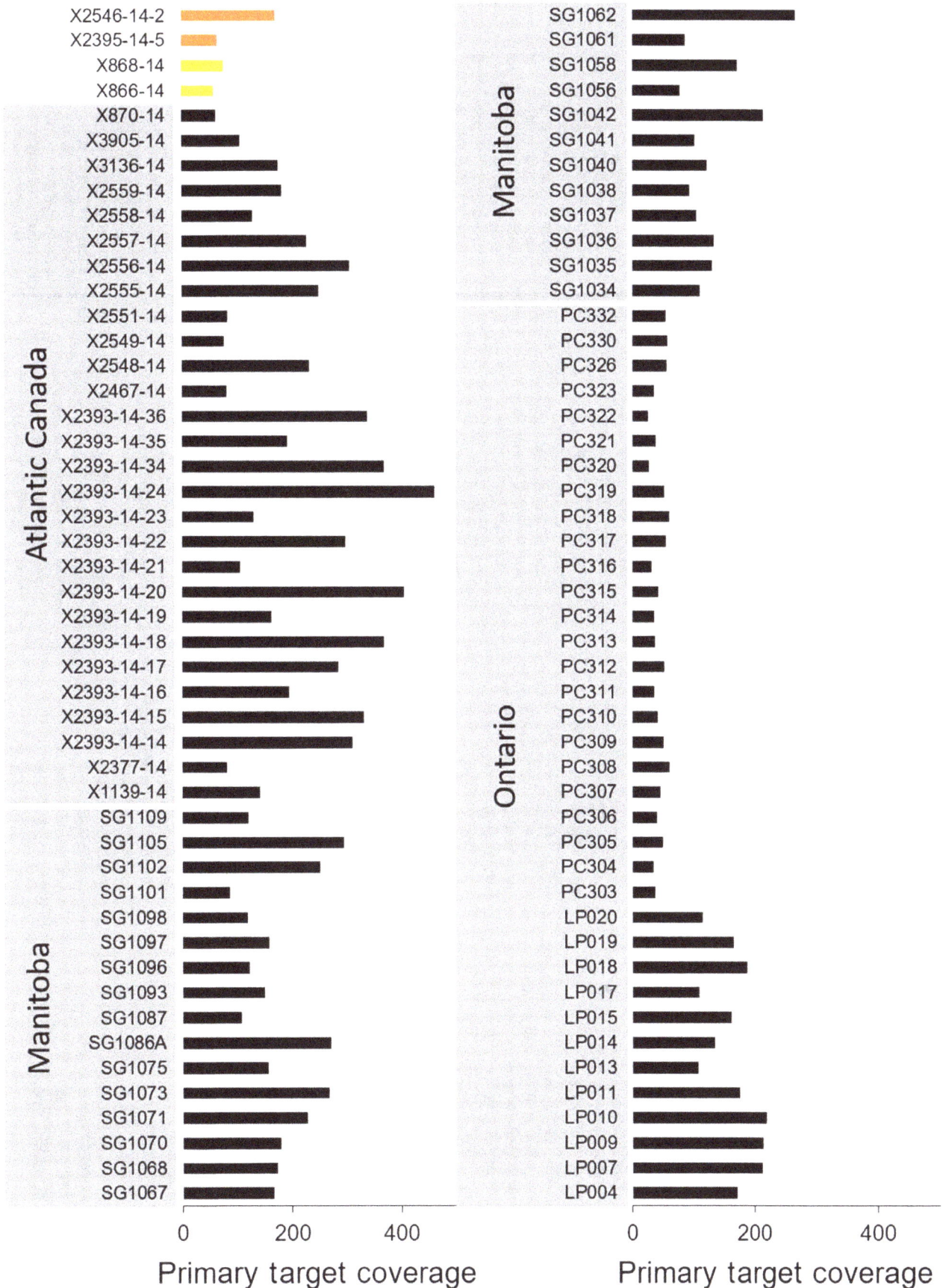

FIGURE 1 Mean depth of coverage for target loci, sorted by sample ID and sampling location. Black bars represent the primary target species, *Myotis lucifugus*. Orange bars: *M. septentrionalis*; yellow bars: *Eptesicus fuscus*

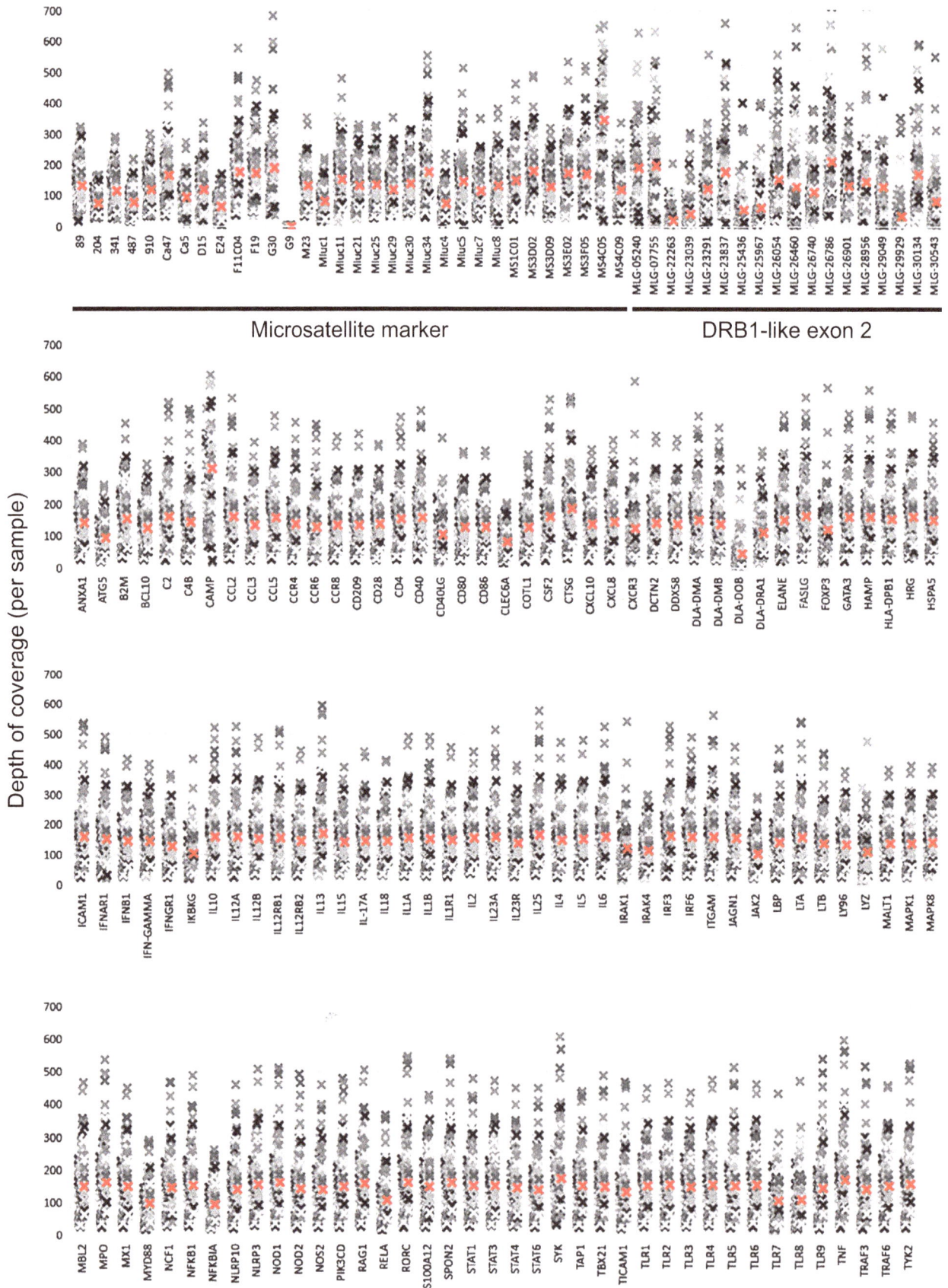

FIGURE 2 Primary target depth of coverage per sample, sorted by locus. Mean depth of coverage for each locus is indicated by a red colored "x." Microsatellite markers and Drb1-like exon 2 targets are marked in the top panel. We condensed the presented data by abbreviating Ensembl-derived *Myotis lucifugus* gene (MLG) identifiers, where "MLG-"="ENSMLUG000000," for the Drb1-like genes that did not have informative Ensembl or GenBank gene names

TABLE 3 Summary of the number of detected, putatively neutral single nucleotide polymorphisms (SNPs) binned by minor allele frequency (MAF)

MAF (%)	SNPs (nr)
2	1038 (111)
5	544 (90)
10	343 (77)
15	236 (68)
20	169 (58)
25	142 (53)

nr = "nonredundant" number of neutral SNP clusters with long-distance (>100-Kbp) SNPs.

hybridization-based method we employed to capture microsatellite loci was successful; however, we were unable to generate genotypes using the reads containing short tandem repeats. To avoid these experimental design and technical problems, we recommend using sequencing technologies that offer longer read lengths. The Illumina MiSeq and Life Technologies PGM System currently yield 300-bp to 400-bp read lengths, which may allow for the microsatellite and flanking regions to be sequenced, and should help microsatellite analysis in nonmodel organisms. Overall, the assay provided

a relatively high number of variable neutral SNPs with frequency differences amenable to population genetic analyses (Table 3) and a large number of F_{ST} outlier SNPs in exons, introns, and regulatory regions (Table 4), several of which were predicted to have important variant effects (Table 5).

4.2 | Immunogenetic diversity and structure

We developed a GBS sequence capture assay to cost-effectively and rapidly reveal genetic diversity in the immunome of endangered *M. lucifugus* threatened by mass die-offs from WNS. The assay characterized neutral population structure to control for stochastic immunogenetic differentiation among sampled areas (Table S5; Figure 3), and also elucidated genetic variation and structure of immune genes via hundreds of SNPs within the exons, introns, and regulatory regions of those genes (Table S6; Figure 4). Preliminary application of the assay to *E. fuscus* and *M. septentrionalis* indicates it may also be an effective tool for these species. By targeting the assay to address specific research questions, our GBS approach can be used across the range of *M. lucifugus* to investigate drivers of genetic, morphological, and behavioral variation.

Our assay revealed subtle immunogenetic variation and structure on a relatively small geographic scale, suggestive of

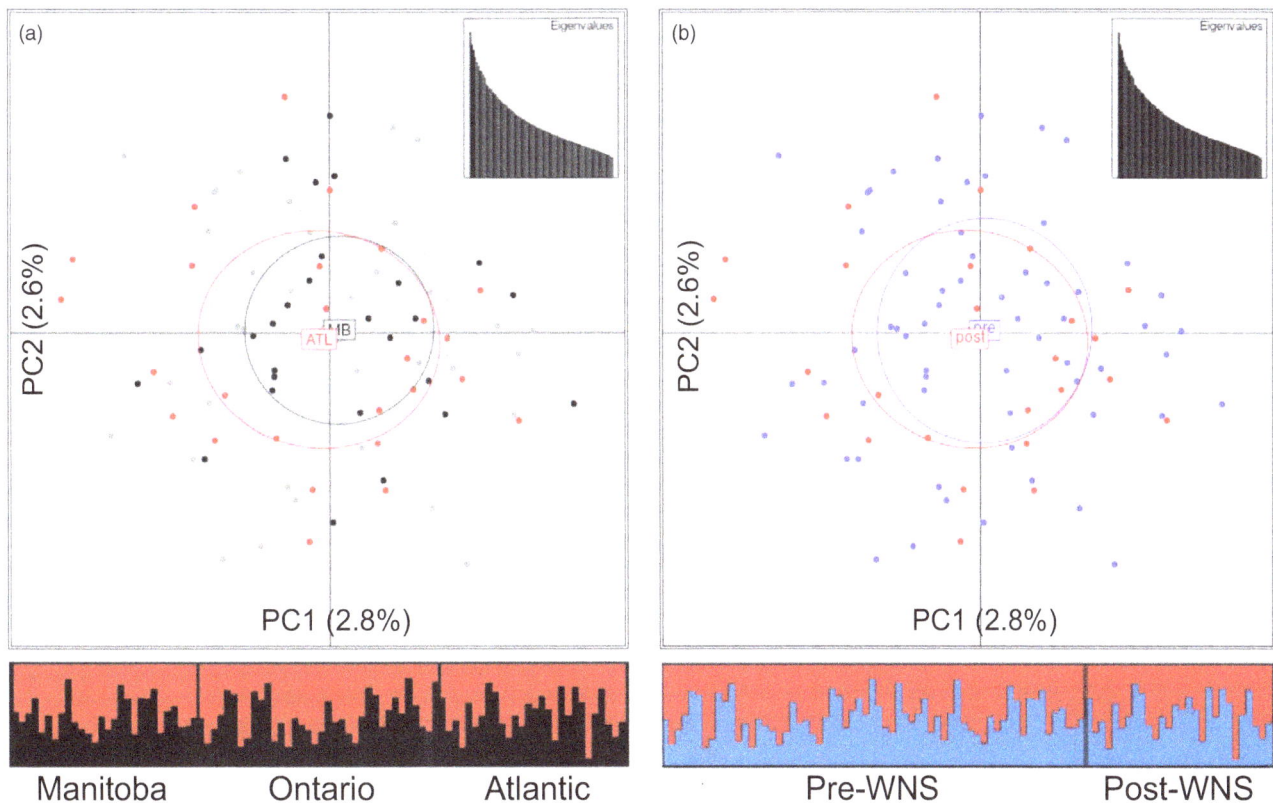

FIGURE 3 Visualizing lack of genetic structure using 1,038 putatively neutral SNPs (>100 kpb from nearest gene) with max-missing genotype of 5% and minor allele frequency of 2% for *Myotis lucifugus* (*n* = 92). Samples were grouped based on (a) geographic location or (b) previous exposure to WNS. Principal component analysis plots were produced using ADEGENET, and the percentage of variation for each axis and a scatter plot of eigenvalues are included for each analysis; barplot shows results of STRUCTURE analysis (*K* = 2). MB = Manitoba (black); ON = Ontario (gray); ATL/post = Atlantic Canada/post-WNS (red); pre = pre-WNS (blue)

TABLE 4　Summary of LOSITAN F_{ST} outliers (FDR < 0.05) in targeted immunome features of *Myotis lucifugus* (n = 92). Directional outliers were used for STRUCTURE and ADEGENET analyses (Figure 4)

Feature type	Geography		WNS	
	Directional outliers (nr)	Balancing outliers (nr)	Directional outliers (nr)	Balancing outliers (nr)
Exon	60 (35)	89 (45)	72 (43)	416 (94)
Intron	159 (62)	249 (71)	131 (54)	1,050 (99)
Regulatory Region	109 (54)	141 (74)	96 (50)	666 (116)
Total	328 (151)	479 (190)	299 (147)	2,132 (309)

nr = "nonredundant" number of genes with F_{ST} outliers.

TABLE 5　Summary of consequences predicted by the variant effect predictor, for directional F_{ST} outliers detected by LOSITAN from immunome sequence capture of *Myotis lucifugus* samples (n = 92)

Feature	Consequence	Geography	WNS
Exon	Synonymous variant	36	43
Exon	Missense variant	21	26
Exon	Missense variant, splice region variant	2	2
Exon	Splice region variant, synonymous variant	1	1
Intron	Intron variant	157	122
Intron	Splice region variant, intron variant	2	8
Intron	Splice donor variant	0	1
Regulatory Region	Upstream gene variant	109	96

local immunogenetic adaptation within an otherwise panmictic population (Figures 3a and 4a; Davy et al., 2017). Comparisons of samples taken before and after the arrival of WNS suggest a nonrandom removal of genetic variants in the immunome by *P. destructans* (Figures 3b and 4b). If similar selection is occurring in *M. lucifugus* that are surviving in WNS-impacted areas, there may be potential for rapid local adaptation to WNS, raising the possibility of evolutionary rescue (Carlson et al., 2016; Maslo & Fefferman, 2015). Conversely, immunogenetic selection by WNS may disrupt previously adaptive patterns of immunogenetic variation as *P. destructans* continues to spread, further complicating the recovery of *M. lucifugus*. Our interpretation of this data is effected by the possibility that the 28 *M. lucifugus* we sampled from Atlantic Canada were not exposed to *P. destructans* during the previous year, and while 27 of these individuals died of WNS in 2014, this might have been their first exposure to an infection with *P. destructans*. To partially address this concern, we note 14 of 28 *M. lucifugus* with sample IDs "X2393-14-N" (where N varies; Table S3) came from a hibernaculum in Prince Edward Island where WNS mortality was identified in the previous winter, 1 year prior to these individuals dying of WNS and being collected for this study.

Immunogenetic diversity in *M. lucifugus* is extremely high. Previous attempts to quantify variation were complicated by duplication of loci in the MHC of *M. lucifugus*, which exhibits up to

24 Drb1-like loci (Davy et al., 2017; Palmer et al., 2016). Our targeted sequence capture assay controls for this gene duplication and allows genotypes to be unambiguously assigned to each individual. We detected functionally significant differentiation in several Drb1-like loci associated with both geography and previous exposure to WNS (Table 6). Exposure to WNS is also associated with a shift in genetic variation at interleukins and Toll-like receptors (Davy et al., 2017; Field et al., 2015; Lilley et al., 2017; Rapin et al., 2014), consistent with the hypothesis that WNS exerts immunogenetic selective pressure on *M. lucifugus*. Our research on the interactions between *M. lucifugus* genetics and *P. destructans* continues to reinforce the need to take both interindividual and inter-regional variation of both the host and pathogen into account when interpreting genetic data. In this study, bats collected from sites <600 km apart in Manitoba and Ontario belong to a panmictic population based on neutral molecular markers, but exhibit local variation in the immunome that may result in different expression of immune genes among sites (Table 6). For example, it is possible that local immunogenetic differentiation between these sites result in different survival rates following the introduction of WNS. Variation in the regulatory regions (Table S6) could also alter the expression of integral immune genes among sites.

High immunogenetic variation in *M. lucifugus* has implications for the interpretation of gene expression studies as well. Bats from different sampling sites may respond differently to immune challenges due to variation in exon and regulatory regions of the immunome. Therefore, experimental gene expression studies related to *P. destructans* or other pathogens should explicitly control for potential geographic variation. Otherwise, observed differences in gene expression cannot be unambiguously attributed to the effects of the pathogen (or other treatments of interest).

The GBS approach used here provides a basis for real-time investigations of evolutionary rescue in populations of bats that persist following initial declines from WNS (Maslo & Fefferman, 2015). The results of our study were based on analysis of a small number of populations, and in the post-WNS population, the *M. lucifugus* we examined did not survive infection. Ideally, future studies will compare immunogenetic variation not only among exposed and unexposed sites, but also among time-series samples taken from bats that have survived multiple selective sweeps from one, two, or more winters in hibernacula containing *P. destructans*.

TABLE 6 LOSITAN-detected F_{ST} outliers from SNP analyses based on a priori grouping by geographic location (Manitoba, Ontario, and Atlantic) or by WNS exposure history (pre-WNS, post-WNS). Only the SNPs that are most likely to have a functional impact by altering amino acids or affecting splice sequences are listed (see Methods and Table S6 for details)

Comparison (#SNPs)	Gene name	Ensembl transcript ID	Brief description	Amino acids
Geographic	CCL3	ENSMLUT00000002888	C-C motif chemokine	A/V
Geographic	CCR4	ENSMLUT00000027956	Chemokine (C-C motif) receptor 4	S/F
Geographic	CD40	ENSMLUT00000006008	CD40 molecule, TNF receptor superfamily member 5	S/N
Geographic	Drb1e2-like-e	ENSMLUT00000027881	DLA class II histocompatibility antigen	N/H
Geographic	Drb1e2-like-f	ENSMLUT00000028450	DLA class II histocompatibility antigen	T/M
Geographic	Drb1e2-like-l	ENSMLUT00000029278	DLA class II histocompatibility antigen	L/R
Geographic	Drb1e2-like-n	ENSMLUT00000030076	DLA class II histocompatibility antigen	E/D
Geographic	Drb1e2-like-r	ENSMLUT00000027745	DLA class II histocompatibility antigen	Q/L[a]
Geographic (2)	HRG	ENSMLUT00000013351	Histidine-rich glycoprotein	K/R, H/Q
Geographic	IFNGR1	ENSMLUT00000008611	Interferon gamma receptor 1	D/E
Geographic (2)	IL12RB1	ENSMLUT00000013802	Interleukin 12 receptor, beta 1	K/R, T/I
Geographic (2)	IL1R1	ENSMLUT00000011035	Interleukin 1 receptor, type I	R/K, E/K
Geographic	IL23A	ENSMLUT00000006770	Interleukin 23, alpha subunit p19	R/T
Geographic	IRF6	ENSMLUT00000004509	Interferon regulatory factor 6	K/N
Geographic (2)	MPO	ENSMLUT00000006099	Myeloperoxidase	Q/L[a], G/R
Geographic	NLRP10	ENSMLUT00000000818	NLR family, pyrin domain containing 10	S/C
Geographic	NOS2	ENSMLUT00000015896	Nitric oxide synthase	G/D
Geographic	RAG1	ENSMLUT00000000542	Recombination activating gene 1	S/N
Geographic	SPON2	ENSMLUT00000017687	Spondin 2, extracellular matrix protein	T/M
WNS	CCR4	ENSMLUT00000027956	Chemokine (C-C motif) receptor 4	I/N
WNS	DDX58	ENSMLUT00000003044	DEAD (Asp-Glu-Ala-Asp) box polypeptide 58	V/I
WNS	DLA-DRA1	ENSMLUT00000027968	DLA class II histocompatibility antigen, DR alpha chain-like	P/T
WNS	Drb1e2-like-i	ENSMLUT00000031273	DLA class II histocompatibility antigen	E/V
WNS (2)	Drb1e2-like-k	ENSMLUT00000023434	DLA class II histocompatibility antigen	D/N, D/E
WNS	Drb1e2-like-p	ENSMLUT00000022698	DLA class II histocompatibility antigen	S/N
WNS (2)	Drb1e2-like-r	ENSMLUT00000027745	DLA class II histocompatibility antigen	Q/L[a], R/H
WNS	HLA-DPB1	ENSMLUT00000016285	Major histocompatibility complex, class II, DP beta 1	[b]
WNS	IFNAR1	ENSMLUT00000025403	Interferon (alpha, beta and omega) receptor 1	S/P
WNS	IL12RB1	ENSMLUT00000013802	Interleukin 12 receptor, beta 1	I/L
WNS	IL12RB2	ENSMLUT00000001415	Interleukin 12 receptor, beta 2	I/V
WNS (2)	IL1R1	ENSMLUT00000011035	Interleukin 1 receptor, type I	L/M, D/G
WNS	IL5	ENSMLUT00000016553	Interleukin 5	K/E
WNS (2)	ITGAM	ENSMLUT00000011332	Integrin, alpha X (complement component 3 receptor 4 subunit)	Q/R, V/L
WNS	MPO	ENSMLUT00000006099	Myeloperoxidase	Q/L[a]
WNS (3)	NOD2	ENSMLUT00000015164	Nucleotide-binding oligomerization domain containing 2	L/V, S/R, S/A
WNS	NOS2	ENSMLUT00000015896	Nitric oxide synthase	A/V
WNS	TBX21	ENSMLUT00000014543	T-box 21	Q/P
WNS	TLR1	ENSMLUT00000008406	Toll-like receptor 1	V/I
WNS	TLR2	ENSMLUT00000012815	Toll-like receptor 2	S/P
WNS (2)	TLR6	ENSMLUT00000008414	Toll-like receptor 6	H/L, I/V
WNS	TLR9	ENSMLUT00000015105	Toll-like receptor 9	A/V

[a]Indicates outlier SNPs were identified in both the geographic and WNS-based comparisons.
[b]Indicates a SNP predicted to have a high impact by altering a splice donor sequence in an intron. The other SNPs listed here are in exons and are predicted have moderate impacts by altering the amino acid sequence.

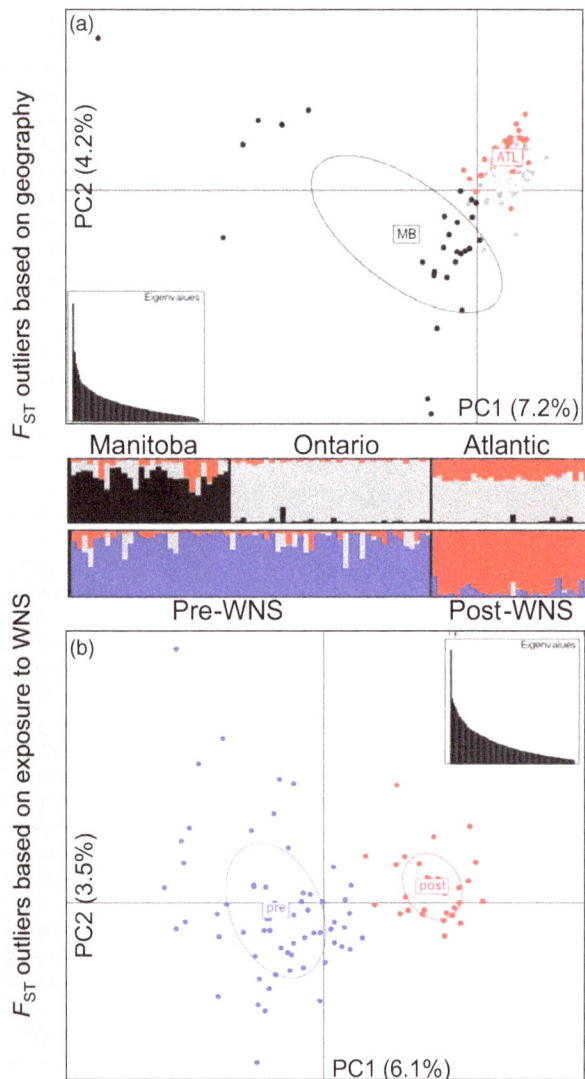

FIGURE 4 Analysis of immunogenetic population structure based on Lositan-detected F_{ST} outliers with max-missing genotype of 5% and minor allele frequency of 2% for *Myotis lucifugus* (*n* = 92). Samples were grouped based on (a) geographic location or (b) previous exposure to WNS. Principal component analysis plots were produced using ADEGENET, and the percentage of variation for each axis and a scatter plot of eigenvalues are included for each analysis; barplot shows results of STRUCTURE analysis (*K* = 3). MB = Manitoba (black); ON = Ontario (gray); ATL/post = Atlantic Canada/post-WNS (red); pre = pre-WNS (blue)

5 | CONCLUSION

We developed a cost-effective high-throughput sequence capture assay to test for immunogenetic shifts in *M. lucifugus* populations following exposure to *P. destructans*. Sequence analysis from 92 *M. lucifugus* identified sequence variation in 138 immune-related genes, their upstream regulatory regions, and 111 putatively neutral regions of the genome. The "one-pot" assay we developed to test for genetic structure and immunogenetic variation identified functional immunogenetic variants in *M. lucifugus* putatively associated with WNS susceptibility, demonstrated a shift in immunogenetic diversity of

populations pre- and post-WNS exposure, and provided preliminary support for a potential evolutionary rescue of *M. lucifugus* in Atlantic Canada given a nonrandom purging of immunogenetic variants in the WNS-susceptible bats. We can use the genetic variants identified in this study as a baseline for future investigations of rangewide immunogenetic adaptation to WNS in little brown myotis. Ultimately, understanding the potential for evolutionary rescue in a species can guide more effective and targeted management actions to mitigate the impacts of WNS on North American bat populations. Overall, this study sets the stage for further research with larger sample sizes and increased population replicates under different types of selective pressure to further understand patterns of local adaptation in this bat species, most importantly in context of WNS exposure and survival.

ACKNOWLEDGEMENTS

This research was funded by a Discovery Grant from the Natural Sciences and Engineering Research Council of Canada (CJK, CKRW), the Liber Ero Fellowship Program (CMD), and the Species at Risk Research Fund for Ontario (CJK, CMD, CKRW). The Canadian Wildlife Health Cooperative provided funding for the WNS surveillance program that enabled the diagnosis of this disease in the bats included from the post-WNS endemic area as well as the collection and shipping of tissues from their carcasses. We thank Katie Dogantzis, Katelyn Jackson (Trent University), and Matt Harnden (Natural Resources DNA Profiling and Forensics Centre) for technical assistance; Mena Farag (Roche) for helping revise the multiplexed sequence capture protocol; Aaron Goldman, Graham Cromar, Roger Shi, and Siwei Chen (The Clinical Genomics Centre at Mount Sinai Hospital, Toronto) for performing the DNA library preparation and Illumina sequencing; Dr. Barry Saville (Trent University) for assistance in analyzing the high-throughput sequence data; and two anonymous reviewers for critique leading to an improved version of our manuscript.

AUTHOR CONTRIBUTIONS

MED, CMD, CKRW, and CJK conceived and designed the experiments. SM, JS, and CKRW collected bat tissue. SM performed postmortem diagnosis of WNS. MED performed the experiments and analyzed the data. CJK contributed reagents/materials/analysis tools. MED, CMD, CKRW, CJK, and SM wrote and revised the manuscript.

REFERENCES

Acevedo-Whitehouse, K., & Cunningham, A. A. (2006). Is MHC enough for understanding wildlife immunogenetics? *Trends in Ecology & Evolution*, *21*, 433–438.

Afgan, E., Baker, D., van den Beek, M., Blankenberg, D., Bouvier, D., Čech, M., … Goecks, J. (2016). The Galaxy platform for accessible, reproducible and collaborative biomedical analyses: 2016 update. *Nucleic Acids Research*, *44*, W3–W10.

Andrews, K. R., Good, J. M., Miller, M. R., Luikart, G., & Hohenlohe, P. A. (2016). Harnessing the power of RADseq for ecological and evolutionary genomics. *Nature Reviews Genetics*, *17*, 81–92.

Antao, T., Lopes, A., Lopes, R. J., Beja-Pereira, A., & Luikart, G. (2008). LOSITAN: A workbench to detect molecular adaptation based on a Fst-outlier method. *BMC Bioinformatics*, 9, 1–5.

Beaumont, M. A., & Nichols, R. A. (1996). Evaluating loci for use in the genetic analysis of population structure. *Proceedings of the Royal Society of London B: Biological Sciences*, 263, 1619–1626.

Blehert, D. S., Hicks, A. C., Behr, M., Meteyer, C.U., Berlowski-Zier, B.M., Buckles, E.L., ... Stone, W.B. (2009). Bat white-nose syndrome: An emerging fungal pathogen? *Science*, 323, 227.

Bonneaud, C., Balenger, S. L., Zhang, J., Edwards, S. V., & Hill, G. E. (2012). Innate immunity and the evolution of resistance to an emerging infectious disease in a wild bird. *Molecular Ecology*, 21, 2628–2639.

Burns, L. E., Broders, H. G., & Frasier, T. R. (2012). Characterization of 11 tetranucleotide microsatellite loci for the little brown bat (*Myotis lucifugus*) based on in silico genome sequences. *Conservation Genetics Resources*, 4, 653–655.

Canadian Wildlife Health Cooperative (2014). *Canadian bat white-nose syndrome necropsy protocol*. 10 pp. http://www.cwhc-rcsf.ca/docs/Canadian%20Bat%20WNS%20Necropsy%20Protocol.pdf. Accessed June 18, 2017.

Carlson, S. M., Cunningham, C. J., & Westley, P. A. H. (2016). Evolutionary rescue in a changing world. *Trends in Ecology & Evolution*, 29, 521–530.

Castella, V., & Ruedi, M. (2000). Characterization of highly variable microsatellite loci in the bat *Myotis myotis* (Chiroptera Vespertilionidae). *Molecular Ecology*, 9, 1000–1002.

Cheng, T. L., Mayberry, H., McGuire, L. P., Hoyt, J. R., Langwig, K. E., Nguyen, H., ... Frick, W. F. (2016). Efficacy of a probiotic bacterium to treat bats affected by the disease white-nose syndrome. *Journal of Applied Ecology*, 54, 701–708.

Chhatre, V. E., & Emerson, K. (2017). StrAuto: Automation and parallelization of STRUCTURE analysis. *BMC Bioinformatics*, 18, 192.

Cornelison, C. T., Gabriel, K. T., Barlament, C., & Crow, S. A. (2014). Inhibition of *Pseudogymnoascus destructans* growth from conidia and mycelial extension by bacterially produced volatile organic compounds. *Mycopathologia*, 177, 1–10.

Cunningham, F., Amode, M. R., Barrell, D., Beal, K., Billis, K., Brent, S., ... Flicek, P. (2015). Ensembl 2015. *Nucleic Acids Research*, 43, D662–D669.

Danecek, P., Auton, A., Abecasis, G., Albers, C.A., Banks, E., DePristo, M.A., ... 1000 Genomes Project Analysis Group. (2011). The variant call format and VCFtools. *Bioinformatics*, 27, 2156–2158.

Davy, C. M., Donaldson, M. E., Kyle, C. J., Saville, B. J., McGuire, L., Mayberry, H., ... Willis, C. K. R. (2017). The Other White-Nose Syndrome Transcriptome: Tolerant and susceptible hosts respond differently to infection with Pseudogymnoascus destructans. *Ecology and Evolution*, http: \\doi.org\10.1002/ece3.3234.

Davy, C. M., Donaldson, M. E., Rico, Y., Lausen, C. L., Dogzantis, K., Ritchie, K., ... Kyle, C. J. (2017). Prelude to a panzootic: Gene flow and continent-wide immunogenetic variation in northern little brown myotis vulnerable to bat white-nose syndrome. *FACETS*, http:\\doi.org\10.1139/facets-2017-0022.

DePristo, M. A., Banks, E., Poplin, R., Garimella, K. V., Maguire, J. R., Hartl, C., ... Daly, M. J.. (2011). A framework for variation discovery and genotyping using next-generation DNA sequencing data. *Nature Genetics*, 43, 491–498.

Dobony, C. A., Hicks, A. C., Langwig, K. E., von Linden, R. I., Okoniewski, J. C., & Rainbolt, R. E. (2011). Little brown myotis persist despite exposure to white-nose syndrome. *Journal of Fish and Wildlife Management*, 2, 190–195.

Earl, D. A., & VonHoldt, B. M. (2012). STRUCTURE HARVESTER: A website and program for visualizing STRUCTURE output and implementing the Evanno method. *Conservation Genetics Resources*, 4, 359–361.

Eizaguirre, C., Lenz, T. L., Kalbe, M., & Milinski, M. (2012). Rapid and adaptive evolution of MHC genes under parasite selection in experimental vertebrate populations. *Nature Communications*, 3, 621.

Evanno, G., Regnaut, S., & Goudet, J. (2005). Detecting the number of clusters of individuals using the software STRUCTURE: A simulation study. *Molecular Ecology*, 14, 2611–2620.

Field, K. A., Johnson, J. S., Lilley, T. M., Reeder, S. M., Rogers, E. J., Behr, M. J., & Reeder, D. M. (2015). The white-nose syndrome transcriptome: Activation of anti-fungal host responses in wing tissue of hibernating little brown myotis. *PLoS Pathogens*, 11, e1005168–e1005168.

Fisher, M. C., Henk, D. A., Briggs, C. J., Brownstein, J. S., Madoff, L. C., McCraw, S. L., & Gurr, S. J. (2012). Emerging fungal threats to animal, plant and ecosystem health. *Nature*, 484, 186–194.

Frank, C. L., Michalski, A., McDonough, A. A., Rahimian, M., Rudd, R. J., & Herzog, C. (2014). The resistance of a North American bat species (*Eptesicus fuscus*) to White-Nose Syndrome (WNS). *PLoS One*, 9, e113958.

Fraser, H. B. (2013). Gene expression drives local adaptation in humans. *Genome Research*, 23, 1089–1096.

Frick, W. F., Pollock, J. F., Hicks, A. C., Langwig, K. E., Reynolds, D. S., Turner, G. G., ... Kunz, T. H. (2010). An emerging disease causes regional population collapse of a common North American Bat species. *Science*, 329, 679–682.

Frick, W. F., Puechmaille, S. J., Hoyt, J. R., Nickel, B. A., Langwig, K. E., Foster, J. T., ... Kilpatrick, A. M. (2015). Disease alters macroecological patterns of North American bats. *Global Ecology and Biogeography*, 24, 741–749.

Fungtammasan, A., Ananda, G., Hile, S. E., Su, M. S.-W., Sun, C., Harris, R., ... Makova, K. D. (2015). Accurate typing of short tandem repeats from genome-wide sequencing data and its applications. *Genome Research*, 25, 736–749.

Gallana, M., Ryser-Degiorgis, M.-P., Wahli, T., & Segner, H. (2013). Climate change and infectious diseases of wildlife: Altered interactions between pathogens, vectors and hosts. *Current Zoology*, 59, 427–437.

Hansen, M. M., Olivieri, I., Waller, D. M., & Nielsen, E. E. (2012). Monitoring adaptive genetic responses to environmental change. *Molecular Ecology*, 21, 1311–1329.

Harrisson, K. A., Pavlova, A., Telonis-Scott, M., & Sunnucks, P. (2014). Using genomics to characterize evolutionary potential for conservation of wild populations. *Evolutionary Applications*, 7, 1008–1025.

Jakobsson, M., & Rosenberg, N. A. (2007). CLUMPP: A cluster matching and permutation program for dealing with label switching and multimodality in analysis of population structure. *Bioinformatics*, 14, 1801–1806.

Johnson, J. B., Roberts, J. H., King, T. L., Edwards, J. W., Ford, W. M., & Ray, D. A. (2014). Genetic structuring of northern myotis (*Myotis septentrionalis*) at multiple spatial scales. *Acta Theriologica*, 59, 223–231.

Jombart, T., & Ahmed, I. (2011). adegenet 1.3-1: New tools for the analysis of genome-wide SNP data. *Bioinformatics*, 27(21), 3070–3071.

Jones, M. R., & Good, J. M. (2016). Targeted capture in evolutionary and ecological genomics. *Molecular Ecology*, 25, 185–202.

Kawakami, T., Backström, N., Burri, R., Husby, A., Olason, P., Rice, A. M., ... Ellegren, H. (2014). Estimation of linkage disequilibrium and inter-specific gene flow in Ficedula flycatchers by a newly developed 50k single-nucleotide polymorphism array. *Molecular Ecology Resources*, 14, 1248–1260.

Knight, J. C. (2013). Genomic modulators of the immune response. *Trends in Genetics*, 29, 74–83.

Kyle, C. J., Rico, Y., Castillo, S., Srithayakumar, V., Cullingham, C. I., White, B. N., & Pond, B. A. (2014). Spatial patterns of neutral and functional genetic variations reveal patterns of local adaptation in raccoon (*Procyon lotor*) populations exposed to raccoon rabies. *Molecular Ecology*, 23, 2287–2298.

Langwig, K. E., Frick, W. F., Bried, J. T., Hicks, A. C., Kunz, T. H., & Kilpatrick, A. M. (2012). Sociality, density-dependence and microclimates determine the persistence of populations suffering from a novel fungal disease, white-nose syndrome. *Ecology Letters*, 15, 1050–1057.

Langwig, K. E., Hoyt, J. R., Parise, K. L., Frick, W. F., Foster, J. T., & Kilpatrick, A. M. (2017). Resistance in persisting bat populations after white-nose syndrome invasion. *Philosophical Transactions of the Royal Society of*

London. Series B, Biological Sciences, 372, 20160044.

Leopardi, S., Blake, D., & Puechmaille, S. J. (2015). White-nose syndrome fungus introduced from Europe to North America. *Current Biology, 25*, R217–R219.

Li, H. (2013). Aligning sequence reads, clone sequences and assembly contigs with BWA-MEM. *arXiv, 1303*, 1–3.

Li, H., Handsaker, B., Wysoker, A., Fennell, T., Ruan, J., Homer, N., ... 1000 Genome Project Data Processing (2009). The sequence alignment/map format and SAMtools. *Bioinformatics, 25*, 2078–2079.

Lilley, T. M., Prokkola, J. M., Johnson, J. S., Rogers, E. J., Gronsky, S., Kurta, A., ... Field, K. A. (2017). Immune responses in hibernating little brown myotis (*Myotis lucifugus*) with white-nose syndrome. *Proceedings of the Royal Society B: Biological Sciences, 284*, 20162232.

Lischer, H. E. L., & Excoffier, L. (2012). PGDSpider: An automated data conversion tool for connecting population genetics and genomics programs. *Bioinformatics, 28*, 298–299.

Lorch, J. M., Palmer, J. M., Lindner, D. L., Ballmann, A. E., George, K. G., Griffin, K., ... Blehert, D. S. (2016). First detection of bat white-nose syndrome in western North America. *mSphere, 1*, e00148-16.

Maslo, B., & Fefferman, N. H. (2015). A case study of bats and white-nose syndrome demonstrating how to model population viability with evolutionary effects. *Conservation Biology, 29*, 1176–1185.

Maslo, B., Valent, M., Gumbs, J. F., & Frick, W. F. (2015). Conservation implications of ameliorating survival of little brown bats with white-nose syndrome. *Ecological Applications, 25*, 1832–1840.

McKenna, A., Hanna, M., Banks, E., Sivachenko, A., Cibulskis, K., Kernytsky, A., ... DePristo, M. A. (2010). The genome analysis toolkit: A MapReduce framework for analyzing next-generation DNA sequencing data. *Genome Research, 20*, 1297–1303.

Moore, M. S., Reichard, J. D., Murtha, T. D., Nabhan, M. L., Pian, R. E., Ferreira, J. S., & Kunz, T. H. (2013). Hibernating little brown myotis (*Myotis lucifugus*) show variable immunological responses to white-nose syndrome. *PLoS One, 8*, e58976.

Morris, K. M., Wright, B., Grueber, C. E., Hogg, C., & Belov, K. (2015). Lack of genetic diversity across diverse immune genes in an endangered mammal, the Tasmanian devil (*Sarcophilus harrisii*). *Molecular Ecology, 24*, 3860–3872.

Narum, S. R., & Hess, J. (2011). Comparison of FST outlier tests for SNP loci under selection. *Molecular Ecology, 11*, 184–194.

Narum, S. R., Buerkle, C. A., Davey, J. W., Miller, M. R., & Hohenlohe, P. A. (2013). Genotyping-by-sequencing in ecological and conservation genomics. *Molecular Ecology, 22*, 2841–2847.

Oyler-McCance, S. J., & Fike, J. A. (2011). Characterization of small microsatellite loci isolated from endangered Indiana bat (*Myotis sodalis*) for use in non-invasive sampling. *Conservation Genetics Resources, 3*, 243–245.

Palmer, J. M., Berkman, L. K., Marquardt, P. E., Donner, D. M., Jusino, M. A., & Lindner, D. L. (2016). Preliminary characterization of little brown bats (*Myotis lucifugus*) immune MHC II DRB alleles using next-generation sequencing. *PeerJ PrePrints, 4*, e1662v1.

Piaggio, A. J., Figueroa, J. A., & Perkins, S. L. (2009). Development and characterization of 15 polymorphic microsatellite loci isolated from Rafinesque's big-eared bat, *Corynorhinus rafinesquii*. *Molecular Ecology Resources, 9*, 1191–1193.

Powell, J. H., Amish, S. J., Haynes, G. D., Luikart, G., & Latch, E. K. (2016). Candidate adaptive genes associated with lineage divergence: Identifying SNPs via next-generation targeted resequencing in mule deer (*Odocoileus hemionus*). *Molecular Ecology Resources, 16*, 1165–1172.

Pritchard, J. K., Stephens, M., & Donnelly, P. (2000). Inference of population structure using multilocus genotype data. *Genetics, 155*, 945–959.

Puechmaille, S. J., Fuller, H., & Teeling, E. C. (2011). Effect of sample preservation methods on the viability of *Geomyces destructans*, the fungus associated with white-nose syndrome in bats. *Acta Chiropterologica, 13*, 217–221.

Puechmaille, S. J., Wibbelt, G., Korn, V., Fuller, H., Forget, F., Mühldorfer, K.,

... Teeling, E. C. (2011). Pan-European distribution of white-nose syndrome fungus (*Geomyces destructans*) not associated with mass mortality. *PLoS One, 6*, e19167.

Purcell, S., Neale, B., Todd-Brown, K., Thomas, L., Ferreira, M. A. R., Bender, D., ... Sham, P. C. (2007). PLINK: A tool set for whole-genome association and population-based linkage analyses. *American Journal of Human Genetics, 81*, 559–575.

Rapin, N., Johns, K., Martin, L., Warnecke, L., Turner, J. M., Bollinger, T. K., ... Misra, V. (2014). Activation of innate immune-response genes in little brown bats (*Myotis lucifugus*) infected with the fungus *Pseudogymnoascus destructans* (J Sun, Ed,). *PLoS One, 9*, e112285.

Reed, K. M., Mendoza, K. M., & Settlage, R. E. (2016). Targeted capture enrichment and sequencing identifies extensive nucleotide variation in the turkey MHC-B. *Immunogenetics, 68*, 219–229.

Rico, Y., Morris-Pocock, J., Zigouris, J., Nocera, J. J., & Kyle, C. J. (2015). Lack of spatial immunogenetic structure among wolverine (*Gulo gulo*) populations suggestive of broad scale balancing selection. *PLoS One, 10*, e0140170.

Roffler, G. H., Amish, S. J., Smith, S., Cosart, T., Kardos, M., Schwartz, M. K., & Luikart, G. (2016). SNP discovery in candidate adaptive genes using exon capture in a free-ranging alpine ungulate. *Molecular Ecology Resources, 16*, 1147–1164.

Rosenberg, N. A. (2004). distruct: A program for the graphical display of population structure. *Molecular Ecology Notes, 4*, 137–138.

Samorodnitsky, E., Jewell, B. M., Hagopian, R., Miya, J., Wing, M. R., Lyon, E., ... Roychowdhury, S. (2015). Evaluation of hybridization capture versus amplicon-based methods for whole-exome sequencing. *Human Mutation, 36*, 903–914.

Savage, A., & Zamudio, K. R. (2011). MHC genotypes associate with resistance to a frog-killing fungus. *Proceedings of the National Academy of Sciences, 108*, 16705–16710.

Schoville, S. D., Bonin, A., François, O., Lobreaux, S., Melodelima, C., & Manel, S. (2012). Adaptive genetic variation on the landscape: Methods and cases. *Annual Review of Ecology, Evolution, and Systematics, 43*, 23–43.

Schweizer, R. M., Robinson, J., Harrigan, R., Silva, P., Galverni, M., Musiani, M., ... Wayne, R. K. (2016). Targeted capture and resequencing of 1040 genes reveal environmentally driven functional variation in gray wolves. *Molecular Ecology, 25*, 357–379.

Smith, A. L., Hewitt, N., Klenk, N., Bazely, D. R., Yan, N., Wood, S., ... Lipsig-Mummé, C. (2012). Effects of climate change on the distribution of invasive alien species in Canada: A knowledge synthesis of range change projections in a warming world. *Environmental Reviews, 20*, 1–16.

Tiffin, P., & Ross-Ibarra, J. (2014). Advances and limits of using population genetics to understand local adaptation. *Trends in Ecology & Evolution, 29*, 673–680.

Trujillo, R. G., & Amelon, S. K. (2009). Development of microsatellite markers in *Myotis sodalis* and cross-species amplification in *M. gricescens, M. leibii, M. lucifugus*, and *M. septentrionalis*. *Conservation Genetics, 10*, 1965–1968.

Van der Auwera, G. A., Carneiro, M. O., Hartl, C., Poplin, R., del Angel, G., Levy-Moonshine, A., ... DePristo, M. A. (2013). From FastQ data to high-confidence variant calls: The Genome Analysis Toolkit best practices pipeline. *Current Protocols in Bioinformatics, 43*, 11.10.1–11.10.33.

Wilcox, A., & Willis, C. K. R. (2016). Energetic benefits of enhanced summer roosting habitat for little brown bats (*Myotis lucifugus*) recovering from white-nose syndrome. *Conservation Physiology, 4*, cov070.

6

Genetic diversity and connectivity within *Mytilus* spp. in the subarctic and Arctic

Sofie Smedegaard Mathiesen[1,2] | Jakob Thyrring[1] | Jakob Hemmer-Hansen[2] | Jørgen Berge[3,4] | Alexey Sukhotin[5,6] | Peter Leopold[3] | Michaël Bekaert[7] | Mikael Kristian Sejr[1,†] | Einar Eg Nielsen[2,†]

[1]Department of Bioscience, Arctic Research Centre, Aarhus University, Aarhus C, Denmark

[2]Section for Marine Living Resources, National Institute of Aquatic Resources, Technical University of Denmark, Silkeborg, Denmark

[3]Faculty of Biosciences, Fisheries and Economics, UiT The Arctic University of Norway, Tromsø, Norway

[4]The University Centre in Svalbard, Longyearbyen, Norway

[5]White Sea Biological Station, Zoological Institute of Russian Academy of Sciences, St. Petersburg, Russia

[6]Invertebrate Zoology Department, St. Petersburg State University, St. Petersburg, Russia

[7]Institute of Aquaculture, University of Stirling, Stirling, UK

Correspondence
Einar Eg Nielsen, Section for Marine Living Resources, National Institute of Aquatic Resources, Technical University of Denmark, Silkeborg, Denmark.
Email: een@aqua.dtu.dk

Abstract

Climate changes in the Arctic are predicted to alter distributions of marine species. However, such changes are difficult to quantify because information on present species distribution and the genetic variation within species is lacking or poorly examined. Blue mussels, *Mytilus* spp., are ecosystem engineers in the coastal zone globally. To improve knowledge of distribution and genetic structure of the *Mytilus edulis* complex in the Arctic, we analyzed 81 SNPs in 534 *Mytilus* spp. individuals sampled at 13 sites to provide baseline data for distribution and genetic variation of *Mytilus* mussels in the European Arctic. *Mytilus edulis* was the most abundant species found with a clear genetic split between populations in Greenland and the Eastern Atlantic. Surprisingly, analyses revealed the presence of *Mytilus trossulus* in high Arctic NW Greenland (77°N) and *Mytilus galloprovincialis* or their hybrids in SW Greenland, Svalbard, and the Pechora Sea. Furthermore, a high degree of hybridization and introgression between species was observed. Our study highlights the importance of distinguishing between congener species, which can display local adaptation and suggests that information on dispersal routes and barriers is essential for accurate predictions of regional susceptibility to range expansions or invasions of boreal species in the Arctic.

KEYWORDS
arctic fauna, bivalves, climate change, glacial refugium, hybrid zone, *Mytilus edulis*, population structure, SNPs

1 | INTRODUCTION

Nowhere else on Earth is the impact of climate change expected to be more severe than in the Arctic. Temperatures in the Arctic are estimated to increase by 4–7°C over the next century, with wide-ranging effects on Arctic species (ACIA 2004; IPCC 2014). This have caused shifts in species' abundances and distributions over the last decades (IPCC 2014; Poloczanska et al., 2013), and future temperature increases are believed to move species distribution limits toward the poles (ACIA 2004). Such effects, however, are nearly impossible to monitor and understand without proper baseline studies of the genetic variation within and between species (Brodersen & Seehausen, 2014). Almost all species investigated, including Arctic, have revealed genetically discrete populations that inhabit a specific subset of the species geographical and environmental range. These populations

†These authors jointly supervised this work.

can exhibit different adaptations and tolerance limits to specific environments (Limborg et al., 2012; Nielsen, Hemmer-Hansen, Larsen, & Bekkevold, 2009; Thyrring, Rysgaard, Blicher, & Sejr, 2015), which make it important to know their current distribution and the connectivity between populations and the processes governing the distribution of genetic variation. Through genetic analysis, it is possible to determine the level of genetic variability within both threatened and newly established populations, the origin of migrating individuals, direction of gene flow, and possible adaptive evolutionary changes associated with climate change (Brodersen & Seehausen, 2014; Hansen, Olivieri, Waller, & Nielsen, 2012; Laikre, Schwartz, Waples, & Ryman, 2010). All key factors needed to make predictions for the likely impact of climate change.

Bivalves of the genus *Mytilus* are frequently used as environmental indicators, as they are semi-sessile, have a relatively long life span, and are widely distributed in coastal regions in both Northern Hemisphere and Southern Hemisphere (Goldberg, 1986; Gosling, 2003; Rainbow, 1995; Thyrring, Juhl, Holmstrup, Blicher, & Sejr, 2015). *Mytilus* spp. are commercially and ecologically important species and often a dominant part of the intertidal and shallow subtidal fauna. Therefore, numerous studies of their responses to various stressors (e.g., temperature, salinity, pollutants) have been performed (Gosling, 2003; Jones, Lima, & Wethey, 2010; Mubiana, Qadah, Meys, & Blust, 2005; Søndergaard, Asmund, Johansen, & Riget, 2011; Wanamaker et al., 2007). Furthermore, *Mytilus* spp. have already demonstrated adaptations to different environments (Blicher, Sejr, & Høgslund, 2013; Thyrring, Rysgaard, et al., 2015) and a shift in their southern geographical range caused by increasing temperatures (Jones et al., 2010), making them an excellent model for inferring how species distributions might change in response to climate change. Additionally, *Mytilus* spp. have been the subjects of genetic studies for decades as the different species are morphologically difficult to distinguish. Consequently, the population structure of individual *Mytilus* species has been difficult to establish. *Mytilus edulis* L. 1758, *Mytilus trossulus* Gould 1850 and *Mytilus galloprovincialis* Lmk. 1819, all belong to the *M. edulis* species complex and are known to coexist and hybridize with conflicting patterns on the fitness for hybrids. Some studies did not observe any depressed fitness (Doherty, Brophy, & Gosling, 2009; Koehn, 1991; Riginos & Cunningham, 2005; Toro, Thompson, & Innes, 2006), while others (Gardner & Thompson, 2001; Toro, Innes, & Thompson, 2004; Toro, Thompson, & Innes, 2002; Tremblay & Landry, 2016) found a difference in fitness between parental types and hybrids and backcrosses. These findings and numerous studies on introgression between them (Fraïsse, Belkhir, Welch, & Bierne, 2016; Roux et al., 2014) have challenged the isolation species concept (White, 1978); however, they are generally considered to be different species, as they remain ecological distinct despite semipermeable barriers for gene flow and introgression (Bierne, Borsa, et al., 2003; Fraïsse, Roux, Welch, & Bierne, 2014; Saarman & Pogson, 2015). *Mytilus trossulus* is thought to have invaded the Arctic Ocean from the Pacific Ocean around 3.5 million years ago (mya) through the Bering Strait (Rawson & Hilbish, 1995, 1998; Vermeij, 1991). As the Bering Strait closed during glacial periods, allopatric speciation resulted in the evolution of

M. edulis in the Atlantic. *Mytilus edulis* has since spread to large parts of the Atlantic and due to apparent low gene flow (at least for some loci); *M. edulis* populations on each side of the Atlantic are genetically distinct (Riginos & Henzler, 2008; Riginos, Hickerson, Henzler, & Cunningham, 2004). Speciation between *M. edulis* and *M. galloprovincialis* most likely occurred through allopatric isolation approximately 2.5 mya (Quesada, Gallagher, Skibinski, & Skibinski, 1998; Rawson & Hilbish, 1995, 1998) with secondary contact and introgression occurring around 0.7 mya (Roux et al., 2014). Between interglacial periods 46,000 and 20,000 years ago, *M. trossulus* reinvaded the Arctic Ocean (Rawson & Harper, 2009). From here, it invaded both sides of the Atlantic founding *M. trossulus*/*M. edulis* hybrid zones along North American and European coasts (Riginos & Cunningham, 2005).

The geographical distribution and genetic population structure of *Mytilus* spp. have been intensively studied in boreal and temperate regions (Bierne, Borsa, et al., 2003; Hilbish, Carson, Plante, Weaver, & Gilg, 2002; Sarver & Foltz, 1993; Väinölä & Strelkov, 2011; Westerbom, Kilpi, & Mustonen, 2002); however, little is known of their distribution and genetic population structure in the Arctic. In the subarctic and Arctic, *M. edulis* is considered the most abundant *Mytilus* species, and it has been recorded in Arctic regions of Russia, along the Norwegian coast, in Iceland and Greenland (Hummel, Colucci, Bogaards, & Strelkov, 2001; Riginos & Henzler, 2008; Sukhotin, Strelkov, Maximovich, & Hummel, 2007; Väinölä & Strelkov, 2011). In Greenland, *Mytilus* spp. populations are found all along the west coast, and southern populations from Tartoq and Narsarsuaq have been shown to be genetically distinct from European *M. edulis* displaying higher resemblance to Canadian and North American *M. edulis* populations (Riginos & Henzler, 2008). Few genetic analyses have been performed on *Mytilus* spp. in Greenland, and most studies have assumed these mussels to be *M. edulis* without genetic verification despite observations of variations in metabolic response to low temperatures between populations from NW and SW Greenland (Thyrring, Rysgaard, et al., 2015). Moreover, in 2004, subtidal *M. edulis* were discovered at the mouth of Isfjorden in Svalbard after 1,000 years of absence (Berge, Johnsen, Nilsen, Gulliksen, & Slagstad, 2005). These mussels were hypothesized to have been transported from Norway by the West Spitsbergen Current in 2002, but their origin has never been confirmed through genetic analysis. *Mytilus trossulus* is less common in Arctic waters. Väinölä and Strelkov (2011) found that *M. trossulus* had a scattered distribution in the White Sea and the Norwegian Sea, and Feder, Norton, and Geller (2003) found live *M. trossulus* in Arctic Alaska in the 1990s. Furthermore, Wenne, Bach, Zbawicka, Strand, and McDonald (2016) has recently reported a NW Greenlandic fjord at Maarmorilik (71°N) to be inhabited by *M. edulis*, *M. trossulus*, and their hybrids. *Mytilus edulis* and *M. trossulus* hybrid zones have also been found and studied on the European and N American Atlantic coasts. Riginos and Cunningham (2005) reviewed the literature on the subject to look at local adaptation and species segregation and found conflicting patterns of species segregation across the Atlantic. In the western Atlantic, *M. trossulus* was found on wave-exposed open coasts, whereas *M. edulis* appeared to dominate in sheltered areas of low salinity. However, European *M. trossulus* populations from the Baltic Sea appeared to be locally adapted to the prevailing low salinities. The latter is in line with the findings of Wenne et al. (2016), who found a higher prevalence of *M. trossulus* in the inner Maarmorilik fjord compared with the more saline outer fjord. *Mytilus galloprovincialis* normally inhabits warmer waters, but in recent years the species and *M. galloprovincialis/M. edulis* hybrids have been observed along the Norwegian coast (Brooks & Farmen, 2013; Riginos & Henzler, 2008). This could be related to human activities like ship traffic in rural areas enabling faster invasion of waters otherwise not directly accessible to them (Anderson, Bilodeau, Gilg, & Hilbish, 2002; Geller, Carlton, & Powers, 1994). Furthermore, it has been demonstrated that *M. galloprovincialis* is capable of tolerating low water temperatures (Inoue et al., 1997), highlighting the potential for this species to occur in the Arctic.

Most studies on *Mytilus* spp. have focused on a few allozymes, mtDNA markers, or microsatellites (Bierne, Daguin, Bonhomme, David, & Borsa, 2003; Brooks & Farmen, 2013; Feder et al., 2003; McDonald, Seed, & Koehn, 1991; Ouagajjou, Presa, Astorga, & Pérez, 2011; Presa, Perez, & Diz, 2002). However, in recent years the use of single nucleotide polymorphisms, SNPs, has become increasingly popular to answer questions about *Mytilus* spp. status, population structure, hybridization, and adaptive variation (Helyar et al., 2011; Saarman & Pogson, 2015; Zbawicka, Drywa, Smietanka, & Wenne, 2012; Zbawicka, Sanko, Strand, & Wenne, 2014).

Utilizing 81 nuclear SNPs, we examined the distribution of *Mytilus* spp. in subarctic and Arctic regions ranging from the eastern Baffin Bay to the Pechora Sea with a special emphasis on the spatiotemporal population structure of *M. edulis*. We further aimed at identifying the source population or populations for the newly discovered *M. edulis* population in Svalbard and whether the observed differences in temperature response found in W Greenland *Mytilus* populations (Thyrring, Rysgaard, et al., 2015) could be caused by genetically based local adaptation.

2 | MATERIALS AND METHODS

2.1 | Study sites and sampling

Nineteen *Mytilus* spp. samples, consisting of 509 individuals in total, were collected from thirteen subarctic and Arctic sites (Table 1 and Fig. 1). Our primary focus was to assure broad geographical coverage and to sample regions where specific hypothesis regarding origin had been generated. The aim was to collect between 30 and 50 individuals from each site. However, as sampling Arctic regions is associated with high logistical costs, we had to rely on already available samples and numbers in some instances. Three samples were collected along the Norwegian NW coast at Tromsø (TRS and TRL) and Lofoten (LOF), and four samples were collected from the Svalbard archipelago (SV1, SV2, SV3, and SV4). Four samples were obtained from the Russian Arctic: two from the White Sea (WS1, WS2) and two from the southeast part of the Barents Sea: Pechora Sea to the east (PSE) and to the west (PSW) of Dolgiy Island. Further, one sample was collected in Iceland: south of Reykjavik (SRI), and six samples were collected in western Greenland: Nuuk (NUS and NUL), Kobbefjord (KOB), Upernavik (UPE),

and Qaanaaq (QAS and QAL). From Tromsø, Nuuk, and Qaanaaq, mussels of different size classes were collected as size can be used as a proxy for age class and hence indicate possible short-term genetic change over time; TRS, NUS, and QAS were smaller mussels in the size range 15–30 mm in length, while TRL, NUL, and QAL being mussels larger than 50 mm.

Samples were stored at −19°C (TRS, TRL, SRI, KOB, UPE, QAS, and QAL) or in 96% ethanol at 4°C (LOF, SV1, SV2, SV3, SV4, WS1, WS2, PSW, and PSE). Measurements of shell length, width, and height of frozen mussels were conducted at the laboratory facilities at DTU Aqua in Silkeborg, Denmark, whereas the measurements of ethanol preserved specimens were taken at the sampling locations.

Reference samples of *M. trossulus* and *M. galloprovincialis* were provided from the Sea of Okhotsk, Russia, and from around Galician Rías in NW Spain, respectively, to evaluate the species status of the sampled mussels and to identify potential hybrids.

2.2 | DNA extraction

A minimum of 30 mg (wet weight) of mantle tissue was dissected from each mussel, and DNA was extracted using the Omega EZNA Tissue DNA kit (Omega Bio-Tek, Norcross, GA, USA) according to the manufacturer's instructions for tissue. DNA content in the extracts was verified on a NanoDrop spectrophotometer (Thermo Scientific, Waltham, MA, USA).

2.3 | SNP genotyping

SNP genotyping was conducted using the Fluidigm Biomark™ HD System. 96.96 Dynamic Array IFCs were read on a real-time PCR system after amplification and scored using Fluidigm SNP Genotyping Analysis software. The samples were genotyped for a panel of 96 SNPs: 19 from previous publications on *Mytilus* spp. genetic structure (Zbawicka et al., 2012, 2014) and 77 new SNPs originated from RAD sequencing of genomic *M. edulis* DNA (EBI Sequence Read Archive (SRA) study ERP006912) at the University of Stirling (Table S1).

2.4 | Summary statistics

Loci with more than 25% missing data across all samples were discarded. Genepop 4.2 (Raymond & Rousset, 1995; Rousset, 2008) was used to test each locus in each sample for departure from Hardy–Weinberg equilibrium (HWE) and linkage disequilibrium (LD) for each locus pair in each sample (10,000 dememorizations, 100 batches, and 5,000 iterations). Within samples, the program diveRsity (Keenan, McGinnity, Cross, Crozier, & Prodohl, 2013) was used to calculate allelic richness and estimate expected (H_e) and observed (H_o) heterozygosities. This was done for both the full data set and for a reduced data set consisting exclusively of inferred *M. edulis* individuals (see explanation in section on *M. edulis* population structure below). Overall and pairwise F_{ST} values for all samples were estimated in Genepop 4.2. This initial sorting and discarding of SNP loci resulted

TABLE 1 Summary of sample information including sampling location and year, sampling code indicating location and possibly size, estimates of expected (H_e), and observed (H_o) heterozygosities and the allelic richness

Country	Location	Latitude	Longitude	Sampling year	Code	N	Habitat type	H_e	H_o	Allelic richness
Greenland	Qaanaaq, 15–30 mm	77.4650	−69.2403	2014	QAS	30	Intertidal zone	0.09	0.06	1.40
	Qaanaaq, >50 mm	77.4650	−69.2403	2014	QAL	30	Intertidal zone	0.07	0.04	1.31
	Upernavik	72.7939	−56.1028	2014	UPE	43	Intertidal zone	0.25	0.14	1.73
	Nuuk, 15–30 mm	64.1968	−51.7104	2014	NUS	30	Intertidal zone	0.25	0.23	1.64
	Nuuk, >50 mm			2014	NUL	24	Intertidal zone	0.24	0.21	1.69
	Kobbefjord	64.1367	−51.3909	2014	KOB	30	Intertidal zone	0.25	0.22	1.65
Iceland	Iceland	62.0261	−22.1594	2014	ICE	45	Intertidal zone	0.27	0.27	1.72
Norway	Lofoten	68.3380	13.8780	2014	LOF	45	Intertidal zone	0.26	0.24	1.76
	Tromsø, 15–30 mm	69.8278	18.9226	2014	TRS	10	Intertidal zone	0.27	0.29	1.76
	Tromsø, >50 mm	69.8278	18.9226	2014	TRL	30	Subtidal zone	0.28	0.23	1.64
	Svalbard									
	Kongsfjorden	79.1123	11.1362	2012	SV1	21	Subtidal zone	0.26	0.23	1.73
	Kongsfjorden	79.1123	11.1362	2013	SV2	13	Subtidal zone	0.29	0.24	1.79
	Kongsfjorden	79.1123	11.1362	2014	SV3	13	Subtidal zone	0.26	0.25	1.70
	Adventfjorden	78.2381	15.6026	2014	SV4	10	Subtidal zone	0.29	0.24	1.78
Russia	Pechora Sea, west of Dolgiy Island	69.3563	58.8393	2014	PSW	18	Subtidal zone	0.29	0.29	1.75
	Pechora Sea, east of Dolgiy Island	69.3204	58.7566	2014	PSE	27	Subtidal zone	0.29	0.29	1.77
	White Sea, Kandalaksha Bay	66.3372	33.6494	2014	WS1	45	Subtidal zone	0.32	0.24	1.86
	White Sea, Onega Bay	64.2079	36.6187	2014	WS2	45	Intertidal zone	0.29	0.27	1.77
	Magadan, Okhotsk Sea[a]			2007	MTR	15		0.11	0.01	1.00
Spain	Galician Rías[a]			2009	MGA	10		0.15	0.11	1.27

[a]These samples were only used for reference i.e. a representative of *Mytilus trossulus* and *Mytilus galloprovincialis*.

in 81 loci being retained for further analyses. As most SNP loci were developed from *M. edulis*, reliable scoring of *M. trossulus* individuals was not possible for three loci (174302_A, 67577_A, and 31051_A), and analyses concerning hybrid identification were performed for 78 SNP loci only.

2.5 | Identification of hybrids

Based on the generated pairwise F_{ST} estimates, the grouping of samples was visualized in a multidimensional scaling plot applying the cmdscale function in R (R Core Team, 2015). Additionally, a principal component analysis scatter plot was created in the R package Adegenet (Jombart, 2008; Jombart & Ahmed, 2011) to illustrate the genetic relationships among individuals across all samples. Structure v2.3.4, utilizing the Bayesian MCMC clustering approach (Pritchard, Stephens, & Donnelly, 2000) was used to visualize species integrity and identify possible hybridization among *Mytilus* spp. using a variable number of predefined clusters (*K*) for grouping individuals. This was also done to positively identify *M. edulis* individuals and subsequently create a reduced data set exclusively aimed at investigating population structure within this species. Considering the close genetic resemblance of *Mytilus* spp. and the assumed low gene flow between geographically distant samples (Riginos & Henzler, 2008), simulations were run for a number of predefined *K* values. Based on

an initial analysis of *K* up to 18, we found the highest likelihoods for *K* = 3–5. Accordingly, we used this as the basis to identify the major groupings within the species complex. For all simulations, a burn-in of 10,000 iterations was used followed by 100,000 MCMC repetitions. To evaluate the power of designating individuals as pure or hybrids, we followed the procedure described in Nielsen, Hansen, Ruzzante, Meldrup, and Grønkjær (2003) using the program Hybridlab (Nielsen, Bach, & Kotlicki, 2006). Briefly, we simulated 1,000 individuals of each of the following classes: parentals, F1/F2, and backcrosses. This was done based on the allele frequencies of the reference samples of *M. trossulus*, *M. galloprovincialis*, and *M. edulis* samples identified by initial Structure runs to likely consist exclusively of *M. edulis* individuals (NUS, KOB and WS2). Separate simulations were conducted for *M. edulis* samples from Greenland (NUS, KOB) and the Eastern Atlantic (WS2). The simulated and real individuals were included in a common Structure run (*K* = 4) and 95% confidence intervals for the inferred ancestry of the simulated individuals were recorded and compared to the real individuals.

2.6 | Population structure of *Mytilus edulis*

A reduced data set was used to assess the population structure in *M. edulis*. Based on the results from the analysis of simulated parentals and hybrids (see results section), we chose to only include individuals

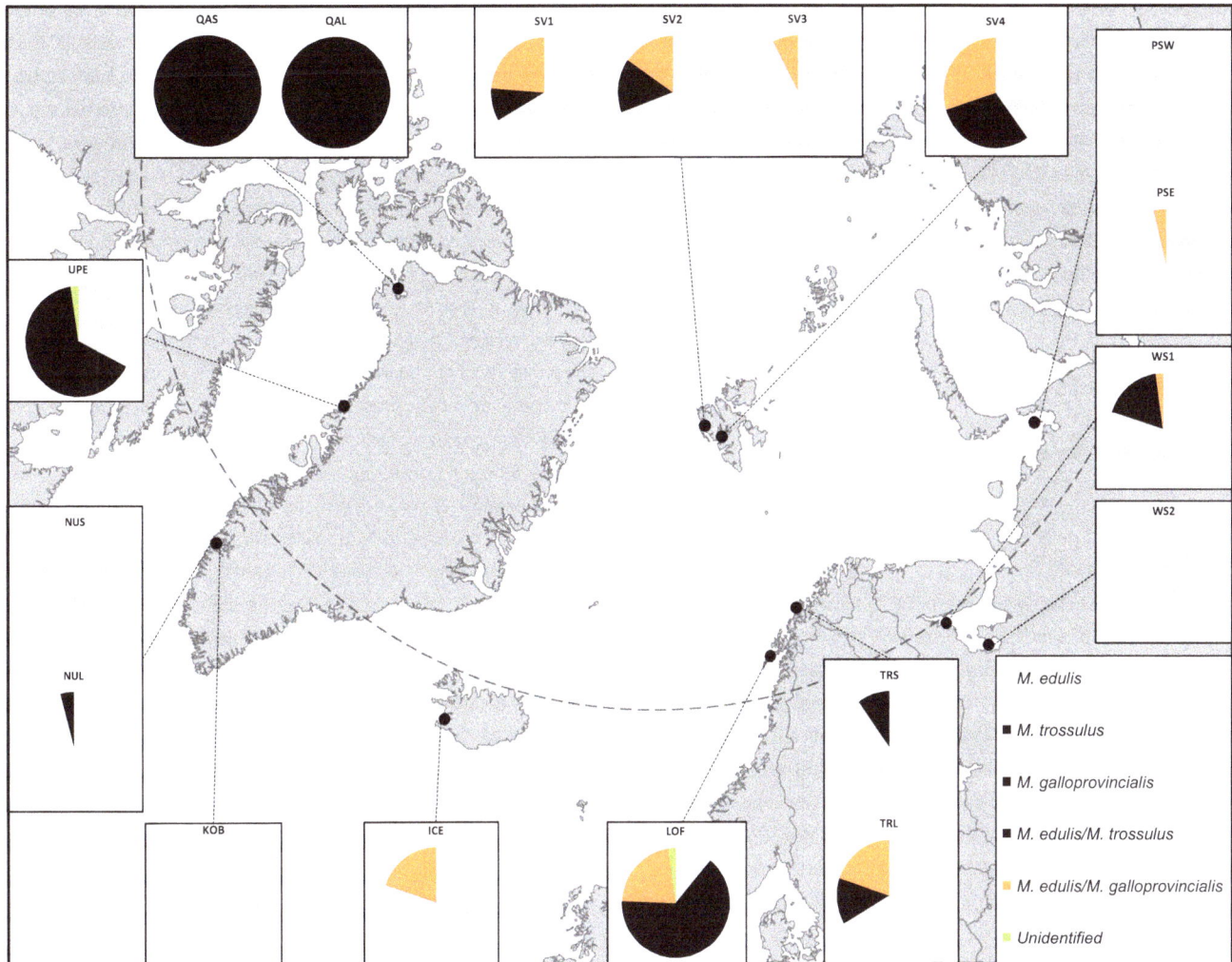

FIGURE 1 Map showing the different sampling locations and the proportion of three different *Mytilus* species and inferred hybrids at each location. For explanation of sample identification codes, see Table 1. Unidentified individuals denote apparent hybrids of all three *Mytilus* spp.

with admixture proportions below 0.2, as estimated by Structure. This was done to avoid extensive influence of hybridization on estimates of population divergence, but at the same time allowing for statistical uncertainty regarding whether individuals were pure *M. edulis* individuals or not. No significant differentiation was found between sampled mussels of different size classes from the same location (Tromsø, Nuuk, and Qaanaaq). Consequently, they were pooled prior to downstream analyses of population structure. Pairwise F_{ST} estimates were generated with Genepop 4.2, while Structure v2.3.4 was used to estimate the most likely number of genetic clusters. A burn-in period of 50,000 iterations was chosen followed by 100,000 MCMC repetitions for K values 2–4. A hierarchical AMOVA was conducted in Arlequin v.3.5.2.2 (Excoffier & Lischer, 2010) to infer the proportion of genetic variance distributed among the different *M. edulis* clusters and among samples within the clusters detected by Structure (see results section). To visualize the genetically based grouping of *M. edulis* population samples, a multidimensional scaling plot was generated, while a principal component analysis, PCA, was used to illustrate the relationships among *M. edulis* individuals in general and specifically for the Norwegian, Svalbard, and Russian samples to infer the likely origin

of Svalbard mussels. The PCA scatter plots were generated in R, using the cmdscale function and the package Adegenet.

2.7 | Outlier analysis

To identify loci potentially under selection in the "*Mytilus edulis*" data set, the joint distribution of F_{ST} and heterozygosity under a hierarchical island model of population structure was examined using Arlequin v.3.5.2.2 (Excoffier & Lischer, 2010) based on the method in Excoffier, Hofer, and Foll (2009). Accounting for the hierarchical population structure reduces the probability of false discoveries (Excoffier et al., 2009). Samples were grouped according to the genetic clustering analyses: (i) Greenlandic samples, (ii) Samples from Norway, the Svalbard archipelago, and Russian waters, and (iii) the Icelandic sample (see the section under Results subsection *Population structure of M. edulis*). The analytical settings for generating 95% and 99% confidence intervals were 20,000 simulations, 100 demes per group, and 10 groups. Loci outside the 95% quantile were considered possible subjects to selection, as these loci deviate more than could be expected under a model of neutral population structure. From this analysis, an exclusive "outlier" data set and a "neutral" data set were created to test the importance of outlier loci for defining the inferred population structure of *M. edulis*; that is, the true connectivity among populations based on neutral processes (drift and migration) could be obscured by loci under divergent selection. For both data sets, overall and pairwise F_{ST} estimates were generated in Genepop 4.2, while Structure v2.3.4 with $K = 2$ (using settings as above) was used to investigate whether the population structure found in *M. edulis* based on all loci could be identified from both the "outlier" and "neutral" data sets, or whether they displayed contrasting patterns.

3 | RESULTS

3.1 | Summary statistics

Three loci (159069_A, 171383_A, and 170478_A) deviated significantly from HWE in ten samples or more, and they were discarded from further analyses.

In total, 73,206 pairwise tests for LD between loci within samples were performed of which 3,194 tests were significant (4.36%). On average, 154 of 3,485 tests were significant within samples (range 0–986 significant tests). Only three SNP pairs displayed significant LD in more than five samples: 137120_A x BM8E (significant in six samples), 100078_A x 40154_A (significant in eight samples), and 175018_A x 241544_A (significant in 12 samples). Subsequently, one locus from each of the coupled SNP pairs was discarded (BM8E, 241544_A and 40154_A) to eliminate effects of linkage on downstream analyses.

Allelic richness ranged from 1.00 to 1.86 (Table 1), with the lowest values in the Qaanaaq samples (QAS and QAL). The levels of H_e and H_o (Table 1) ranged from 0.14 to 0.32 in all samples except for QAS and QAL, which had particularly low values ranging from 0.04 to 0.09. In all samples, H_o was close to H_e except for the Upernavik sample (UPE), where H_e and H_o was 0.25 and 0.14, respectively, and one of the White Sea samples (WS1) with H_e and H_o of 0.32 and 0.24. In the reduced data set, with samples consisting only of inferred *M. edulis*, individuals provided estimates of allelic richness between 1.57 and 1.67 (Appendix 1) and H_e/H_o values ranging between 0.22 and 0.30.

The overall F_{ST} across samples was 0.273. The pairwise F_{ST} values ranged between 0 and 0.860 (Table S2) with the highest pairwise F_{ST} value between the *M. trossulus* and *M. galloprovincialis* reference samples. Further, high values were found between *M. galloprovincialis* and the N Greenland samples from Qaanaaq; QAS and QAL (0.738 and 0.774) and between *M. trossulus* and all other samples except the three N Greenland samples (QAS, QAL, and UPE).

3.2 | Identification of hybrids

The multidimensional scaling plot (Fig. 2A) visualizes the genetic differentiation among all samples including the reference samples for *M. trossulus* (MTR) and *M. galloprovincialis* (MGA). The majority of samples clustered together in a "*M. edulis*" cluster. However, the QAS and QAL samples clustered with the *M. trossulus* reference sample, while samples UPE, WS1, and Lofoten (LOF) were located between the

three main "species" clusters. UPE and WS1 appeared to be distributed between the *M. trossulus* and *M. edulis* clusters, while LOF was situated between the "*M. edulis*" and "*M. galloprovincialis*" clusters. The clustering of QAS and QAL with MTR, and the inferred separation of UPE, WS1, and LOF from the "*M. edulis*" cluster were further supported by the principal component analysis scatter plot of individual genotypes (Fig. 3A). Most individuals clustered together as a "*M. edulis*" cluster, except for individuals from the UPE and WS1 samples, which appeared to contain individuals distributed between the "*M. trossulus*" and "*M. edulis*" clusters, suggesting that these individuals may be hybrids. The *M. galloprovincialis* reference sample clustered in the proximity of the "*M. edulis*" samples in the multidimensional scaling plot; however, a clear separation between the *M. edulis* and *M. galloprovincialis* samples was still apparent (Fig. 2A).

The Structure clustering analysis for $K = 4$ (Fig. 4, for $K = 3$ and 5 see Appendix 2) showed that the clusters make biologically sense as they corresponded to *M. trossulus*, *M. galloprovincialis*, Greenlandic *M. edulis*, and other *M. edulis*. This configuration also allowed the identification of *M. edulis*/*M. galloprovincialis* or *M. edulis*/*M. trossulus* hybrids. This was supported by the Structure analysis including the simulated parentals and hybrids, which showed relatively narrow 95% confidence intervals for the simulated *M. edulis* parentals regardless of their geographical origin (0.89–0.99 *M. edulis* ancestry for Greenland and 0.92–0.99 for the other *M. edulis*, see Appendix 3). Likewise, the simulated *M. trossulus*, *M. galloprovincialis* parentals suggested very high power for identifying *M. edulis*/*M. galloprovincialis* or *M. edulis*/*M. trossulus* hybrids. However, as only two relatively small samples of *M. trossulus* and *M. galloprovincialis* provided the foundation for the simulations, we conservatively chose an admixture level of 20% as the cutoff point between *Mytilus* spp. parentals and *M. edulis*/*M. galloprovincialis* or *M. edulis*/*M. trossulus* hybrids. This was done in order to allow for uncertainty caused by population structure and missing genotypes within the samples of real individuals. This approach enabled the construction of an exclusive "*Mytilus edulis*" data set. When using $K = 3$ the analysis was unable to split the samples into the three a priori defined species groups (Appendix 2B).

The inferred proportion (using the 20% criterion) of the different *Mytilus* species and hybrids in each of the geographical samples (Fig. 1) show that *M. edulis* is the most common species within the sampled subarctic and Arctic populations, where pure *M. edulis* specimens constitute approximately 66% of all sampled individuals. Pure *M. edulis* were present in all samples except for QAS and QAL, which were mainly *M. trossulus* (87%–90%), with few individuals (3–4) showing evidence of *M. edulis* hybridization. Only two samples, UPE and WS1, contained both pure *M. edulis* and *M. trossulus* individuals. The UPE sample contained approximately 51% *M. trossulus* and 33% *M. edulis* and 14% *M. edulis*/*M. trossulus* hybrids, while the WS1 sample was comprised of 9% *M. trossulus*, 80% *M. edulis*, 9% *M. edulis*/*M. trossulus* hybrids, and 2% *M. edulis*/*M. galloprovincialis* hybrids. The distribution of *M. galloprovincialis* individuals is mainly restricted to samples from the Norwegian coast and the Svalbard archipelago (34 and 4 individuals, respectively). A single apparent *M. galloprovincialis* individual was found in the sample of large mussels from Nuuk (NUL).

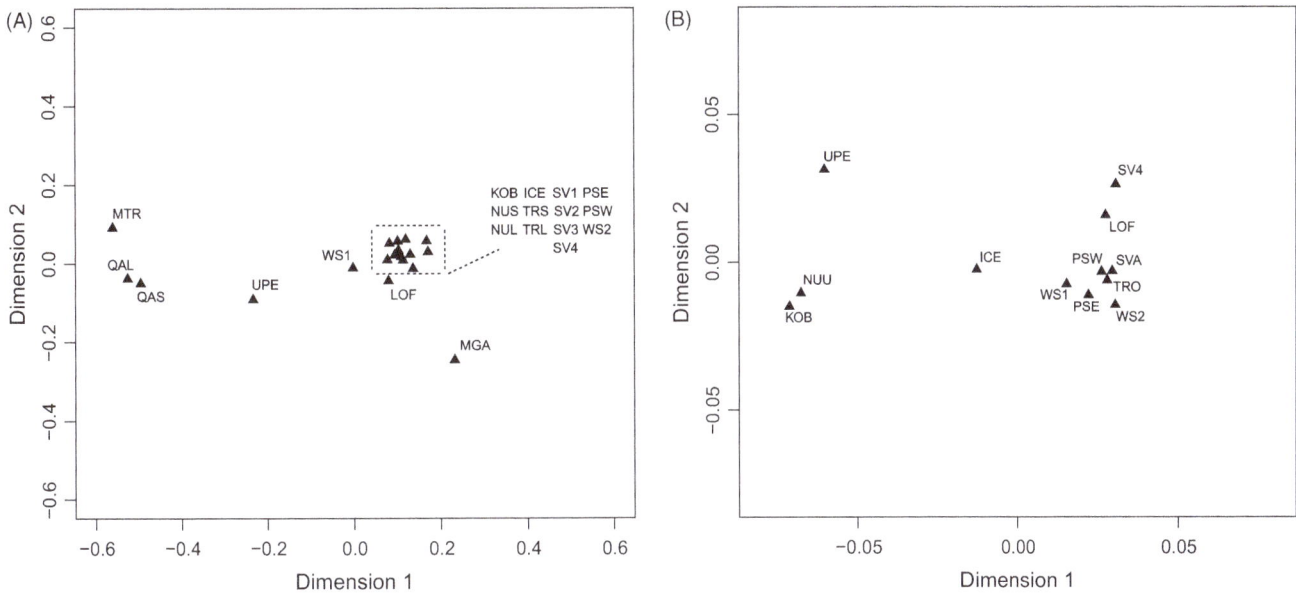

FIGURE 2 Multidimensional scaling plot of (A) all samples and (B) designated *Mytilus edulis* samples based on pairwise genetic distances among samples. For explanation of sample identification codes, see Table 1. Further codes: NUU comprise of NUS and NUL, TRO comprise of TRS and TRL, and SVA comprise of SV1, SV2, and SV3

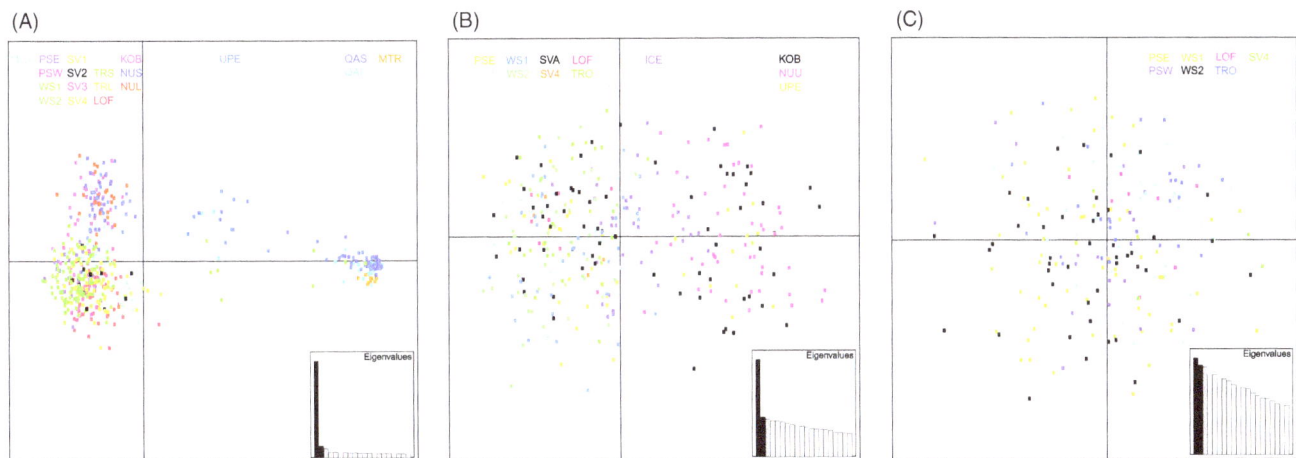

FIGURE 3 Principal component scatter plot of individual genotypes for (A) all samples (B) *Mytilus edulis* samples, and (C) Norwegian, Svalbard, and Russian samples. For explanation of sample identification codes, see Table 1. Further codes: NUU comprise of NUS and NUL, TRO comprise of TRS and TRL, and SVA comprise of SV1, SV2, and SV3

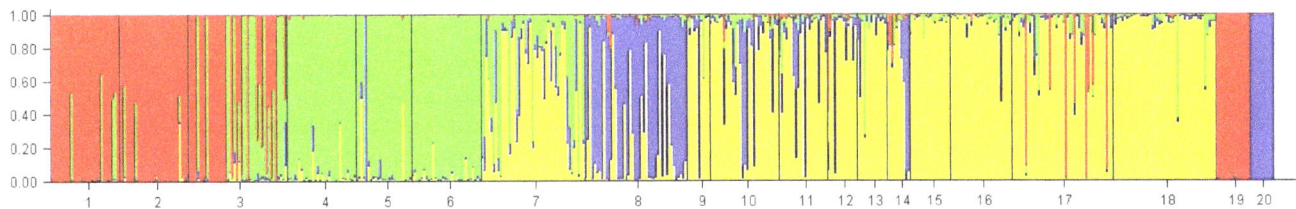

FIGURE 4 Results of Structure clustering analyses for the full data set for $K = 4$. Samples are 1: QAS, 2: QAL, 3: UPE, 4: NUS, 5: NUL, 6: KOB, 7: ICE, 8: LOF, 9: TRS, 10: TRL, 11: SV1, 12: SV2, 13: SV3, 14: SV4, 15: PSW, 16: PSE, 17: WS1, 18: WS2, 19: MTR, and 20: MGA. For explanation of sample identification codes, see Table 1

The LOF sample contained the highest number of *M. galloprovincialis* observed—64% and further 11% *M. edulis* and 22% *M. edulis/M. galloprovincialis* hybrids. In cases where more than a few hybrids were found, the distribution of admixture estimates of real individuals was compared to the simulated hybrids. In all cases, different classes of hybrids (F1, F2, and backcrosses) were suggested. However, as explained above the comparison of real and simulated individuals should be interpreted with caution.

3.3 | Population structure of *Mytilus edulis*

The overall F_{ST} for all samples identified as *M. edulis* was 0.048. Pairwise F_{ST} values ranged from 0 to 0.113 with the highest values between the Greenlandic samples and the Norwegian, Svalbard, and Russian samples (Table S3). The lowest F_{ST} values were found between geographically proximate samples such as the two White Sea samples (WS1 and WS2) and the two sampling sites in Svalbard (SVA and SV4). For sites with samples of different size classes, F_{ST} estimates ranged between 0.001 for the Nuuk samples (NUS and NUL) and 0.018 for the Tromsø samples (TRS and TRL) (Supplementary Table S3). The low F_{ST} for Nuuk samples indicates short-term temporal stability of genetic population structure. The higher F_{ST} estimate for Tromsø mussels was not significant, thus allowing the pooling of size classes for the downstream analyses.

The cluster analysis of the "*Mytilus edulis*" data set (K = 2–4) showed a clear clustering of samples, essentially separating the Greenlandic samples from the other samples (Fig. 5). The likelihood of K = 2 was highest splitting the *M. edulis* samples into two groups; the Greenlandic samples versus the Norwegian, Svalbard, and Russian samples and identifying the Icelandic sample a mixture of eastern and western Atlantic gene pools (Fig. 5A). The plots for K = 3 and K = 4 added no additional biologically sensible information.

The hierarchical AMOVA for the three groups (Greenlandic, Icelandic, and Norwegian/Svalbard/Russian) provided an estimated variance of 5.78% among groups and 0.43% among samples within groups. The multidimensional scaling plot of population samples (Fig. 2B) and the principal component analysis scatter plot of individual genotypes (Fig. 3B) further supported the population structure of *M. edulis* inferred by Structure with three groups: (i) Greenlandic samples, (ii) Norwegian, Svalbard, and Russian samples, and (iii) the Icelandic sample found between the two main clusters inferred by axis 1.

The principal component analysis scatter plot including only Norwegian, Svalbard, and Russian samples (Fig. 3C) did not provide a clear separation of individuals as these individuals were scattered with no apparent pattern.

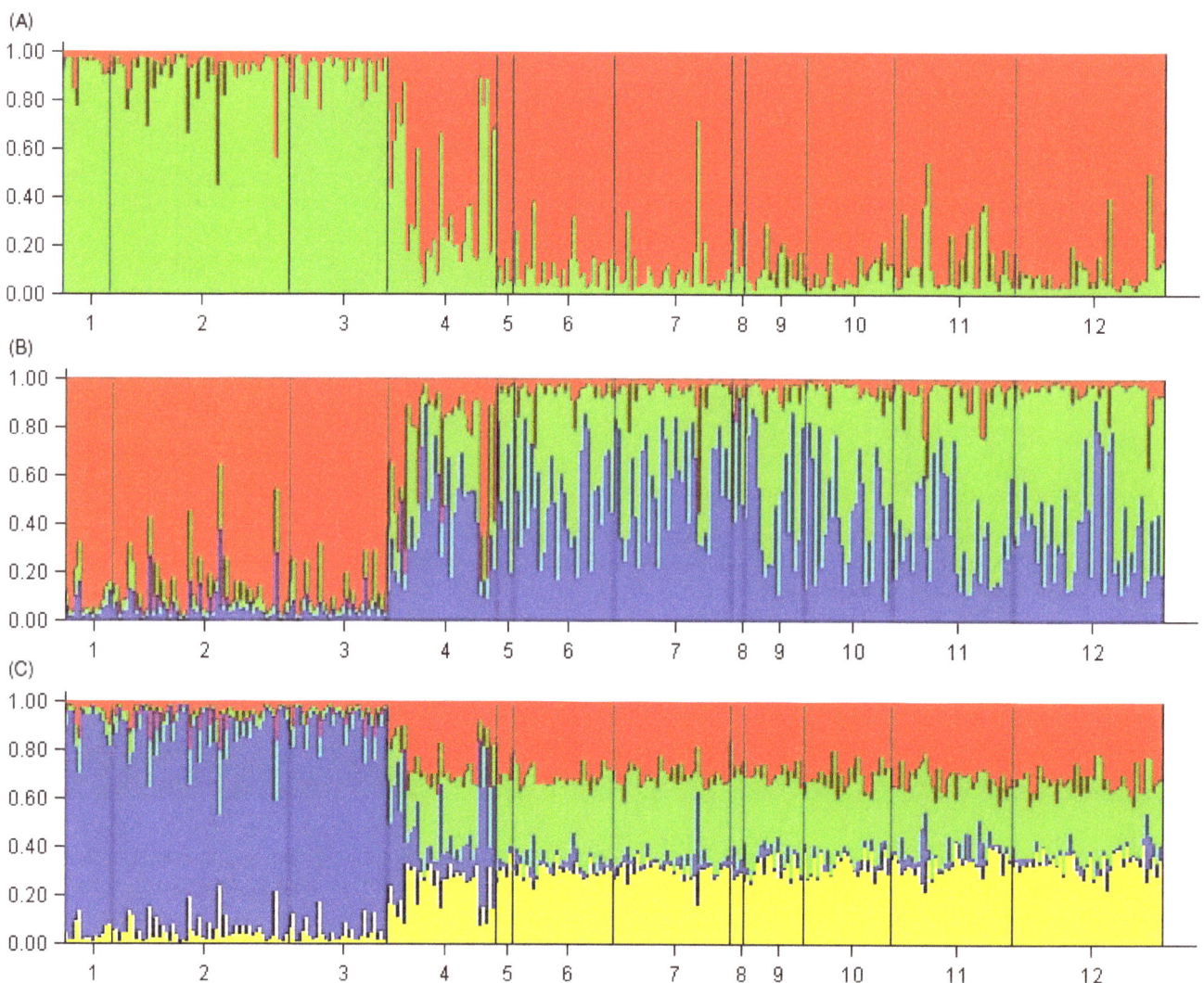

FIGURE 5   Results from clustering analyses of the "*Mytilus edulis*" data set with: (A) K = 2, (B) K = 3, and (C) K = 4. Samples are as follows: 1: UPE, 2: NUU comprising of NUS and NUL, 3: KOB, 4: ICE, 5: LOF, 6: TRO comprising of TRS and TRL, 7: SVA comprising of SV1, SV2, and SV3, 8: SV4, 9: PSW, 10: PSE, 11: WS1, and 12: WS2. For explanation of sample identification codes, see Table 1

3.4 | Outlier analysis

The outlier tests identified six loci as F_{ST} outliers, with six loci significant at the 5% level and three at the 1% level. All of these outliers are high F_{ST} outliers (Fig. 6) indicating diversifying selection (Beaumont & Nichols, 1996), although a few of them could represent the upper tail of the neutral F_{ST} distribution. Also, a strong genetic cline as observed here is known to sometimes overestimate the number of loci under diversifying selection (Strand, Williams, Oleksiak, & Sotka, 2012). Furthermore, introgression between Mytilus spp. has been found to cause high F_{ST} outliers (Gosset and Bierne 2012). Pairwise F_{ST} values ranged from 0 to 0.059 for the "neutral" data set and from 0 to 0.474 for the "outlier" data set (Tables S4 and S5). The Structure analyses for both the "neutral" and "outlier" data set also supported the initial population structure separating Greenlandic samples from the Norwegian, Svalbard, and Russian samples and with the Icelandic sample of admixed origin (Appendices 4 and 5).

4 | DISCUSSION

4.1 | Distribution of Mytilus spp. in the Arctic

Baseline information of species distribution and their genetic composition is imperative in order to quantify the impacts of climate change on species distribution ranges, biodiversity, and the effects of hybridization between species and populations (Gardner, Zbawicka, Westfall, & Wenne, 2016). Molecular genetic knowledge is a key

measure to identify the distribution of invasive congener species (Geller, Darling, & Carlton, 2010), which may cause cascading ecosystem effects. Despite congener species appearing morphologically similar, interspecific variation in ecology and physiology may impact population fitness (Fly & Hilbish, 2013; Fraïsse et al., 2016; Somero, 2005). In the Arctic, baselines studies on genetic variation and species abundance are largely absent but urgently needed (Bluhm et al., 2011; Wassmann, Duarte, Agusti, & Sejr, 2011). Pioneer work should therefore focus on keystone model species (such as Mytilus), because of their disproportionally large effect on their environment.

Mytilus spp. were found pan-Arctic (although only one individual of M. galloprovincialis was identified in Greenland). Generally, M. edulis was the most common species making up approximately 66% of all sampled individuals. The biogeographic structures of the three Mytilus spp. reflect the major current systems of the region. Pure populations were mainly found in regions (such as W Greenland and the Pechora Sea) with a lower influence of Pacific and Atlantic water, than other sampling sites. Northwards currents from boreal waters facilitate larvae dispersal from southern populations (Berge et al., 2005; Renaud, Sejr, Bluhm, Sirenko, & Ellingsen, 2015). For instance, the northward flowing current regimes (such as the Norwegian Current) allows non-Arctic species to extend their range into the Arctic from the Atlantic or Pacific Ocean (Bluhm et al., 2011; Fetzer & Arntz, 2008; Sirenko & Gagaev, 2007). Ocean currents also explain why Mytilus spp. remain absent in NE Greenland despite the environmental resemblance of NW Greenland with regard to temperatures and ice conditions (Sejr, Blicher, & Rysgaard, 2009). In general, the NE Greenland shelf

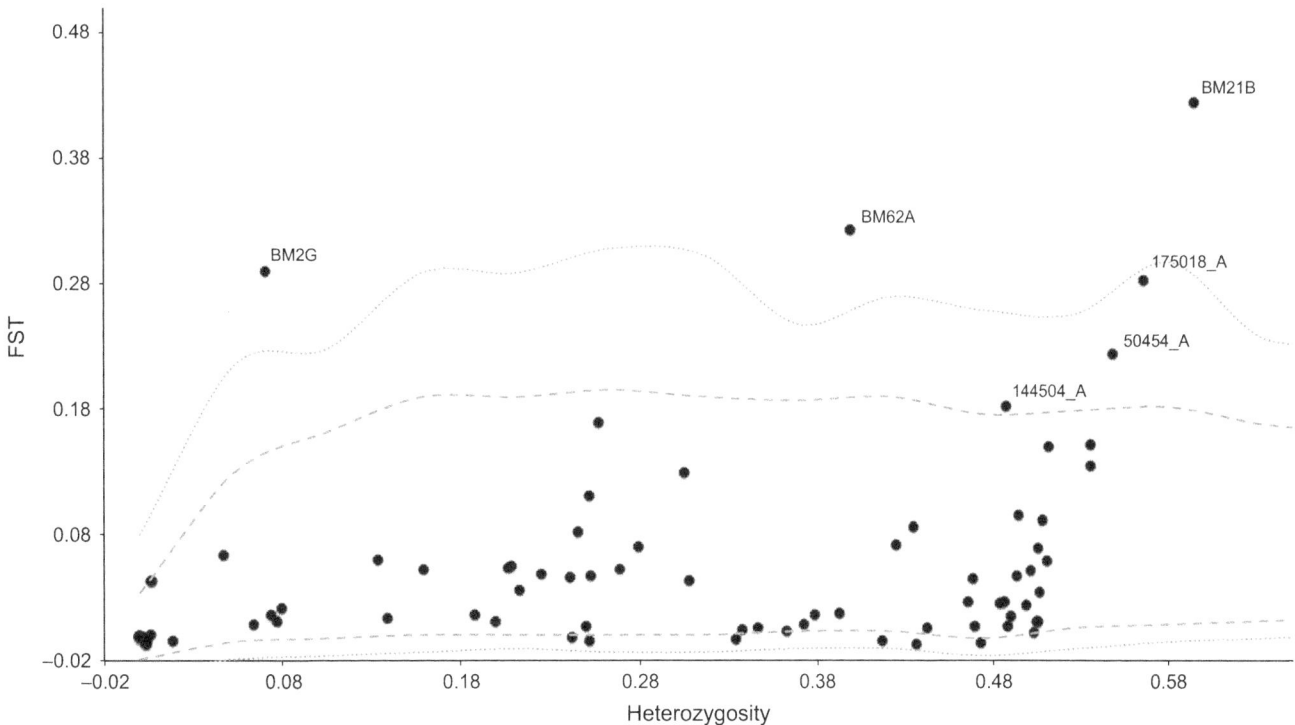

FIGURE 6 F_{ST} outlier analyses in Arlequin v3.5.1.3 utilizing the hierarchical island model. *Black solid dots* denote loci, and *gray dashed* and *dotted lines* indicate 95% and 99% confidence intervals, respectively. Loci outside the 95% and 99% confidence intervals are suggested to be under selection

is considered biogeographically different from the rest of Greenland (Piepenburg et al., 2011). The absence of *Mytilus* mussels in NE Greenland is likely a result of dispersal barriers due to the lack of an downstream source population, as the East Greenland Current flows from north to south, exemplifying how outflow shelves may respond slowly to climatic changes (Renaud et al., 2015). This is further supported by the presence of *Mytilus* mussels in SE Greenland, at Tasiilaq (Ammassalik, 65°N) (Ockelmann, 1958), which is influenced by a branch of the Irminger Current from the Atlantic Ocean.

The present study highlights the need for further genetic studies in the region as a *M. trossulus* population was found in the most northern sampled region of NW Greenland (77°N) with *M. edulis* populations residing in SW Greenland (64°N). This discovery was unexpected, as a seemingly established *M. trossulus* population has not been found in the high Arctic prior to this study. Several possible mechanisms could explain the presence of *M. trossulus* in Qaanaaq and Upernavik. First, these populations could have survived in a refugium near NW Greenland during the last glacial period. Glacial refugia are known from North Atlantic temperate regions and evidence suggests that *M. edulis* may have survived north of the ice margin (Maggs et al., 2008; Riginos & Henzler, 2008). Second, there could be a contemporary spread of *M. trossulus* from the Pacific Ocean. Jones et al. (2003) found that waters around NW Greenland contained high levels of phosphate indicating Pacific water being transported into this area. Also, there are a few reports of live *M. trossulus* in Arctic Alaska and Canada (Feder et al., 2003), so the spread of planktonic larvae from the Canadian Arctic could be possible. A third scenario could be that *M. trossulus* spread to Arctic Greenland from the East coast of Canada. However, as the West Greenland Current moves along the coast from south to north, and *Mytilus* mussels are expected to disperse with rather than against currents, this scenario seems unlikely (McQuaid & Phillips, 2000). Finally, *Mytilus* spp. are known to disperse by human activities and can survive long distances and fluctuating temperatures (Lee & Chown, 2007). Qaanaaq is situated less than 150 km from the US Thule Air Base, which receives supplies by US ships; this is providing an alternative dispersal route of *M. trossulus* from the north Pacific.

The invasive *M. galloprovincialis* appeared widespread from Greenland to the Pechora Sea. In Norway, *M. galloprovincialis* appears common along the coastline (Brooks & Farmen, 2013), and the discovery of *M. galloprovincialis* in Svalbard suggests colonization by ocean currents as hypothesized by Berge et al. (2005) or ship traffic from the Norwegian mainland (Ware et al., 2014).

4.2 | *Mytilus* hybrid zones in the Arctic

Most sampling locations displayed varying degrees of hybridization and introgression between different *Mytilus* spp. and only four locations contained apparently pure populations (Fig. 1). Introgression can affect a population's fitness and vulnerability to climate change. In the study region, hybrid zones were found in Norway, Svalbard, and Greenland, with the highest abundance of the invasive *M. galloprovincialis* found along the Norwegian coast, especially in Lofoten (68°N) further supporting the findings by Brooks and Farmen (2013)

and Riginos and Henzler (2008). Additionally, a surprisingly high amount of *M. galloprovincialis* was found at Svalbard. We also found evidence of limited introgression of *M. galloprovincialis* in the Russian and Icelandic samples, and the ecological consequences of invasive mussels in these regions need to be studied further. In the White Sea, *M. trossulus* individuals were only recorded in one of two locations. This small-scale regional variation in species composition was also observed by Väinölä and Strelkov (2011), who also found *M. trossulus* and *M. edulis*/*M. trossulus* hybrids but to a much lesser extent than *M. edulis*. It is believed that the expansion of *M. trossulus* in the White Sea is most likely facilitated by ships (Väinölä & Strelkov, 2011). This explains the fact that populations dominated by *M. trossulus* are confined to sites with harbors and seaports, while *M. edulis* inhabit all the coastline of the White Sea, where the substrates are appropriate. In the present study, the sample WS1 that contained *M. trossulus* and their hybrids were collected directly in the area of the White Sea Biological Station Kartesh, which has a regular ship connection with Chupa, a small town in Kandalaksha Bay. Recently, *M. trossulus* was found in the Chupa harbor (Katolikova, Khaitov, Vänölä, Gantsevich, & Strelkov, 2016), where ship traffic from the Barents Sea has been relatively intensive. In contrast, the WS2 site with pure *M. edulis* in the sample is located on an uninhabited island Kondostrov in the Onega Bay, which is far from the towns with intensive ship traffic.

4.3 | Population structure of *Mytilus edulis*

The genetic structure of the *M. edulis* populations in this study revealed a significant split between *M. edulis* samples from each side of the Atlantic, with Icelandic *M. edulis* appearing as an admixture of the two gene pools. This divergence of W and E Atlantic populations is in line with the findings of Riginos and Henzler (2008) and Waltari and Hickerson (2013), who suggested that *M. edulis* survived in a W Atlantic glacial refugium. Furthermore, Riginos et al. (2004) found low gene flow across the Atlantic, providing an explanation for the continuing divergence of *M. edulis* populations from W and E Atlantic coasts. These studies primarily looked at mitochondrial DNA, but their results are strongly supported by the SNP analysis presented here. This, however, contrasts to the meta-population analysis of polychaete and echinoderm populations in the Arctic showing high gene flow between populations (Hardy et al., 2011). This difference in gene flow patterns between different species with long planktonic larval stage further highlights the necessity of understanding the population structure within species to best conserve biodiversity in the Arctic.

In general, F_{ST} values between samples from Norway, Svalbard, and Russia and the Icelandic sample are lower than between the Icelandic sample and Greenlandic samples. Śmietanka, Burzyński, Hummel, and Wenne (2014) suggested a single glacial Atlantic refugium founding European *M. edulis*. However, our studied sample from Iceland suggests the population to consist of individuals of mixed ancestry. Further analyses of their origin/history could be elucidated by conducting additional analysis of samples from both sides of the Atlantic. Considering that the major North Atlantic Current reaches Iceland from the east, it is perhaps more likely that Iceland would be recruiting spat from East Atlantic populations. This is also inferred by Riginos and Henzler (2009), who found postcolonization gene flow from northern Europe to Iceland.

The outlier tests identified six loci as F_{ST} outliers at the 5% significance levels. All of these outliers are high F_{ST} outliers (Fig. 6) indicating diversifying selection (Beaumont & Nichols, 1996). However, a strong genetic cline as observed here is known to sometimes overestimate the number of loci under diversifying selection (Strand et al., 2012). Furthermore, introgression between Mytilus spp. have been found to cause high F_{ST} outliers (Gosset & Bierne, 2012), and this result should be interpreted with some caution. Still, we find that the pattern of population structure is the same for the "neutral" and the "outlier" data sets (Appendices 4 and 5), suggesting that patterns of neutral population structure is correlated with adaptive evolution in response to divergent local environmental conditions. Temperature influences the large-scale geographical distribution of species (Sunday, Bates, & Dulvy, 2011); however, on a local scale other factors including predation, the presence of sea ice, suitable habitats, water current, and salinity can influence the distribution of intertidal species (Høgslund, Sejr, Wiktor, Blicher, & Wegeberg, 2014; Kroeker et al., 2016; Paine, 1974), and these conditions are very different between W Greenland and the other sampling sites (Rayner et al., 2003). Still, the high divergence between samples from the Eastern Atlantic and Greenland cannot be explained alone by loci subject to selection. F_{ST} values for the "neutral" data set are still high (Table S4) suggesting a high degree of isolation between groups. This isolation in turn may have facilitated local adaptation at this rather large geographical scale. For more specific insights on the environmental factors responsible for local adaptation, the geographical scale, and its genomewide significance, a more elaborate sampling design is warranted including more regional samples and a higher degree of genomic coverage.

4.4 | Implications for conservation of marine species in the face of climate change

The effects of global warming increase the spread and associated threat of nonindigenous species across the globe (Gardner et al., 2016; Hellmann, Byers, Bierwagen, & Dukes, 2008; Saarman & Pogson, 2015). A study by Wisz et al. (2015) predicted that continued warming of the Arctic could open the Bering Strait and thus facilitate a Pacific–Arctic exchange of nonindigenous species, which could have adverse impact on Arctic biodiversity. Moreover, human activities are short-cutting natural dispersal barriers for nonindigenous species (Carlton & Geller, 1993), posing a global risk of spreading these to novel regions. In this regard, especially ship traffic facilitates dispersal (e.g. in ballast water and hull fouling; Chan, MacIsaac, & Bailey, 2015; Geller et al., 1994; Ware et al., 2014). Such intrusions of nonindigenous species into the Arctic have already occurred (e.g. Pacific king crabs Paralithodes camtschaticus and bluefin tuna Thunnus thynnus; CAFF 2013; MacKenzie, Payne, Boje, Hoyer, & Siegstad, 2014; Oug, Cochrane, Sundet, Norling, & Nilsson, 2011), and Saarman and Pogson (2015) found that the nonindigenous M. galloprovincialis pose an ecological threat to M. trossulus along the Californian coast as it

had displaced and continues to displace the native M. trossulus. The surprisingly broad distribution of M. galloprovincialis in the Arctic therefore highlights the benefit of using genetic tools and stresses the need for developing measures to detect and identify nonindigenous species and pathways of introduction, to understand and reduce the threat of invasive species in the Arctic.

Prior to the current investigation, multiple studies have assumed Mytilus mussels in the Arctic to be exclusively M. edulis (Berge et al., 2005; Hansen, Hanken, Nielsen, Nielsen, & Thomsen, 2011; Jensen, 1905; Strand & Asmund, 2003). The identification of three Mytilus spp. across the Arctic has implications for ecological and ecotoxicological research in the region. Mytilus mussels are extensively used in biological monitoring programs (Wenne et al., 2016). However, interspecific differences in physiology and responses to environmental pollutants have been reported (Brooks, Farmen, Heier, Blanco-Rayon, & Izagirre, 2015; Fly & Hilbish, 2013), and thus, the lack of genetic knowledge could seriously affect the conclusions of ongoing biological monitoring. We therefore emphasize the importance of applying genetic tools to document species status, when conducting ecological, ecotoxicological, and physiological studies.

Moreover, assuming that the distribution and genetic connectivity between regions observed in this study is to be a first approximation representative for benthic invertebrates in general, several important observations were made related to quantifying changes in species distribution in a warmer Arctic. A number of congener species exists, which display different responses to changes in temperature. The genetic connectivity and inferred gene flow are closely linked to major ocean currents, which means that predicting range changes purely based on future climate predictions without considering dispersal potential or barriers can be misleading. In fact, changes in ocean currents and thereby in supply of potential colonizers may be a more important driver of change than warming per se. This has previously been demonstrated by the species changes observed during the large northward expansion of Atlantic water in the Barents Sea and along the W Greenland coast in the 1930s (Drinkwater, 2006). Genetically isolated areas like outflow shelves without upstream source populations (such as, NE Greenland) appear to be especially vulnerable to human vectors (such as shipping) as the absence of several species here likely reflects lacking postglacial invasion rather than adverse climatic conditions. Finally, NW Greenland M. trossulus populations with an affinity to the Pacific suggest that exchange of species from the Pacific across the Arctic and into the Atlantic is taking place. However, all of these factors should be further validated through urgently needed studies documenting current distribution and genetic composition of marine species in the Arctic.

ACKNOWLEDGEMENTS

We wish to thank Dr. Risto Väinölä and Prof. Paulino Martinez Portela for providing mussel tissue for reference samples. Dorte Meldrup is acknowledged for laboratory support. Sampling in Nuuk was conducted as part of the Greenland Ecosystem Monitoring (GEM) program, and we also appreciate Snorri Gunnarson and Dr. Dorte

Krause-Jensen for collecting mussels. The project was funded by the 15th of June Foundation, the Program of Russian Academy of Sciences "Fundamental Research to the Development of Arctic", the Framcentre flagship Fjord and Coast project "Life at the Edge", and a Norwegian Research Council project (project nr 225044). This work is a contribution to the Arctic Science Partnership (ASP), asp-net.org.

REFERENCES

ACIA. (2004). *Impacts of a warming arctic: Arctic climate impact assessment.* ACIA Overview report (140 pp.). Cambridge: Cambridge University Press.

Anderson, A. S., Bilodeau, A. L., Gilg, M. R., & Hilbish, T. J. (2002). Routes of introduction of the Mediterranean mussel (*Mytilus galloprovincialis*) to Puget Sound and Hood Canal. *Journal of Shellfish Research, 21,* 75–79.

Beaumont, M. A., & Nichols, R. A. (1996). Evaluating loci for use in the genetic analysis of population structure. *Proceedings of the Royal Society of London B: Biological Sciences, 263,* 1619–1626.

Berge, J., Johnsen, G., Nilsen, F., Gulliksen, B., & Slagstad, D. (2005). Ocean temperature oscillations enable reappearance of blue mussels *Mytilus edulis* in Svalbard after a 1000 year absence. *Marine Ecology Progress Series, 303,* 167–175.

Bierne, N., Borsa, P., Daguin, C., Jollivet, D., Viard, F., Bonhomme, F., & David, P. (2003). Introgression patterns in the mosaic hybrid zone between *Mytilus edulis* and *M. galloprovincialis*. *Molecular Ecology, 12,* 447–461.

Bierne, N., Daguin, C., Bonhomme, F., David, P., & Borsa, P. (2003). Direct selection on allozymes is not required to explain heterogeneity among marker loci across a *Mytilus* hybrid zone. *Molecular Ecology, 12,* 2505–2510.

Blicher, M. E., Sejr, M. K., & Høgslund, S. (2013). Population structure of *Mytilus edulis* in the intertidal zone in a sub-Arctic fjord, SW Greenland. *Marine Ecology Progress Series, 487,* 89–100.

Bluhm, B. A., Gebruk, A. V., Gradinger, R., Hopcroft, R. R., Huettmann, F., Kosobokova, K. N., ... Weslawski, J. M. (2011). Arctic marine biodiversity: An update of species richness and examples of biodiversity change. *Oceanography, 24,* 232–248.

Brodersen, J., & Seehausen, O. (2014). Why evolutionary biologists should get seriously involved in ecological monitoring and applied biodiversity assessment programs. *Evolutionary Applications, 7,* 968–983.

Brooks, S. J., & Farmen, E. (2013). The distribution of the mussel *Mytilus* species along the Norwegian coast. *Journal of Shellfish Research, 32,* 265–270.

Brooks, S. J., Farmen, E., Heier, L. S., Blanco-Rayon, E., & Izagirre, U. (2015). Differences in copper bioaccumulation and biological responses in three *Mytilus* species. *Aquatic Toxicology, 160,* 1–12.

CAFF. 2013. *Arctic biodiversity assessment—Status and trends in Arctic biodiversity* (678 pp.). Akureyri, Iceland: Author.

Carlton, J. T., & Geller, J. B. (1993). Ecological roulette—The global transport of nonindigenous marine organisms. *Science, 261,* 78–82.

Chan, F. T., MacIsaac, H. J., & Bailey, S. A. (2015). Relative importance of vessel hull fouling and ballast water as transport vectors of nonindigenous species to the Canadian Arctic. *Canadian Journal of Fisheries and Aquatic Sciences, 72,* 1230–1242.

Doherty, S. D., Brophy, D., & Gosling, E. (2009). Synchronous reproduction may facilitate introgression in a hybrid mussel (*Mytilus*) population. *Journal of Experimental Marine Biology and Ecology, 378,* 1–7.

Drinkwater, K. F. (2006). The regime shift of the 1920s and 1930s in the North Atlantic. *Progress in Oceanography, 68,* 134–151.

Excoffier, L., Hofer, T., & Foll, M. (2009). Detecting loci under selection in a hierarchically structured population. *Heredity, 103,* 285–298.

Excoffier, L., & Lischer, H. E. L. (2010). Arlequin suite ver 3.5: A new series of programs to perform population genetics analyses under Linux and Windows. *Molecular Ecology Resources, 10,* 564–567.

Feder, H. M., Norton, D. W., & Geller, J. B. (2003). A review of apparent 20th century changes in the presence of mussels (*Mytilus trossulus*) and macroalgae in Arctic Alaska, and of historical and paleontological evidence used to relate mollusc distributions to climate change. *Arctic, 56,* 391–407.

Fetzer, I., & Arntz, W. E. (2008). Reproductive strategies of benthic invertebrates in the Kara Sea (Russian Arctic): Adaptation of reproduction modes to cold water. *Marine Ecology Progress Series, 356,* 189–202.

Fly, E. K., & Hilbish, T. J. (2013). Physiological energetics and biogeographic range limits of three congeneric mussel species. *Oecologia, 172,* 35–46.

Fraïsse, C., Belkhir, K., Welch, J. J., & Bierne, N. (2016). Local interspecies introgression is the main cause of extreme levels of intraspecific differentiation in mussels. *Molecular Ecology, 25,* 269–286.

Fraïsse, C., Roux, C., Welch, J. J., & Bierne, N. (2014). Gene-flow in a mosaic hybrid zone: Is local introgression adaptive? *Genetics, 197,* 939–951.

Gardner, J. P. A., & Thompson, R. J. (2001). The effects of coastal and estuarine conditions on the physiology and survivorship of the mussels *Mytilus edulis, M. trossulus* and their hybrids. *Journal of Experimental Marine Biology and Ecology, 265,* 119–140.

Gardner, J. P. A., Zbawicka, M., Westfall, K. M., & Wenne, R. (2016). Invasive blue mussels threaten regional scale genetic diversity in mainland and remote offshore locations: The need for baseline data and enhanced protection in the Southern Ocean. *Global Change Biology, 22,* 3182–3195.

Geller, J. B., Carlton, J. T., & Powers, D. A. (1994). PCR based detection of mtDNA haplotypes of native and invading mussels on the northeastern Pacific coast—Latitudinal pattern of invasion. *Marine Biology, 119,* 243–249.

Geller, J. B., Darling, J. A., & Carlton, J. T. (2010). Genetic perspectives on marine biological invasions. *Annual Review of Marine Science, 2,* 367–393.

Goldberg, E. D. (1986). The mussel watch concept. *Environmental Monitoring and Assessment, 7,* 91–103.

Gosling, E. (2003). *Bivalve molluscs: Biology, ecology and culture* (456 pp.). Bodmin, Cornwall, UK: Fishing News Books.

Gosset, C. C., & Bierne, N. (2012). Differential introgression from a sister species explains high FST outlier loci within a mussel species. *Journal of Evolutionary Biology, 26,* 14–26.

Hansen, J., Hanken, N. M., Nielsen, J. K., Nielsen, J. K., & Thomsen, E. (2011). Late Pleistocene and Holocene distribution of *Mytilus edulis* in the Barents Sea region and its palaeoclimatic implications. *Journal of Biogeography, 38,* 1197–1212.

Hansen, M. M., Olivieri, I., Waller, D. M., Nielsen, E. E., & T. G. W. Group. (2012). Monitoring adaptive genetic responses to environmental change. *Molecular Ecology, 21,* 1311–1329.

Hardy, S. M., Carr, C. M., Hardman, M., Steinke, D., Corstorphine, E., & Mah, C. (2011). Biodiversity and phylogeography of Arctic marine fauna: Insights from molecular tools. *Marine Biodiversity, 41,* 195–210.

Hellmann, J. J., Byers, J. E., Bierwagen, B. G., & Dukes, J. S. (2008). Five potential consequences of climate change for invasive species. *Conservation Biology, 22,* 534–543.

Helyar, S. J., Hemmer-Hansen, J., Bekkevold, D., Taylor, M. I., Ogden, R., Limborg, M. T., ... Nielsen, E. E. (2011). Application of SNPs for population genetics of nonmodel organisms: New opportunities and challenges. *Molecular Ecology Resources, 11,* 123–136.

Hilbish, T. J., Carson, E. W., Plante, J. R., Weaver, L. A., & Gilg, M. R. (2002). Distribution of *Mytilus edulis, M. galloprovincialis,* and their hybrids in open-coast populations of mussels in southwestern England. *Marine Biology, 140,* 137–142.

Høgslund, S., Sejr, M. K., Wiktor Jr, J., Blicher, M. E., & Wegeberg, S. (2014). Intertidal community composition along rocky shores in South-west Greenland: A quantitative approach. *Polar Biology, 37,* 1549–1561.

Hummel, H., Colucci, P., Bogaards, R. H., & Strelkov, P. (2001). Genetic traits in the bivalve *Mytilus* from Europe, with an emphasis on Arctic populations. *Polar Biology, 24,* 44–52.

Inoue, K., Odo, S., Noda, T., Nakao, S., Takeyama, S., Yamaha, E., … Harayama, S. (1997). A possible hybrid zone in the Mytilus edulis complex in Japan revealed by PCR markers. Marine Biology, 128, 91–95.

IPCC. (2014). Summary for policymakers. In C. B. Field, V. R. Barros, D. J. Dokken, K. J. Mach, M. D. Mastrandrea, T. E. Bilir, M. Chatterjee, K. L. Ebi, Y. O. Estrada, R. C. Genova, B. Girma, E. S. Kissel, A. N. Levy, S. MacCracken, P. R. Mastrandrea, & L. L. White (Eds.), Climate change 2014: Impacts, adaptation, and vulnerability. Part A: Global and sectoral aspects. Contribution of working group II to the fifth assessment report of the intergovernmental panel on climate change (pp. 1–32). Cambridge, UK and New York, NY: Cambridge University Press.

Jensen, A. S. (1905). On the mollusca of East Greenland. In Meddelelser om Grønland. Copenhagen, Denmark: Kommissionen for videnskabelige undersøgelser i Grønland.

Jombart, T. (2008). adegenet: A R package for the multivariate analysis of genetic markers. Bioinformatics, 24, 1403–1405.

Jombart, T., & Ahmed, I. (2011). adegenet 1.3-1: New tools for the analysis of genome-wide SNP data. Bioinformatics, 27, 3070–3071.

Jones, S. J., Lima, F. P., & Wethey, D. S. (2010). Rising environmental temperatures and biogeography: Poleward range contraction of the blue mussel, Mytilus edulis L., in the western Atlantic. Journal of Biogeography, 37, 2243–2259.

Jones, E. P., Swift, J. H., Anderson, L. G., Lipizer, M., Civitarese, G., Falkner, K. K., … McLaughlin, F. (2003). Tracing Pacific water in the North Atlantic Ocean. Journal of Geophysical Research-Oceans, 108, 3116.

Katolikova, M., Khaitov, V., Vänölä, R., Gantsevich, M., & Strelkov, P. (2016). Genetic, ecological and morphological distinctness of the blue mussels Mytilus trossulus Gould and M. edulis L. in the White Sea. PLoS One, 11, e0152963.

Keenan, K., McGinnity, P., Cross, T. F., Crozier, W. W., & Prodohl, P. A. (2013). diveRsity: An R package for the estimation and exploration of population genetics parameters and their associated errors. Methods in Ecology and Evolution, 4, 782–788.

Koehn, R. K. (1991). The genetics and taxonomy of species in the genus Mytilus. Aquaculture, 94, 125–145.

Kroeker, K. J., Sanford, E., Rose, J. M., Blanchette, C. A., Chan, F., Chavez, F. P., … Washburn, L. (2016). Interacting environmental mosaics drive geographic variation in mussel performance and predation vulnerability. Ecology Letters, 19, 771–779.

Laikre, L., Schwartz, M. K., Waples, R. S., & Ryman, N. (2010). Compromising genetic diversity in the wild: Unmonitored large-scale release of plants and animals. Trends in Ecology & Evolution, 25, 520–529.

Lee, J. E., & Chown, S. L. (2007). Mytilus on the move: Transport of an invasive bivalve to the Antarctic. Marine Ecology Progress Series, 339, 307–310.

Limborg, M. T., Helyar, S. J., de Bruyn, M., Taylor, M. I., Nielsen, E. E., Ogden, R., … FPT Consortium. (2012). Environmental selection on transcriptome-derived SNPs in a high gene flow marine fish, the Atlantic herring (Clupea harengus). Molecular Ecology, 21, 3686–3703.

MacKenzie, B. R., Payne, M. R., Boje, J., Hoyer, J. L., & Siegstad, H. (2014). A cascade of warming impacts brings bluefin tuna to Greenland waters. Global Change Biology, 20, 2484–2491.

Maggs, C. A., Castilho, R., Foltz, D., Henzler, C., Jolly, M. T., Kelly, J., … Wares, J. (2008). Evaluating signatures of glacial refugia for North Atlantic benthic marine taxa. Ecology, 89, 108–122.

McDonald, J. H., Seed, R., & Koehn, R. K. (1991). Allozymes and morphometric characters of three species of Mytilus in the Northern and Southern Hemispheres. Marine Biology, 111, 323–333.

McQuaid, C. D., & Phillips, T. E. (2000). Limited wind-driven dispersal of intertidal mussel larvae: In situ evidence from the plankton and the spread of the invasive species Mytilus galloprovincialis in South Africa. Marine Ecology Progress Series, 201, 211–220.

Mubiana, V. K., Qadah, D., Meys, J., & Blust, R. (2005). Temporal and spatial trends in heavy metal concentrations in the marine mussel Mytilus edu-

lis from the Western Scheldt estuary (The Netherlands). Hydrobiologia, 540, 169–180.

Nielsen, E. E., Bach, L. A., & Kotlicki, P. (2006). Hybridlab (version 1.0): A program for generating simulated hybrids from population samples. Molecular Ecology Notes, 6, 971–973.

Nielsen, E. E., Hansen, M. M., Ruzzante, D. E., Meldrup, D., & Grønkjær, P. (2003). Evidence of a hybrid-zone in Atlantic cod (Gadus morhua) in the Baltic and the Danish Belt Sea revealed by individual admixture analysis. Molecular Ecology, 12, 1497–1508.

Nielsen, E. E., Hemmer-Hansen, J., Larsen, P. F., & Bekkevold, D. (2009). Population genomics of marine fishes: Identifying adaptive variation in space and time. Molecular Ecology, 18, 3128–3150.

Ockelmann, W. K. (1958). The zoology of East Greenland: Marine lamellibranchiata. In Meddelelser om Grønland. Copenhagen, Denmark: Kommissionen for videnskabelige undersøgelser i Grønland.

Ouagajjou, Y., Presa, P., Astorga, M., & Pérez, M. (2011). Microsatellites of Mytilus chilensis: A genomic print of its taxonomic status within Mytilus sp. Journal of Shellfish Research, 30, 325–330.

Oug, E., Cochrane, S. K. J., Sundet, J. H., Norling, K., & Nilsson, H. C. (2011). Effects of the invasive red king crab (Paralithodes camtschaticus) on soft-bottom fauna in Varangerfjorden, northern Norway. Marine Biodiversity, 41, 467–479.

Paine, R. T. (1974). Intertidal community structure—Experimental studies on relationship between a dominant competitor and its principal predator. Oecologia, 15, 93–120.

Piepenburg, D., Archambault, P., Ambrose Jr, W., Blanchard, A., Bluhm, B., Carroll, M., … Włodarska-Kowalczuk, M. (2011). Towards a pan-Arctic inventory of the species diversity of the macro- and megabenthic fauna of the Arctic shelf seas. Marine Biodiversity, 41, 51–70.

Poloczanska, E. S., Brown, C. J., Sydeman, W. J., Kiessling, W., Schoeman, D. S., Moore, P. J., … Richardson, A. J. (2013). Global imprint of climate change on marine life. Nature Climate Change, 3, 919–925.

Presa, P., Perez, M., & Diz, A. P. (2002). Polymorphic microsatellite markers for blue mussels (Mytilus spp.). Conservation Genetics, 3, 441–443.

Pritchard, J. K., Stephens, M., & Donnelly, P. (2000). Inference of population structure using multilocus genotype data. Genetics, 155, 945–959.

Quesada, H., Gallagher, C., Skibinski, D. A. G., & Skibinski, D. O. F. (1998). Patterns of polymorphism and gene flow of gender-associated mitochondrial DNA lineages in European mussel populations. Molecular Ecology, 7, 1041–1051.

R Core Team. (2015). R: A language and environment for statistical computing. Vienna, Austria: R Foundation for Statistical Computing.

Rainbow, P. S. (1995). Biomonitoring of heavy metal availability in the marine environment. Marine Pollution Bulletin, 31, 183–192.

Rawson, P. D., & Harper, F. M. (2009). Colonization of the northwest Atlantic by the blue mussel, Mytilus trossulus postdates the last glacial maximum. Marine Biology, 156, 1857–1868.

Rawson, P. D., & Hilbish, T. J. (1995). Evolutionary relationships among the male and female mitochondrial-DNA lineages in the Mytilus edulis species complex. Molecular Biology and Evolution, 12, 893–901.

Rawson, P. D., & Hilbish, T. J. (1998). Asymmetric introgression of mitochondrial DNA among European populations of blue mussels (Mytilus spp.). Evolution, 52, 100–108.

Raymond, M., & Rousset, F. (1995). Genepop (Version-1.2)—population genetics software for exact tests and ecumenicism. Journal of Heredity, 86, 248–249.

Rayner, N. A., Parker, D. E., Horton, E. B., Folland, C. K., Alexander, L. V., Rowell, D. P., … Kaplan, A. (2003). Global analyses of sea surface temperature, sea ice, and night marine air temperature since the late nineteenth century. Journal of Geophysical Research-Atmospheres, 108, 4407.

Renaud, P. E., Sejr, M. K., Bluhm, B. A., Sirenko, B., & Ellingsen, I. H. (2015). The future of Arctic benthos: Expansion, invasion, and biodiversity. Progress in Oceanography, 139, 244–257.

Riginos, C., & Cunningham, C. W. (2005). Local adaptation and species segregation in two mussel (Mytilus edulis x Mytilus trossulus) hybrid zones.

Molecular Ecology, 14, 381–400.

Riginos, C., & Henzler, C. M. (2008). Patterns of mtDNA diversity in North Atlantic populations of the mussel *Mytilus edulis*. *Marine Biology, 155*, 399–412.

Riginos, C., & Henzler, C. M. (2009). Patterns of mtDNA diversity in North Atlantic populations of the mussel *Mytilus edulis* (vol 155, pg 399, 2008). *Marine Biology, 156*, 2649.

Riginos, C., Hickerson, M. J., Henzler, C. M., & Cunningham, C. W. (2004). Differential patterns of male and female mtDNA exchange across the Atlantic Ocean in the blue mussel, *Mytilus edulis*. *Evolution, 58*, 2438–2451.

Rousset, F. (2008). GENEPOP'007: A complete re-implementation of the GENEPOP software for Windows and Linux. *Molecular Ecology Resources, 8*, 103–106.

Roux, C., Fraïsse, C., Castric, V., Vekemans, X., Pogson, G. H., & Bierne, N. (2014). Can we continue to neglect genomic variation in introgression rates when inferring the history of speciation? A case study in a *Mytilus* hybrid zone. *Journal of Evolutionary Biology, 27*, 1662–1675.

Saarman, N. P., & Pogson, G. H. (2015). Introgression between invasive and native blue mussels (genus *Mytilus*) in the central California hybrid zone. *Molecular Ecology, 24*, 4723–4738.

Sarver, S. K., & Foltz, D. W. (1993). Genetic population structure of a species' complex of blue mussels (*Mytilus* spp.). *Marine Biology, 117*, 105–112.

Sejr, M. K., Blicher, M. E., & Rysgaard, S. (2009). Sea ice cover affects inter-annual and geographic variation in growth of the Arctic cockle *Clinocardium ciliatum* (Bivalvia) in Greenland. *Marine Ecology Progress Series, 389*, 149–158.

Sirenko, B. I., & Gagaev, S. Y. (2007). Unusual abundance of macrobenthos and biological invasions in the Chukchi Sea. *Russian Journal of Marine Biology, 33*, 355–364.

Śmietanka, B., Burzyński, A., Hummel, H., & Wenne, R. (2014). Glacial history of the European marine mussels *Mytilus*, inferred from distribution of mitochondrial DNA lineages. *Heredity, 113*, 250–258.

Somero, G. (2005). Linking biogeography to physiology: Evolutionary and acclimatory adjustments of thermal limits. *Frontiers in Zoology, 2*, 1.

Søndergaard, J., Asmund, G., Johansen, P., & Riget, F. (2011). Long-term response of an arctic fiord system to lead-zinc mining and submarine disposal of mine waste (Maarmorilik, West Greenland). *Marine Environmental Research, 71*, 331–341.

Strand, J., & Asmund, G. (2003). Tributyltin accumulation and effects in marine molluscs from West Greenland. *Environmental Pollution, 123*, 31–37.

Strand, A. E., Williams, L. M., Oleksiak, M. F., & Sotka, E. E. (2012). Can diversifying selection be distinguished from history in geographic clines? A population genomic study of killifish (*Fundulus heteroclitus*). *PLoS One, 7*, e45138.

Sukhotin, A. A., Strelkov, P. P., Maximovich, N. V., & Hummel, H. (2007). Growth and longevity of *Mytilus edulis* (L.) from northeast Europe. *Marine Biology Research, 3*, 155–167.

Sunday, J. M., Bates, A. E., & Dulvy, N. K. (2011). Global analysis of thermal tolerance and latitude in ectotherms. *Proceedings of the Royal Society of London B: Biological Sciences, 278*, 1823–1830.

Thyrring, J., Juhl, B. K., Holmstrup, M., Blicher, M. E., & Sejr, M. (2015). Does acute lead (Pb) contamination influence membrane fatty acid composition and freeze tolerance in intertidal blue mussels in arctic Greenland? *Ecotoxicology, 24*, 2036–2042.

Thyrring, J., Rysgaard, S., Blicher, M. E., & Sejr, M. K. (2015). Metabolic cold adaptation and aerobic performance of blue mussels (*Mytilus edulis*) along a temperature gradient into the high arctic region. *Marine Biology, 162*, 235–243.

Toro, J., Innes, D. J., & Thompson, R. J. (2004). Genetic variation among life-history stages of mussels in a *Mytilus edulis, M.trossulus* hybrid zone. *Marine Biology, 145*, 713–725.

Toro, J. E., Thompson, R. J., & Innes, D. J. (2002). Reproductive isolation and reproductive output in two sympatric mussel species (*Mytilus edulis, M. trossulus*) and their hybrids from Newfoundland. *Marine Biology, 141*, 897–909.

Toro, J. E., Thompson, R. J., & Innes, D. J. (2006). Fertilization success and early survival in pure and hybrid larvae of *Mytilus edulis* (Linnaeus, 1758) and *M. trossulus* (Gould, 1850) from laboratory crosses. *Aquaculture Research, 37*, 1703–1708.

Tremblay, R., & Landry, T. (2016). The implication of metabolic performance of *Mytilus edulis, Mytilus trossulus* and hybrids for mussel aquaculture in eastern canadian waters. *Journal of Marine Biology and Aquaculture, 2*, 1–7.

Väinölä, R., & Strelkov, P. (2011). *Mytilus trossulus* in Northern Europe. *Marine Biology, 158*, 817–833.

Vermeij, G. J. (1991). Anatomy of an invasion: The trans-Arctic interchange. *Paleobiology, 17*, 281–307.

Waltari, E., & Hickerson, M. J. (2013). Late Pleistocene species distribution modelling of North Atlantic intertidal invertebrates. *Journal of Biogeography, 40*, 249–260.

Wanamaker, A. D., Kreutz, K. J., Borns, H. W., Introne, D. S., Feindel, S., Funder, S., ... Barber, B. J. (2007). Experimental determination of salinity, temperature, growth, and metabolic effects on shell isotope chemistry of *Mytilus edulis* collected from Maine and Greenland. *Paleoceanography, 22*, PA2217.

Ware, C., Berge, J., Sundet, J. H., Kirkpatrick, J. B., Coutts, A. D. M., Jelmert, A., ... Alsos, I. G. (2014). Climate change, non-indigenous species and shipping: Assessing the risk of species introduction to a high-Arctic archipelago. *Diversity and Distributions, 20*, 10–19.

Wassmann, P., Duarte, C. M., Agusti, S., & Sejr, M. K. (2011). Footprints of climate change in the Arctic marine ecosystem. *Global Change Biology, 17*, 1235–1249.

Wenne, R., Bach, L., Zbawicka, M., Strand, J., & McDonald, J. H. (2016). A first report on coexistence and hybridization of *Mytilus trossulus* and *M. edulis* mussels in Greenland. *Polar Biology, 39*, 343–355.

Westerbom, M., Kilpi, M., & Mustonen, O. (2002). Blue mussels, *Mytilus edulis* at the edge of the range: Population structure, growth and biomass along a salinity gradient in the north-eastern Baltic Sea. *Marine Biology, 140*, 991–999.

White, M. J. D. (1978). *Modes of speciation*. San Francisco, CA: W. H. Freeman.

Wisz, M. S., Broennimann, O., Grønkjær, P., Moller, P. R., Olsen, S. M., Swingedouw, D., ... Pellissier, L. (2015). Arctic warming will promote Atlantic-Pacific fish interchange. *Nature Climate Change, 5*, 261–265.

Zbawicka, M., Drywa, A., Smietanka, B., & Wenne, R. (2012). Identification and validation of novel SNP markers in European populations of marine *Mytilus* mussels. *Marine Biology, 159*, 1347–1362.

Zbawicka, M., Sanko, T., Strand, J., & Wenne, R. (2014). New SNP markers reveal largely concordant clinal variation across the hybrid zone between *Mytilus* spp. in the Baltic Sea. *Aquatic Biology, 21*, 25–36.

APPENDIX 1 H_e, H_o and allelic richness for the "Mytilus edulis" data set.

Estimates of expected (H_e) and observed (H_o) heterozygosities and the allelic richness for the samples only including *M. edulis*.

Code	H_e	H_o	Allelic richness
UPE	00.24	00.24	1.54
NUS	00.25	00.23	1.53
NUL	00.23	00.22	1.54
KOB	00.25	00.22	1.54
ICE	00.28	00.27	1.60
LOF	00.26	00.28	1.58
TRS	00.27	00.29	1.64
TRL	00.28	00.24	1.50
SV1	00.27	00.25	1.59
SV2	00.27	00.27	1.62
SV3	00.26	00.25	1.58
SV4	00.27	00.25	1.60
PSW	00.30	00.29	1.66
PSE	00.30	00.29	1.67
WS1	00.28	00.25	1.62
WS2	00.29	00.27	1.66

APPENDIX 2 Structure analysis for the full data set.

Results of Structure clustering analyses for the full data set with: (a) $K = 3$ and (b) $K = 5$. Samples are 1: QAS, 2: QAL, 3: UPE, 4: NUS, 5: NUL, 6: KOB, 7: ICE, 8: LOF, 9: TRS, 10: TRL, 11: SV1, 12: SV2, 13: SV3, 14: SV4, 15: PSW, 16: PSE, 17: WS1, 18: WS2, 19: MTR, and 20: MGA. For explanation of sample identification codes, see Table 1.

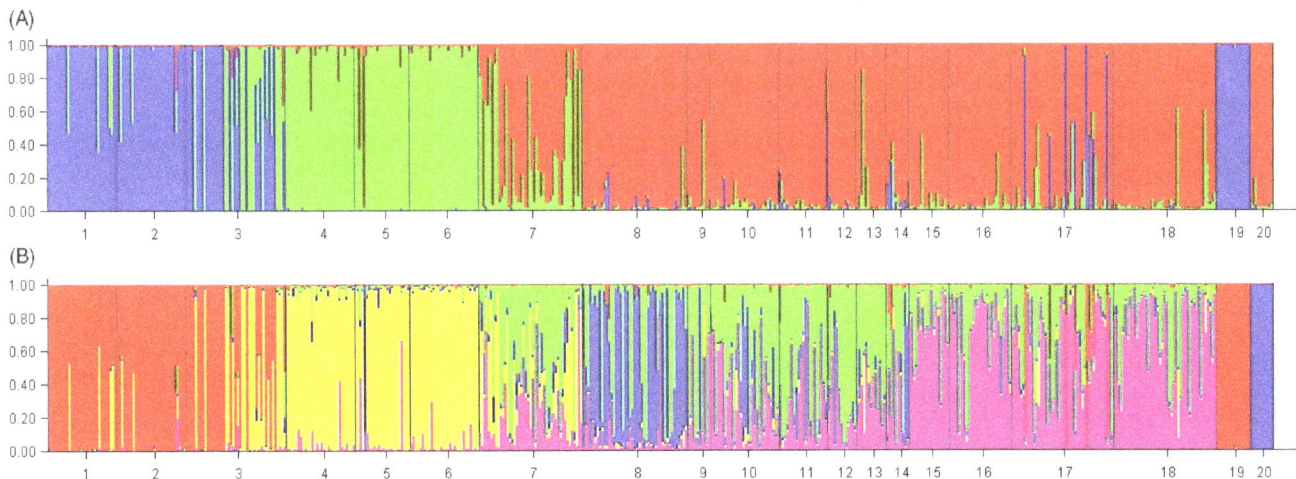

APPENDIX 3 Confidence intervals

95% confidence intervals for the inferred ancestry for simulated parentals and hybrids analyzed with Structure (*K* = 4). Ancestry estimates for the two inferred *Mytilus edulis* clusters (Greenland and Eastern Atlantic) are pooled. See text for explanation.

Simulated individuals	*Mytilus edulis*	*Mytilus trossulus*	*Mytilus galloprovincialis*
Greenland			
M. edulis	0.89–0.99	0.00–0.02	0.01–0.10
M. trossulus	0.00–0.01	0.99–1.00	0.00–0.01
M. edulis/ M. trossulus F1	0.34–0.54	0.44–0.56	0.01–0.10
M. edulis/ M. trossulus F2	0.36–0.59	0.39–0.60	0.01–0.09
M. edulis/M. trossulus backcross M. edulis	0.60–0.82	0.15–0.34	0.01–0.10
M. edulis/M. trossulus backcross M. trossulus	0.14–0.31	0.67–0.82	0.01–0.07
Eastern Atlantic			
M. edulis	0.92–0.99	0.00–0.02	0.01–0.07
M. trossulus	0.00–0.01	0.99–1.00	0.00–0.01
M. galloprovincialis	0.01–0.02	0.00–0.01	0.97–0.99
M. edulis/M. trossulus F1	0.43–0.55	0.43–0.54	0.01–0.06
M. edulis/M. trossulus F2	0.38–0.60	0.37–0.59	0–01–0.09
M. edulis/M. trossulus backcross M. edulis	0.70–0.76	0.20–0.25	0.04–0.05
M. edulis/M. trossulus backcross M. trossulus	0.17–0.33	0.65–0.80	0.01–0.08
M. edulis/M. galloprovincialis F1	0.37–0.68	0.00–0.02	0.31–0.62
M. edulis/M. galloprovincialis F2	0.32–0.73	0.00–0.03	0.26–0.67
M. edulis/M. galloprovincialis backcross M. edulis	0.61–0.96	0.00–0.02	0.03–0.39
M. edulis/M. galloprovincialis backcross M. galloprovincialis	0.14–0.39	0.00–0.02	0.60–0.86

APPENDIX 4 Structure analysis for the "Mytilus edulis" "neutral" data set.

Results from clustering analysis (*K* = 2) for the "*Mytilus edulis*" data set only including SNPs not documented as outliers. Samples are 1: UPE, 2: NUU comprising of NUS and NUL, 3: KOB, 4: ICE, 5: LOF, 6: TRO comprising of TRS and TRL), 7: SVA comprising of SV1, SV2 and SV3, 8: SV4, 9: PSW, 10: PSE, 11: WS1, and 12: WS2. For explanation of sample identification codes, see Table 1.

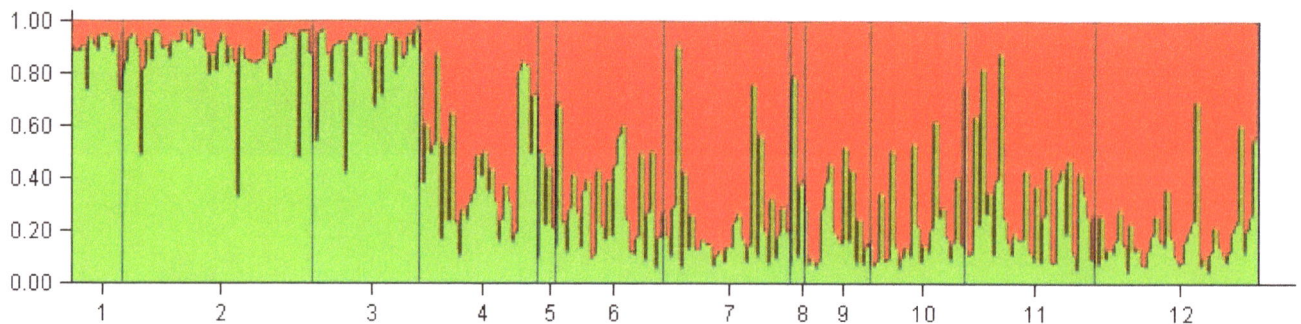

APPENDIX 5 Structure analysis for the "*Mytilus edulis*" "outlier" data set.

Results from clustering analysis (*K* = 2) for the "*Mytilus edulis*" data set only including the six outlier SNPs Samples are 1: UPE, 2: NUU comprising of NUS and NUL, 3: KOB, 4: ICE, 5: LOF, 6: TRO comprising of TRS and TRL), 7: SVA comprising of SV1, SV2, and SV3, 8: SV4, 9: PSW, 10: PSE, 11: WS1, and 12: WS2. For explanation of sample identification codes, see Table 1.

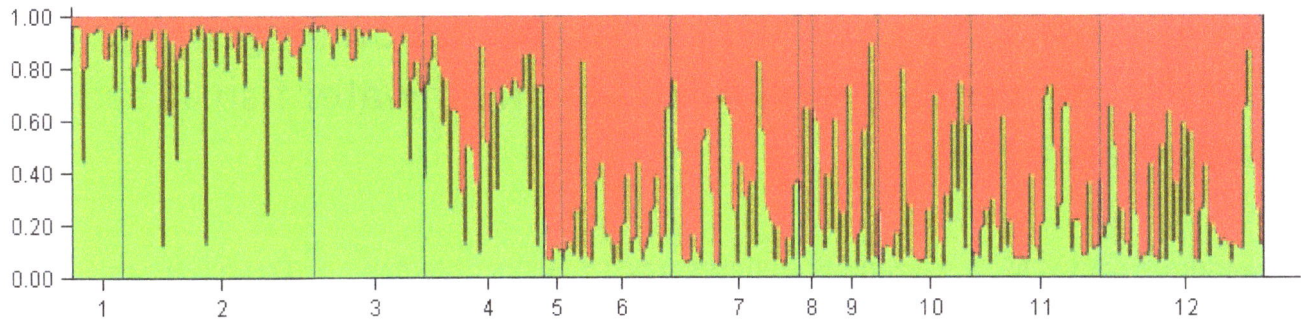

7

Severe consequences of habitat fragmentation on genetic diversity of an endangered Australian freshwater fish: A call for assisted gene flow

Alexandra Pavlova[1] (iD) | Luciano B. Beheregaray[2] | Rhys Coleman[3] | Dean Gilligan[4] | Katherine A. Harrisson[1,5,6] | Brett A. Ingram[7] | Joanne Kearns[5] | Annika M. Lamb[1] | Mark Lintermans[8] | Jarod Lyon[5] | Thuy T. T. Nguyen[9] | Minami Sasaki[2] | Zeb Tonkin[5] | Jian D. L. Yen[10] | Paul Sunnucks[1]

[1]School of Biological Sciences, Clayton Campus, Monash University, Clayton, VIC, Australia

[2]School of Biological Sciences, Flinders University, Adelaide, SA, Australia

[3]Applied Research, Melbourne Water, Docklands, VIC, Australia

[4]Freshwater Ecosystems Research, NSW Department of Primary Industries – Fisheries, Batemans Bay, NSW, Australia

[5]Department of Environment, Land Water and Planning, Arthur Rylah Institute, Land, Fire and Environment, Heidelberg, VIC, Australia

[6]Department of Ecology Environment and Evolution, School of Life Sciences, La Trobe University, Bundoora, Victoria, 3083, Australia

[7]Department of Economic Development, Jobs, Transport and Resources, Fisheries Victoria, Alexandra, VIC, Australia

[8]Institute for Applied Ecology, University of Canberra, Canberra, ACT, Australia

[9]Agriculture Victoria, AgriBio, Centre for AgriBioscience, Bundoora, VIC, Australia

[10]School of Physics and Astronomy, Clayton Campus, Monash University, Clayton, VIC, Australia

Correspondence
Alexandra Pavlova, School of Biological Sciences, Clayton Campus, Monash University, Clayton, VIC, Australia.
Email: alexandra.pavlova@monash.edu

Funding information
Australian Research Council, Grant/Award Number: LP110200017; VIC Department of Sustainability and Environment (now DELWP-Department of Environment, Land, Water & Planning); Icon Water (formerly ACTEW Corporation); Melbourne Water; Fisheries Victoria (now within DEDJTR- Department of Economic Development, Jobs, Transport and Resources); Grant/Award Number: Recreational Fishing Grants program; NSW DPI- Fisheries; Goulburn Broken Catchment Management Authority

Abstract

Genetic diversity underpins the ability of populations to persist and adapt to environmental changes. Substantial empirical data show that genetic diversity rapidly deteriorates in small and isolated populations due to genetic drift, leading to reduction in adaptive potential and fitness and increase in inbreeding. Assisted gene flow (e.g. via translocations) can reverse these trends, but lack of data on fitness loss and fear of impairing population "uniqueness" often prevents managers from acting. Here, we use population genetic and riverscape genetic analyses and simulations to explore the consequences of extensive habitat loss and fragmentation on population genetic diversity and future population trajectories of an endangered Australian freshwater fish, Macquarie perch *Macquaria australasica*. Using guidelines to assess the risk of outbreeding depression under admixture, we develop recommendations for population management, identify populations requiring genetic rescue and/or genetic restoration and potential donor sources. We found that most remaining populations of Macquarie perch have low genetic diversity, and effective population sizes below the threshold required to retain adaptive potential. Our simulations showed that under management inaction, smaller populations of Macquarie perch will face inbreeding depression within a few decades, but regular small-scale translocations will rapidly rescue populations from inbreeding depression and increase adaptive potential through genetic restoration. Despite the lack of data on fitness loss, based on our genetic data for Macquarie perch populations, simulations and empirical results from other systems, we recommend regular and frequent translocations among remnant populations within catchments. These translocations will emulate the effect of historical gene flow and improve population persistence through decrease in demographic and genetic stochasticity. Increasing population genetic connectivity within each catchment will help to maintain large effective population sizes and maximize species adaptive potential. The approach proposed here could be readily applicable to genetic management of other threatened species to improve their adaptive potential.

KEYWORDS

adaptive potential, effective population size, genetic rescue, genetic restoration, inbreeding depression, Macquarie perch *Macquaria australasica*, management, population persistence

1 | INTRODUCTION

A primary goal of conservation management is to improve the adaptive potential of populations, that is, the ability of populations to adapt and persist in the face of environmental changes (Sgrò, Lowe, & Hoffmann, 2011). Genetic diversity underpins adaptive potential, but is lost in small populations through genetic drift, which can lead to loss of fitness, accumulation of genetic load and ultimately population extinction (Harrisson, Pavlova, Telonis-Scott, & Sunnucks, 2014; Reed & Frankham, 2003; Willi, Van Buskirk, & Hoffmann, 2006). Through reduction or cessation of gene flow, habitat fragmentation reduces species ranges to small populations at high extinction risk, contributed to by environmental, demographic and genetic factors (e.g. inbreeding depression, genetic load and inability to rapidly adapt to environmental changes) and their interactions (Benson et al., 2016; Fountain, Nieminen, Sirén, Wong, & Hanski, 2016; Frankham, 2005; Lopez, Rousset, Shaw, Shaw, & Ronce, 2009). Given that taxa of conservation concern have typically undergone habitat fragmentation and/or strong population contractions, genetic issues should be addressed in recovery or management plans, although this is often not done (Pierson et al., 2015, 2016).

Effective and efficient allocation of conservation resources is best assisted by information about the distribution of genetic variation across landscapes (Hoffmann et al., 2015). Spatial patterns of genetic diversity and genetic differentiation reflect stochastic and environmental effects on key demographic and evolutionary processes (e.g. population size, gene flow, adaptive potential) linked to viability. Thus, genetic data can inform about the conservation status of populations, improve predictions about population responses to environmental change and management interventions, and support the development of effective conservation strategies for enhancing population viability (Willi & Hoffmann, 2009).

For a given population or species, a direct demonstration of genetic problems, such as loss of fitness due to decline in genetic diversity or inbreeding, presents a strong case for genetic management. In contrast,

unavailability of evidence for genetic problems in a specific case often prompts an argument against genetic intervention until such evidence is collected. This position ignores evidence acquired from numerous wild and captive systems that small and isolated populations will most likely be facing genetic problems with negative demographic consequences (Keller & Waller, 2002; Saccheri et al., 1998; Woodworth, Montgomery, Briscoe, & Frankham, 2002), reflecting cultural rather than evidence-based decisions concerning genetic management (Love Stowell, Pinzone, & Martin, 2017). Obtaining data on population fitness is challenging and expensive and may take years. Postponing decision-making until such data are available, or disregarding genetic problems completely, is likely to result in managing small populations to extinction (Frankham et al., in press; Love Stowell et al., 2017; Weeks, Stoklosa, & Hoffmann, 2016). Instead, the augmentation of gene flow is a powerful conservation management option for counteracting loss of genetic diversity resulting from drift in small populations, with the potential to promote positive demographic outcomes (Frankham, 2015; Hufbauer et al., 2015; Whiteley, Fitzpatrick, Funk, & Tallmon, 2015). Genetic augmentation to alleviate detrimental effects of inbreeding and/or genetic load in small, isolated populations (i.e. genetic rescue) and/or increase levels of genetic diversity and adaptive potential (i.e. genetic restoration) can be achieved through reconnection of habitat or human-assisted translocations (Attard et al., 2016; Weeks et al., 2011; Whiteley et al., 2015). Strong precedents supporting the overwhelmingly positive and long-lasting effects of genetic rescue on fitness exist across a broad range of taxa (Frankham, 2015, 2016). By bringing in new genetic diversity, genetic augmentation is expected to improve adaptive potential and probability of persistence of small and isolated populations, even if they do not yet suffer from inbreeding depression and genetic load (Willi et al., 2006).

Another common argument against genetic augmentation is that gene flow between divergent populations may "swamp" local adaptation and distinctiveness, or contribute to outbreeding depression (Frankham et al., 2011; Le Cam, Perrier, Besnard, Bernatchez, &

Evanno, 2015; Love Stowell et al., 2017). Such concerns frequently lead to recommendations to maintain isolation of differentiated populations, but without appropriate assessment of the risks and benefits of gene flow (Farrington, Lintermans, & Ebner, 2014; Frankham et al., 2011; Roberts, Baker, & Perrin, 2011). Genetic differentiation per se is not a clear indicator of risk that populations are differently adapted and gene flow might be harmful: some differentiation is inevitable under restricted dispersal, particularly among small populations, and is often caused by recent human impacts (Cole et al., 2016; Coleman, Weeks, & Hoffmann, 2013; Faulks, Gilligan, & Beheregaray, 2011). Even where genetic differentiation is associated with adaptive divergence, locally adapted traits can still be maintained in a population in the presence of gene flow from a differently adapted population (Fitzpatrick, Gerberich, Kronenberger, Angeloni, & Funk, 2015), unless the level of gene flow is so large that it overwhelms natural selection (Lowe & Allendorf, 2010). The potential benefits of gene flow are large, and any risks of outbreeding depression must be weighed against the risk of extinction due to inaction, for example from inbreeding depression, genetic load or inability to adapt to a novel selective pressure (Becker, Tweedie, Gilligan, Asmus, & Whittington, 2013; Fisher, Garner, & Walker, 2009; Frankham et al., in press; Harrisson, Pavlova, et al., 2016; Lintermans, 2013).

In developing plans for genetic rescue and/or restoration, four key questions must be considered:

1. What level of genetic variation should trigger genetic rescue/ restoration?
2. Which populations can be used as sources for translocations?
3. What is the risk of outbreeding depression? and
4. How much gene flow is required to rescue the populations at risk and restore their adaptive potential?

Recent reviews provide guidelines:

1. Effective population size (N_e) ≥100 is required to prevent inbreeding depression in the short-term and limit loss of fitness to ≤10% over 5 generations. A much larger global effective population size, N_e ≥ 1,000, is required to retain adaptive potential, by maintaining an equilibrium between the accumulation of genetic variation for a selectively neutral trait by mutation, and the loss variation by random genetic drift (Frankham, Bradshaw, & Brook, 2014). Based on a combination of theory and empirical evidence, the numbers from previously proposed 50/500 rule (Franklin, 1980) were found to be insufficient to prevent inbreeding depression/avoid loss of genetic variation for fitness (Frankham et al., 2014).
2. Any population only recently diverged from a genetically depleted one can be used as a source, but the magnitude of genetic rescue is greater if the source population is outbred and/or differentiated (Frankham, 2015).
3. The risk of outbreeding depression will be low if populations have the same karyotype, were isolated for <500 years and are adapted to similar environments (Frankham et al., 2011).

4. If the risk of outbreeding depression is low, then up to 20% gene flow from the source population will improve population adaptive potential; ongoing monitoring within an adaptive management framework (i.e. iterative process of decision-making based on monitoring of outcomes) is recommended to avoid genetic swamping and outbreeding depression (Hedrick, 1995; Weeks et al., 2011; Whiteley et al., 2015).

The Macquarie perch *Macquaria australasica* Cuvier 1830 is an endangered Australian endemic freshwater fish species that has undergone substantial population decline and range contraction and fragmentation. Prior to European colonization in the late 18th century, the Macquarie perch was widespread, occurring in the inland Murray–Darling Basin (MDB), and across three coastal drainage basins, Hawkesbury–Nepean (HNB), Shoalhaven and Georges (Figure 1). Although distinct taxa within *Macquaria australasica* are not formally recognized, morphological, allozyme and DNA data all indicate that the inland and coastal forms of Macquarie perch diverged in the Pleistocene and probably represent different species (Dufty, 1986; Faulks, Gilligan, & Beheregaray, 2010a, 2015; Faulks et al., 2011; Knight & Bruce, 2010; Lintermans & Ebner, 2010; Pavlova et al., 2017). Over the last two centuries, Macquarie perch suffered strong declines, particularly in the lowland reaches of rivers, and as a result the species is now restricted to isolated locations in the headwaters of Lachlan, Murrumbidgee and Murray tributaries in the MDB, Hawkesbury and Nepean rivers in the HNB, and Georges River in the Georges Basin (Figure 1) (Gilligan, McGarry, & Carter, 2010; Knight & Bruce, 2010; Lintermans, 2007). A highly differentiated form in the Shoalhaven Basin is thought to have become extinct by 1998, which underscores the urgent need to understand and manage remaining genomic diversity within the species (Faulks et al., 2010a). Many anthropogenic factors have been implicated in the species' decline across its range, including loss of 95% of habitat, fragmentation by barriers to movement, increased river sedimentation, alien fish species, overexploitation and altered flow regimes reducing previously connected habitats to small fragments (Cadwallader, 1978; Ingram, Douglas, & Lintermans, 2000; Ingram et al., 1990; Pollard, Ingram, Harris, & Reynolds, 1990). Loss of physical and genetic connectivity across the species' range is expected to have contributed to its ongoing decline, although establishing that directly is challenging given the lack of historical genetic samples.

In attempts to reverse its decline, the Macquarie perch has been subject to various management actions including habitat restoration, threat amelioration, wild-to-wild translocations and captive breeding with subsequent stocking of captive-bred fish to re-establish extinct populations, augment threatened ones and support recreational fisheries (Ho & Ingram, 2012; Lintermans, 2012; Lintermans, Lyon, Hammer, Ellis, & Ebner, 2015). Despite substantial financial investment and effort, these management actions have so far failed to reverse the decline of the species. Indeed, water scarcity under recent droughts has made several of the remaining populations exceedingly small and vulnerable to environmental, demographic and genetic stochasticity. Historically, vulnerable or extinct populations would have been recolonized from neighbouring populations when flow returned, but this is no longer possible due to insufficient physical connectivity

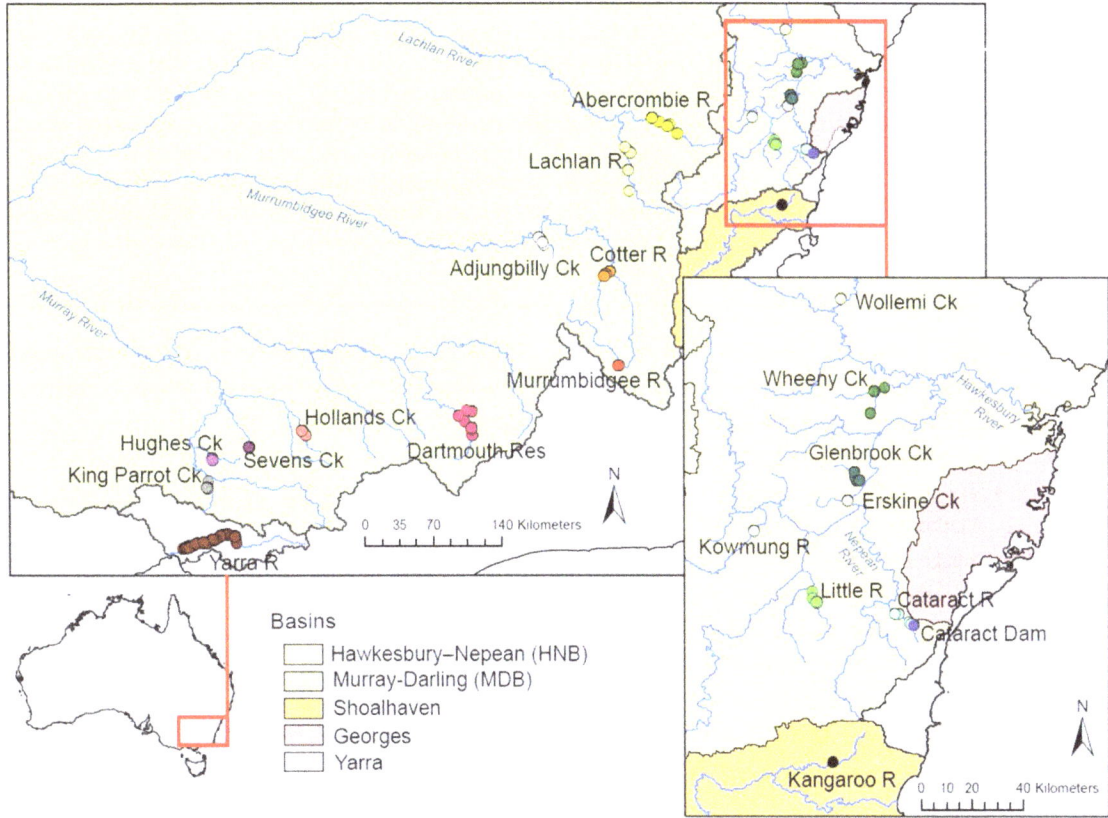

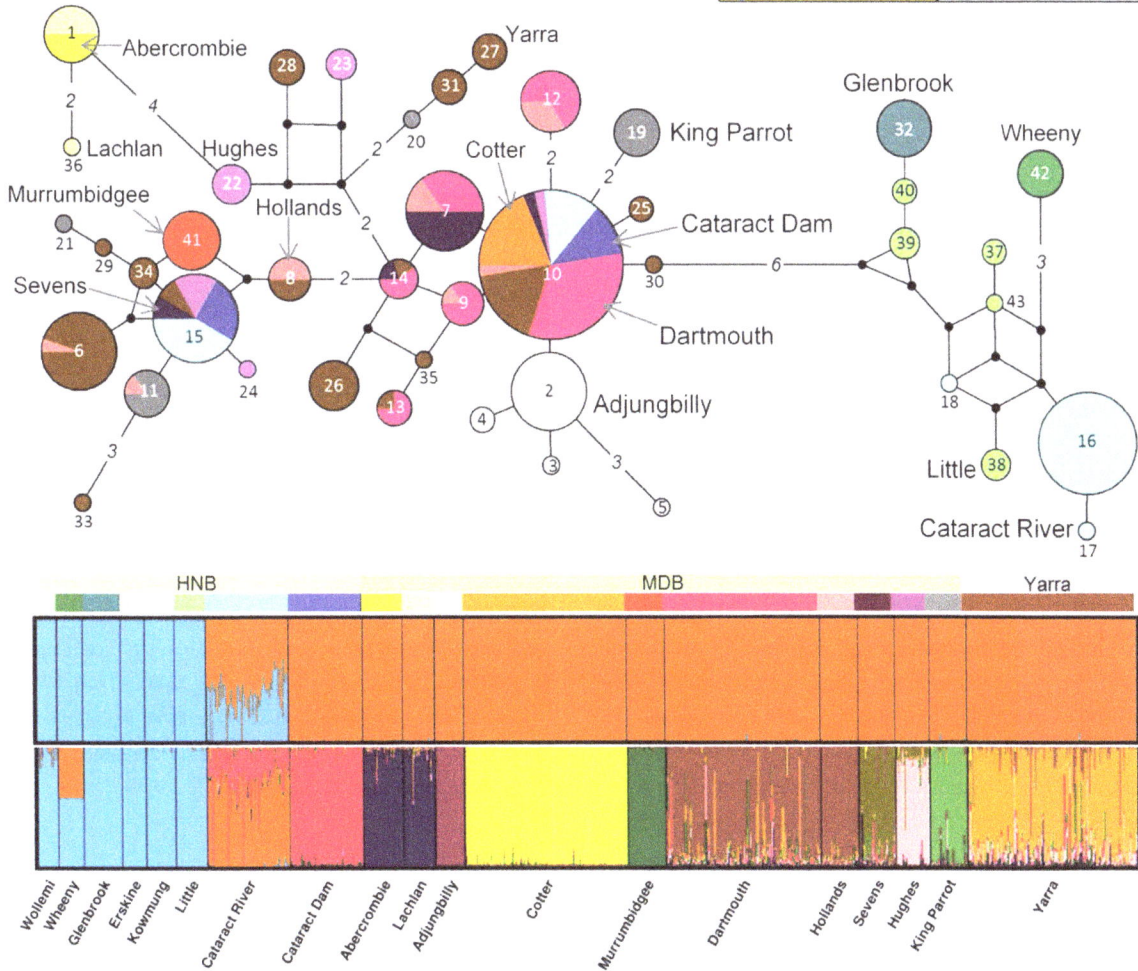

FIGURE 1 Geographic distribution of Macquarie perch samples analysed for this study (top), mitochondrial control region haplotype network showing distribution of haplotypes across populations (middle; Shoalhaven haplotype not included) and individual memberships in two or 12 genetic clusters inferred from microsatellite data (major structure inferred by $K = 2$ and $K = 12$ Structure analysis of all HNB and MDB samples; see Appendix S9; Shoalhaven individual not included). Two colour stripes above Structure plots are colour codes for basins (top) and populations (bottom), as on map. Yarra and Cataract Dam populations are translocated from the MDB; Cataract River comprises hybrids between endemic HNB lineage and fish dispersed from Cataract Dam (details in Appendix S9). On the haplotype network, populations are coloured as on the map (population labels are also given for each colour); each circle is a unique haplotype (number within-haplotype ID, small black circle – missing haplotype); the size of each circle is proportional to the haplotype frequency; connections between circles are single substitutions unless marked with a number (of substitutions). In five locations, a single haplotype is fixed (1 in Abercrombie, 10 in Cotter, 41 in Murrumbidgee, 32 in Glenbrook, 42 in Wheeny)

between remnant habitat patches. As others have stated (Love Stowell et al., 2017; Whiteley et al., 2015), we argue that although the main reasons for population decline (e.g. habitat availability and degradation) should be addressed by conservation management, boosting genetic diversity through assisted gene flow might be necessary to improve population fitness and/or adaptive potential and decrease extinction probability of small populations. Management decisions need to be made now based on the best available information. When direct information on genetic problems in a specific case is unavailable, drawing on accumulated evidence from empirical, theoretical and simulation studies of other systems is a more evidence-based approach than making the unvalidated assumption that there are no genetic problems (Frankham et al., in press; Love Stowell et al., 2017; Weeks et al., 2016).

Here, we explore patterns of genetic diversity and genetic differentiation across the entire range of the Macquarie perch and expand on earlier genetic work that revealed deep divisions within the named species, hybridization among inland and coastal forms, strong population substructure and environmental factors affecting the distribution of population genetic diversity (details in Appendix S1; Faulks et al., 2010a, 2011; Pavlova et al., 2017). For the first time, we estimate effective population sizes across the species range, evaluate them against established thresholds (Frankham et al., 2014) and demonstrate by simulations that within a few decades, smaller Macquarie perch populations are likely to suffer erosion of genetic diversity and inbreeding unless genetic diversity is restored and maintained by translocations. In addition, we identify novel environmental variables associated with individual-based heterozygosity and improve earlier estimates of population history, key divergence times, population genetic diversity and differentiation, and population structure at basin- and catchment scales by analysing additional samples and populations, additional microsatellite loci and longer mitochondrial sequences. Our results will contribute to developing separate management plans for appropriate evolutionary units, the selection of broodstock source populations for the current hatchery breeding programmes and the identification of critical locations for stocking and translocations. The processes of habitat loss, degradation and fragmentation that have affected the previously common and widespread Macquarie perch are similar to those affecting many species including, but not limited to, other threatened MDB species that have experienced recent population extinctions, for example trout cod *Maccullochella macquariensis*, southern pygmy perch *Nannoperca australis* and southern-purple spotted gudgeon *Mogurnda adspersa* (Hammer et al., 2015; Lintermans, 2007; Trueman,

2011); hence, the workflow here could be readily applicable to genetic management of other threatened species to improve their adaptive potential.

2 | MATERIALS AND METHODS

2.1 | Population sampling, tissue sampling and DNA extraction

Macquarie perch samples were collected predominantly between 2002 and 2014 from 20 populations (Figure 1, Appendix S2), including ten MDB, seven HNB (note that the Cataract River population consists of endemic HNB as well as introduced MDB fish), two coastal populations that have originated from translocation of MDB individuals (Cataract Dam in HNB and the Yarra River in the Yarra River Basin) and a Shoalhaven Basin population (a single known genetic sample from Kangaroo River). Individuals were captured using various electrofishing, netting or angling techniques. Most fish were measured for length and weight and immediately released at the site of capture after collection of fin clips. Fin tissue was preserved in 100% ethanol and stored at −20°C. DNA was extracted using a Qiagen DNeasy tissue extraction kit, unless available from other projects (Faulks et al., 2010a, 2011; Nguyen, Ingram, Lyon, Guthridge, & Kearns, 2012).

2.2 | Mitochondrial control region sequencing

Sequences of an 844-bp mitochondrial DNA fragment including the complete control region were analysed for 339 individuals from 17 populations (Appendix S2 Table S2A). For five samples, sequences were extracted from complete mitogenomes (GenBank accessions KR152235, KR152240, KR152241, KR152248 and KR152253) (Pavlova et al., 2017). For the remaining samples, mtDNA fragments were amplified and sequenced using primers A (Lee, Conroy, Howell, & Kocher, 1995) and Dmod (GCCCATCTTAACATCTTCAGTG) (Appendix S3). Chromatograms were edited and aligned in Geneious Pro 6.1.3 (Drummond, Ashton, et al., 2012). Sequences were uploaded to GenBank (accession KT626048–KT626381).

2.3 | Microsatellite genotyping

Individuals were genotyped for 19 microsatellite loci (Appendix S3). Twenty individuals with scores missing for more than three loci were excluded, as well as a single Dartmouth individual that returned a triploid genotype (confirmed by repeated genotyping of re-extracted DNA). Final analyses included genotyping scores for 871 individuals from 20 populations (Appendix S2; Supporting Information); only 0.3% of scores were missing. Tests for deviation from Hardy–Weinberg and linkage equilibria were performed using Genepop 4.2 (Rousset, 2008) (Appendix S3).

2.4 | Estimating population history and key divergence times

To evaluate the level of mitochondrial haplotype-sharing across populations, indicative of past gene flow, a median-joining network (Bandelt, Forster, & Roehl, 1999) was constructed in Network v4.6.1.0 (www.fluxus-engineering.com). Times of divergence among major mitochondrial lineages were previously estimated from complete mitogenomes (Pavlova et al., 2017; summarized in Appendix S1); however, some lineages were not represented. Using mitochondrial haplotypes, we estimated the times of lineage divergence with Beast 1.8.2 (Drummond, Suchard, Xie, & Rambaut, 2012), assuming the HKY+G+I substitution model, coalescent tree prior and Bayesian skyline (with 10 groups and piecewise-constant skyline model) as population size prior. A strict molecular clock was used, based on results of complete mitogenome analysis (Pavlova et al., 2017); the control region clock rate, estimated based on a wide clock rate prior in that analysis, was used here as a prior (i.e. lognormal clock rate with mean (in real space) = 0.05 and SD = 0.5; this corresponds to 95% of rates being sampled between 0.017 and 0.118 substitutions per site per million years). Three replicates were run for 100 million steps sampling every 100,000; runs were checked for convergence, combined and analysed in Tracer 1.6.0 (Rambaut & Drummond, 2007) after discarding first 10% of samples from each run as burn-in.

2.5 | Genetic differentiation and tests for evolutionary timescale of divergence

Arlequin 3.5 (Excoffier & Lischer, 2010) was used to calculate pairwise population F_{ST}, R_{ST} (microsatellites) and Φ_{ST} (mtDNA sequences) between catchments. SPAGeDi ver 1.5 (Hardy & Vekemans, 2002) with 10,000 permutations of microsatellite allele sizes was used to test if populations have been evolving independently long enough to evolve new microsatellite alleles through mutation; locus AB009 was excluded, because it was monomorphic in all but one population.

2.6 | Assessing population structure at basin- and catchment scales

Structure (Pritchard, Stephens, & Donnelly, 2000) was used to identify genetic clusters from microsatellite genotypes (data on population of origin were not applied). We first tested for general population structure across the species range using all 870 MDB and HNB individuals. Twenty replicates of 10^6 burn-in iterations followed by 3×10^6 iterations were run for each K from 1 to 16 (the maximum was set according to preliminary runs showing no meaningful structure above

K = 12) using the admixture model with correlated allele frequencies. Runs were summarized using the web server Clumpak (Kopelman, Mayzel, Jakobsson, Rosenberg, & Mayrose, 2015). The largest number of genetic clusters resulting in coherent geographic groups was considered to best represent population structure.

Three subsets of the data were further analysed using Structure: (i) a data set of 671 individuals of MDB origin (10 MDB populations, Cataract Dam and Yarra) was run for K_{max} = 16 to detect major population subdivisions within MDB; (ii) a data set of 375 individuals from the Murray and Yarra catchments (Dartmouth, Hollands, Sevens, Hughes, King Parrot and Yarra populations) was run for K_{max} = 7 to explore the origin of two previously detected genetic clusters in Dartmouth and Yarra (Nguyen et al., 2012); and (iii) a data set of 134 individuals from six HNB populations (all except the introduced Cataract River and Cataract Dam samples) was run for K_{max} = 7 to test for presence of structure within the HNB.

2.7 | Analyses of genetic diversity, proxy for historical effective population sizes

For mtDNA sequencing data, Arlequin was used to calculate haplotype diversity (Hd), nucleotide diversity (π), the number of segregating sites (S) as well as Tajima's D (Tajima, 1996) and Fu's Fs (Fu, 1997) tests for selective neutrality or population size changes. For microsatellite data, Arlequin was used to calculate expected heterozygosity (He); Genalex (Peakall and Smouse 2006), the number of private alleles (alleles restricted to a single location); and Fstat (Goudet, 2001), allelic richness (AR) standardized to the minimum sample size of 14 individuals. He and AR reflect coalescent effective population sizes (N_e) under mutation–drift equilibrium and may not reliably estimate contemporary N_e associated with inbreeding and the probability of population persistence (Hare et al., 2011).

2.8 | Analyses of contemporary effective population sizes

We estimated N_e using two single-sample methods, both of which assume that population size is stable over a few generations. First, we used the approximate Bayesian computation method based on eight summary statistics (including linkage disequilibrium, LD) implemented in OneSamp (Tallmon, Koyuk, Luikart, & Beaumont, 2008); 50,000 populations were simulated using a prior N_e range of 4–1,000 individuals. OneSamp was also run with a prior N_e range of 4–500, to confirm estimate consistency. Second, an LD-based method was used (LDNe; Waples & Do, 2008), implemented in NeEstimator V2.0 (Do et al., 2014); random mating model and P_{Crit} = 0.02 were applied. LDNe estimates were interpreted only for samples of ≥50 individuals, as smaller samples are not informative when true N_e > 100 (Tallmon et al., 2010). When estimated from a single cohort of an iteroparous species, an LD-based estimate of N_e reflects the harmonic mean of the effective number of breeders in one reproductive cycle (N_b) and effective size per generation (N_e), but when estimated from mixed-age sample, it approximates effective size per generation, albeit consistently downwardly biased (50–90% of true N_e, least biased when the number of cohorts in a sample is close to the length of the generation time) (Waples, Antao, & Luikart, 2014). Whereas N_b reflects short-term effective population size relevant to inbreeding, N_e is responsible for shaping long-term evolutionary processes (Waples et al., 2014); thus, both would be required for evaluating two parts of the 100/1,000 N_e threshold (Frankham et al., 2014). Using simulations, Waples, Luikart, Faulkner, and Tallmon (2013) showed that the N_b/N_e ratio can be approximated using two life-history traits. According to their formulae, for Macquarie perch $N_b/N_e = 1.156$ (assuming age at maturity of 3 years and adult life span of 23 years (Appleford, Anderson, & Gooley, 1998; Lintermans & Ebner, 2010)); thus, an LD-based estimate of N_e should reasonably approximate both N_b and N_e, provided that other biases can be accounted for. Macquarie perch samples including ≥7 cohorts would be least biased, assuming a generation time of 7 years (calculated in Vortex (Lacy, Miller, & Traylor-Holzer, 2015), below). According to the range of individual fish lengths in each population (Supporting Information) and the modelled relationship between length and age for Macquarie perch (Figure 3 of Todd & Lintermans, 2015), our Macquarie perch population samples included from ~3 to >7 cohorts, and we expect the least downward bias in LD-based estimates of N_e for the seven populations with a range of sampled lengths >200 mm (Abercrombie, Adjungbilly, Dartmouth, Hughes, King Parrot, Lachlan, Sevens, Yarra). For species with N_b/N_e ~1, LD-based estimates of N_b (and N_e) are expected to be unaffected by two-loci Wahlund effect arising when parents from different cohorts are combined in a single sample (Waples et al., 2014).

2.9 | Identifying environmental variables associated with individual-based genetic diversity

We used an individual-based Bayesian modelling approach to identify environmental variables that were correlated with spatial distribution of genetic variation (Harrisson, Yen, et al., 2016). The model was based on the 16 populations from the MDB and the HNB (translocated Yarra and Cataract Dam and admixed Cataract River samples were excluded).

The response variable was homozygosity-by-locus (HL), calculated for each individual using the Rhh package (Alho, Valimaki, & Merilä, 2010) in R 3.2.0 (R Development Core Team 2014). HL measures the proportion of homozygous loci per individual, attributing greater weight to homozygosity at more variable loci (Aparicio, Ortego, & Cordero, 2006). Low HL of an individual indicates low genetic similarity of its parents, which is expected in populations with high genetic diversity. Using an individual-based measure of genetic diversity, rather than a population-based one (such as AR), enabled us to consider environmental variation within populations (Appendix S4) rather than population averages, which should resolve environmental-genetic associations at finer spatial scales.

Eleven environmental variables reflecting flow regime, connectivity, habitat and climate, previously shown to be indicators of Macquarie perch presence and population health (Appendix S4), were originally considered for inclusion in the model as predictors of

individual genetic diversity (Table 1). All but one were sourced from the National Environmental Stream Attributes Database v1.1.5 (http://www.ga.gov.au/metadata-gateway/metadata/record/gcat_75066) (Geoscience Australia 2012; Stein, 2006) linked via stream segment number to the Geofabric stream network layer (Bureau of Meterology 2012); the data on mean November temperature were downloaded from http://www.bom.gov.au. Environmental data were extracted using ArcGIS 10.2 (Environmental Systems Research Institute 1999-2014) for each sampling site (1–8 sites per population for 16 MDB and HNB populations; Supporting Information). Environmental predictor variables were standardized to zero mean and unit variance, and Pearson's correlations among variables were calculated (using a single record per sampling site). Only seven variables that were not highly correlated (Pearson's $|r| < 0.7$) were included in the final model (Table 1; Appendices S4 and S5).

We used a hierarchical Bayesian regression model to estimate the relationships between HL and environmental variables (Appendix S5). We used a reversible-jump Markov chain Monte Carlo (MCMC) algorithm to perform model selection. Basin and site were included as clustering variables (given exchangeable priors, equivalent to random effects in a standard mixed model) to account for spatial clustering of the samples. The model was fitted using WinBUGS 1.4 (Lunn, Best, & Whittaker, 2009; Lunn, Thomas, Best, & Spiegelhalter, 2000; Lunn, Whittaker, & Best, 2006). Outputs were managed in R 3.1.2 (R Development Core Team 2014). All parameters were assigned vague prior distributions. The primary model outputs are estimates of the probability of inclusion for each environmental variable and the fitted associations between HL and each environmental variable. The prior probability of variable inclusion was 0.5, so that posterior probabilities of inclusion >0.5 provide evidence in favour of variable inclusion. Posterior probabilities of inclusion >0.75 provide strong evidence for variable inclusion (odds ratio >3). Full model details and model code are in Appendix S5.

Fivefold cross-validation was used to estimate the predictive capacity of the fitted model and to test whether the model was likely to be identifying true relationships. To account for possible correlation of HL among individuals from the same genetic clusters due to shared history, cross-validation test data sets comprised nine population clusters, with approximately 120 individuals in each test data set (1–5 population clusters; Appendix S5). We used microsatellite clusters, rather than phylogenetic groups, because recent history (e.g. drift in small populations) could have a strong effect on HL, which can differ across distinct genetic clusters. Model predictions for cross-validation were based on environmental variables only; clustering variables (basin and site) were used in model-fitting but not to make predictions.

2.10 | Simulating genetic rescue/genetic restoration of isolated Macquarie perch populations

To demonstrate genetic rescue and/or genetic restoration effects of translocations to small populations of Macquarie perch, we simulated the outcomes of two management scenarios (do-nothing vs. 50 years of translocation) for two pairs of populations using an age-structured

TABLE 1 Details of the environmental model predicting homozygosity-by-locus (HL, an inverse of genetic diversity): final set of seven variables (first column) and variables highly correlated to them (second column), predicted relationships with HL, probability of inclusion in the model (value >0.7 indicate a strong association with HL, shown in bold) and direction of the effect (positive means that a variable increases HL; negative means that a variable decreases HL). Environmental variables were sourced from National Environmental Stream Attributes Database. Environs are valley bottoms associated with the stream

Environmental predictors	Correlated variables	Predicted relationships with HL	Probability of inclusion	Direction of effect
Flow regime disturbance index calculated for period 1970–2000	Annual mean accumulated soil water surplus, Stream and environs average hottest month maximum temperature, Coefficient of variation of monthly totals of accumulated soil water surplus	Higher genetic diversity (lower HL) within larger and more permanent streams and/or more variance in genetic diversity in small rivers	0.28	Neutral
Barrier-free flow-path length		Higher genetic diversity (lower HL) in larger river fragments	**1**	Positive
Maximum barrier-free flow-path length upstream	Annual mean accumulated soil water surplus	Increase in genetic diversity (decrease in HL) with distance to upstream dam	**0.89**	Negative
Stream segment slope		Higher genetic diversity (lower HL) in streams with higher slope	0.49	Neutral
Stream and valley percentage extant woodland and forest cover		Higher genetic diversity (lower HL) in streams with more vegetated banks	0.19	Positive
Stream and environs average coldest month minimum temperature	Mean segment elevation	Lower genetic diversity (higher HL) in warmer streams	**0.73**	Positive
Mean November temperature		Higher genetic diversity (lower HL) in streams with cooler temperatures during start of the breeding season	0.47	Positive

population model in Vᴏʀᴛᴇx 10.1.6 (Lacy & Pollak, 2014). Vᴏʀᴛᴇx simulates effects of deterministic forces, and those of demographic, environmental and genetic stochasticity on population parameters, such as probability of survival, time to extinction, population size, allelic diversity, heterozygosity and inbreeding (approximated as homozygosity at modelled loci). Age-specific mortality was modelled based on Todd and Lintermans (2015) (details in Appendix S6). Inbreeding depression was modelled by assuming that each founder individual has unique recessive alleles (6.29 lethal equivalents per individual, the empirically derived Vᴏʀᴛᴇx default value). Fifty percent of inbreeding depression was assumed to be due to recessive lethal alleles: inbred offspring with two copies of the same recessive alleles die before reproduction. The translocation scenario assumed that six individuals (3 of each sex) were moved from larger to smaller populations each year for 50 years to approximate the impacts of repeated, modest-sized translocations; designing the most cost-effective translocation programme is beyond the scope of this paper. All translocated adults were assumed to survive translocation. Genetic results were summarized for 1 mitochondrial and 19 microsatellite markers simulated based on observed allele frequencies. The first pair of populations had initial population sizes of 3,000 and 500 (modelled using allele frequencies of Dartmouth and King Parrot, respectively), and the second pair had initial population sizes of 300 and 100 (modelled using Cataract Dam and Murrumbidgee). Five hundred forward-in-time simulations were run for 100 years for each scenario using parameters

outlined in Appendix S6. To compensate for the inability of Vᴏʀᴛᴇx to model large female fecundities, calculations were restricted to adults (Lacy et al., 2015).

3 | RESULTS

3.1 | Population history, divergence time estimates and genetic differentiation

The mitochondrial haplotype network was consistent with historical isolation between the MDB and the HNB, historical isolation of the Lachlan catchment from the rest of the MDB, isolation of Adjungbilly—the only known remnant Macquarie perch population in the lower Murrumbidgee catchment—from other populations, and isolation among HNB populations (Figure 1). Lack of haplotype-sharing among any HNB populations (Appendix S7) suggests that strong genetic drift operates in this basin, consistent with earlier findings from a mitogenome study (Pavlova et al., 2017). In contrast, haplotype-sharing across Murray River tributaries (Dartmouth, Hollands, Sevens, Hughes, King Parrot) indicated historical connectivity within the southern MDB, consistent with a continuous reconstructed historical distribution in the southern MDB (Trueman, 2011). Similarly, haplotype-sharing between Lachlan and Abercrombie (northern MDB) supported previously inferred contemporary connectivity between these two populations (Faulks et al., 2011). The translocated fish from the MDB population into Cataract Dam apparently bore two haplotypes (that were the most common in southern MDB), both of which appear to have escaped into the Cataract River, where three endemic HNB haplotypes also occur. The translocated Yarra population shared haplotypes with most populations from the southern MDB and also had several haplotypes not sampled elsewhere.

Phylogenetic analysis in BEAST (Appendix S8) showed estimates of lineage divergence times consistent with those from complete mitogenomes (Pavlova et al., 2017): time to the most recent common ancestor (TMRCA) of Shoalhaven and HNB+MDB lineages was estimated at 1.061 (95% HPD 0.204–2.840) million years ago, of MDB and HNB lineages (which formed clades with posterior probabilities (PP) of 0.85 and 1, respectively) at 264 (59–674) KY (thousand years) ago, of all HNB haplotypes at 90 (16–240) KY and of all MDB haplotypes at 177 (45–464) KY. The estimates of TMRCA of Lachlan+Abercrombie clade (PP = 1) at 27 (2–88) KY bring forward the previous estimate of isolation of the Lachlan catchment from the rest of the MDB based on partial control region sequence (310 KY; Faulks et al., 2010a). Historical isolation of the lower Murrumbidgee catchment (Adjungbilly) from the upper Murrumbidgee for 36 (8–100) KY (TMRCA of Adjungbilly clade; PP = 0.83) is a novel result for a previously unstudied population.

Strong population differentiation among most HNB populations, all three MDB catchments and three populations of the Murrumbidgee catchment (Adjungbilly, Cotter and Murrumbidgee) was evident from generally high pairwise population Φ_{ST} (mtDNA) and F_{ST} (microsatellites) values (details in Appendices S9 and S10), consistent with Faulks, Gilligan, & Beheregaray (2010b), Faulks et al., (2010a, 2011). An evolutionary timescale of isolation within the HNB was supported by strong mitochondrial divergence among HNB populations and by SPAGeDi analysis, which showed that evolution of microsatellite allele sizes has contributed to divergence among northern HNB (Wollemi and Wheeny) versus southern HNB populations (rest) and within northern HNB (as well as between the HNB and MDB), but not within the MDB (Appendix S10). Large, significant ($p < 0.001$) microsatellite F_{ST} values across almost all pairwise comparisons supported contemporary isolation of populations and strong drift effects, which is a novel result.

3.2 | Assessing population genetic structure at basin- and catchment scales

For all 19 MDB and HNB populations ($N = 870$), STRUCTURE analysis assuming $K = 2$ (Figure 1) was consistent with the HNB-MDB divergence and hybridization of translocated MDB fish with endemic HNB fish in Cataract River (Faulks et al., 2011). Our analysis of a large Cataract River sample showed that individual memberships (Q) in the endemic HNB cluster ranged from 0.68 to 0.001 and were not associated with mtDNA lineage, suggesting that gene flow from Cataract Dam might be ongoing and that hybridization in Cataract River is not limited to the first generation and is not sex-biased (Appendix S11). All other STRUCTURE analyses supported predominantly geographic structure and at least partial contemporary genetic isolation of populations within basins. Analysis of all individuals ($N = 870$) assuming $K = 12$ yielded geographically meaningful genetic clusters (Figure 1),

supported by separate analyses of 12 populations of MDB origin ($N = 671$) and 6 populations from the southern MDB ($N = 375$). Analysis of six HNB populations (all except Cataract River and Cataract Dam; $N = 134$) assigned all HNB populations to different genetic clusters, except a single individual in the Little River had Q > 90% in the Kowmung cluster, suggesting a rare long-distance dispersal event or undocumented translocation (details of all analyses are in Appendix S11). STRUCTURE analyses did not support the previously reported two clusters within Dartmouth and Yarra (Nguyen et al., 2012).

3.3 | Analyses of genetic diversity and effective population sizes

Mitochondrial (Hd and π) and nuclear (AR and He) genetic diversity ranged widely across populations in both basins (Figure 2; Appendix S2). Generally, populations with low nuclear genetic diversity (Wollemi, Whenny, Glenbrook in the HNB and Adjungbilly, Cotter and Murrumbidgee in the MDB) also had low mtDNA diversity (Wollemi was not sequenced here, but had low Hd and π in the study of Faulks et al., 2010a), but mtDNA diversity was also low for some other populations (Abercrombie, Lachlan). Tajima's D was significantly negative for Glenbrook and Adjungbilly, suggesting purifying selection and/or past population growth. In comparison, both translocated populations (Cataract Dam and Yarra) had relatively high levels of genetic diversity (Figure 2). The Shoalhaven individual was homozygous at 17 of 19 loci.

Linkage disequilibrium-based estimates of N_e (LDNe) for five populations with N > 50 were smallest in Cotter (mean N_e=63, confidence interval 35–134), slightly higher in Cataract River (164, 103–349) and Cataract Dam (128, 68–501) and moderate in Dartmouth (307, 201–593) and Yarra (344, 228–642; Appendix S2). ONESAMP N_e estimates for the same five populations were ~2.8× (range 1.5–4.3) lower than LDNe estimates. ONESAMP N_e estimates were low (mean N_e < 100) for all populations except Yarra (mean N_e=132; Figure 2), consistently so for both sets of priors (Appendix S2).

3.4 | Estimated relationships between individual genetic diversity and environmental variables

Approximately 66% of the variation in HL could be explained by environmental and clustering variables (model $r^2 = 0.66$). Environmental variables could predict only c. 17% of the variation in HL (cross-validated $r^2 = 0.17$), suggesting that stochastic processes (e.g. genetic drift in small populations) have a strong effect on genetic diversity of Macquarie perch and/or that one or more key variable impacting HL was not included in the model (e.g. EHN virus; Becker et al., 2013). Three of the seven environmental variables had probabilities of inclusion in the fitted model >0.7, indicating a strong association with HL (odds ratios > 2.3, Table 1, Appendix S5). Maximum barrier-free flow-path length upstream had a negative association with HL (i.e. distance from the upstream dam was associated with higher individual genetic diversity), whereas barrier-free flow-path length, and minimum temperature of the coldest month, had positive associations with HL (Table 1, Figure 3).

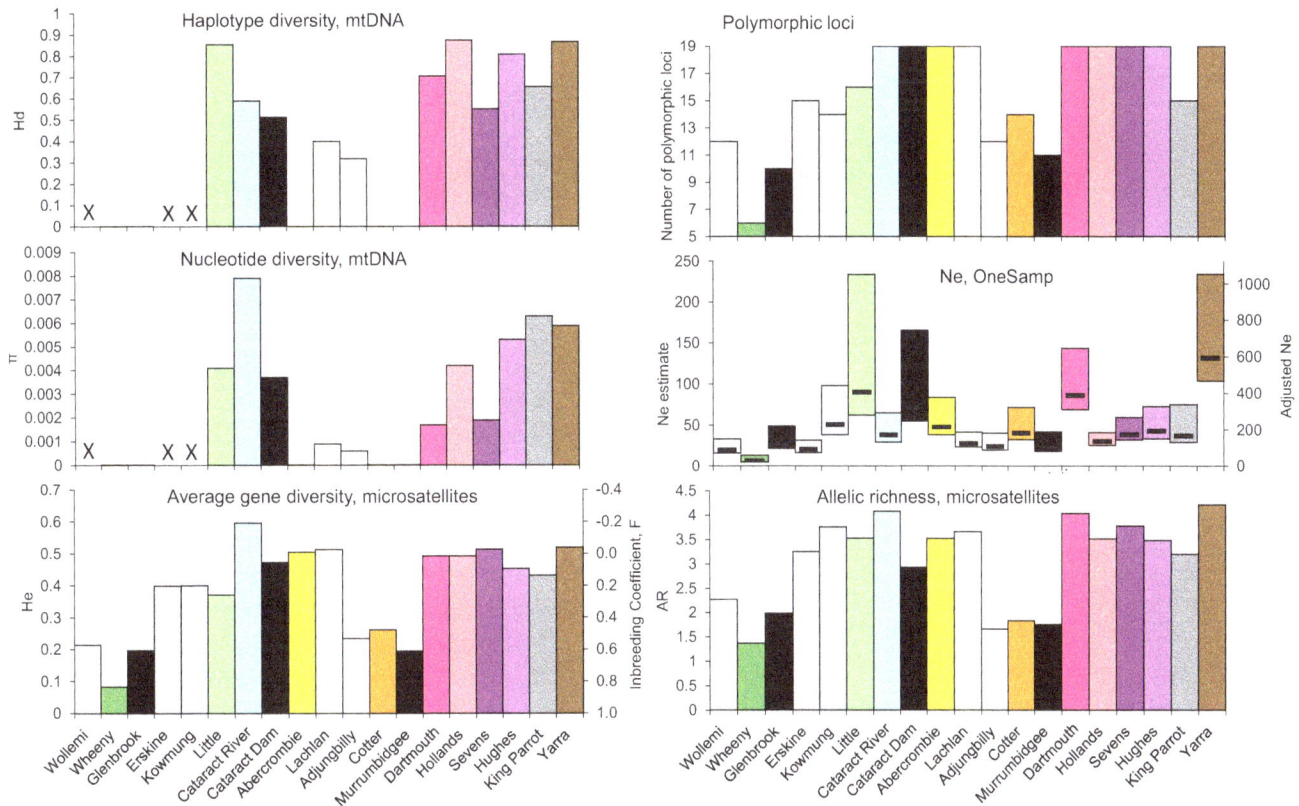

FIGURE 2   Estimates of mitochondrial and nuclear genetic diversity and effective population sizes (N_e, ONESAMP: lower–upper 95% confidence limits, bars show mean N_e) for 19 populations (coloured as on Figure 1). Crosses indicate populations for which mtDNA was not sequenced here (estimates from partial control region (Faulks et al., 2010a) were the following: Wollemi Hd = 0.111, π = 0.0006; Erskine Hd = 0, π = 0; Kowmung Hd = 0.708, π = 0.0076). Inbreeding coefficient F_e for each population is calculated as $F_e = 1 - He/He_{Dartmouth}$; adjusted N_e is calculated as 4.5× ONESAMP N_e (see Section 4; lower and upper bounds are not adjusted)

3.5 | Simulating genetic rescue/genetic restoration of small isolated Macquarie perch populations

For do-nothing scenarios run over 100 years, models for all four populations showed overall decrease in survival probability, decrease in genetic diversity and increase in inbreeding and inbreeding depression over time (Appendix S6). With decreasing initial population sizes (3,000, 500, 300 and 100 for Dartmouth, King Parrot, Cataract Dam and Murrumbidgee, respectively) extinction probability increased, time to the first extinction decreased, inbreeding depression increased (i.e. the number of lethal alleles per individual at year 100 decreased) and proportion of retained heterozygosity for extant populations decreased ($He_{year100}/He_{year0}$: 0.98, 0.89, 0.85 and 0.79; Table 2). Final N_e-values estimated based on allele frequency changes resulted in a mean N_e/N ratio of 0.17 (range 0.1–0.26; Table 2).

Results of 50-years-of-translocation scenarios showed that addition of individuals into King Parrot and Murrumbidgee, compared to do-nothing scenarios, decreased population probability of extinction 35- and 26-fold, respectively, increased time to first extinction and decreased inbreeding depression 1.1- and 1.5-fold, and for populations extant at year 100, increased population sizes (2- and 3.8-fold), heterozygosity (1.2- and 2.2-fold) and number of alleles (2.4- and 2.5-fold; Table 2, Figure 4). The genetic restoration effect was rapid: in the first

10 years of translocations, the number of alleles in King Parrot reached 90%, and in Murrumbidgee >100% of those of the source populations (Appendix S6). Genetic rescue from inbreeding depression (i.e. larger number of lethal alleles per individual under 50-years-of-translocation compared to the do-nothing scenario; Figure 2) was apparent for the small Murrumbidgee population at year 10 and for the larger King Parrot at year 40. Genetic rescue/restoration effects were long-lasting: genetic diversity remained higher and inbreeding lower than in do-nothing scenario even 50 years after cessation of translocations. Effect of harvest (for translocation) on population size, heterozygosity, number of alleles or inbreeding of both source populations was minimal, but harvest increased the probability of extinction and decreased mean time to extinction for the smaller Cataract Dam (and to a lesser extent, the larger population at Dartmouth; Table 2, Appendix S6).

4 | DISCUSSION

4.1 | The need for specieswide genetic management to prevent further loss of genetic diversity and boost adaptive potential

Using novel genetic data, we investigated whether genetic augmentation is likely to be beneficial and necessary for long-term

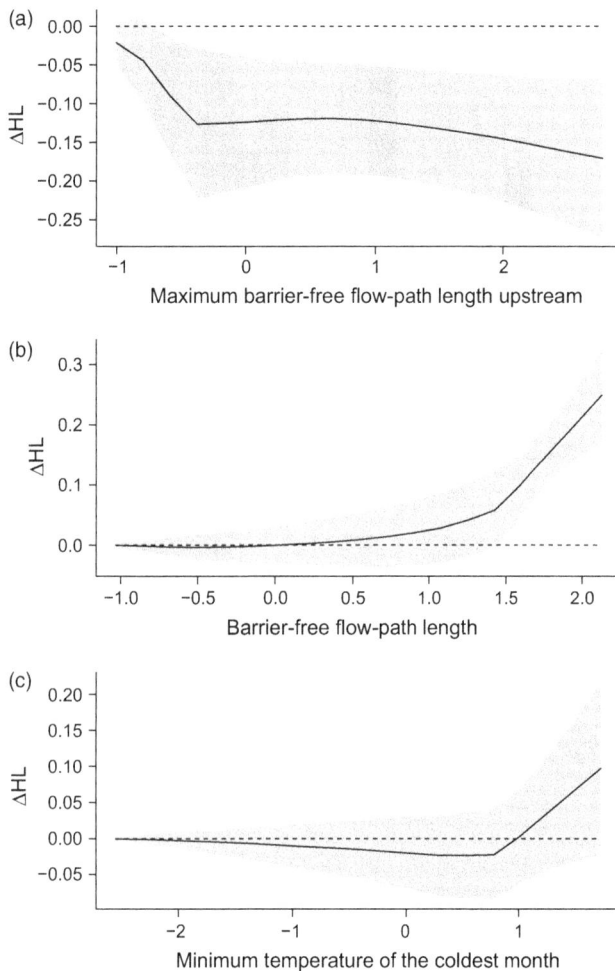

FIGURE 3 Fitted relationships between HL and environmental variables for variables with probability of inclusion greater than 0.7. The x-axis shows standardized values for each variable (e.g. a value of 1 means 1 standard deviation above the mean value for that variable), and the y-axis shows deviation from the mean HL value (in units of HL) for a given standardized value of the predictor variable. Grey shading is one standard deviation of the fitted effect. Models included basin and site as clustering variables, so these fitted effects account for differences among sites or between basins

survival of an endangered freshwater fish, the Macquarie perch. The species has experienced dramatic declines in range and numbers of most populations, intense habitat fragmentation, levels of inbreeding at which inbreeding depression is often observed in wild populations and continuing loss of populations despite intensive management. Assuming levels of inbreeding depression similar to those reported for other wild populations, our simulations support a decline in population viability unless there is active genetic management. We believe that these combined lines of evidence provide sufficient support for genetic augmentation to be implemented now, despite a lack of data on fitness decline, which may take too long to collect in order to be useful for preventing extinctions of many populations (Weeks et al., 2016).

Significant investment in habitat restoration and protection has failed to lead to detectable recovery of some Macquarie perch populations (Mark Turner, GBCMA pers. comm.). Although no data on

fitness loss exist, it is reasonable to assume that inbreeding depression might have contributed to this failure. Our simulations suggest that inbreeding depression may contribute to population declines and impede population recovery in the near future. Our estimates of N_e suggest that most populations will have limited capacity to adapt to rapid environmental changes. Thus, genetic problems should be addressed in conjunction with other key threats to population persistence. Restoring the historical, within-basin genetic connectivity of Macquarie perch via assisted gene flow will improve genetic diversity and adaptive potential across the species' range. Our simulations indicate that small-scale translocations (6 adults/year) will result in rapid and long-lasting increase in genetic diversity (limited only by genetic diversity and divergence of the source(s)) and decrease in inbreeding and probability of population extinction. Although large source populations should not be impacted by appropriately limited harvest (e.g. for translocation or captive breeding), smaller sources may become more vulnerable to extinction, loss of genetic diversity and inbreeding. Therefore, where possible, management should consider reciprocal translocations and multiple sources.

Our Vortex model has a number of limitations: (i) offspring that do not survive to reproduction were not modelled (due to software limitation), but, given extreme fecundity (up to 110,000 eggs/female; Cadwallader & Rogan, 1977), strong purifying selection acting on juveniles could purge deleterious mutations, reducing inbreeding depression; (ii) individuals at the start of simulations were assumed to be unrelated, which, if incorrect, artificially increases the time when inbreeding depression starts to impact viability; (iii) mutations were not modelled, and thus, novel genetic diversity was not accounted for; (iv) adaptive capacity was not addressed; (v) there are large uncertainties around environmental effects and population parameters, including an assumption that all individuals survive translocation. Nevertheless, our simulations are in agreement with previous simulation studies (Allendorf & Ryman, 2002; Rieman & Allendorf, 2001) that under management inaction, populations of 100–500 adults will lose genetic diversity over time (>10% of alleles and >10% of heterozygosity in 100 years), and the loss will be faster in smaller populations. Inference of low genetic diversity in Macquarie perch (mean observed heterozygosity across populations Ho = 0.371, estimated from 19 loci with 13.5 alleles/locus on average) was consistent with similar inferences for other threatened Australian freshwater fishes, Yarra pygmy perch *Nannoperca obscura* (Ho = 0.318; 14 loci, 11.9 alleles/locus; Brauer, Unmack, Hammer, Adams, & Beheregaray, 2013), southern pygmy perch *N. australis* (Ho = 0.47; 12 loci, 10.3 alleles/locus; Cole et al., 2016) and dwarf galaxias *Galaxiella pusilla* (Ho = 0.395; 11 loci, 16.3 alleles/locus; Coleman et al., 2010), although estimates of Ho themselves might not be strictly comparable when estimated from marker sets of different variability (Hedrick, 1999). Nevertheless, higher diversity reported for the golden perch *Macquaria ambigua* (Ho = 0.52; 8 loci, 15.3 alleles/locus; Faulks et al., 2010b), the more common and abundant Macquarie perch congener, is consistent with an earlier finding of a mean decrease of 35% in heterozygosity in 170 threatened taxa compared to their nonthreatened relatives (Spielman, Brook, & Frankham, 2004).

TABLE 2 Summary of Vortex simulations of two management scenarios. In 50-years-of-translocation scenarios, King Parrot and Murrumbidgee populations are supplemented by individuals from Dartmouth and Cataract Dam, respectively

Scenario	Do-nothing		50-years-of-translocation		Do-nothing		50-years-of-translocation	
Populations	Dartmouth	King Parrot	Dartmouth	King Parrot	Cataract Dam	Murrumbidgee	Cataract Dam	Murrumbidgee
Initial population size, N	3,000	500	3,000	500	300	100	300	100
Probability of extinction	0.002	0.21	0.022	0.006	0.36	0.828	0.776	0.032
Time to first extinction, years	85	82	76	92	78	63	42	90
N_e at Year 100	340	61	257	−221	44	31	42	−18
N (extant) at Year 100	3,269	337	3,743	680	337	118	478	450
N_e/N	0.104	0.181	Not estimated	Not estimated	0.131	0.26	Not estimated	Not estimated
He at year 0	0.491	0.426	0.491	0.426	0.472	0.19	0.472	0.19
He at year 100	0.481	0.379	0.478	0.442	0.401	0.151	0.398	0.331
Number of alleles at year 0	5.68	3.52	5.68	3.52	3.36	1.8	3.36	1.8
Number of alleles at year 100	5.03	2.92	4.95	4.26	2.6	1.5	2.6	2.86
Number of haplotypes at year 0	6	4	6	4	2	1	2	1
Number of haplotypes at year 100	5.53	2.53	5.36	6.11	1.73	1	1.75	2.51
Number of lethal alleles/individual at year 0	3.15	3.15	3.15	3.15	3.15	3.15	3.15	3.15
Number of lethal alleles/individual at year 100	3	2.53	2.94	2.8	2.28	1.83	2.44	2.72

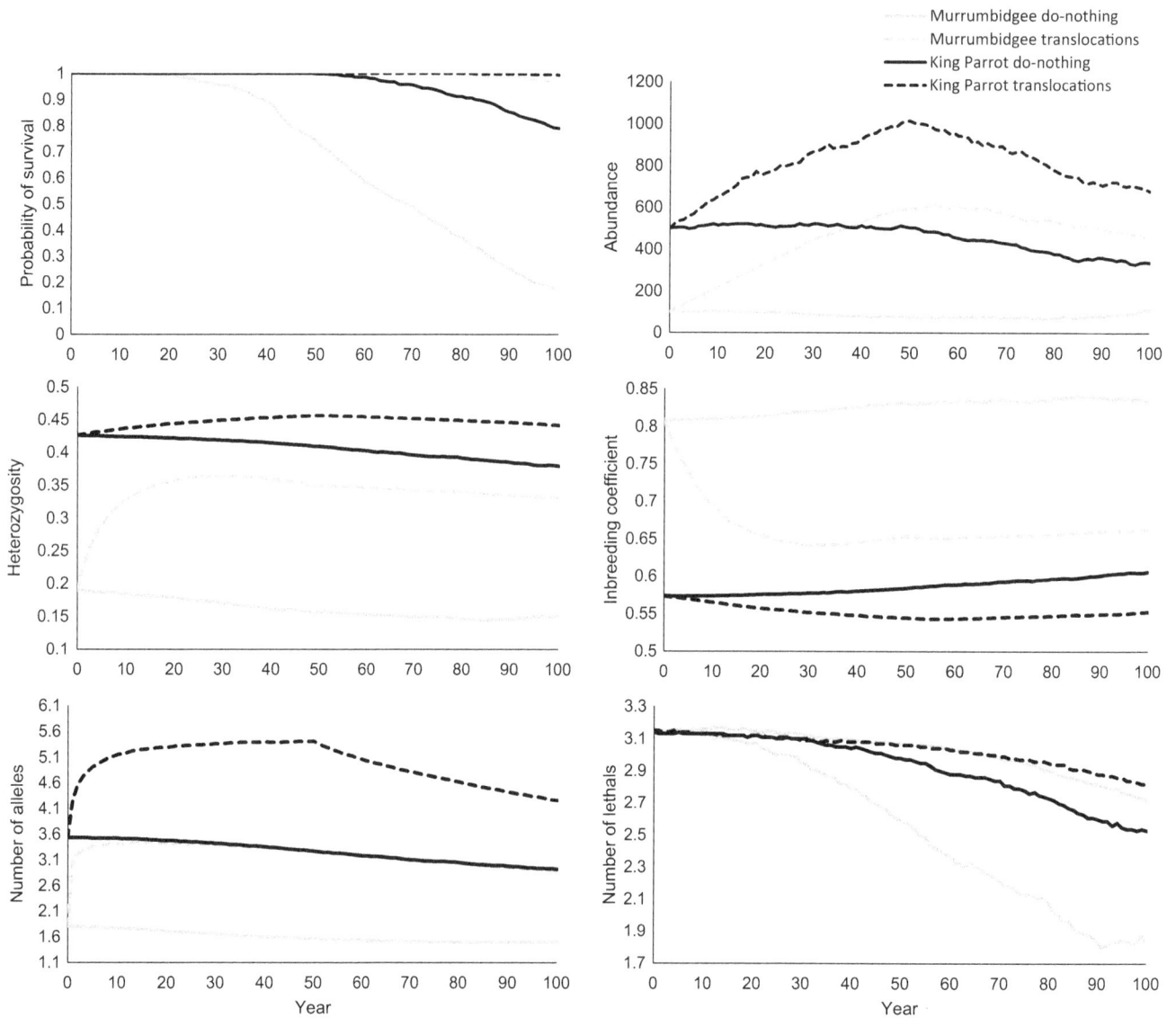

FIGURE 4   Results of Vortex simulations for King Parrot (initial population size *N* = 500; black line) and Murrumbidgee (*N* = 100; grey line) under do-nothing (solid line) or 50-years-of-translocation (dashed line) scenarios

Although in simulations inbreeding depression did not appear to impact the largest population (initial *N* = 3,000, VORTEX N_e = 257) for at least 50 years (~7 Macquarie perch generations), it started to impact three smaller populations (initial *N* ≤ 500, VORTEX N_e < 100) within the first 10 years (i.e. <2 generations), supporting the short-term inbreeding threshold N_e of 100 (Frankham et al., 2014). Inbreeding can be approximated for a small population by scaling its genetic diversity by the diversity of a known outbred population using the effective inbreeding coefficient F_e: $F_e = 1 - He_{inbred}/He_{outbred}$, where He_{inbred} is heterozygosity (for neutral variation) of a population in question and $He_{outbred}$ is heterozygosity of an outbred population (Frankham, 1998). If Dartmouth, one of the largest natural populations of Macquarie perch in the MDB, is used as an outbred reference, then F_e coefficients for Wollemi, Wheeny, Glenbrook, Adjungbilly, Cotter and upper Murrumbidgee are >0.45, and for Little is 0.25 (i.e. is as high as F_e for the offspring of full siblings; Figure 2). At F_e of 0.2, inbreeding

depression is typically observed for populations of naturally outcrossing species (Frankham, 1995; Szulkin & Sheldon, 2007; Walling et al., 2011; Woodworth et al., 2002). Consistently, our simulations show that within a decade, inbreeding depression can be expected to act in Macquarie perch populations of ≤500 adults, unless translocations are performed. Although theoretically one effective migrant per generation may be sufficient to provide inbreeding connectivity (i.e. significantly reduce harmful effects of inbreeding; Mills & Allendorf, 1996), as well as adaptive connectivity (i.e. potential of highly advantageous alleles to spread in a recipient population; Rieseberg & Burke, 2001), it is not expected to drive F_e below 0.2 (Lowe & Allendorf, 2010). In any case, well before F_e climbs to 0.2, many populations will already be suffering inbreeding depression. A long-term study of inbreeding depression in the wild using genomic estimates of inbreeding detected major negative impacts on lifetime reproductive success at *F* even below 0.1 (Huisman, Kruuk, Ellis, Clutton-Brock, & Pemberton, 2016). Thus, more than one effective migrant per generation may be required to drop F_e values sufficiently (e.g. below 0.1) to rescue genetically depauperate populations from strong inbreeding depression (Frankham et al., in press; Weeks et al., 2011).

Contemporary N_e related to inbreeding is best estimated using two or more temporal genetic samples (inbreeding N_e; Luikart, Ryman, Tallmon, Schwartz, & Allendorf, 2010), or using a sample from a single-cohort and adjusting estimates using life-history traits (Waples et al., 2014). N_e relevant to adaptive potential is best estimated from two or more genetic samples taken several generations apart and measuring the change in allele frequencies over time (variance N_e; Luikart et al., 2010). Our data were not suitable for temporal analyses: although sampling was spread over years, temporal samples spanning generations at individual sites were lacking, but sampling was random, comprising multiple generations, and overlapped 15-year drought, during which many populations ceased regular breeding. Because N_e and population size are key parameters for evaluating existing thresholds and predicting future population trajectories, here we used single-sample N_e estimates corrected for biases, along with N_e/N ratios from simulations based on the observed allele frequencies, to estimate bounds on population sizes of Macquarie perch. Although we believe that our estimates of N_e and N should be sufficiently accurate for the purposes stated above, they are associated with a number of untested or violated assumptions (see below) and hence should be interpreted with caution and re-examined when genomic data become available (Hollenbeck, Portnoy, & Gold, 2016). If we assume that (i) ONESAMP N_e estimate is systematically downwardly biased and is ~2.8× (range 1.5–4.3) lower than LDNe N_e (based on five samples of >50 individuals with both estimates; Appendix S2), (ii) LDNe N_e estimate is ~1.6× (1.1–2) lower than the true N_e (Waples et al., 2014) and (iii) N_e is ~5.9× (3.8–9.6) lower than the number of adult Macquarie perch N (Table 2), then (i) the true N_e for each population could be ~4.5× (1.7–8.6) higher than ONESAMP N_e estimate (adjusted N_e; Figure 2), and (ii) adult population size N could be ~44.4× (6.27–82.6) higher than ONESAMP N_e estimate (Appendix S2).

Given that inbreeding N_e is similar to variance N_e for this species (see Section 2), comparisons of the above approximations with 100/1,000 minimum N_e thresholds of Frankham et al. (2014) suggest that (i) the majority of populations have mean $N_e < 600$ (lower bound of $N_e < 100$, upper bound of $N_e < 1,000$), implying that genetic restoration is required to boost population adaptive potential and avoid inbreeding depression within a few decades; (ii) Wollemi, Wheeny and Erskine have mean $N_e < 100$, implying they might experience inbreeding depression within a few generations even if they do not already; (iii) the majority of populations have mean sizes <3,000 (lower bound <500, upper bound <10,000), which our simulations suggest may lead to genetic problems in the next decade; (iv) Little, Cataract Dam, Dartmouth and Yarra have the largest N_e and mean sizes >3,000 and thus are the best sources for translocations. We note that for four populations with sample sizes of 30 individuals (Upper Murrumbidgee, Hollands, Sevens and King Parrot) LDNe estimates of N_e had means and ranges <100 (e.g. outside the range of Tallmon et al. (2010)'s simulations) and were smaller than ONESAMP estimates, suggesting that

for small populations these ONESAMP-based approximations could be strongly upwardly biased (Appendix S2). For example, the number of adults monitored at the Upper Murrumbidgee site is estimated at ~20–50 (Lintermans, unpublished data), whereas our ONESAMP-based approximation provides a much more optimistic value of 113–3,500 (the LDNe-based approximation is closer, 13–307; Appendix S2).

For an ideal Wright–Fisher population of a stable size, different methods of estimating N_e (e.g. from LD, loss of genetic diversity over generation, drift among replicates over generations and/or inbreeding coefficients) should yield similar estimates. However, when conditions of the Wright–Fisher population model are violated (e.g. by the effects of unequal sex-ratio, variance in family sizes and/or fluctuations in population size over generations), different methods yield estimates related to different time frames and spatial scales (Crow & Kimura, 1970; Wang, 2005). The calculations above do not correct for biases related to violations of the Wright–Fisher model. In declining populations, estimates of N_e are expected to be upwardly biased, reflecting a previously large N_e one to two generations earlier (for LDNe) or several generations in the case of ONESAMP, and declines are observed in some Macquarie perch populations (e.g. Yarra; Tonkin, unpublished data). Despite much uncertainty around our approximations of population sizes, many of our results point to the need to restore genetic diversity by translocations in most Macquarie perch populations. Reconnection of several populations by gene flow might be required to achieve a global $N_e > 1,000$ necessary to maintain population adaptive potential and evolvability in the face of environmental change (Weeks et al., 2011). Measuring N_e dynamics using multiple temporal genetic samples will allow managers to monitor the effects of assisted gene flow on N_e. When genomic data including those on physical linkage or genomic position of genetic markers become available for Macquarie perch, contemporary N_e and its recent change could be estimated from contemporary samples (Hollenbeck et al., 2016).

4.2 | Designing translocation strategies while considering risks of outbreeding depression and loss of local adaptation

Selecting appropriate sources of admixture requires simultaneous consideration of the timescale of population differentiation and the risks of inbreeding versus outbreeding depression and loss of local adaptation (Frankham et al., 2011; Love Stowell et al., 2017; Weeks et al., 2011, 2016). Long-term divergence implies that reproductive isolation may have evolved through adaptive differentiation and/or drift; hence, admixture between regions showing histories of long-term divergence and/or local adaptation should be undertaken with caution (Frankham et al., 2011). Long-term divergence of the MDB and HNB Macquarie perch lineages (119–385 KY; Pavlova et al., 2017) in different environments (e.g. Appendix S4) might have resulted in adaptive differences between basins. However, analyses of complete mitogenomes suggest that genetic drift was the prominent force driving mitochondrial divergence; drift was stronger in the HNB either due to smaller historical effective population sizes (approximated by allelic richness or haplotype diversity) or weaker environmental pressures (Pavlova et al., 2017). In accordance with mitochondrial differentiation, the microsatellite genotype of the single individual from the Shoalhaven Basin population differed markedly from those of the HNB and MDB lineages (Appendices S8 and S9). The fact that the Shoalhaven individual was homozygous at 90% of loci suggests that loss of genetic variation and individual inbreeding accompanied extinction of the endemic Shoalhaven lineage. The widespread and not sex-limited hybridization beyond first generation of HNB and MDB lineages in lower Cataract River provides no evidence for outbreeding depression, but evidence for absence of genomic incompatibilities and unimpeded fitness of hybrids is lacking (Verhoeven, Macel, Wolfe, & Biere, 2011). Separate management of Macquarie perch lineages from different basins, which is current practice, is warranted until an informed risk–benefit analysis suggests otherwise (Frankham et al., 2011). Likewise, long-term (58–191 KY; Pavlova et al., 2017) divergence between the northern (here: Wollemi, Wheeny) and southern HNB populations (here: the remaining HNB populations) suggests that genetic rescue/restoration be conducted independently within these regions, unless an informed risk–benefit analysis indicates otherwise.

In some cases, nearby (presumably similarly adapted) source populations do not exist, so there may be no option but to source individuals for translocation from a more strongly divergent population (Love Stowell et al., 2017; Weeks et al., 2011), as suggested for threatened southern pygmy perch from the MDB based on genotype–environment association analyses (Brauer, Hammer, & Beheregaray, 2016). In the Lachlan catchment (MDB), two sampled Macquarie perch populations are recent derivatives from the same source (and are connected by unidirectional gene flow from the Abercrombie to Lachlan; Faulks et al., 2011); thus, little genetic rescue/restoration effect would be expected from their admixture (Frankham, 2015). However, the estimate of divergence of the Lachlan catchment from the other MDB catchments (2–88 KY ago) is not recent enough to dismiss the possibility of outbreeding depression under admixture with southern MDB populations under recent divergence criteria (Frankham et al., 2011). The range of likely adjusted N_e values (41–360) and approximate population sizes (150–3,500) for the two Lachlan catchment populations suggests that under a do-nothing scenario these populations will face inbreeding and loss of genetic diversity within a few generations. Although knowing with certainty the fitness consequences of cross-catchment admixtures would be desirable, the need for management intervention to prevent loss of genetic diversity and inbreeding in the near future may necessitate gene flow from larger populations of fish of MDB genetic origins (e.g. Dartmouth or Yarra) before the risk of outbreeding depression can be assessed through breeding experiments (Frankham, 2015; Frankham et al., in press). This suggestion is in line with calls for focussing on preservation of genetic diversity at the species level, rather than population, subspecies or evolutionary significant unit (Love Stowell et al., 2017; Weeks et al., 2016).

In cases when it is unclear whether mitochondrial divergence confers selective advantage on local mitogenomes (such as Adjungbilly, inferred to have diverged from the other populations of the Murrumbidgee catchment ~8–100 KY ago), genetic rescue could involve within-catchment translocation of males. Assuming that on this timescale mitochondrial divergence did not drive mito-nuclear co-evolution (Wolff, Ladoukakis, Enriquez, & Dowling, 2014), this translocation approach would preserve any potential selective advantage of local mtDNA while providing nuclear gene flow for increasing N_e (Weeks & Corrigan, 2011).

In contrast to cases of long-term divergence, admixture of populations that have diverged only recently (including due to recent anthropogenic barriers to movement) should be associated with very low risk of outbreeding depression or loss of local adaptation (Frankham, 2015), whereas managing these populations separately will decrease their adaptive potential and increase species extinction risk (Coleman et al., 2013; Weeks et al., 2016). In the HNB, extreme divergence between the two northern Macquarie perch populations (Wollemi and Wheeny) is likely due to recent isolation by anthropogenic barriers (Faulks et al., 2011) and strong effects of genetic drift in very small populations (mean adjusted N_e < 100). In the MDB, historical records indicate that prior to arrival of Europeans in Australia, Macquarie perch were common along the length of the Murrumbidgee River and in many tributaries of the Murray River (including lowland zone of the Central Murray River), suggesting that the Murrumbidgee and Murray catchments were previously connected by gene flow (Trueman, 2011). Historical cross-catchment connectivity is supported by the most common MDB haplotype (10 on Figure 1) being shared between Murrumbidgee and Murray catchments. Current differentiation among populations within the Murrumbidgee and Murray catchments (likely resulting from very recent isolating effects of large dams and impoundments) should be reversed by within-catchment admixtures. Occasional cross-catchment translocations should also be considered to improve population adaptive potential. Designing cross-catchment translocations (e.g. from the Murray catchment to the Lachlan or Murrumbidgee catchments) would benefit from further modelling and captive breeding experiments measuring relative fitness of intercatchment crosses. Similar recommendations were suggested for other threatened Australian fishes with anthropogenically impeded genetic connectivity (Brauer et al., 2013, 2016; Cole et al., 2016).

Historically translocated populations, that have been isolated for sufficient generations to accumulate allelic differentiation, represent viable sources for translocations back into their population of origin if they contain diversity lost from the recipient population (Frankham, 2015). Two translocated populations of Macquarie perch, Cataract Dam, established from the Murrumbidgee River in 1916 (Legislative Assembly of New South Wales 1916) and the Yarra, sourced between 1907 and 1943 from multiple southern MDB populations including King Parrot Creek, Broken River and the Goulburn River (where the species is now locally extinct; Cadwallader, 1981; Trueman, 2011; Ho & Ingram, 2012), could be used as source populations for low-risk translocations to Murrumbidgee and Murray catchments, respectively. Given their high genetic diversity, Cataract Dam and Yarra populations may constitute particularly viable translocation options, provided their N_e and N can support it.

Finally, our simulations suggest that in cases when conditions of a remnant population are unlikely to improve enough to facilitate long-term sustainability (as in creeks lacking connectivity with more stable river systems, and with insufficient water availability under climate change scenarios), translocating all remaining individuals into a more sustainable system might be more beneficial than attempting to maintain the original populations, for the overall chance of species survival and adaptive potential. Successful examples of fish rescue during extreme drought and bushfire runoff already exist (including Macquarie perch rescue; Kearns, 2009; Hammer et al., 2013, 2015). Management plans incorporating such actions could have prevented high mortality and loss of genetic diversity of the Broken River population of Macquarie perch during the Millennium Drought of 1995–2009; only three individuals have been detected in the reach since 2011, whereas prior to the drought this population was considered stable (Kearns, unpublished data).

4.3 | Environmental correlates of genetic diversity

Environmental modelling suggested that Macquarie perch were more genetically diverse (i.e. had lower HL) if they were (i) further downstream from a barrier (dam wall, spillway or large dam) or had larger flow-path lengths upstream, (ii) in a short-to-medium, but not long, stretch of a barrier-free fragment of the stream, and (iii) in streams that are colder in winter. Lower genetic diversity in sites immediately downstream from dams could indicate that cold-water pollution and/or other aspects of water supply releases (e.g. changes in habitat quality/availability) have a long-lasting effect on Macquarie perch populations, as was previously inferred for Murray cod (Todd, Ryan, Nicol, & Bearlin, 2005). Indeed, the Macquarie perch population in the Mitta Mitta River downstream of Dartmouth Dam in Victoria was extirpated following the dam's construction, with cold-water pollution the most likely cause (Koehn, Doeg, Harrington, & Milledge, 1995). Alternatively, the correlation between barrier-free flow-path length upstream and mean annual flow ($r = 0.78$, Appendix S5) suggests that larger streams (with higher mean annual flow) might be harbouring larger populations, for example by providing more habitat and/or by being more resilient to droughts or other disturbances. Unfortunately, few reaches with these attributes remain, with the overwhelming majority of mid- and upland reaches of large rivers where the species were once abundant (Trueman, 2011) now containing, or regulated heavily by, major weirs and dams. High levels of correlation among environmental variables make it challenging to identify underlying causal drivers.

Our second result, that individuals are less diverse in very long connected stream fragments (>1.5 standard deviations from the mean length; Figure 3), is counterintuitive, as one would expect longer barrier-free paths to provide more habitat. The longest barrier-free flow-path lengths are restricted to three MDB populations: Adjungbilly, upper Murrumbidgee and Sevens (Appendix S4), and are largest (>2 standard deviations) in Adjungbilly and upper Murrumbidgee, the two populations with low genetic diversity. Thus, a significant relationship between barrier-free flow-path length with HL could be spurious and driven by population history or other correlated variables (e.g. limited access to suitable spawning sites or presence of natural

barriers that subdivide population – that is, waterfalls, not considered for this metric). The final finding that individuals are less diverse in streams that are warmer in winter raises the possibility that some life stages of Macquarie perch could be sensitive to warmer winters and that climate warming may further exacerbate loss of genetic diversity. Although we found some potentially meaningful environmental correlates of individual genetic diversity, the data were limited in their ability to distinguish whether environmental factors or population history (genetic drift/gene flow/mutation) are the main determinants of genetic diversity, mainly because the species is limited to very few small and isolated upstream reaches, without much environmental variation across sites in each (Appendix S4).

4.4 | Summary of management recommendations

We found that most Macquarie perch populations are associated with low levels of genetic diversity and effective population size. Given that Macquarie perch continue to experience ongoing declines in spite of habitat restoration and threat amelioration efforts, these populations are at risk of extinction, at least in part due to negative genetic effects (e.g. inbreeding depression, low adaptive potential). To elevate genetic diversity, N_e and population sizes, we recommend reconnecting populations *within* the following regions with assisted gene flow: (i) northern HNB; (ii) southern HNB; (iii) the Murray catchment, where Yarra is recommended as an additional source of translocations; and (iv) the Murrumbidgee catchment, where Cataract Dam is recommended as an additional source. Several lines of evidence suggest that the mixing recommended above poses a negligible risk of outbreeding depression and swamping of local adaptation: (i) many populations are derived from a larger ancestral population with genetic diversity primarily lost through drift in populations that are small through human actions; (ii) these populations have only recently been isolated; thus, genomic incompatibilities are unlikely; (iii) the admixed Yarra population appears to be healthy and self-sustainable. Managing populations within these regions as a metapopulation is likely to be beneficial for long-term population viability.

Genetic rescue/restoration is a powerful management tool for a number of reasons: (i) it can have a large and long-lasting impact on genetic diversity and adaptive potential for relatively little additional work and expense (Frankham, 2015), (ii) it does not necessarily require ongoing intervention, although if populations remain small, assisted gene flow will need to be performed periodically (Hedrick, Peterson, Vucetich, Adams, & Vucetich, 2014), (iii) it could work at a range of scales from single species rescue (Hedrick & Fredrickson, 2010) to landscape reconnections (Smith, van der Ree, & Rosell, 2015). Population viability analysis can assist in developing objective and measurable criteria of metapopulation connectivity (Carroll, Fredrickson, & Lacy, 2014; Lacy, 2000), provided sufficient knowledge of species biology exists. Empirical results from other systems could be used to help provide priors for unknown parameters (e.g. inbreeding depression) in population viability models used to predict population persistence under management inaction and various scenarios of genetic intervention. Translocations of juveniles, in addition to adults, over several consecutive years have been advocated when there are limited adult fish available in a source populations (Todd & Lintermans, 2015). For these cases, we argue for translocating genetically diverse sets of individuals, achieved by collecting a small number of juveniles from each of many sampling sites. Numbers of translocated juveniles would need to account for high juvenile mortality (e.g. for Macquarie perch ~56 1-year-old juveniles would be required to produce 6 adults; Appendix S6; Todd and Lintermans (2015)).

Even if direct evidence of genetic problems (fitness loss) in specific cases is regarded as desirable, it should not be required to justify genetic intervention for small populations. Genetic augmentation alone often may not be sufficient to reverse population declines, but it must accompany other threat management actions for small populations (Frankham et al., in press; Love Stowell et al., 2017). Performed and monitored carefully within a risk-assessment framework (e.g. after considering the risk of outbreeding depression), in most instances genetic augmentation will be beneficial, and not harmful, for long-term population viability.

ACKNOWLEDGEMENTS

This work was supported by ARC Grant LP110200017 to Monash University, Flinders University of South Australia and the University of Canberra, with Partner Organization University of Montana. Funding and other support were contributed by industry Partner Organizations Icon Water (formerly ACTEW Corporation), Melbourne Water, Department of Sustainability and Environment (Victoria) and Fisheries Victoria (now within DEDJTR - Department of Economic Development, Jobs, Transport and Resources). Goulburn Broken Catchment Management Authority (GBCMA) provided the funds to undertake fish surveys and collect genetic material. Additional funding was provided to AP and PS by the Fisheries Victoria Recreational Fishing Grants programme (DEDJTR) and to DG by NSW DPI—Fisheries. Computationally intensive analyses were performed on a Monash computer cluster. We thank M. Asmus, R. Ayres, P. Boyd, B. Broadhurst, A. Bruce, T. Cable, R. Clear, J. Doyle, B. Ebner, L. Farrington, L. Faulks, D. Hartwell, P. Heath, J. Knight, J. Mahoney, A. McDonald, T. McGarry, J. O'Mahony, A. Pickworth, K. Pitman, S. Raymond, M. Rodgers, J. Stanger, D. Stoessel, B. Talbot, M. Timmins and P. Unmack for sample collection, M. Roitman, A. Sunnucks, R. Albury for laboratory assistance, D. Tallmon for provision of the locally installable version of ONESAMP, L. Faulks for curatorial assistance of earlier samples and J. Thomson for statistical advice. We are grateful to the Chief Editor Prof. Louis Bernatchez, Associate Editor Prof. Craig Primmer and three anonymous reviewers, whose comments helped us to improve the manuscript.

standardized environmental variables for 16 populations). Sequence data are also deposited to GenBank (accession KT626048–KT626381).

REFERENCES

Alho, J. S., Valimaki, K., & Merilä, J. (2010). Rhh: An R extension for estimating multilocus heterozygosity and heterozygosity-heterozygosity correlation. *Molecular Ecology Resources, 10*, 720–722.

Allendorf, F. W., & Ryman, N. (2002). *The role of genetics in population viability analysis. Population viability analysis* (pp. 50–85). Chicago: University of Chicago Press.

Aparicio, J. M., Ortego, J., & Cordero, P. J. (2006). What should we weigh to estimate heterozygosity, alleles or loci? *Molecular Ecology, 15*, 4659–4665.

Appleford, P., Anderson, T., & Gooley, G. (1998). Reproductive cycle and gonadal development of Macquarie perch, *Macquaria australasica* Cuvier (Percichthyidae), in Lake Dartmouth and tributaries of the Murray–Darling Basin, Victoria, Australia. *Marine & Freshwater Research, 49*, 163–169.

Attard, C., Möller, L., Sasaki, M., Hammer, M., Bice, C., Brauer, C., ... Beheregaray, L. (2016). A novel holistic framework for genetic-based captive-breeding and reintroduction programs. *Conservation Biology, 30*, 1060–1069.

Bandelt, H.-J., Forster, P., & Roehl, A. (1999). Median-joining networks for inferring intraspecific phylogenies. *Molecular Biology & Evolution, 16*, 37–48.

Becker, J. A., Tweedie, A., Gilligan, D., Asmus, M., & Whittington, R. J. (2013). Experimental infection of Australian freshwater fish with epizootic haematopoietic necrosis virus (EHNV). *Journal of Aquatic Animal Health, 25*, 66–76.

Benson, J. F., Mahoney, P. J., Sikich, J. A., Serieys, L. E., Pollinger, J. P., Ernest, H. B., & Riley, S. P. (2016). Interactions between demography, genetics, and landscape connectivity increase extinction probability for a small population of large carnivores in a major metropolitan area. *Proceedings of the Royal Society B, 283*, 20160957.

Brauer, C. J., Hammer, M. P., & Beheregaray, L. B. (2016). Riverscape genomics of a threatened fish across a hydroclimatically heterogeneous river basin. *Molecular Ecology, 25*, 5093–5113.

Brauer, C. J., Unmack, P. J., Hammer, M. P., Adams, M., & Beheregaray, L. B. (2013). Catchment-scale conservation units identified for the threatened Yarra Pygmy Perch (*Nannoperca obscura*) in highly modified river systems. *PLoS One, 8*, e82953.

Bureau of Meterology. (2012). *Australian hydrological geospatial fabric version 2, Bureau of Meterology.* Retrieved from http://www.bom.gov.au/water/geofabric

Cadwallader, P. L. (1978). Some causes of the decline in range and abundance of native fish in the Murray–Darling river system. *Proceedings of the Royal Society of Victoria, 90*, 211–224.

Cadwallader, P. (1981). Past and present distributions and translocations of Macquarie perch *Macquaria australasica* (Pisces: Percichthyidae), with particular reference to Victoria. *Proceedings of the Royal Society of Victoria, 93*, 23–30.

Cadwallader, P., & Rogan, P. (1977). The Macquarie perch, *Macquria australasica* (Pisces: Percichthyidae), of Lake Eildon, Victoria. *Australian Journal of Ecology, 2*, 409–418.

Carroll, C., Fredrickson, R. J., & Lacy, R. C. (2014). Developing metapopulation connectivity criteria from genetic and habitat data to recover the endangered Mexican wolf. *Conservation Biology, 28*, 76–86.

Cole, T. L., Hammer, M. P., Unmack, P. J., Teske, P. R., Brauer, C. J., Adams, M., & Beheregaray, L. B. (2016). Range-wide fragmentation in a threatened fish associated with post-European settlement modification in the Murray–Darling Basin, Australia. *Conservation Genetics, 17*, 1377.

Coleman, R., Pettigrove, V., Raadik, T., Hoffmann, A., Miller, A., & Carew, M. (2010). Microsatellite markers and mtDNA data indicate two distinct groups in dwarf galaxias, *Galaxiella pusilla* (Mack)(Pisces: Galaxiidae), a threatened freshwater fish from south-eastern Australia. *Conservation Genetics, 11*, 1911–1928.

Coleman, R., Weeks, A., & Hoffmann, A. (2013). Balancing genetic uniqueness and genetic variation in determining conservation and translocation strategies: A comprehensive case study of threatened dwarf galaxias, *Galaxiella pusilla* (Mack) (Pisces: Galaxiidae). *Molecular Ecology, 22*, 1820–1835.

Crow, J. F., & Kimura, M. (1970). *An introduction to population genetic theory.*

New York, NY: Harper & Row.

Do, C., Waples, R. S., Peel, D., Macbeth, G., Tillett, B. J., & Ovenden, J. R. (2014). NeEstimator v2: Re-implementation of software for the estimation of contemporary effective population size (Ne) from genetic data. *Molecular Ecology Resources*, *14*, 209–214.

Drummond, A., Ashton, B., Buxton, S., Cheung, M., Cooper, A., Duran, C., ... Wilson, A. (2012). *Geneious v5.6*. Retrieved from http://www.geneious.com

Drummond, A. J., Suchard, M. A., Xie, D., & Rambaut, A. (2012). Bayesian phylogenetics with BEAUti and the BEAST 1.7. *Molecular Biology and Evolution*, *29*, 1969–1973.

Dufty, S. (1986). Genetic and morphological divergence between populations of Macquarie perch (*Macquaria australasica*) east and west of the Great Dividing Range. Unpublished BSc.(Hons) thesis, University of New South Wales, Kensington.

Environmental Systems Research Institute. (1999-2014). *ArcGIS version 10.2* in R. California, ed. Redlands California.

Excoffier, L., & Lischer, H. E. L. (2010). Arlequin suite ver 3.5: A new series of programs to perform population genetics analyses under Linux and Windows. *Molecular Ecology Resources*, *10*, 564–567.

Farrington, L. W., Lintermans, M., & Ebner, B. C. (2014). Characterising genetic diversity and effective population size in one reservoir and two riverine populations of the threatened Macquarie perch. *Conservation Genetics*, *15*, 707–716.

Faulks, L. K., Gilligan, D. M., & Beheregaray, L. B. (2010a). Evolution and maintenance of divergent lineages in an endangered freshwater fish, *Macquaria australasica*. *Conservation Genetics*, *11*, 921–934.

Faulks, L. K., Gilligan, D. M., & Beheregaray, L. B. (2010b). Islands of water in a sea of dry land: Hydrological regime predicts genetic diversity and dispersal in a widespread fish from Australia's arid zone, the golden perch (*Macquaria ambigua*). *Molecular Ecology*, *19*, 4723–4737.

Faulks, L. K., Gilligan, D. M., & Beheregaray, L. B. (2011). The role of anthropogenic vs. natural in-stream structures in determining connectivity and genetic diversity in an endangered freshwater fish, Macquarie perch (*Macquaria australasica*). *Evolutionary Applications*, *4*, 589–601.

Faulks, L., Gilligan, D., & Beheregaray, L. B. (2015). 'Ragged mountain ranges, droughts and flooding rains': The evolutionary history and conservation of Australian freshwater fishes. In A. Stow, N. Maclean, & G. I. Holwell (Eds.), *Austral ark: The state of wildlife in Australia and New Zealand*. Cambridge: Cambridge University Press.

Fisher, M. C., Garner, T. W., & Walker, S. F. (2009). Global emergence of *Batrachochytrium dendrobatidis* and amphibian chytridiomycosis in space, time, and host. *Annual Review of Microbiology*, *63*, 291–310.

Fitzpatrick, S., Gerberich, J., Kronenberger, J., Angeloni, L., & Funk, W. (2015). Locally adapted traits maintained in the face of high gene flow. *Ecology Letters*, *18*, 37–47.

Fountain, T., Nieminen, M., Sirén, J., Wong, S. C., & Hanski, I. (2016). Predictable allele frequency changes due to habitat fragmentation in the Glanville fritillary butterfly. *Proceedings of the National Academy of Sciences*, *113*, 2678–2683.

Frankham, R. (1995). Inbreeding and extinction: A threshold effect. *Conservation Biology*, *9*, 792–799.

Frankham, R. (1998). Inbreeding and extinction: Island populations. *Conservation Biology*, *12*, 665–675.

Frankham, R. (2005). Genetics and extinction. *Biological Conservation*, *126*, 131–140.

Frankham, R. (2015). Genetic rescue of small inbred populations: Meta-analysis reveals large and consistent benefits of gene flow. *Molecular Ecology*, *24*, 2610–2618.

Frankham, R. (2016). Genetic rescue benefits persist to at the F3 generation, based on a meta-analysis. *Biological Conservation*, *195*, 33–36.

Frankham, R., Ballou, J. D., Eldridge, M. D. B., Lacy, R. C., Ralls, K., Dudash, M. R., & Fenster, C. B. (2011). Predicting the probability of outbreeding depression. *Conservation Biology*, *25*, 465–475.

Frankham, R., Ballou, J. D., Ralls, K., Eldridge, M. D. B., Dudash, M. R.,

Fenster, C. B., ... Sunnucks, P. (in press). *Genetic management of fragmented animal and plant populations*. Oxford: Oxford University Press.

Frankham, R., Bradshaw, C. J. A., & Brook, B. W. (2014). Genetics in conservation management: Revised recommendations for the 50/500 rules, Red List criteria and population viability analyses. *Biological Conservation*, *170*, 56–63.

Franklin, I. R. (1980). Evolutionary change in small populations. In M. E. Soule, & B. A. Wilcox (Eds.), *Conservation biology: An evolutionary-ecological perspective* (pp. 135–148). Sunderland, MA: Sinauer.

Fu, Y.-X. (1997). Statistical tests of neutrality of mutations against population growth, hitchhiking and background selection. *Genetics*, *147*, 915–925.

Geoscience Australia. (2012). *National environmental stream attributes database v1.1.5*. Geoscience Australia. Retrieved from http://www.ga.gov.au/metadata-gateway/metadata/record/gcat_75066

Gilligan, D., McGarry, T., & Carter, S. (2010). A scientific approach to developing habitat rehabilitation strategies in aquatic environments: A case study on the endangered Macquarie perch (*Macquaria australasica*) in the Lachlan catchment. A report to the Lachlan Catchment Management Authority. Department of Industry and Investment (Industry & Investment NSW).

Goudet, J. (2001). *FSTAT, a program to estimate and test gene diversities and fixation indices (version 2.9.3)*.

Hammer, M. P., Bice, C. M., Hall, A., Frears, A., Watt, A., Whiterod, N. S., ... Zampatti, B. P. (2013). Freshwater fish conservation in the face of critical water shortages in the southern Murray–Darling Basin, Australia. *Marine & Freshwater Research*, *64*, 807–821.

Hammer, M. P., Goodman, T. S., Adams, M., Faulks, L. F., Unmack, P. J., Whiterod, N. S., & Walker, K. F. (2015). Regional extinction, rediscovery and rescue of a freshwater fish from a highly modified environment: The need for rapid response. *Biological Conservation*, *192*, 91–100.

Hardy, O. J., & Vekemans, X. (2002). SPAGeDi: A versatile computer program to analyse spatial genetic structure at the individual or population levels. *Molecular Ecology Notes*, *2*, 618–620.

Hare, M. P., Nunney, L., Schwartz, M. K., Ruzzante, D. E., Burford, M., Waples, R. S., ... Palstra, F. (2011). Understanding and estimating effective population size for practical application in marine species management. *Conservation Biology*, *25*, 438–449.

Harrisson, K. A., Pavlova, A., Gonçalves da Silva, A., Rose, R., Bull, J. J., Lancaster, M., ... Sunnucks, P. (2016). Scope for genetic rescue of an endangered subspecies though re-establishing natural gene flow with another subspecies. *Molecular Ecology*, *25*, 1242–1258.

Harrisson, K. A., Pavlova, A., Telonis-Scott, M., & Sunnucks, P. (2014). Using genomics to characterize evolutionary potential for conservation of wild populations. *Evolutionary Applications*, *7*, 1008–1025.

Harrisson, K. A., Yen, J. D. L., Pavlova, A., Rourke, M. L., Gilligan, D. M., Ingram, B., ... Sunnucks, P. (2016). Identifying environmental correlates of intra-specific genetic variation. *Heredity*, *117*, 155–164.

Hedrick, P. W. (1995). Gene flow and genetic restoration: The Florida panther as a case study. *Conservation Biology*, *9*, 996–1007.

Hedrick, P. W. (1999). Perspective: Highly variable loci and their interpretation in evolution and conservation. *Evolution*, *53*, 313–318.

Hedrick, P. W., & Fredrickson, R. (2010). Genetic rescue guidelines with examples from Mexican wolves and Florida panthers. *Conservation Genetics*, *11*, 615–626.

Hedrick, P. W., Peterson, R. O., Vucetich, L. M., Adams, J. R., & Vucetich, J. A. (2014). Genetic rescue in Isle Royale wolves: Genetic analysis and the collapse of the population. *Conservation Genetics*, *15*, 1111–1121.

Ho, H. K., & Ingram, B. A. (2012). *Genetic risk assessment for stocking Macquarie perch into Victorian waterways 2011*. Fisheries Victoria Internal Report No. 45 (p. 24). Alexandra: Department of Primary industries, Fisheries Victoria.

Hoffmann, A., Griffin, P., Dillon, S., Catullo, R., Rane, R., Byrne, M., ... Sgrò, C. (2015). A framework for incorporating evolutionary genomics into biodiversity conservation and management. *Climate Change Responses, 2*, 1–24.

Hollenbeck, C., Portnoy, D., & Gold, J. (2016). A method for detecting recent changes in contemporary effective population size from linkage disequilibrium at linked and unlinked loci. *Heredity, 117*, 207–216.

Hufbauer, R. A., Szucs, M., Kasyon, E., Youngberg, C., Koontz, M. J., Richards, C., … Melbourne, B. A. (2015). Three types of rescue can avert extinction in a changing environment. *Proceedings of the National Academy of Sciences, 112*, 10557–10562.

Huisman, J., Kruuk, L. E., Ellis, P. A., Clutton-Brock, T., & Pemberton, J. M. (2016). Inbreeding depression across the lifespan in a wild mammal population. *Proceedings of the National Academy of Sciences, 113*, 3585–3590.

Ingram, B., Barlow, C., Burchmore, J., Gooley, G., Rowland, S., & Sanger, A. (1990). Threatened native freshwater fishes in Australia—Some case histories. *Journal of Fish Biology, 37*, 175–182.

Ingram, B. A., Douglas, J. W., & Lintermans, M. (2000). Threatened fishes of the world: *Macquaria australasica* Cuvier, 1830 (Percichthyidae). *Environmental Biology of Fishes, 59*, 68–68.

Kearns, J. (2009). *Translocation of Macquarie Perch, Macquaria australasica, from King Parrot Creek and Hughes Creek, Victoria 2009. Confidential client report prepared for the Murray–Darling Basin Authority (MDBA).* Heidelberg, VIC: Department of Sustainability and Environment.

Keller, L. F., & Waller, D. M. (2002). Inbreeding effects in wild populations. *Trends in Ecology & Evolution, 17*, 230–241.

Knight, J., & Bruce, A. (2010). Threatened fish profile: 'Eastern' Macquarie perch *Macquaria australasica* Cuvier 1830. *Australian Society for Fish Biology Newsletter, 40*(2), 76–78.

Koehn, J. D., Doeg, T. J., Harrington, D. J., & Milledge, G. A. (1995). *The effects of Dartmouth Dam on the aquatic fauna of the Mitta Mitta River.* Report to Murray–Darling Basin Commission (p. 151). Victoria: Department of Conservation and Natural Resources.

Kopelman, N. M., Mayzel, J., Jakobsson, M., Rosenberg, N. A., & Mayrose, I. (2015). Clumpak: A program for identifying clustering modes and packaging population structure inferences across K. *Molecular Ecology Resources, 15*, 1179–1191.

Lacy, R. C. (2000). Structure of the VORTEX simulation model for population viability analysis. *Ecological Bulletins, 48*, 191–203.

Lacy, R. C., Miller, P. S., & Traylor-Holzer, K. (2015). *Vortex 10 user's manual. 15 April 2015 update.* IUCN SSC Conservation Breeding Specialist Group, and Chicago Zoological Society Apple Valley, Minnesota, USA.

Lacy, R., & Pollak, J. (2014). *VORTEX: A stochastic simulation of the extinction process, version 10.0.* Brookfield, IL: Chicago Zoological Society.

Le Cam, S., Perrier, C., Besnard, A.-L., Bernatchez, L., & Evanno, G. (2015). Genetic and phenotypic changes in an Atlantic salmon population supplemented with non-local individuals: A longitudinal study over 21 years. *Proceedings of the Royal Society B: Biological Sciences, 282*, 20142765.

Lee, W.-J., Conroy, J., Howell, W. H., & Kocher, T. (1995). Structure and evolution of teleost mitochondrial control regions. *Journal of Molecular Evolution, 41*, 54–66.

Legislative Assembly of New South Wales (1916). *Report on the fisheries of New South Wales for the year 1915.*

Lintermans, M. (2007). *Fishes of the Murray–Darling Basin: An introductory guide.* Canberra, ACT: Murray–Darling Basin Commission.

Lintermans, M. (2012). Managing potential impacts of reservoir enlargement on threatened *Macquaria australasica* and *Gadopsis bispinosus* in southeastern Australia. *Endangered Species Research, 16*, 1–16.

Lintermans, M. (2013). The rise and fall of a translocated population of the endangered Macquarie perch, *Macquaria australasica*, in south-eastern Australia. *Marine & Freshwater Research, 64*, 838–850.

Lintermans, M., & Ebner, B. (2010). Threatened fish profile: 'Western' Macquarie perch *Macquaria australasica* Cuvier 1830. *Australian Society*

for Fish Biology Newsletter, 40(2), 76–78.

Lintermans, M., Lyon, J. P., Hammer, M. P., Ellis, I., & Ebner, B. C. (2015). Underwater, out of sight: Lessons from threatened freshwater fish translocations in Australia. In D. P. Armstrong, M. W. Hayward, D. Moro, & P. J. Seddon (Eds.), *Advances in reintroduction biology of Australian and New Zealand Fauna* (pp. 237–253). Collingwood: CSIRO Publishing.

Lopez, S., Rousset, F., Shaw, F. H., Shaw, R. G., & Ronce, O. (2009). Joint effects of inbreeding and local adaptation on the evolution of genetic load after fragmentation. *Conservation Biology, 23*, 1618–1627.

Love Stowell, S. M., Pinzone, C. A., & Martin, A. P. (2017). Overcoming barriers to active interventions for genetic diversity. *Biodiversity and Conservation.* https://doi.org/10.1007/s10531-017-1330-z.

Lowe, W. H., & Allendorf, F. W. (2010). What can genetics tell us about population connectivity? *Molecular Ecology, 19*, 3038–3051.

Luikart, G., Ryman, N., Tallmon, D. A., Schwartz, M. K., & Allendorf, F. W. (2010). Estimation of census and effective population sizes: The increasing usefulness of DNA-based approaches. *Conservation Genetics, 11*, 355–373.

Lunn, D. J., Best, N., & Whittaker, J. C. (2009). Generic reversible jump MCMC using graphical models. *Statistics and Computing, 19*, 395–408.

Lunn, D. J., Thomas, A., Best, N., & Spiegelhalter, D. (2000). WinBUGS—A Bayesian modelling framework: Concepts, structure, and extensibility. *Statistics and Computing, 10*, 325–337.

Lunn, D. J., Whittaker, J. C., & Best, N. (2006). A Bayesian toolkit for genetic association studies. *Genetic Epidemiology, 30*, 231–247.

Mills, S. L., & Allendorf, F. W. (1996). The one-migrant-per-generation rule in conservation and management. *Conservation Biology, 10*, 1509–1518.

Nguyen, T. T. T., Ingram, B. A., Lyon, J., Guthridge, K., & Kearns, J. (2012). *Genetic diversity of populations of Macquarie perch, Macquaria australasica, in Victoria.* Fisheries Victoria Internal Report No. 40 (p. 14). Queenscliff: Fisheries Victoria.

Pavlova, A., Gan, H. M., Lee, Y. P., Austin, C. M., Gilligan, D., Lintermans, M., & Sunnucks, P. (2017). Purifying selection and genetic drift shaped Pleistocene evolution of the mitochondrial genome in an endangered Australian freshwater fish. *Heredity, 118*, 466–476. https://doi.org/10.1038/hdy.2016.120

Peakall, R., & Smouse, P. E. (2006). GenAlEx 6: Genetic analysis in Excel. Population genetic software for teaching and research. *Molecular Ecology Notes 6*, 288–295.

Pierson, J. C., Beissinger, S. R., Bragg, J. G., Coates, D. J., Oostermeijer, J. G. B., Sunnucks, P., … Young, A. G. (2015). Incorporating evolutionary processes into population viability models. *Conservation Biology, 29*, 755–764.

Pierson, J. C., Coates, D., Oostermeijer, J. G. B., Beissinger, S. R., Bragg, J. G., Sunnucks, P., … Young, A. G. (2016). Genetic factors in threatened species recovery plans on three continents. *Frontiers in Ecology and the Environment, 14*, 433–440.

Pollard, D., Ingram, B., Harris, J., & Reynolds, L. (1990). Threatened fishes in Australia—An overview. *Journal of Fish Biology, 37*, 67–78.

Pritchard, J. K., Stephens, M., & Donnelly, P. (2000). Inference of population structure using multilocus genotype data. *Genetics, 155*, 945–959.

R Development Core Team (2014). *R: A language and environment for statistical computing.* Vienna, Austria: R Foundation for Statistical Computing. Retrieved from http://www.R-project.org/ ISBN 3-900051-07-0.

Rambaut, A., & Drummond, A. J. (2007). *Tracer v1.4.* Retrieved from http://beast.bio.ed.ac.uk/Tracer

Reed, D. H., & Frankham, R. (2003). Correlation between fitness and genetic diversity. *Conservation Biology, 17*, 230–237.

Rieman, B. E., & Allendorf, F. (2001). Effective population size and genetic conservation criteria for bull trout. *North American Journal of Fisheries Management, 21*, 756–764.

Rieseberg, L. H., & Burke, J. (2001). A genic view of species integration. *Journal of Evolutionary Biology, 14*, 883–886.

Roberts, D. G., Baker, J., & Perrin, C. (2011). Population genetic structure of the endangered Eastern Bristlebird, *Dasyornis brachypterus*; implications for conservation. *Conservation Genetics, 12,* 1075–1085.

Rousset, F. (2008). genepop'007: A complete re-implementation of the genepop software for Windows and Linux. *Molecular Ecology Resources, 8,* 103–106.

Saccheri, I., Kuussaari, M., Kankare, M., Vikman, P., Fortelius, W., & Hanski, I. (1998). Inbreeding and extinction in a butterfly metapopulation. *Nature, 392,* 491–494.

Sgrò, C. M., Lowe, A. J., & Hoffmann, A. A. (2011). Building evolutionary resilience for conserving biodiversity under climate change. *Evolutionary Applications, 4,* 326–337.

Smith, D. J., van der Ree, R., & Rosell, C. (2015). *Wildlife crossing structures. Handbook of road ecology* (pp. 172–183). Chichester: John Wiley & Sons, Ltd.

Spielman, D., Brook, B. W., & Frankham, R. (2004). Most species are not driven to extinction before genetic factors impact them. *Proceedings of the National Academy of Sciences of the United States of America, 101,* 15261–15264.

Stein, J. L. (2006). *A continental landscape framework for systematic conservation planning for Australian rivers and streams.* Canberra: Australian National University.

Szulkin, M., & Sheldon, B. C. (2007). The environmental dependence of inbreeding depression in a wild bird population. *PLoS One, 2,* e1027.

Tajima, F. (1996). The amount of DNA polymorphism maintained in a finite population when the neutral mutation rate varies among sites. *Genetics, 143,* 1457–1465.

Tallmon, D. A., Gregovich, D., Waples, R. S., Scott Baker, C., Jackson, J., Taylor, B. L., ... Schwartz, M. K. (2010). When are genetic methods useful for estimating contemporary abundance and detecting population trends? *Molecular Ecology Resources, 10,* 684–692.

Tallmon, D. A., Koyuk, A., Luikart, G., & Beaumont, M. A. (2008). ONeSAMP: A program to estimate effective population size using approximate Bayesian computation. *Molecular Ecology Resources, 8,* 299–301.

Todd, C. R., & Lintermans, M. (2015). Who do you move? A stochastic population model to guide translocation strategies for an endangered freshwater fish in south-eastern Australia. *Ecological Modelling, 311,* 63–72.

Todd, C. R., Ryan, T., Nicol, S. J., & Bearlin, A. R. (2005). The impact of cold water releases on the critical period of post-spawning survival and its implications for Murray cod (*Maccullochella peelii peelii*): A case study of the Mitta Mitta River, southeastern Australia. *River Research and Applications, 21,* 1035–1052.

Trueman, W. T. (2011). *True tales of the trout cod: River histories of the Murray–Darling Basin,* MDBA Publication No. 215/11.

Verhoeven, K. J., Macel, M., Wolfe, L. M., & Biere, A. (2011). Population admixture, biological invasions and the balance between local adaptation and inbreeding depression. *Proceedings of the Royal Society of London B: Biological Sciences, 278,* 2–8.

Walling, C. A., Nussey, D. H., Morris, A., Clutton-Brock, T. H., Kruuk, L. E., & Pemberton, J. M. (2011). Inbreeding depression in red deer calves. *BMC Evolutionary Biology, 11,* 318.

Wang, J. (2005). Estimation of effective population sizes from data on genetic markers. *Philosophical Transactions of the Royal Society B: Biological Sciences, 360,* 1395–1409.

Waples, R. S., Antao, T., & Luikart, G. (2014). Effects of overlapping generations on linkage disequilibrium estimates of effective population size. *Genetics, 197,* 769–780.

Waples, R. S., & Do, C. (2008). LDNE: A program for estimating effective population size from data on linkage disequilibrium. *Molecular Ecology Resources, 8,* 753–756.

Waples, R. S., Luikart, G., Faulkner, J. R., & Tallmon, D. A. (2013). Simple life-history traits explain key effective population size ratios across diverse taxa. *Proceedings of the Royal Society of London B: Biological Sciences, 280,* 20131339.

Weeks, A. R., & Corrigan, T. (2011). *Translocation strategy for the Mount Buller population of the mountain pygmy possum.* Melbourne: Victorian Government Department of Sustainability and Environment (DSE).

Weeks, A. R., Sgro, C. M., Young, A. G., Frankham, R., Mitchell, N. J., Miller, K. A., ... Hoffmann, A. A. (2011). Assessing the benefits and risks of translocations in changing environments: A genetic perspective. *Evolutionary Applications, 4,* 709–725.

Weeks, A. R., Stoklosa, J., & Hoffmann, A. A. (2016). Conservation of genetic uniqueness of populations may increase extinction likelihood of endangered species: The case of Australian mammals. *Frontiers in Zoology, 13,* 31.

Whiteley, A. R., Fitzpatrick, S. W., Funk, W. C., & Tallmon, D. A. (2015). Genetic rescue to the rescue. *Trends in Ecology & Evolution, 30,* 42–49.

Willi, Y., & Hoffmann, A. A. (2009). Demographic factors and genetic variation influence population persistence under environmental change. *Journal of Evolutionary Biology, 22,* 124–133.

Willi, Y., Van Buskirk, J., & Hoffmann, A. A. (2006). Limits to the adaptive potential of small populations. *Annual Review of Ecology, Evolution, and Systematics, 37,* 433–458.

Wolff, J. N., Ladoukakis, E. D., Enriquez, J. A., & Dowling, D. K. (2014). Mitonuclear interactions: Evolutionary consequences over multiple biological scales. *Philosophical Transactions of the Royal Society of London. Series B, Biological Sciences, 369,* 20130443.

Woodworth, L. M., Montgomery, M. E., Briscoe, D. A., & Frankham, R. (2002). Rapid genetic deterioration in captive populations: Causes and conservation implications. *Conservation Genetics, 3,* 277–288.

8

Using fine-scale spatial genetics of Norway rats to improve control efforts and reduce leptospirosis risk in urban slum environments

Jonathan L. Richardson[1] | Mary K. Burak[1] | Christian Hernandez[2] |
James M. Shirvell[2] | Carol Mariani[2] | Ticiana S. A. Carvalho-Pereira[3] |
Arsinoê C. Pertile[3] | Jesus A. Panti-May[3] | Gabriel G. Pedra[3] | Soledad Serrano[3] |
Josh Taylor[3] | Mayara Carvalho[3] | Gorete Rodrigues[4] | Federico Costa[5] |
James E. Childs[6] | Albert I. Ko[3,6] | Adalgisa Caccone[2]

[1]Department of Biology, Providence College, Providence, RI, USA

[2]Department of Ecology & Evolutionary Biology, Yale University, New Haven, CT, USA

[3]Centro de Pesquisas Gonçalo Moniz, Fundação Oswaldo Cruz, Ministério da Saúde, Salvador, Brazil

[4]Centro de Controle de Zoonoses, Secretaria Municipal de Saúde, Ministério da Saúde, Salvador, Brazil

[5]Instituto de Saúde Coletiva, Universidade Federal da Bahia, UFBA, Salvador, Brazil

[6]Department of Epidemiology of Microbial Disease, Yale School of Public Health, New Haven, CT, USA

Correspondence
Jonathan L. Richardson, Department of Biology, Providence College, Providence, RI, USA.
Email: jrichardson@providence.edu

Funding information
Oswaldo Cruz Foundation; Secretariat of Health Surveillance; Brazilian Ministry of Health; Wellcome Trust, Grant/Award Number: 102330/Z/13/Z; NSF-NIH, Grant/Award Number: 5 R01 AI052473, 5 U01 AI088752, 1 R25 TW009338, 1 R01 AI121207, F31 AI114245, R01 AI052473, U01 AI088752, R01 TW009504 and R25 TW009338; Ecology and Evolution of Infection Diseases (EEID), Grant/Award Number: 1 R01 TW009504.

Abstract

The Norway rat (*Rattus norvegicus*) is a key pest species globally and responsible for seasonal outbreaks of the zoonotic bacterial disease leptospirosis in the tropics. The city of Salvador, Brazil, has seen recent and dramatic increases in human population residing in slums, where conditions foster high rat density and increasing leptospirosis infection rates. Intervention campaigns have been used to drastically reduce rat numbers. In planning these interventions, it is important to define the eradication units - the spatial scale at which rats constitute continuous populations and from where rats are likely recolonizing, post-intervention. To provide this information, we applied spatial genetic analyses to 706 rats collected across Salvador and genotyped at 16 microsatellite loci. We performed spatially explicit analyses and estimated migration levels to identify distinct genetic units and landscape features associated with genetic divergence at different spatial scales, ranging from valleys within a slum community to city-wide analyses. Clear genetic breaks exist between rats not only across Salvador but also between valleys of slums separated by <100 m—well within the dispersal capacity of rats. The genetic data indicate that valleys may be considered separate units and identified high-traffic roads as strong impediments to rat movement. Migration data suggest that most (71–90%) movement is contained within valleys, with no clear source population contributing to migrant rats. We use these data to recommend eradication units and discuss the importance of carrying out individual-based analyses at different spatial scales in urban landscapes.

KEYWORDS
epidemiology, favela, individual-based sampling, intervention, landscape genetics, population genetics, public health, reservoir host, spatial scale, urban ecology, vector control

1 | INTRODUCTION

Over the last century, there has been a marked increase in the proportion of the human population living in urban areas, from <14% in 1900 to more than 54% in 2015 (UN; WHO). As a result, during this period urban landscapes have been expanding while human population densities have increased. Despite this trend, urban habitat has generally been considered depauperate in biodiversity and received little attention from ecologists and population geneticists relative to its increasing scale on the landscape. However, urban landscapes can have large impacts on wildlife (Mcdonald, Kareiva, & Forman, 2008). The negative impacts are well-appreciated, and declines in population size and barriers to movement have been documented in species sensitive to habitat modification (Geslin et al., 2016; Krausman, Derbridge, & Merkle, 2008; Noël, Ouellet, Galois, & Lapointe, 2007; Wood & Pullin, 2002). However, there are a number of species that persist and even thrive in urban habitats. Some species can inhabit small patches of remnant habitat within an urbanized area, while other species characterized as "urban exploiters" can take advantage of the modified habitat and increased food resources (Blair, 2001; Hulme-Beaman et al., 2016; McKinney, 2002).

To date, very few studies have investigated how species that thrive in urban habitats disperse through these heavily developed landscapes. Importantly, many of the species best adapted to living alongside urban human populations are considered nuisance species capable of damaging property, ruining food stocks, and transmitting disease (Lyytimäki, Petersen, Normander, & Bezák, 2008). In USA, the costs of rats to agriculture alone are estimated at 19 billion dollars (Pimentel, Zuniga, & Morrison, 2005). The global costs of eradication and control efforts are uncertain but considerable, particularly in urban habitats. It is critical to understand how the urban landscape shapes dispersal patterns in urban species—particularly rat pests—so that public health and urban infrastructure officials can pursue strategies that control or restrict the movement of these species.

The Norway rat (*Rattus norvegicus*, also called the brown rat) is an urban exploiting pest species that has expanded to a near-global range over the last 300 years. Originally native to forests and brushy habitat in northern China, *R. norvegicus* is now most commonly associated with human-dominated landscapes and thrives in areas of high human population density (Feng & Himsworth, 2013). As a result, *R. norvegicus* is one of the most important nuisance species globally, responsible for significant agricultural losses, negative impacts on native ecosystems, and disease transmission as the reservoir host of major zoonotic pathogens (Antoniou et al., 2010; Capizzi, Bertolino, & Mortelliti, 2014; Pimentel et al., 2005). In particular, urban areas with low socioeconomic conditions and high human densities in developing countries face the greatest risk of rat-associated diseases that are linked with increasing rat populations (Gratz, 1994; Himsworth, Parsons, Jardine, & Patrick, 2013; Ko, Reis, Dourado, Johnson, & Riley, 1999; Lau, Smythe, Craig, & Weinstein, 2010). The rapid increase in people living in slum communities with subadequate housing, poor sanitation, and limited access to health care is considered to be one of

the major challenges to public health (Riley, Ko, Unger, & Reis, 2007; Sclar, Garau, & Carolini, 2005; Un-Habitat, 2004).

The city of Salvador, Brazil (pop. 2.9 million), has seen a 500% increase in human population size over the last 60 years, with most new settlement occurring in urban slum, or favela, neighborhoods within the city (Reis et al., 2008). These densely populated and low socioeconomic areas often have limited access to municipal services and experience derelict housing and poor sanitation (Felzemburgh et al., 2014; Hagan et al., 2016; Reis et al., 2008). These conditions support rodent activity and *R. norvegicus* infestation with several pathogens that cause spillover infections to humans (Costa et al., 2014, 2015; Glass, Childs, Korch, & LeDuc, 1989; Ko, Goarant, & Picardeau, 2009). There has also been a corresponding rise in the incidence of leptospirosis in this area—a potentially life-threatening disease affecting respiratory, renal, and liver functioning in infected humans (Ko et al., 1999; Lau et al., 2010). The disease is zoonotic and caused by spirochete bacteria in the same family as the agents of syphilis and Lyme disease. Rats are chronically infected with the bacteria, and the infectious leptospires are shed in rat urine creating an environmental reservoir for the pathogen (Costa et al., 2015; Lau et al., 2010). The leptospires can survive in soil and aquatic environments for weeks to months, leading to increased human outbreaks during the rainy season.

In cities throughout Brazil, which, as in Salvador, have annual rainfall-associated outbreaks of leptospirosis, public health officials have initiated intervention campaigns to sharply reduce rat numbers within these urban habitats at increased risk. Control efforts consist of rodenticide poisoning and environmental interventions (e.g., covering free-flowing sewers) in areas with high leptospirosis incidence. However, rat numbers rebound quickly with population sizes reaching pre-intervention densities within six months (de Masi, Vilaça, & Razzolini, 2009). Therefore, data on rat dispersal patterns are critical for designing more efficient rodent control programs, including where to target resources for intervention campaigns across large sections of urban habitat to impede re-colonization of treated areas (Abdelkrim, Pascal, & Samadi, 2007). Groups of vector organisms that are in close proximity and interconnected are defined as "eradication units." Pest populations within these units must be controlled simultaneously to prevent reinvasion and augment long-term success of control campaigns (Abdelkrim et al., 2007; Robertson & Gemmell, 2004; Russell et al., 2010).

Distinguishing dispersal events from population rebound due to animals surviving interventions is critical for effective control. While data on movement patterns are difficult to gather efficiently, or ethically untenable in the case of a zoonotic disease vector, using traditional mark–recapture and radio tracking methods (Davis, Emlen, & Stokes, 1948; Fenn, Tew, & Macdonald, 1987; Glass et al., 1989; Peakall, Ebert, Cunningham, & Lindenmayer, 2006; Taylor, 1978; Taylor & Quy, 1978; Wilson, Efford, Brown, Williamson, & McElrea, 2007), genetic data can be used to estimate rates of gene flow and provide valuable insights into the degree of connectivity, routes of dispersal, and spatial networks linking organisms across a complex urban landscape (Munshi-South, 2012). Researchers have used genetic data to determine relatedness of rats within prescribed locations and to estimate connectivity across urban areas (Gardner-Santana et al., 2009; Kajdacsi et al., 2013). However, even these genetic studies suffer from modest sample sizes (n = 146–277 rats) distributed across large urban regions, where low sampling density limits the resolution of spatial genetic data and restricts our ability to detect patterns across complex urban landscapes. Moreover, the distribution of sampled rats tends to be clumped at a limited number of sites, reducing the spatial coverage of the study to the areas sampled. While genetic data are a valuable proxy for individual movement and dispersal, it should be noted that these data are providing insight into the subset of all dispersal events that result in breeding (Lowe & Allendorf, 2010). This functional connectivity may not represent all individual movements, for which local factors may inhibit or promote assimilation into the local breeding cohort (Fraser, Banks, & Waters, 2015).

In this study, we evaluated genetic connectivity among more than 700 urban Norway rats captured from Salvador, Brazil, to identify eradication units that will be used to direct future rat control campaigns in Salvador. We also used our large sampling numbers to advance the spatial scope and analytical framework beyond what has been used to date in urban population genetics. We focused on three hierarchical spatial scales of analyses, from a local slum neighborhood to city-wide sampling, to evaluate how spatial genetic patterns differ based on the scale and resolution of sampling. We combined geographically distributed rat genotype data with individual-based analyses, providing much greater resolution of spatial genetic patterns than could be afforded from combining individuals in arbitrarily assigned sampling sites—particularly at the fine spatial scales that correspond to heterogeneous urban habitat structure. We use spatially explicit analyses that increase our power to detect genetic structure and link this directly to spatial location and connections on the landscape (Guillot, Santos, & Estoup, 2008; Jombart, Devillard, Dufour, & Pontier, 2008). We also use two methods that are free from the limiting assumptions of many population genetic models and that bypass the issue of nonindependence among pairwise genetic distances (Galpern, Peres-Neto, Polfus, & Manseau, 2014; Jombart, Devillard, & Balloux, 2010).

To achieve these goals, we collected rats from across the city of Salvador, with an emphasis on Pau da Lima, a slum neighborhood that has been the focus of long-term epidemiological studies related to leptospirosis (Figure 1) (Felzemburgh et al., 2014; Hagan et al., 2016; Reis et al., 2008). We genotyped rats at 16 variable microsatellite loci, and used these data to identify spatial genetic patterns and levels of connectivity that promote or impede the movement of rats at three spatial scales, ranging from city wide to areas within and around Pau da Lima. We also used spatial genetic structure to delineate practical eradication units for future interventions. Our study takes advantage of the largest sample size ever included in a study of urban rat ecology and their spatial genetic structure. This includes many samples collected within the dispersal range observed for *R. norvegicus*—distances within which any potential genetic subdivision must be caused by ecological or habitat features rather than movement abilities.

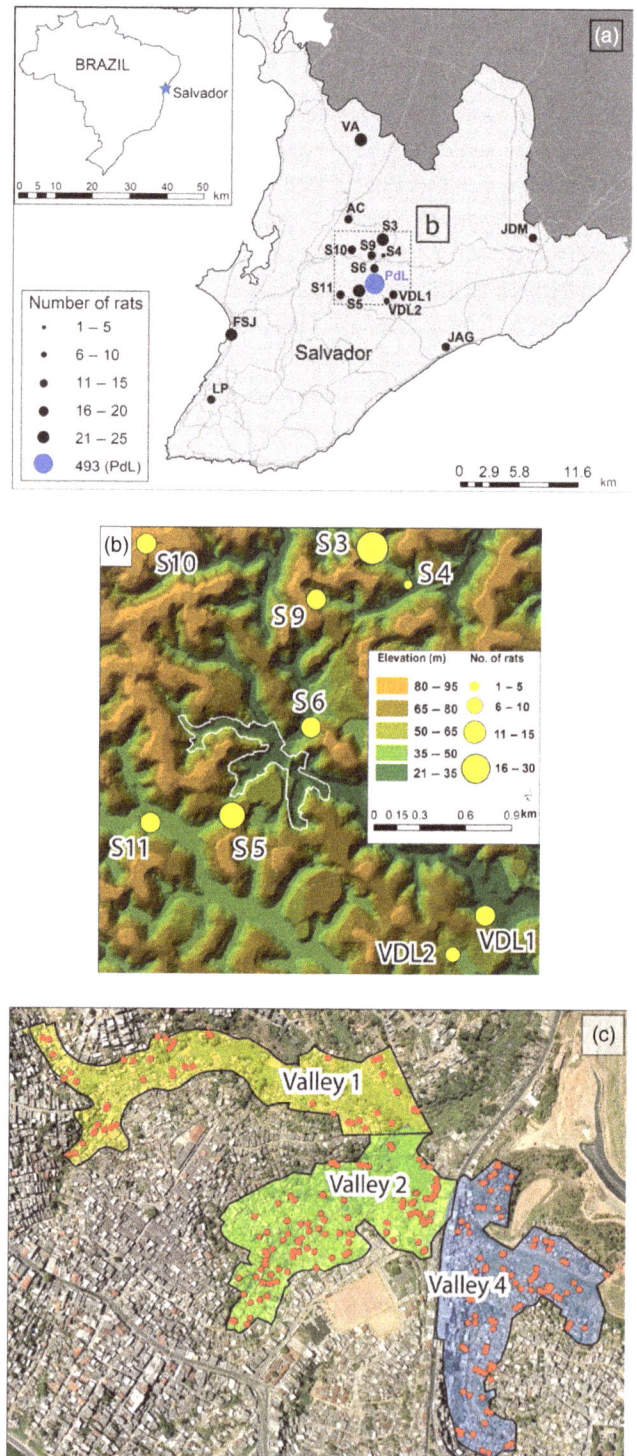

FIGURE 1 (a) Map of the study area within Salvador, Brazil. Black circles (and size) indicate sampled sites and the number of rat genotypes analyzed at each site at the city-wide scale. (b) The intermediate-scale analyses included nine satellite sites within 1 km of the Pau da Lima slum. Yellow dots are sampled sites and are scaled to the number of rats. (c) An aerial image of the Pau da Lima slum, with the three main valleys delineated. Red dots represent the sampling locations of the 493 rats analyzed at this slum scale. Note that in some cases, multiple rats were collected from the same geographic coordinate, corresponding to a building or structure

2 | MATERIALS AND METHODS

2.1 | Study area and sampling

We sampled 706 rats at 19 geographically separate regions across the city of Salvador using both live and snap traps (Figure 1a). These 706 rats greatly expand the number of areas and samples analyzed by Kajdacsi et al. (2013). Trap sites were geo-referenced, and tail tissue was collected and stored in ethanol at −80°C until DNA was extracted. Nearly 60% of the human population of Salvador lives within slum developments (Moreira and Pereira 2008). For that reason, we sampled most intensively across one of them, Pau da Lima. This slum is located in the center of Salvador and includes four connected valleys (valleys 1–4; Figure 1c). Pau da Lima was selected as the smallest scale with increased sampling based on the high density of rats and the high incidence human leptospirosis (Costa et al., 2014; Felzemburgh et al., 2014). The samples in this study were collected from three of the valleys (valleys 1, 2, and 4; Figure 1); Valley 3 has not been part of the larger long-term eco-epidemiological study. The 493 samples collected from within Pau da Lima constituted the finest spatial scale, referred to as "slum scale." The next spatial scale included 122 samples collected from eight "satellite" sites located within 1 km of Pau de Lima; we refer to the combination of these "satellite" locations and Pau da Lima ones as "intermediate scale." Lastly, 91 rats were collected beyond this intermediate scale in seven additional areas of the city (2–21 km from Pau da Lima); we refer to this scale as "city wide" in our three-tier hierarchy of sampling (Figure 1).

2.2 | DNA extraction and genotyping

DNA was extracted from 2–5 mm of tail tissue using standard kit-based extraction protocols (Qiagen and ZyGEM). We then amplified 16 microsatellite loci previously identified as polymorphic in *Rattus* spp. using a touchdown PCR protocol and reaction conditions following Kajdacsi et al. (2013). PCR amplicons were identified using capillary electrophoresis on an ABI 3730 DNA sequencer. GeneMarker software was used to score alleles, and Microsatellite Toolkit v3.1 (Park 2001) was used to check for scoring errors.

Although most of our analyses have no requirements of Hardy–Weinberg equilibrium and the loci were selected based on chromosome location to ensure linkage equilibrium among loci, we evaluated any deviations from these two common population genetic models. Independence of loci was evaluated by testing for linkage disequilibrium in FSTAT v2.9.3 (Goudet, 1995). Hardy–Weinberg equilibrium (HWE) for each locus was also estimated, with significance estimated using 10,000 randomizations. Significance values were adjusted for multiple testing using the sequential Bonferroni method (Holm 1979). Null allele frequencies were obtained using Microchecker v2.2.3 (van Oosterhout, Hutchinson, Wills, & Shipley, 2004). Loci exhibiting null allele frequencies >0.20 were excluded from later analyses (Chapuis & Estoup, 2007). Results for these marker tests and summary statistics can be found in the online Supporting Information.

2.3 | Individual-based genetic analyses

Rather than arbitrarily grouping rats for population-level genetic analyses, we took advantage of our high-resolution sampling (i.e., many rats over a large area) to perform individual-based analyses. Using individuals as the unit of analysis rather than populations allows for the evaluation of much finer-scale genetic patterns because information is not lost by averaging within groups (Richardson, Brady, Wang, & Spear, 2016). Most importantly, individual-based approaches can be more powerful in detecting small-scale genetic patterns (Luximon, Petit, & Broquet, 2014; Prunier et al., 2013) and are at least complementary to the population-based models that most population genetic analyses are based on. We used the program SPAGeDi to calculate genetic distances between every pair of individuals in the dataset (Hardy & Vekemans 2002). To look for a correlation between genetic differentiation and geographic distance between individuals, we performed a Mantel test of matrix correlation with 10,000 permutations using the vegan package within R (Oksanen et al., 2016).

2.4 | Spatially explicit ordination of the genetic data (spatial PCA)

Principal component analyses (PCA) are an effective method to create new synthetic variables that maximize the variance among genotypic data (Patterson, Price, & Reich, 2006). More recently, extensions of PCA have been developed to include spatial data in this ordination approach (Jombart et al., 2008). We performed spatial principal component analysis (sPCA) within the adegenet package of R (Jombart et al., 2016). We selected the parameters used based on initial insights from eigenvalue decomposition and recommendations from the program authors. We used an inverse distance-weighted connection network with no exponent because of the large number of samples and a minimum distance of one to maximize sample inclusion.

sPCA uses the spatial coordinates of each sample to estimate the degree and direction of spatial autocorrelation in the genetic data using Moran's *I* index. Global genetic structure evaluates the level of positive spatial autocorrelation or whether a cline exists where genetic similarity between individuals decreases as the distance between samples increases. Strong local-scale structure indicates negative spatial autocorrelation, and sharp genetic breaks between samples located close together. We used the eigenvalue variance and spatial components of the sPCA analysis to select the number of global and local axes to retain and visualize (Jombart et al., 2016). We used the sPCA-specific multivariate significance test to determine the strength of the global and local genetic structure across genotypes. These tests use Moran eigenvector maps (MEM) created by decomposing the connection network matrix and estimating the r^2 correlation coefficients of alleles from the global and local MEMs. The significance was determined using simulated *p*-values across 10,000 permutations where each row in the allele frequency matrix is randomized (Jombart et al., 2008). This allows us to test the null hypothesis that allele frequencies are distributed randomly across the connection network, with no spatial structure.

2.5 | Discriminant analysis of genetic structure

We used discriminant analysis of principal components (DAPC) to evaluate (A) the degree of genetic similarity among individuals in the dataset and (B) the patterns of genetic similarity among individuals from different areas of Salvador and within Pau da Lima. DAPC uses coefficients of allele loadings in linear combinations to maximize the between-groups variance while minimizing within-group variances in these loadings (Jombart et al., 2010). Studies using simulated and empirical data have found that DAPC performs as well or better than other individual-based clustering methods (e.g., STRUCTURE; Pritchard, Stephens, & Donnelly, 2000), particularly when complex processes are operating (Jombart et al., 2010; Klaassen, Gibbons, Fedorova, Meis, & Rokas, 2012). Prior to DAPC analyses, we added location tags for each rat corresponding to the geographic area that it was collected to help visualize the patterns in discriminant function space. We used guidelines in the *adegenet* package in R 2.14 to conduct DAPC and k-means clustering while retaining the best supported numbers of principal components and eigenvalues (R Development Core Team 2015; Jombart et al., 2016).

2.6 | Mapping spatially explicit genetic discontinuity

We used a fourth method that is based on Hardy–Weinberg equilibrium to look for convergence among different approaches characterizing spatial genetic structure. This approach, implemented in the Geneland package in R, is a Bayesian model using both genotypic data and spatial coordinates to identify areas of genetic continuity, where Hardy–Weinberg disequilibrium is minimized (Guillot, Mortier, & Estoup, 2005). Results are then visualized using Voronoi tessellation and membership coefficients. Studies indicate that Geneland performs better than other Bayesian clustering approaches in terms of correctly assigning individuals into groups (François & Durand, 2010). We used 100,000 Markov chain Monte Carlo iterations of the model with a thinning interval of 100. Uncorrelated allele frequencies and geographic coordinates without error were used to parameterize the model run. We used posterior probabilities to determine the number of genetic groups that were best supported across these runs. Three independent runs were used to confirm convergent results after the 100,000 iterations.

2.7 | Moran's eigenvector mapping to account for spatial nonindependence

Much criticism has been raised regarding population genetic analyses based on pairwise distance data (e.g., genetic and geographic distance), which are often used in Mantel tests of isolation by distance (Legendre, Fortin, & Borcard, 2015). This is because the pairwise values are not independent when each site is associated with multiple distances in the matrix. Moran's eigenvector mapping (MEM) has been developed to evaluate the spatial relationship among locations for any variable of interest. The MEM approach creates new uncorrelated (i.e., orthogonal) variables describing the spatial autocorrelation,

and hence spatial relationships, among sampling sites (Dray, Pierre, & Peres-Neto, 2006).

We performed a MEM analysis to identify spatial genetic patterns among rats using our geographic and genotype data within the MEMGENE package of R (Galpern et al., 2014). This approach produces new spatially independent variables that summarize the spatial relationship of genetic differences between sampled rats (Peres-Neto & Galpern, 2015). The proportion of shared alleles was used as the measure of genetic distance between individual rats (Bowcock et al., 1994). We extracted new MEM variables and visualized patterns via spatially explicit mapping of these MEM variable scores.

2.8 | Migration rates across Salvador

We estimated the rates of recent and ongoing genetic migration between the different sampling areas using the default parameters within the program BayesAss v3.0 (Wilson & Rannala, 2003). This approach uses Bayesian Markov chain Monte Carlo resampling to estimate posterior probabilities of ancestry within the last several generations. It also allows deviations from Hardy–Weinberg equilibrium, while assuming that loci are in linkage equilibrium and null alleles are not present. Immigrants are assumed to exhibit short-term linkage disequilibrium relative to the recipient population. This analysis estimates migration for all pairwise connections (i.e., [$n(n-1)/2$], and to make this computationally feasible, we grouped the 706 rats into 28 geographic areas based on topography and distance from other sampling areas (Fig. S1).

3 | RESULTS

3.1 | Spatially explicit ordination of the genetic data (spatial PCA)

We found significant global structure in the rat genetic data at all three spatial scales (slum scale $p = .002$, intermediate scale $p = .008$, city-wide scale $p = .002$). At each scale, Valley 2 within Pau da Lima (Figure 1b) was clearly identified as genetically dissimilar from the adjacent valleys (Figures 2a, 3a, and 4a). At the slum scale within the Pau da Lima community, the PCA-based genetic signature of Valley 1 is intermediate between valleys 2 and 4, a pattern also seen in the DAPC analysis (Figure 2a,b). There were no significant patterns of local-scale genetic structure at any of the three spatial scales, meaning that there was no negative autocorrelation in the genetic data across the study area (city-wide scale $p = .23$, intermediate scale $p = .19$, slum scale $p = .27$). One to three principal components explained the most variation at all scales, and those eigenvalues were used to plot the sPCA data (Jombart et al., 2016). These patterns of global structure can result from a number of processes, including isolation of patches, adaptive divergence in response to spatially autocorrelated natural selection, or clinal isolation by distance (Jombart et al., 2008; Wagner & Fortin, 2005). We detected no significant local genetic structure with sPCA, which can arise when individuals avoid breeding with other members of their population/deme (e.g., inbreeding avoidance,

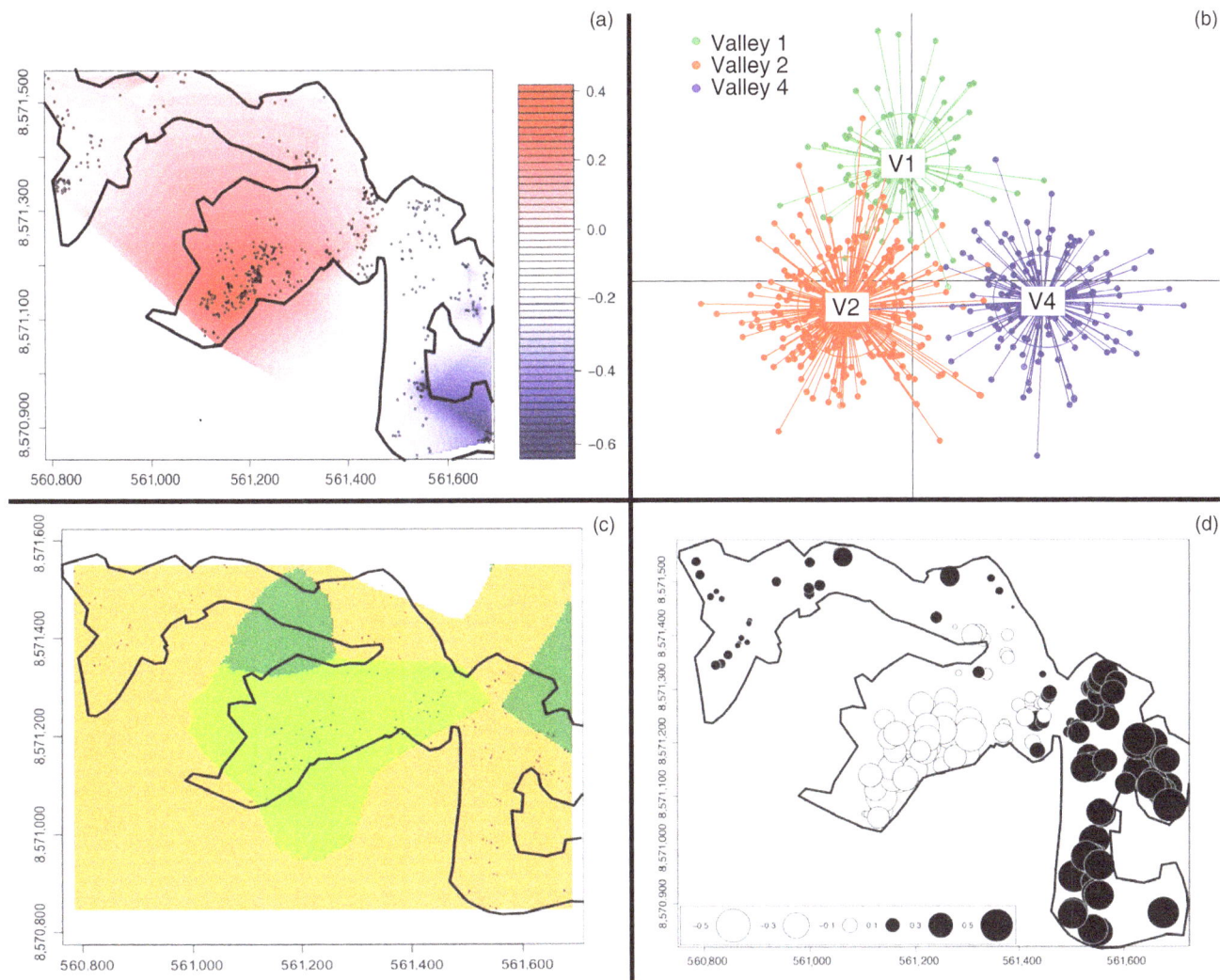

FIGURE 2 Analyses from the smallest spatial scale within the Pau da Lima slum show strong and consistent genetic segregation between valleys 2 and 4. Points in panels a, c, and d are the 493 spatially referenced rat sampling sites, overlaid with the outline of Pau da Lima (x- and y-axes are spatial coordinates). (a) Spatially explicit principal component (sPCA) vector scores interpolated across the sampling area show a sharp genetic break between valleys 2 and 4, while Valley 1 is admixed but more similar to Valley 2. (b) Discriminant analysis of principal components (DAPC) within Pau da Lima shows genetic separation between all three valleys in discriminant function space; each rat genotype is represented by a point with a line connecting it to the centroid ellipse of that group. Axes are discriminant function variables. (c) The spatial Bayesian model also indicates strong divergence between valleys 2 and 4, represented by the different tessellation colors. Each color unit has at least one rat contained within. (d) Consistent with the other analyses, Moran's eigenvector mapping (MEM) shows clear divergence between valleys 2 and 4, with circle color and size representing genetic similarity along the first MEM variable axis. Valley 1 is more similar to Valley 4 in the Bayesian and MEM analyses

repulsion), or where adaptive divergence occurs at small spatial scales in response to sharp gradients in natural selection (Richardson, Urban, Bolnick, & Skelly, 2014).

3.2 | Discriminant analysis of genetic structure

For each DAPC run, we retained the number of principal components necessary to explain between 80 and 90% of the variance, which ranged from 75 to 100 PCs. There was clear genetic grouping of individuals across all three spatial scales, including the finest scale within Pau da Lima. At the slum scale, all three valleys within Pau da Lima were genetically distinct, with more overlap between valleys 1 and 2 than Valley 4 (Fig. 2b). At the intermediate scale, Valley 4 was still clearly distinct,

while valleys 1 and 2 overlapped extensively with the nearby satellite populations, with the exception of site S10 (Figure 3b). When all 706 rats across Salvador were considered at the city-wide scale, Valley 4 again was distinct from the rest of Pau da Lima and the nearby satellite sampling areas, while the outlying sampled areas across Salvador were genetically segregated from the three Pau da Lima valleys (Figure 4b).

3.3 | Mapping spatially explicit genetic discontinuity

Results from each Geneland run converged between 10,000 and 40,000 iterations, well before the completed 100,000 runs. Four, five, and eight were the most likely number of clusters for the slum, intermediate, and city-wide scales, respectively. At the slum scale, most

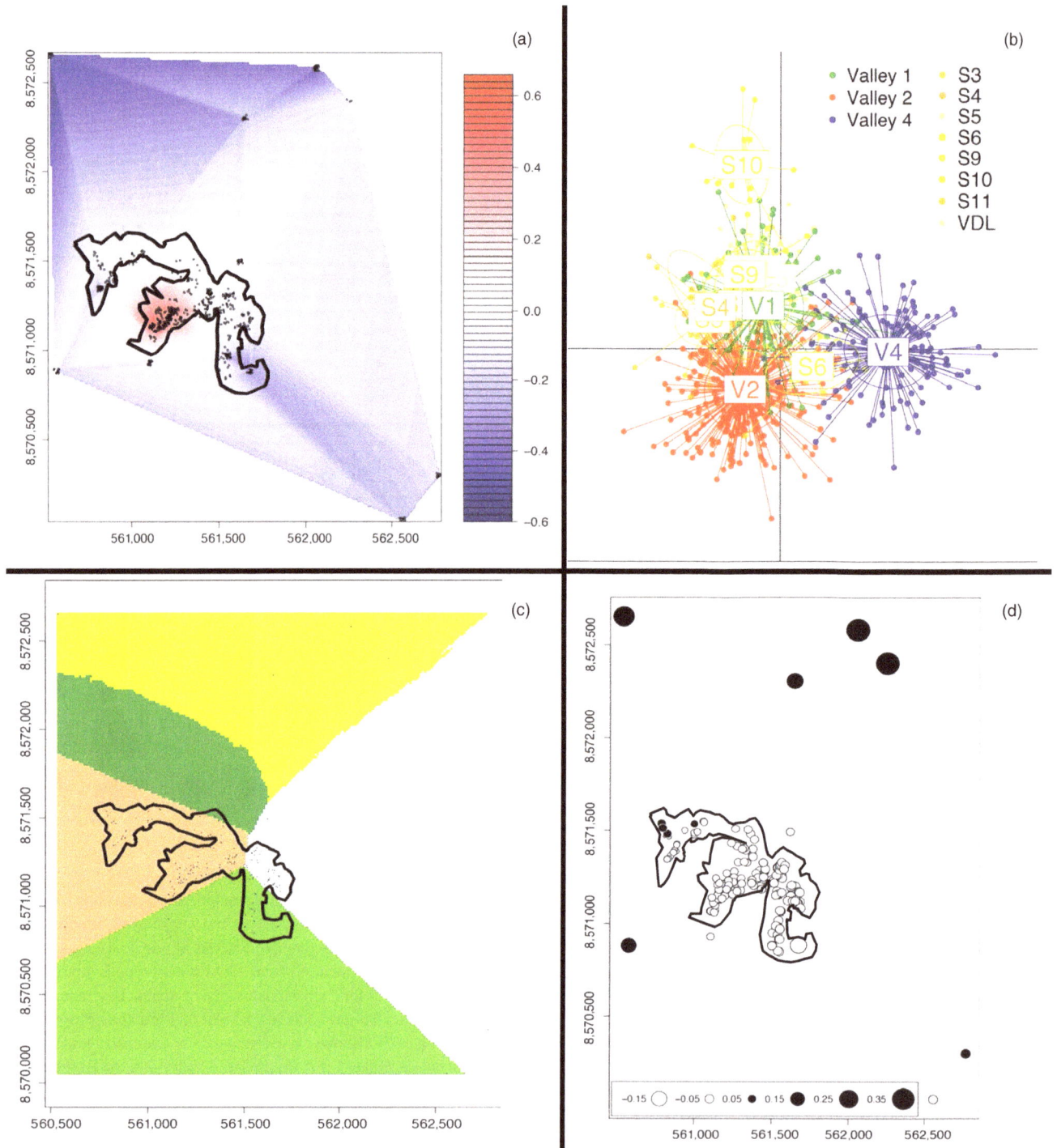

FIGURE 3   Analyses from the intermediate spatial scale, including Pau da Lima (black polygon outline) and all satellite sites within 1 km, show strong genetic divergence between valleys 2 and 4, with the exception of the MEM (D). Points in a, c, and d are the 614 spatially referenced rat samples. (a) sPCA vector scores interpolated across the sampling area show a sharp genetic break between Valley 2 and the rest of the sampling area, while Valley 1 is admixed but more similar to Valley 4. (b) DAPC indicates that most of the sampling sites cluster together, with the exception of clear divergence of Valley 4 and satellite site S10 in discriminant function space. (c) The spatial Bayesian model also shows strong divergence between valleys 2 and 4, represented by the different tessellation colors. At this scale, valleys 1 and 2 are grouped together and genetically distinct from all other sample areas. (d) MEM at the intermediate scale is the only analysis that does not exhibit divergence between valleys 2 and 4, as indicative by symbol color and size. Most satellite sites away from Pau da Lima appear as separate genetic group (black circles)

individuals in valleys 1 and 4 were clustered together, while Valley 2 individuals were clearly separated into a distinct group (Figure 2c). At the intermediate scale, the patterns differed slightly; Valley 4 was subdivided into two groups that were clustered away from valleys 1 and 2 (Figure 3c). Some satellite sites were grouped with other satellites, while others were included with the Pau da Lima valley clusters.

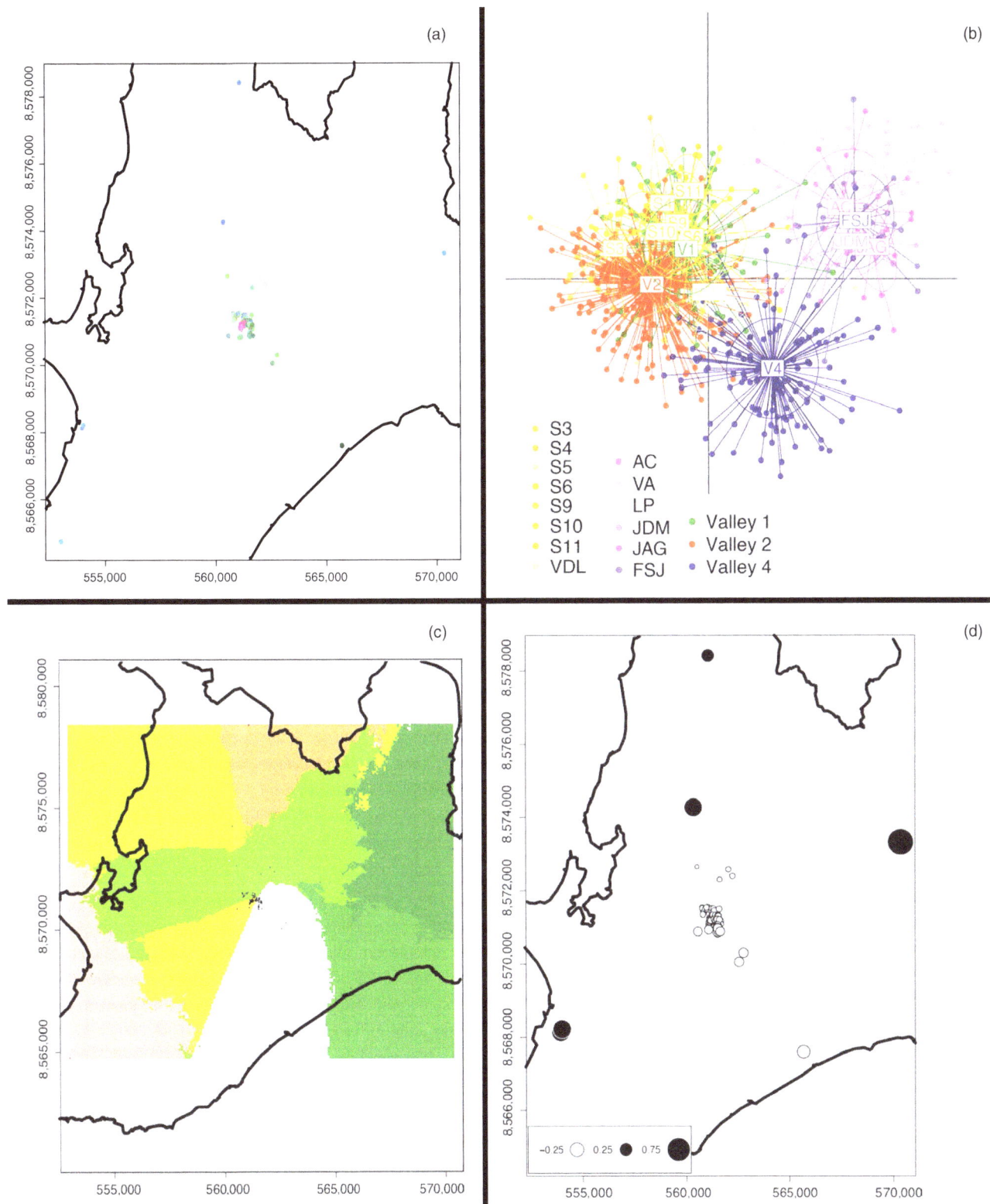

FIGURE 4   Analyses at the largest city-wide scale from all sites in Salvador also show strong and consistent genetic segregation between valleys 2 and 4. Most outlying sampling sites around the city cluster together relative to the Pau da Lima area. Points in a, c, and d are the 706 spatially referenced rat sampling sites. (a) The sPCA color plot displays genetic similarity as similarly colored points; the interpolated scores used in Figures 2 and 3 are less tractable at this large scale. Salvador-wide data exhibit a sharp genetic break between valleys 2 and 4, while the outlying sites are clustered as one group. (b) DAPC indicates three primary genetic groups—Valley 4, all outlying sites, and the rest of the Pau da Lima and nearby satellite sites. (c) The spatial Bayesian model shows strong divergence between all three valleys within Pau da Lima, while the outlying sites are mostly represented by a unique genetic group based on geography (represented by the different tessellation colors). (d) MEM exhibits strong divergence between the outlying sites around Salvador and the Pau da Lima area and their nearby satellite sampling areas

At the broadest city-wide scale, the samples from the three Pau da Lima valleys were genetically isolated to their own valley (Figure 4c). The two sites in the southwestern part of the city (FSJ, LP) cluster together around the port, while other satellite sites represent their own group or cluster with either Valley 1 or Valley 4 (Figure 4c).

3.4 | Moran's eigenvector mapping to account for spatial nonindependence

Moran's eigenvector analysis indicated that the proportion of all genetic variation attributed to spatial patterns differed across scales. The smallest scale (within slum) had the most genetic variation that was explained by spatial patterns (adjusted $r^2 = .124$), representing a surprisingly high degree of spatial genetic structure at such a small scale, given that the area is only 0.37 km^2. Genetic variation at the intermediate ($r^2 = .051$) and city-wide scales ($r^2 = .06$) was less strongly shaped by spatial orientation. At all scales, the first MEM variable, representing the principal coordinate eigenvector, explained more than 30% of the spatial genetic variation (proportional variances: 0.30 for the slum scale, 0.33 for the intermediate scale, and 0.41 for the city-wide scale). Visualizing this first MEM variable score within Pau da Lima indicates that valleys 2 and 4 are strongly divergent despite their adjacent locations (Figure 2d). Valley 1 is genetically intermediate, with MEM scores closer to Valley 4 than Valley 2 (Figure 2d). MEM scores at the intermediate and city-wide scales highlight the genetic distinction between Pau da Lima rats and all other sites across Salvador (Figures 3, and 4d).

3.5 | Migration rates across Salvador

As expected, most of the recent gene flow (i.e., over the last several generations) occurred within sites, rather than among sampled areas. An average of 79% of "migration" occurred internally within each valley, satellite site, or sampling area. Valley 2 showed the greatest genetic insulation with 90% of individuals identified as originating from within Valley 2. In Valley 4, 84% of individuals were estimated to have originated within this valley. Consistent with the other spatial genetic analyses, Valley 1 showed the highest proportion of migrants from outside this valley (71% from within Valley 1). The source of migration estimated to have come into Pau da Lima, and other sampling areas, was distributed fairly evenly across the other sampling areas (i.e., no single area contributed disproportionately high migration rates into any other population). A Mantel matrix correlation test found no significant relationship between migration rate and geographic distance among sampled areas ($r = -.116$, $p = .96$), or pairwise migration rate and genetic distance ($r = -.13$, $p = .98$). There was a significant association between geographic and genetic distance among sampling groups across the city, indicating isolation by distance in this system ($r = .451$, $p = .008$).

4 | DISCUSSION

Norway rats within the Pau da Lima slum showed strong patterns of genetic divergence within a very small area, despite modest divergence across the entire city of Salvador, Brazil. Importantly, this pattern of fine-scale divergence was consistent using four very different analytical approaches and spatial scales. Within Pau da Lima, a sharp break in genetic similarity distinguished valleys 2 and 4, two geographically distinct micro-districts within this densely inhabited slum settlement. Our migration data also indicate that rat movements occur mostly within their home valley or sampling neighborhoods and that there may not be clear "source" populations contributing to the bulk of migrants. Together, we use these spatial genetic data to identify units of eradication that are practical for future intervention campaigns.

4.1 | Genetic patterns at three spatial scales

Analyses at the smallest scale incorporated all 493 rats sampled within the Pau da Lima community (Figure 1). The most striking pattern at this spatial scale is the large degree of genetic divergence among rats located within half a kilometer of each other. The sharpest genetic breaks occur among rats in valleys 2 and 4, separated by <50 m (Figure 2a–d). This is remarkable considering that the average dispersal distance for urban *R. norvegicus* is 150 m (Glass et al., 1989; Traweger, Travnitzky, Moser, Walzer, & Bernatzky, 2006; Richardson unpublished data) and more than 1 km for *R. norvegicus* in rural areas (Macdonald & Fenn, 1995; Taylor & Quy, 1978). Given these dispersal abilities, our data indicate that there is a strong barrier to movement and gene flow between valleys 2 and 4. These two valleys are separated by a modest saddle, representing 50 m of elevation gain, as well as a high-traffic road connecting two neighborhoods on either side of Pau da Lima. Rat movements are likely impeded by the combination of this road and topographic relief.

The intermediate-scale analyses included 615 rats from Pau da Lima and the eight satellite sites sampled within 1 km of Pau da Lima. The strongest signal of genetic divergence at this scale was again between valleys 2 and 4 within Pau da Lima (Figure 3a–c). The association between Valley 1 and the rest of Pau da Lima was less consistent, with Valley 1 more similar to Valley 2 for the DAPC and Bayesian model approaches (Figure 3b,c), but showing admixture more closely associated with Valley 4 in the sPCA (Figure 3a). Interestingly, the level of admixture for Valley 1 switched from being more associated with Valley 2 at the slum scale or Valley 4 at this intermediate scale. This was true for both the sPCA (Figures 2a, and 3a) and spatial Bayesian model (Figures 2c, and 3c). These results indicate that valleys 2 and 4 are strongly diverged genetically, while Valley 1 is admixed with affinities to both valleys 2 and 4. As noted above, valleys 2 and 4 are separated by an altitudinal saddle and a busy road, but the immediately adjacent valleys 1 and 2 are separated by only a low-lying area, inundated with water during the rainy winter months, with few human settlements (Figure 1b). A model-based approach (e.g., approximate Bayesian computation) may be able to distinguish between true admixture and other mechanisms, such as drift, that could create the inconsistent pattern observed for Valley 1. The intermediate-scale MEM analysis showed very little divergence among all Pau da Lima samples, but separation between Pau da Lima and the nearby satellite sites (Figure 3d). MEM analyses are expected to be particularly useful at small scales where higher gene flow occurs, such as within genetically distinct groups (Galpern et al., 2014). The lower sensitivity of MEM between genetically divergent groups may explain why MEM is consistent with the other analyses at the Pau da Lima scale (Figure 2), but does not distinguish the separate valleys at the two larger scales where the genetic divergence is much greater between Pau da Lima and the satellite sites (Figures 3d, and 4d, Tables S1 and S2).

When analyzing all 706 rat samples from the city-wide scale, which included rats from 17 sites across Salvador, the sites situated further from the intermediate-scale sampling area were more similar to each other than to the sites at the smaller spatial scales (Figure 4a,b,d). Consistent with the results for the other two spatial scales analyzed, Valley 2 is genetically distinct from valleys 1 and 4 within Pau da Lima in the sPCA (Figure 4a), with the DAPC analysis clearly segregating Valley 4 from all other sites (Figure 4b). The approach using a spatially explicit Bayesian model did identify distinct genetic groups among the city-wide sampling sites (Figure 4c), where sites located outside the Pau da Lima slum and intermediate scales were each assigned to distinct genetic groups, with the exception of two sites in the southwestern area of the city (FSJ and LP). These sites are the closest pair among the outlying Salvador sites, located <3 km apart. Additionally, they are both in close proximity (within 3 km) to Salvador's main port—the most likely point of entry for additional genotypes of newly introduced rats. This Bayesian model also identified the micro-scale differentiation within the Pau da Lima slum (Figure 4c).

4.2 | Using spatial genetic patterns to guide public health interventions

Our study provides important insights into the movement and spatial genetic patterns of Norway rats in Salvador, Brazil. These insights are timely, as the ministry of health is planning intervention campaigns to reduce the prevalence of leptospirosis in Pau da Lima and other sections of Salvador by wide-scale application of rodenticides. Similar genetic data have been used to outline eradication units in studies concerned with controlling island pest species (Adams, van Heezik, Dickinson, & Robertson, 2014; Piertney et al., 2016; Ragionieri et al., 2013; Robertson & Gemmell, 2004). More broadly, the methods used in the current study provide important advances in both sampling and analysis that can benefit future studies of urban epidemiology and vector control.

Strong spatial genetic patterns can be used to infer levels of movement and connectivity across a study area, and delineate unique genetic groups that can be treated as focal units during intervention campaigns. Epidemiologists and public health officials often define "eradication units" as interconnected groups of vector organisms that are targeted during control efforts. Defining eradication units also makes evaluating the effectiveness of an intervention campaign possible in a tractable way (Robertson & Gemmell, 2004). For example, rigorous estimates of rat population sizes, densities, and removal rates can be estimated for each of the defined units to monitor the efficacy of control campaigns (Hacker et al., 2016). Genetic data can provide another important facet of eradication success when samples can be compared before and after intervention campaigns to evaluate the degree of genetic variation that is lost or reduced via postintervention genetic bottlenecks (Abdelkrim, Pascal, Calmet, & Samadi, 2005; Russell et al., 2010; Veale et al., 2013).

In the current study, we found strong genetic divergence between valleys in Pau da Lima. This divergence was strong and consistent between rats in valleys 2 and 4, located <50 m apart. Valley 1 was also divergent from the other two valleys, but this pattern differed across analyses and scales suggesting higher admixture in Valley 1. These spatial genetic patterns support defining eradication units based on these valley divisions within Pau da Lima. Specifically, the genetic data strongly support valleys 2 and 4 being delineated as separate units. The fine-scale DAPC and MEM analyses also support the segregation of Valley 1 from both other valleys as another eradication unit (Figure 2b,d). Moreover, we identified that high-traffic roads and modest topography likely play a role in impeding rat movement and that both features can be used to define eradication units in the context of Salvador. Additionally, the observed migration rates indicate that most rat movements (71–90%) are contained within valleys, providing further support for creating eradication units based on valleys. It is important to note that using topography alone would not be sufficient, as valleys 1 and 2 are connected via a low-elevation area. The spatial genetic data provide clear and consistent results across analyses to use the valley boundaries to define eradication units and design future intervention campaigns. However, an explicit landscape genetics analysis is needed to identify the exact features shaping the spatial genetic patterns observed.

We focused genetic sampling on the areas within and around Pau da Lima because this slum has been the focus of long-term epidemiological research (Felzemburgh et al., 2014; Hagan et al., 2016; Maciel et al., 2008). Pau da Lima has high human and rat densities, poor municipal infrastructure, and high incidence of leptospirosis cases corresponding to elevated levels of *Leptospira* infection exceeding 80% among Norway rats in this area (Costa et al., 2014; Felzemburgh et al., 2014). *Rattus norvegicus* is the primary host reservoir, and control campaigns are conducted regularly to target rats for removal. However, rat populations rebound in size within 6–12 months, as indicated by surveys of rat activity and abundance (Hacker et al., 2016). Therefore, identifying the contributions of remnant eradication survivors vs. immigrant re-colonizers to this rapid recovery is a central goal in the public health strategies designed to reduce leptospirosis in urban centers. Distinguishing between these sources is critical to determining the efficacy of the control campaign, and if future improvement is more likely to come from increased local interventions or continuous targeting of dispersal corridors (Russell et al., 2010). Our data indicate that spatial genetics and migration assignment techniques can be used to differentiate the source of new animals postintervention. While the current study does not focus on pre- and postintervention sampling, our migration data suggest that most individuals originated from within their specific sampling area. Further, there does not appear to be a clear source population contributing any sizeable proportion of identified migrant rats. However, future studies would need to explicitly sample for pre- and postintervention genetic patterns to suitably address this within Pau da Lima.

There are practical and immediate applications of these data in designing future intervention campaigns in Salvador. These data provide a better understanding of where rats are moving in a dense urban landscape and which areas constitute genetically, and presumably demographically, isolated units. In the case of Pau da Lima and the surrounding area, we recommend units corresponding to valley designation and satellite sampling neighborhoods, based on our data. The ministry of health in Salvador is using these units as the focal areas to delineate the boundaries of future intervention campaigns. This means that instead of spreading resources (i.e., traps, manpower) across all of Pau da Lima, particular valleys can be targeted individually. Moreover, continual trapping efforts can be used to limit rats migrating between valleys 1 and 2, even outside of intervention campaigns. Our data suggest that these permanent trapping efforts are not needed between valleys 2 and 4. Importantly, verifying the success of this modified intervention strategy will require evaluating rat numbers and genetic signatures before and after control campaigns, as described in the previous paragraph.

4.3 | Advancing spatial genetic studies in urban contexts

Our study demonstrates the power and utility of combining dense sampling with spatially explicit analyses to characterize the genetic structure and likely movement patterns of wildlife in complex urban landscapes. These patterns are difficult to study, but Norway rats are an excellent species to investigate urban ecology and movement. *Rattus norvegicus* is a globally distributed pest species invasive to most cities because it is so adept at exploiting urban habits and resources (Long, 2003). This species has been the focus of many ecological and epidemiological studies in cities around the world (Capizzi et al., 2014); however, few studies have looked at the genetic patterns of movement and gene flow within an urban context. Gardner-Santana et al. (2009) evaluated genetic structure at 10 microsatellite loci for 277 rats from 11 sites within ~84 km^2 of Baltimore, Maryland, USA. Three genetic groups were distinguished based on Bayesian clustering and high assignment probabilities, which suggested low levels of dispersal among the 11 separate sampling locations, despite examples of rare long-distance dispersal (Gardner-Santana et al., 2009). A previous study of 146 Norway rats captured within and around Pau da Lima found genetic divergence among three identified genetic clusters from nine sites (Kajdacsi et al., 2013). The current study extends the findings of Kajdacsi et al. (2013) by including many more rats (n = 706) captured over a much wider area across Salvador. These previous studies from Baltimore and Salvador support our findings that genetic heterogeneity exists within urban landscapes, even at very small scales. However, the two aforementioned studies were limited by the low density of samples across complex urban landscapes.

Our use of individual-based analyses is an important advance in spatial population genetics. Individual-based approaches treat individual samples as the unit of analysis, negating the need to group samples *a priori* for analysis, which is often arbitrary with respect to biology (Waples & Gaggiotti, 2006). With sufficient sampling density, this analytical framework can also be more powerful in detecting genetic patterns at small spatial scales (Luximon et al., 2014; Prunier et al., 2013). This increased power to detect fine-scale genetic patterns is particularly useful when investigating urban landscapes, where habitat complexity such as density of humans and the presence of physical barriers (e.g., roads and topographical features) can be much higher than exurban areas (Pickett et al., 2008). Population densities of urban species can also be inflated relative to natural densities, facilitating the sampling density needed for individual-based analyses. To this effect, we were able to collect and analyze a large numbers of samples (n = 493) within a relatively small area within the Pau da Lima neighborhood (0.37 km^2), which vastly exceeds previous efforts. Such high-resolution spatial genetic data provide newly available and powerful information for the public health initiatives to control Norway rats and advance urban population genetics more broadly.

Very few studies have explicitly investigated how scale can influence observed spatial genetic patterns. Those that have evaluated scale found similar patterns to what we observed in Salvador rats, including different patterns of genetic divergence in simulated (Cushman & Landguth, 2010) and empirical data using amphibians (Angelone, Kienast, & Holderegger, 2011) and insects (Keller, Holderegger, & van Strien, 2013). However, none of these previous studies investigated urban landscapes, where high habitat heterogeneity exists at much finer scales (Hacker et al., 2013; Pickett et al., 2008). In our study, most analyses retained patterns of spatial genetic divergence across valleys at different scales. The specific genetic groupings of rats in Salvador varied among the different scales of analysis based on the analytical methods. For example, sPCA and the Bayesian analyses have Valley 1 associated with Valley 2 at one scale and with Valley 4 at another scale (Figure 2a,c and Figure 3a,c). However, one pattern that was consistent across all scales and analyses was the clear divergence between valleys 2 and 4. These multiscale analyses illustrate the importance of either defining the scale of investigation based on detailed knowledge or specific hypotheses about mechanisms of genetic divergence over space (Richardson et al., 2016) or explicitly evaluating patterns at multiple spatial scales to identify how patterns differ. Either approach requires a sampling design with sufficient power to detect spatial genetic patterns (Oyler-McCance, Fedy, & Landguth, 2013).

It should be noted that genetics is a powerful proxy for dispersal and movement of individual across a landscape; however, genetic data only represent migrants and movements that result in successful breeding. Therefore, individual movements that do not lead to breeding and genetic exchange are not detected by genetic data, biasing connectivity estimates downward (Lowe & Allendorf, 2010). The dense sampling employed in the current study is one way to minimize the risk of not detecting nonbreeding dispersers. We also estimated migration rates, which are designed to detect the movement of first-generation migrants regardless of their breeding status. So a scenario where a large proportion of rats are moving across Salvador or Pau da Lima but not breeding (e.g., if high densities of rats with strong priority effects are preventing migrants from penetrating new areas; Fraser et al., 2015) is possible, but very unlikely given our study design and sharp patterns of genetic divergence. In fact, the sharp genetic boundaries we detected are all the more striking if previous intervention campaigns disrupted priority effects and competitive exclusion of migrants by resident rats.

4.4 | Potential role of evolution in urban rat populations

The observed patterns of strong genetic divergence at very fine spatial scales may arise not only from restricted movement and gene flow, but also from evolutionary divergence at microgeographic scales. This could occur if environmental conditions differ substantially within a small area such as Pau da Lima's valleys. While speculative for rats in Salvador, evidence from other studies indicates that complex urban habitats often impose strong and diverse natural selection on urban-dwelling species (Donihue & Lambert, 2015). Dissimilar habitats can promote divergent natural selection on rats occupying different parts of the environmental space, creating the potential for adaptive divergence within a small area (Richardson et al., 2014). For example, strong divergence in urban rodents has been linked to environmental differences in New York City, where genetic differences linked to metabolic function and dietary specialization have arisen within decades, rather than centuries (Harris & Munshi-South, 2016; Harris, Munshi-South, Obergfell, & O'Neill, 2013).

We do not currently know enough about the evolutionary selective environment for urban rats, and how this differs across Salvador, to ascribe the observed genetic patterns to evolutionary divergence. However, the sharp break in genetic similarity between Valley 2 and Valley 4 indicates the potential for such adaptive divergence to either arise or explain the observed genetic break. Outbreeding avoidance may also contribute to this divergence, where rats are more likely to breed with more closely related individuals coming from within the local population, rather than immigrants from other areas of Pau da Lima or Salvador (Costa et al., 2016). There are an increasing number of examples of evolutionary divergence at microgeographic scales, including in response to complex urban environments and their natural selection regimes (Richardson et al., 2014; Saccheri, Rousset, Watts, Brakefield, & Cook, 2008; Selander & Kaufman, 1975).

5 | CONCLUSIONS

High-resolution spatial genetic data can provide important information for epidemiological studies on the areas of movement, gene flow, and potential routes of recolonization after interventions to control vector species. Here, we used such data on over 700 rats in Salvador Brazil to (A) define eradication units based on genetic divergence among valleys in the Pau da Lima slum community, (B) evaluate migration rates to determine that no clear source area is responsible for recolonization after intervention campaigns that reduce rat populations, and (C) advance the methods associated with urban ecology using spatially explicit and individual-based approaches to evaluate genetic patterns at multiple spatial scales. The data from this study are particularly useful when combined with other epidemiological interventions, including reductions in the number of environmental reservoirs supporting the parasite, and implementing changes in human behavior that can reduce the prevalence of disease. These approaches are currently being employed in urban centers in Brazil and other developing countries to reduce the risk of leptospirosis, which has emerged due to the expansion of slum settlements and rat populations that populate such communities. The approaches used in this study also provide a framework for future work on urban ecology and landscape genetics, as well as the potential for microgeographic evolution in urban habitats that vary dramatically within very small spatial scales, in terms of both landscape features and genetic divergence.

ACKNOWLEDGEMENTS

We thank the staff of Fiocruz and the Zoonosis Control Center in Salvador for their assistance in conducting the study. We also thank Priscilla Machado for assistance with database processing and geo-referenced maps. We thank the collaborative groups from the Urban Health Council of Pau da Lima, including residents and community leaders, as well as the Oswaldo Cruz Foundation. This work was funded by the Oswaldo Cruz Foundation and Secretariat of Health Surveillance, Brazilian Ministry of Health; the Wellcome Trust (102330/Z/13/Z); and a NSF-NIH grant from the Ecology and Evolution of Infection Diseases (EEID) program (1 R01 TW009504) and NIH grants (5 R01 AI052473, 5 U01 AI088752, 1 R25 TW009338, 1 R01 AI121207 (F31 AI114245, R01 AI052473, U01 AI088752, R01 TW009504, R25 TW009338). Comments from three anonymous reviewers greatly improved the manuscript.

REFERENCES

Abdelkrim, J., Pascal, M., Calmet, C., & Samadi, S. (2005). Importance of assessing population genetic structure before eradication of invasive species: Examples from insular Norway rat populations. *Conservation Biology, 19,* 1509–1518.

Abdelkrim, J., Pascal, M., & Samadi, S. (2007). Establishing causes of eradication failure based on genetics: Case study of ship rat eradication in Ste, Anne archipelago. *Conservation Biology, 21,* 719–730.

Adams, A. L., van Heezik, Y., Dickinson, K. J. M., & Robertson, B. C. (2014). Identifying eradication units in an invasive mammalian pest species. *Biological Invasions, 16,* 1481–1496.

Angelone, S., Kienast, F., & Holderegger, R. (2011). Where movement happens: Scale-dependent landscape effects on genetic differentiation in the European tree frog. *Ecography, 34,* 714–722.

Antoniou, M., Psaroulaki, A., Toumazos, P., Mazeris, A., Ioannou, I., Papaprodromou, M., Georgiou, K., Hristofi, N., Patsias, A., Loucaides, F., Moschandreas, J., Tsatsaris, A., & Tselentis, Y. (2010). Rats as indicators of the presence and dispersal of pathogens in Cyprus: Ectoparasites, parasitic helminths, enteric bacteria, and encephalomyocarditis virus. *Vector-Borne and Zoonotic Diseases, 10,* 867–873.

Blair, R. B. (2001). Birds and butterflies along urban gradients in two ecoregions of the U.S. In J.L. Lockwood & M.L. McKinney (Eds.), *Biotic homogenization* (pp. 33–56). Norwell, MA: Kluwer Academic Publishers.

Bowcock, A. M., Ruiz-Linares, A., Tomfohrde, J., Minch, E., Kidd, J. R., & Cavalli-Sforza, L. L. (1994). High resolution of human evolutionary trees with polymorphic microsatellites. *Nature*, *368*, 455–457.

Capizzi, D., Bertolino, S., & Mortelliti, A. (2014). Rating the rat: Global patterns and research priorities in impacts and management of rodent pests. *Mammal Review*, *44*, 148–162.

Chapuis, M.-P., & Estoup, A. (2007). Microsatellite null alleles and estimation of population differentiation. *Molecular Biology and Evolution*, *24*, 621–631.

Costa, F., Ribeiro, G. S., Felzemburgh, R. D. M., Santos, N., Reis, R. B., Santos, A. C., Fraga, D. B. M., Araujo, W. N., Santana, C., Childs, J. E., Reis, M. G., & Ko, A. I. (2014). Influence of household rat infestation on leptospira transmission in the urban slum environment. *PLoS Neglected Tropical Diseases*, *8*, e3338.

Costa, F., Richardson, J. L., Dion, K., Mariani, C., Pertile, A. C., Burak, M. K., Childs, J. E., Ko, A. I., & Caccone, A. (2016). Multiple paternity in the Norway rat, *Rattus norvegicus*, from urban slums in Salvador, Brazil. *The Journal of Heredity*, *107*, 181–186.

Costa, F., Wunder, E. A., De Oliveira, D., Bisht, V., Rodrigues, G., Reis, M. G., Ko, A. I., Begon, M., & Childs, J. E. (2015). Patterns in leptospira shedding in Norway rats (*Rattus norvegicus*) from Brazilian slum communities at high risk of disease transmission. *PLoS Neglected Tropical Diseases*, *9*, e0003819.

Cushman, S. A., & Landguth, E. L. (2010). Scale dependent inference in landscape genetics. *Landscape Ecology*, *25*, 967–979.

Davis, D. E., Emlen, J. T., & Stokes, A. W. (1948). Studies on home range in the brown rat. *Journal of Mammalogy*, *29*, 207–225.

Donihue, C. M., & Lambert, M. R. (2015). Adaptive evolution in urban ecosystems. *Ambio*, *44*, 194–203.

Dray, S., Pierre, L., & Peres-Neto, P. R. (2006). Spatial modelling: A comprehensive framework for principal coordinate analysis of neighbour matrices (PCNM). *Ecological Modelling*, *196*, 483–493.

Felzemburgh, R. D. M., Ribeiro, G. S., Costa, F., et al. (2014). Prospective study of leptospirosis transmission in an urban slum community: Role of poor environment in repeated exposures to the Leptospira agent. *PLoS Neglected Tropical Diseases*, *8*, e2927.

Feng, A. Y. T., & Himsworth, C. G. (2013). The secret life of the city rat: A review of the ecology of urban Norway and black rats (*Rattus norvegicus* and *Rattus rattus*). *Urban Ecosystems*, *17*, 149–162.

Fenn, M. G. P., Tew, T. E., & Macdonald, D. W. (1987). Rat movements and control on an Oxfordshire farm. *Journal of Zoology*, *213*, 745–749.

François, O., & Durand, E. (2010). Spatially explicit Bayesian clustering models in population genetics. *Molecular Ecology Resources*, *10*, 773–784.

Fraser, C. I., Banks, S. C., & Waters, J. M. (2015). Priority effects can lead to underestimation of dispersal and invasion potential. *Biological Invasions*, *17*, 1–8.

Galpern, P., Peres-Neto, P., Polfus, J., & Manseau, M. (2014). MEMGENE: Spatial pattern detection in genetic distance data. *Methods in Ecology and Evolution*, *5*, 1116–1120.

Gardner-Santana, L. C., Norris, D. E., Fornadel, C. M., Hinson, E. R., Klein, S. L., & Glass, G. E. (2009). Commensal ecology, urban landscapes, and their influence on the genetic characteristics of city-dwelling Norway rats (*Rattus norvegicus*). *Molecular Ecology*, *18*, 2766–2778.

Geslin, B., Le Féon, V., Folschweiller, M., Flacher, F., Carmignac, D., Motard, E., Perret, S., & Dajoz, I. (2016). The proportion of impervious surfaces at the landscape scale structures wild bee assemblages in a densely populated region. *Ecology and Evolution*, *6*, 6599–6615.

Glass, G., Childs, J., Korch, G., & LeDuc, J. (1989). *Comparative ecology and social interactions of Norway rat (Rattus norvegicus) populations in Baltimore, Maryland*. Lawrence, Kansas: Museum of Natural History, the University of Kansas.

Goudet, J. (1995). FSTAT (version 1.2): A computer program to calculate F-statistics. *Journal of Heredity*, *86*, 485–486.

Gratz, N. G. (1994). Rodents as carriers of disease. In A. P. Buckle, & R. H. Smith (Eds.), *Rodent pests and their control* (pp. 85–108). Wallingford, UK: CAB International.

Guillot, G., Mortier, F., & Estoup, A. (2005). GENELAND: A computer package for landscape genetics. *Molecular Ecology Notes*, *5*, 712–715.

Guillot, G., Santos, F., & Estoup, A. (2008). Analysing georeferenced population genetics data with Geneland: A new algorithm to deal with null alleles and a friendly graphical user interface. *Bioinformatics*, *24*, 1406–1407.

Hacker, K. P., Minter, A., Begon, M., Diggle, P. J., Serrano, S., Reis, M. G., Childs, J. E., Ko, A. I., & Costa, F. (2016). A comparative assessment of track plates to quantify fine scale variations in the relative abundance of Norway rats in urban slums. *Urban Ecosystems*, *19*, 561–575.

Hacker, K. P., Seto, K. C., Costa, F., Corburn, J., Reis, M. G., Ko, A. I., & Diuk-Wasser, M. A. (2013). Urban slum structure: Integrating socioeconomic and land cover data to model slum evolution in Salvador, Brazil. *International Journal of Health Geographics*, *12*, 45.

Hagan, J. E., Moraga, P., Costa, F., Capian, N., Ribeiro, G. S., Wunder, E. A., Felzemburgh, R. D. M., Reis, R. B., Nery, N., Santana, F. S., Fraga, D., Dos Santos, B. L., Santos, A. C., Queiroz, A., Tassinari, W., Carvalho, M. S., Reis, M. G., Diggle, P. J., & Ko, A. I. (2016). Spatiotemporal determinants of urban leptospirosis transmission: Four-year prospective cohort study of slum residents in Brazil. *PLoS Neglected Tropical Diseases*, *10*, e0004275.

Hardy, O. J., & Vekemans, X. (2002). SPAGeDi: a versatile computer program to analyse spatial genetic structure at the individual or population levels. *Molecular Ecology Notes*, *2*, 618–620.

Harris, S. E., & Munshi-South, J. (2016). Scans for positive selection reveal candidate genes and local adaptation of *Peromyscus leucopus* populations to urbanization. *bioRxiv*, 038141.

Harris, S. E., Munshi-South, J., Obergfell, C., & O'Neill, R. (2013). Signatures of rapid evolution in urban and rural transcriptomes of white-footed mice (*Peromyscus leucopus*) in the New York metropolitan area. *PLoS ONE*, *8*, e74938.

Himsworth, C. G., Parsons, K. L., Jardine, C., & Patrick, D. M. (2013). Rats, cities, people, and pathogens: A systematic review and narrative synthesis of literature regarding the ecology of rat-associated zoonoses in urban centers. *Vector Borne and Zoonotic Diseases*, *13*, 349–359.

Holm, S. (1979). A simple sequentially rejective multiple test procedure. *Scandinavian Journal of Statistics*, *6*, 65–70.

Hulme-Beaman, A., Dobney, K., Cucchi, T., & Searle, J. B. (2016). An ecological and evolutionary framework for commensalism in anthropogenic environments. *Trends in Ecology and Evolution*, *31*, 633–645.

Jombart, T., Devillard, S., & Balloux, F. (2010). Discriminant analysis of principal components: A new method for the analysis of genetically structured populations. *BMC Genetics*, *11*, 94.

Jombart, T., Devillard, S., Dufour, A.-B., & Pontier, D. (2008). Revealing cryptic spatial patterns in genetic variability by a new multivariate method. *Heredity*, *101*, 92–103.

Jombart, T., Kamvar, Z. N., & Lustrik, R. Collins, C., Beugin, M. P., Knaus, B., Solymos, P., Schliep, K., Ahmed, I., Cori, A., & Calboli, F. (2016). adegenet: an R package for the Exploratory Analysis of Genetic and Genomic Data. *R package version 2.0.1.*

Kajdacsi, B., Costa, F., Hyseni, C., Porter, F., Brown, J., Rodrigues, G., Farias, H., Reis, M. G., Childs, J. E., Ko, A. I., & Caccone, A. (2013). Urban population genetics of slum-dwelling rats (*Rattus norvegicus*) in Salvador, Brazil. *Molecular Ecology*, *22*, 5056–5070.

Keller, D., Holderegger, R., & van Strien, M. J. (2013). Spatial scale affects landscape genetic analysis of a wetland grasshopper. *Molecular Ecology*, *22*, 2467–2482.

Klaassen, C. H. W., Gibbons, J. G., Fedorova, N. D., Meis, J. F., & Rokas, A. (2012). Evidence for genetic differentiation and variable recombination rates among Dutch populations of the opportunistic human pathogen *Aspergillus fumigatus*. *Molecular Ecology*, *21*, 57–70.

Ko, A. I., Goarant, C., & Picardeau, M. (2009). Leptospira: The dawn of the molecular genetics era for an emerging zoonotic pathogen. *Nature Reviews. Microbiology*, *7*, 736–747.

Ko, A. I., Reis, M. G., Dourado, C. M. R., Johnson, W. D., & Riley, L. W. (1999). Urban epidemic of severe leptospirosis in Brazil. *The Lancet, 354*, 820-825.

Krausman, P. R., Derbridge, J., & Merkle, J. A. (2008). *Suburban and exurban influences on wildlife and fish.* MT, USA: Helena.

Lau, C. L., Smythe, L. D., Craig, S. B., & Weinstein, P. (2010). Climate change, flooding, urbanisation and leptospirosis: Fuelling the fire? *Transactions of the Royal Society of Tropical Medicine and Hygiene, 104*, 631-638.

Legendre, P., Fortin, M.-J., & Borcard, D. (2015). Should the Mantel test be used in spatial analysis? (P Peres-Neto, Ed,). *Methods in Ecology and Evolution, 6*, 1239-1247.

Long, J. L. (2003). *Introduced mammals of the world.* Collingwood, Australia: CSIRO Publishing.

Lowe, W. H., & Allendorf, F. W. (2010). What can genetics tell us about population connectivity? *Molecular Ecology, 19*, 3038-3051.

Luximon, N., Petit, E. J., & Broquet, T. (2014). Performance of individual vs. group sampling for inferring dispersal under isolation-by-distance. *Molecular Ecology Resources, 14*, 745-752.

Lyytimäki, J., Petersen, L. K., Normander, B., & Bezák, P. (2008). Nature as a nuisance? Ecosystem services and disservices to urban lifestyle. *Environmental Sciences, 5*, 161-172.

Macdonald, D. W., & Fenn, M. G. P. (1995). Rat ranges in arable areas. *Journal of Zoology, 236*, 349-353.

Maciel, E. A. P., de Carvalho, A. L. F., Nascimento, S. F., et al. (2008). Household transmission of leptospira infection in urban slum communities. *PLoS Neglected Tropical Diseases, 2*, e154.

de Masi, E., Vilaça, P. J., & Razzolini, M. T. P. (2009). Evaluation on the effectiveness of actions for controlling infestation by rodents in Campo Limpo region, Sao Paulo Municipality, Brazil. *International Journal of Environmental Health Research, 19*, 291-304.

Mcdonald, R. I., Kareiva, P., & Forman, R. T. T. (2008). The implications of current and future urbanization for global protected areas and biodiversity conservation. *Biological Conservation, 141*, 1695-1703.

McKinney, M. L. (2002). Urbanization, biodiversity, and conservation. *BioScience, 52*, 883-890.

Moreira, I. M., & Pereira, G. C. (2008). *Salvador and Metropolitan Region* (p. 228). Salvador, Brazil: Edufba

Munshi-South, J. (2012). Urban landscape genetics: Canopy cover predicts gene flow between white-footed mouse (*Peromyscus leucopus*) populations in New York City. *Molecular Ecology, 21*, 1360-1378.

Noël, S., Ouellet, M., Galois, P., & Lapointe, F.-J. (2007). Impact of urban fragmentation on the genetic structure of the eastern red-backed salamander. *Conservation Genetics, 8*, 599-606.

Oksanen, J., Blanchet, F. G., Kindt, R., Legendre, P., Minchin, P. R., O'Hara, B., Simpson, G. L., Solymos, P., Stevens, M. H. H., & Wagner, H. (2016). Vegan: community ecology package. *R package version 2.3.*

van Oosterhout, C., Hutchinson, W. F., Wills, D. P., & Shipley, P. (2004). MICRO-CHECKER: Software for identifying and correcting genotyping errors in microsatellite data. *Molecular Ecology Notes, 4*, 535-538.

Oyler-McCance, S. J., Fedy, B. C., & Landguth, E. L. (2013). Sample design effects in landscape genetics. *Conservation Genetics, 14*, 275-285.

Park, S. (2001). *The Excel Microsatellite Toolkit (v3.1).* Ireland: Animal Genomics Laboratory, UCD.

Patterson, N., Price, A. L., & Reich, D. (2006). Population structure and eigenanalysis. *PLoS Genetics, 2*, e190.

Peakall, R., Ebert, D., Cunningham, R., & Lindenmayer, D. (2006). Mark-recapture by genetic tagging reveals restricted movements by bush rats (*Rattus fuscipes*) in a fragmented landscape. *Journal of Zoology, 268*, 207-216.

Peres-Neto, P., & Galpern, P. (2015). Spatial pattern detection in genetic distance data using Moran's Eigenvector Maps. *R package version 1.0.*

Pickett, S. T. A., Cadenasso, M. L., Grove, J. M., Nilon, C. H., Pouyat, R. V., Zipperer, W. C., & Costanza, R. (2008). Urban ecological systems: Linking terrestrial ecological, physical, and socioeconomic components of metropolitan areas. In J. M. Marzluff, E. Shulenberger, W. Endlicher, et al. (Eds.), *Urban ecology* (pp. 99-122). Boston, MA, USA: Springer.

Piertney, S. B., Black, A., Watt, L., Christie, D., Poncet, S., & Collins, M. A. (2016). Resolving patterns of population genetic and phylogeographic structure to inform control and eradication initiatives for brown rats *Rattus norvegicus* on South Georgia (C Bieber, Ed,). *Journal of Applied Ecology, 53*, 332-339.

Pimentel, D., Zuniga, R., & Morrison, D. (2005). Update on the environmental and economic costs associated with alien-invasive species in the United States. *Ecological Economics, 52*, 273-288.

Pritchard, J. K., Stephens, M., & Donnelly, P. (2000). Inference of population structure using multilocus genotype data. *Genetics, 155*, 945-959.

Prunier, J. G., Kaufmann, B., Fenet, S., Picard, D., Pompanon, F., Joly, P., & Lena, J. P. (2013). Optimizing the trade-off between spatial and genetic sampling efforts in patchy populations: Towards a better assessment of functional connectivity using an individual-based sampling scheme. *Molecular Ecology, 22*, 5516-5530.

Ragionieri, L., Cutuli, G., Sposimo, P., Spano, G., Navone, A., Capizzi, D., Baccetti, N., Vannini, M., & Fratini, S. (2013). Establishing the eradication unit of Molara Island: A case of study from Sardinia, Italy. *Biological Invasions, 15*, 2731-2742.

R Development Core Team (2015) R: A Language and Environment for Statistical Computing. http://www.r-project.org.

Reis, R. B., Ribeiro, G. S., Felzemburgh, R. D. M., Santana, F. S., Mohr, S., Melendez, A. X. T. O., Queiroz, A., Santos, A. C., Ravines, R. R., Tassinari, W. S., Carvalho, M. S., Reis, M. G., & Ko, A. I. (2008). Impact of environment and social gradient on Leptospira infection in urban slums. *PLoS Neglected Tropical Diseases, 2*, e228.

Richardson, J. L., Brady, S. P., Wang, I. J., & Spear, S. F. (2016). Navigating the pitfalls and promise of landscape genetics. *Molecular Ecology, 25*, 849-863.

Richardson, J. L., Urban, M. C., Bolnick, D. I., & Skelly, D. K. (2014). Microgeographic adaptation and the spatial scale of evolution. *Trends in Ecology and Evolution, 29*, 165-176.

Riley, L. W., Ko, A. I., Unger, A., & Reis, M. G. (2007). Slum health: Diseases of neglected populations. *BMC International Health and Human Rights, 7*, 2.

Robertson, B. C., & Gemmell, N. J. (2004). Defining eradication units to control invasive pests. *Journal of Applied Ecology, 41*, 1042-1048.

Russell, J. C., Miller, S. D., Harper, G. A., MacInnes, H. E., Wylie, M. J., & Fewster, R. M. (2010). Survivors or reinvaders? Using genetic assignment to identify invasive pests following eradication. *Biological Invasions, 12*, 1747-1757.

Saccheri, I. J., Rousset, F., Watts, P. C., Brakefield, P. M., & Cook, L. M. (2008). Selection and gene flow on a diminishing cline of melanic peppered moths. *Proceedings of the National Academy of Sciences of the United States of America, 105*, 16212-16217.

Sclar, E. D., Garau, P., & Carolini, G. (2005). The 21st century health challenge of slums and cities. *Lancet (London, England), 365*, 901-903.

Selander, R. K., & Kaufman, D. W. (1975). Genetic structure of populations of brown snail (*Helix aspersa*) I. Microgeographic variation. *Evolution, 29*, 385-401.

Taylor, K. D. (1978). Range of movement and activity of common rats (*Rattus norvegicus*) on agricultural land. *Journal of Applied Ecology, 15*, 663-677.

Taylor, K. D., & Quy, R. J. (1978). Long distance movements of a common rat (*Rattus norvegicus*) revealed by radio-tracking. *Mammalia, 42*, 63-72.

Traweger, D., Travnitzky, R., Moser, C., Walzer, C., & Bernatzky, G. (2006). Habitat preferences and distribution of the brown rat (*Rattus norvegicus* Berk.) in the city of Salzburg (Austria): Implications for an urban rat management. *Journal of Pest Science, 79*, 113-125.

Un-Habitat, (2004). The challenge of slums: Global report on human settlements 2003. *Management of Environmental Quality, 15*, 337-338.

Veale, A. J., Edge, K.-A., McMurtrie, P., Fewster, R. M., Clout, M. N., & Gleeson, D. M. (2013). Using genetic techniques to quantify reinvasion, survival and in situ breeding rates during control operations. *Molecular Ecology*, *22*, 5071–5083.

Wagner, H. H., & Fortin, M.-J. (2005). Spatial analysis of landscapes: Concepts and statistics. *Ecology*, *86*, 1975–1987.

Waples, R. S., & Gaggiotti, O. (2006). What is a population? An empirical evaluation of some genetic methods for identifying the number of gene pools and their degree of connectivity. *Molecular Ecology*, *15*, 1419–1439.

Wilson, D., Efford, M., Brown, S., Williamson, J., & McElrea, G. (2007). Estimating density of ship rats in New Zealand forests by capture-mark-recapture trapping. *New Zealand Journal of Ecology*, *31*, 47–59.

Wilson, G. A., & Rannala, B. (2003). Bayesian inference of recent migration rates using multilocus genotypes. *Genetics*, *163*, 1177–1191.

Wood, B. C., & Pullin, A. S. (2002). Persistence of species in a fragmented urban landscape: The importance of dispersal ability and habitat availability for grassland butterflies. *Biodiversity and Conservation*, *11*, 1451–1468.

9

Genetic and epigenetic variation in *Spartina alterniflora* following the *Deepwater Horizon* oil spill

Marta Robertson[1] (iD) | Aaron Schrey[2] | Ashley Shayter[3] | Christina J Moss[4] | Christina Richards[1] (iD)

[1]Department of Integrative Biology, University of South Florida, Tampa, FL, USA

[2]Department of Biology, Armstrong State University, Savannah, GA, USA

[3]Rehabilitation Institute, Southern Illinois University, Carbondale, IL, USA

[4]Department of Cell Biology, Microbiology and Molecular Biology, University of South Florida, Tampa, FL, USA

Correspondence
Marta Robertson, University of South Florida, Department of Integrative Biology, Tampa, FL, USA.
Email: mhr@mail.usf.edu

Funding information
This work was partially supported by funding from a University of South Florida New Researcher Award and the National Science Foundation (U.S.A.) through DEB-1419960 (to CLR).

Abstract

Catastrophic events offer unique opportunities to study rapid population response to stress in natural settings. In concert with genetic variation, epigenetic mechanisms may allow populations to persist through severe environmental challenges. In 2010, the *Deepwater Horizon* oil spill devastated large portions of the coastline along the Gulf of Mexico. However, the foundational salt marsh grass, *Spartina alterniflora*, showed high resilience to this strong environmental disturbance. Following the spill, we simultaneously examined the genetic and epigenetic structure of recovering populations of *S. alterniflora* to oil exposure. We quantified genetic and DNA methylation variation using amplified fragment length polymorphism and methylation sensitive fragment length polymorphism (MS-AFLP) to test the hypothesis that response to oil exposure in *S. alterniflora* resulted in genetically and epigenetically based population differentiation. We found high genetic and epigenetic variation within and among sites and found significant genetic differentiation between contaminated and uncontaminated sites, which may reflect nonrandom mortality in response to oil exposure. Additionally, despite a lack of genomewide patterns in DNA methylation between contaminated and uncontaminated sites, we found five MS-AFLP loci (12% of polymorphic MS-AFLP loci) that were correlated with oil exposure. Overall, our findings support genetically based differentiation correlated with exposure to the oil spill in this system, but also suggest a potential role for epigenetic mechanisms in population differentiation.

KEYWORDS
AFLP, *Deepwater Horizon*, DNA methylation, environmental stressors, epigenetics, MS-AFLP, *Spartina alterniflora*

1 | INTRODUCTION

Ecological theory predicts that adaptation to local conditions can result when populations harbor heritable phenotypic variation for traits that increase tolerance to local conditions. Classic population genetics studies demonstrate that natural selection in different microhabitats can result in associations of genotypes, or alleles of candidate genes, with habitat type (e.g., Hamrick & Allard, 1972; Salzman, 1985; Schmidt & Rand, 1999; Schmidt et al., 2008). In concert with other evolutionary mechanisms, disturbance events may also create population genetic structure, by diminishing standing genetic diversity through mortality (Hermisson & Pennings, 2005; Orr & Betancourt, 2001). These classic predictions are intuitive and often supported empirically (e.g., Clausen, Keck, & Hiesey, 1948). However, in some cases, data across a diversity of taxa show either no association of genetic differences with habitat (e.g., Richards, Hamrick, Donovan, & Mauricio, 2004; Foust et al., 2016; examples in Schmidt et al., 2008) or that low levels of molecular diversity are not associated with decreased phenotypic variation (Dlugosch & Parker, 2008; Richards et al., 2008). The disconnect between empirical findings and ecological theory suggests the possibility of additional, underexplored molecular mechanisms, such as epigenetic modifications, that mediate the relationship between phenotype and environment.

The recent application of molecular techniques to ecological questions has revealed that epigenetic regulatory mechanisms, such as DNA methylation, may respond dynamically and independently to sudden changes in the environment (e.g., Gugger, Fitz-Gibbon, Pellegrini, & Sork, 2016; Trucchi et al., 2016). Although there are several epigenetic mechanisms that can alter gene expression (e.g., chromatin remodeling, histone modifications, small interfering RNAs), DNA methylation of cytosines is the most widely studied (Schrey et al., 2013; Verhoeven, Vonholdt, & Sork, 2016) and can have important ecological effects. For example, studies in *Taraxacum officinale* show that when DNA methylation machinery is disrupted, flowering time differences among populations of these plants are removed (Wilschut, Oplaat, Snoek, Kirschner, & Verhoeven, 2016). Additionally, natural populations typically harbor high amounts of epigenetic variation (Keller, Lasky, & Yi, 2016; Paun et al., 2010; Richards, Schrey, & Pigliucci, 2012), which can be structured by local environmental conditions along with genetic variation. Variation in DNA methylation is correlated with habitat type in mangroves (Lira-Medeiros et al., 2010) and knotweed (Richards et al., 2012), herbivory in viola (Herrera & Bazaga, 2010), and climate in natural accessions of *Arabidopsis thaliana* (Keller et al., 2016). This association between DNA methylation and plant ecology may reflect the modulation of gene expression (Bewick et al., 2016; Zilberman, Gehring, Tran, Ballinger, & Henikoff, 2007) or recombination rates (Mirouze et al., 2012), the release of transposable elements (Dowen et al., 2012), or other regulatory processes in response to environmental conditions in addition to covariance with genetic structure. In some cases, epigenetic variation can be restructured during periods of environmental stress and these changes can persist after the stress is relieved (Verhoeven, Jansen, van Dijk, Biere, 2010; Verhoeven, Van Dijk, Biere, 2010; Dowen et al., 2012 but see Wibowo et al., 2016). These findings suggest that epigenetic mechanisms may allow for rapid modification of phenotype in response to immediate and acute stressors (Rapp & Wendel, 2005).

In this study, we simultaneously examined genetic and epigenetic patterns in populations of *S. alterniflora* along the Gulf Coast that were exposed to heavy oiling following the *Deepwater Horizon* (*DWH*) oil spill ("heavy" sensu Lin et al., 2016; Nixon et al., 2016). In 2010, 4.9 million barrels of oil spilled into the Gulf of Mexico over a period of 3 months, with devastating effects on coastal ecology and salt marsh ecosystems (Lin & Mendelssohn, 2012; Lin et al., 2016; Silliman et al., 2012; Whitehead et al., 2012). As the dominant plant on the leading edge of salt marshes, many *S. alterniflora* populations across the northern Gulf of Mexico were negatively impacted by the *DWH* oil spill. Despite large die-off of aboveground biomass and

reduced carbon fixation and transpiration in heavily oiled populations, *S. alterniflora* showed high resilience to the hydrocarbon exposure (Lin & Mendelssohn, 2012; Lin et al., 2016), and aboveground biomass and live stem density levels recovered to the same level as uncontaminated reference marshes within 18 months (Lin et al., 2016). However, while these and other studies support that *S. alterniflora* is resilient to hydrocarbon stress, the extent of intraspecific variation in resilience is uncertain, and it remains unknown whether there was differential mortality among *S. alterniflora* genotypes in natural populations exposed to the *DWH* oil spill. We measured genetic and epigenetic variation using amplified fragment length polymorphism (AFLP) and methylation sensitive fragment length polymorphism (MS-AFLP) to test the hypothesis that oil exposure in *S. alterniflora* resulted in genetic and epigenetic signatures of population differentiation. As in previous studies of *S. alterniflora* (Edwards, Travis, & Proffitt, 2005; Foust et al., 2016; Hughes & Lotterhos, 2014; Richards et al., 2004; Travis, Proffitt, & Ritland, 2004), we expected to see high levels of genetic and epigenetic variation. However, we anticipated that moderate, nonrandom differential mortality in response to oil exposure would result in genetic differentiation of oil-exposed populations from unexposed populations. Further, we anticipated a concurrent but stronger epigenetic signature of oil exposure, given its reflection of gene expression and physiological response to environmental stimuli (Dowen et al., 2012; Verhoeven, Jansen, et al., 2010; Verhoeven, Van Dijk, et al. 2010; Xie et al., 2015).

2 | MATERIALS AND METHODS

2.1 | Sample collection

Spartina alterniflora is a clonal halophyte, native to the east coast of the United States and invasive in coastlines around the world (Ainouche et al., 2009; Ayres, Smith, Zaremba, Klohr, & Strong, 2004; Pennings & Bertness, 2001). *Spartina alterniflora* displays diverse phenotypes in response to the natural environmental gradients in marshes, producing less aboveground biomass in response to increasingly saline soil (Richards, Pennings, & Donovan, 2005). Populations of *S. alterniflora* display high genetic diversity (Edwards et al., 2005; Foust et al., 2016; Hughes & Lotterhos, 2014; Richards et al., 2004; Travis et al., 2004) and substantial resilience to both natural variation in the salt marsh (Pennings & Bertness, 2001) and anthropogenic stressors, such as crude oil (Lin & Mendelssohn, 2012; Lin et al., 2016).

We collected leaf tissue from *S. alterniflora* stems at approximately ten-meter intervals along the shoreline from three oil-contaminated and four uncontaminated reference sites along the Gulf Coast in August 2010, while oil was still standing on the soil surface at contaminated sites (Table 1; Figure 1). Oil contamination was defined by visually confirmed presence of crude oil in the sediment and complete aboveground dieback of *S. alterniflora* in populations on the leading edge of the marsh. The only visible live tissue was the regrowth of stems from rhizomes through the wrack of dead aboveground *S. alterniflora* (Figure 2) from which we collected leaf tissue. Contamination levels were later confirmed through Natural Resource

TABLE 1 GPS coordinates of seven study sites

Population	N	Coordinates Longitude	Latitude
Oil-contaminated			
GIO1	6	29°26′42.8″N	89°55′45.7″W
GIO2	7	29°26′11.2″N	89°54′35.9″W
MSO	8	30°15′29.1″N	89°24′45.6″W
Unaffected			
GIN1	9	29°10′09.2″N	90°09′05.7″W
GIN2	8	29°10′49.4″N	90°06′31.6″W
MSN	8	30°20′21.1″N	89°21′15.3″W
AR	10	28°13′00.3″N	96°59′16.8″W

Damage Assessment databases (2014; Figure 1). Shoreline Cleanup Assessment Technique categories delineate oil contamination into five categories from "no oil observed" to "heavy oiling" (Nixon et al., 2016; Zengel et al., 2016). Our three contaminated sites fit the description of heavily oiled marshes, whereas the four uncontaminated sites had no visible oiling or impacted vegetation at the time samples were collected, and were not annotated as contaminated in the oil assessment databases. Samples were collected from the middle of the so-called tall plant zone near the leading edge of the marsh (sensu "low-salt habitat" in Foust et al., 2016). From each plant, we collected the 3rd fully expanded leaf to standardize age and minimize developmental bias in sampling. The contaminated sites were Grand Isle, LA

oiled site 1 (GIO1) (n = 6); Grand Isle, LA oiled site 2 (GIO2) (n = 7); and Bay St. Louis, MS oiled (MSO) (n = 8). Nearby uncontaminated reference sites were Grand Isle, LA no-oil site 1 (GIN1) (n = 9), Grand Isle, LA no-oil site 2 (GIN2) (n = 10), and Bay St. Louis, MS no-oil (MSN) (n = 8). Because the minimum number of populations required to detect differences between two groups at the level of alpha = 0.05 is suggested to be n = 7 (Fitzpatrick, 2009), we also sampled one additional reference site, Aransas, TX (AR) (n = 10), which was not affected by the *DWH* oil spill (Table 1). Sites in Mississippi and Louisiana were separated by a minimum of 10 km and maximum of 35 km, and AR was 775 km from Mississippi. Tissue samples were snap-frozen in liquid nitrogen and stored at −80°C.

2.2 | AFLP genotyping

We used AFLP to assess genetic variation between the field sites using a standard protocol described in Richards et al. (2012). Briefly, we isolated DNA in duplicate from leaf tissue with the Qiagen DNeasy Plant Mini Kit according to the manufacturer's recommended protocol (Qiagen, Valencia, CA) and conducted the entire protocol on duplicate reactions to ensure the consistent scoring of fragments and control for the potential error rate of AFLP markers. For selective PCR, we used fluorescently labeled primers *Eco*RI + AGC (6-FAM) and +ACG (HEX) and unlabeled *Mse*I + CAC primers. We sent selective PCR products to the DNA Facility at Iowa State University, IA, USA, where they were electrophoresed on an ABI 3130XL. We scored resulting fragments in duplicate as "1" for present and "0" for absent using Peak

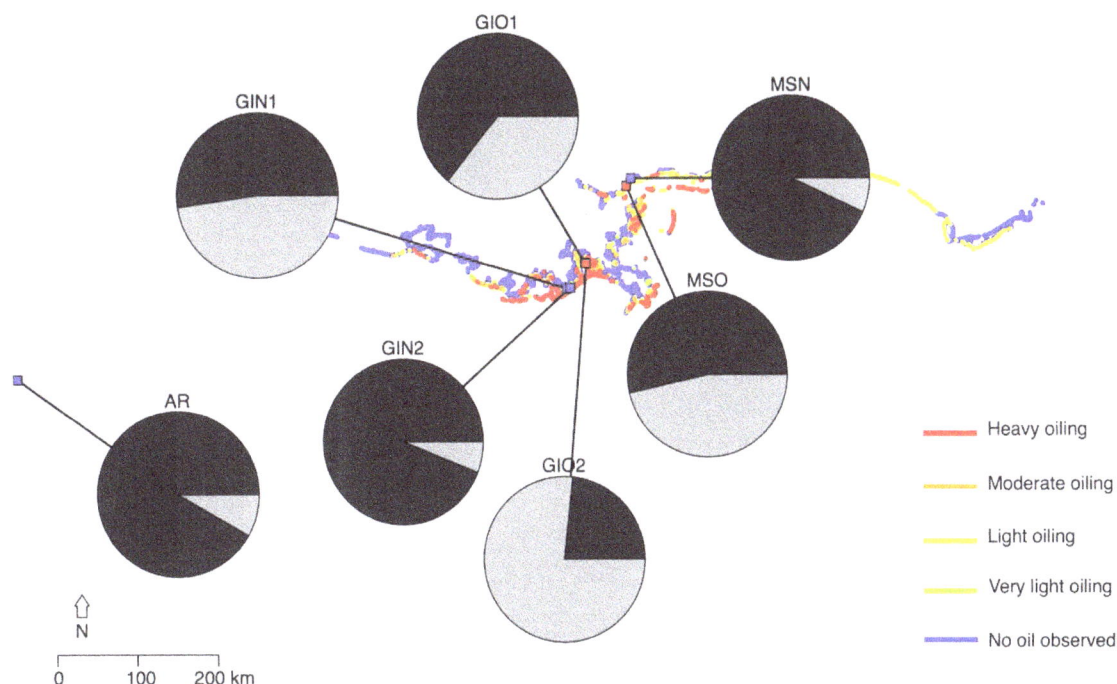

FIGURE 1 Map of seven study sites and atheir relative locations in the Gulf Coast, with site-specific oil intensity following the *Deepwater Horizon* (*DWH*) oil spill, according to NRDA databases, and the results of Bayesian clustering. Population assignment to two groups is indicated by the shaded portion of the circle for each species. Group 1 = dark gray, group 2 = light gray

FIGURE 2 Examples of (a) noncontaminated Grand Isle, LA no-oil site 1, (GIN1) and (b) contaminated sites Grand Isle, LA site 1 (GIO1) in the Gulf Coast following the *Deepwater Horizon* (DWH) oil spill. Oil was present on the soil surface at the time of sampling, and plants experienced substantial dieback. New growth sampled for this study (arrow) can be seen emerging from ramets under the soil surface through the dead wrack aboveground

Scanner (Thermo Fisher Scientific) and excluded markers that were not supported in duplicate.

2.3 | MS-AFLP epigenotyping

We used MS-AFLP to assess genomewide DNA methylation on the same duplicate DNA extractions used in the AFLP protocol (Reyna-Lopez, Simpson, & Ruiz-Herrera, 1997). We used *MspI* and *HpaII* restriction enzymes, which have different sensitivities to cytosine methylation of the same CCGG sequence (Reyna-Lopez et al., 1997; Salmon, Clotault, Jenczewski, Chable, & Manzanares-Dauleux, 2008). DNA extracts were digested with both *EcoRI/MspI* and *EcoRI/HpaII* enzyme combinations independently for each individual, and selective PCR was run with fluorescently labeled primers *EcoRI* + AGC (6-FAM) and +ACG (HEX) and unlabeled primers *HpaII/MspI* +TCAC and *HpaII/MspI* + TCAT. We sent selective PCR products to the DNA Facility at Iowa State University, IA, USA, where they were analyzed on an ABI3130XL. We visualized the resulting electropherograms using Peak Scanner and scored fragments as "1" when present and "0" when absent.

Together, *MspI* and *HpaII* produce four types of evaluative variation (Salmon et al., 2008). *MspI* does not cut when the external cytosines are fully or hemimethylated, and *HpaII* does not cut when either the internal or external cytosines are methylated on both strands. Likewise, cleaving by both enzymes is blocked when both cytosines are methylated. The resulting fragments can be classified as either type I when the corresponding sequence restriction site is nonmethylated and fragments occur in both digests, type II when fragments are absent in *EcoRI* + *HpaII* digests but present in *MspI*, type III when fragments are absent in *EcoRI* + *MspI* digests only, or type IV when no fragments occur in either digest. We treated type IV variation as missing data, because the methylation state cannot be specified (Salmon et al., 2008). Although some advocate for discriminating between type II and type III methylation as these types are expected to capture methylation in CG versus CHG contexts (Medrano, Herrera, & Bazaga, 2014; Schulz, Eckstein, & Durka, 2014), type II variation and type III variation cannot simply be interpreted as CG versus CHG methylation as apparent CHG methylation can be caused by the nesting of internal restriction sites within MS-AFLP fragments that exhibit differential CG methylation (Fulneček & Kovařík, 2014). Therefore, we combined type II variation and type III variation to represent the presence of DNA methylation in any context. Throughout this manuscript, we use "locus" to indicate a specific fragment size in the AFLP and MS-AFLP results. We use "haplotype" to indicate the binary variable positions (dominant genotypes) for each individual's collection of AFLP loci, and "epigenotype" to indicate the collection of binary variable positions of MS-AFLP loci.

2.4 | Data analysis

To identify the number of different genetic groups represented in our collection independent of sampling location in our populations, we performed Bayesian clustering of the genetic data only using Structure v.2.3.4 (Falush, Stephens, & Pritchard, 2003, 2007; Hubisz, Falush, Stephens, & Pritchard, 2009; Pritchard, Stephens, & Donnelly, 2000). Our previous work has shown population structure within native *S. alterniflora* populations (Foust et al., 2016; Richards et al., 2004). Although we designed our sampling to avoid subpopulation structure in this study by only sampling near the leading edge of the marsh, we tested for the possibility of finding more populations than expected. We tested ten populations (k = 1–10) with ten independent runs at each k. We performed analyses with 50,000 burn-in sweeps and 1,000,000 postburn-in sweeps, assuming admixture and without including sample location, or any geographic information as priors in the analysis. We estimated the number of clusters represented by the data using Evanno's delta K (Evanno, Regnaut, & Goudet, 2005).

We used GenAlEx version 6.41 (Peakall & Smouse, 2012) to estimate the haplotype and epigenotype diversity (h-AFLP and h-MS-AFLP). We also used GenAlEx to calculate estimates of genetic differentiation over all AFLP and MS-AFLP loci with a hierarchical AMOVA, nesting study sites within oil exposure to compare genetic variation among oil-contaminated and uncontaminated sites (Φ_{RT}), among sites within contamination level (Φ_{PR}), and within sites (Φ_{PT}). We also used GenAlEx to conduct a locus-by-locus AMOVA to characterize genetic and epigenetic differentiation at each locus, using the same hierarchical design. Finally, we performed pairwise AMOVA comparisons to determine which populations were differentiated. For all AMOVA analyses, we used 9,999 permutations to estimate statistical significance and adjusted for multiple comparisons using the sequential Bonferroni method whenever multiple tests were performed.

In addition to AMOVA, we tested for the effect of oil on AFLP and MS-AFLP multilocus marker profiles via permutational multivariate analysis of variance (perMANOVA), which allows for comparison of nested terms within hierarchical experimental design. Using the Adonis function within the Vegan package of R (Oksanen et al., 2017), we derived p-values based on 9,999 permutations within populations using the following formula: Adonis (AFLP genetic distance matrix~ oil exposure, strata = population, permutations = 10,000). In each analysis, variation in marker profiles was represented by the Euclidean distance matrices as calculated from the binary AFLP and MS-AFLP methylation data (with interpolation of missing values) generated by GenAlEx 6.41. We also used the RDA function within the Vegan package of R (Oksanen et al., 2017) to conduct a partial redundancy analysis of the relationship between contamination level (presence or absence) and MS-AFLP, while removing the effects of AFLP. We used the following formula: RDA [x ~ y + z] where x = the Euclidean epigenetic distance matrix generated by GenAlEx, y = site condition (presence or absence of oil), and z = the Euclidean genetic distance matrix generated by GenAlEx. To create the site condition matrix, we used zero to indicate uncontaminated sites and one to indicate contaminated sites. This strategy makes the assumption that differences between contaminated and uncontaminated populations will be essentially the same magnitude regardless of individual population differences.

3 | RESULTS

3.1 | Genetic diversity and structure

A power analysis indicated that we could detect an effect of oil contamination among seven groups using our sample sizes (Fitzpatrick, 2009), and previous work reports population differentiation in hierarchical analyses is detectable with as few as five individuals per population (Nelson & Anderson, 2013). We found 71 polymorphic loci, which is well above the minimum of 30 markers reported in previous work

to be required to detect significant patterns of differentiation (Nelson & Anderson, 2013). Of these loci, six were present or absent in only one sample. We ran these analyses with and without including these single-variable loci and found no substantial differences in the results. The data presented here are based on the complete set of 71 polymorphic loci. Although a modest data set, our markers identified that genetic diversity was high (h-AFLP ranged from 0.103 to 0.206), and 55 of 56 individuals displayed a unique genotype. There was no difference in genotype diversity between contaminated and uncontaminated sites ($p = .262$). Bayesian clustering identified two genetic groups ($\Delta K = 1,517.81$); however, these groups did not clearly reflect either differentiation by oil contamination or geographic separation (Figure 1).

Hierarchical AMOVA revealed significant variation between contaminated and unaffected populations (explaining 6% of the genetic variance), and among populations within site type (explaining 16% of the genetic variance; Table 3), as well as most (66%) pairwise comparisons between sites (Table 4), indicating the presence of population structure between contamination types and among populations. These results were supported by perMANOVA, which showed a significant effect of oil contamination on multilocus genetic marker profiles ($F = 0.092$, $p = .017$). Locus-by-locus AMOVA revealed 17 loci that varied significantly between oil-contaminated and unaffected sites (Figure 3).

3.2 | Epigenetic diversity and structure

We found 39 polymorphic epigenetic loci from 71 observed. Of these loci, seven were present or absent in only one sample. We ran these analyses with and without including these single-variable loci and found no substantial differences in the results. The data presented here are based on the complete set of 39 polymorphic loci. Epigenotype diversity was high (h-MSAFLP ranged from 0.089 to 0.222), and each individual displayed a unique epigenotype. Like the estimates for genetic patterns, there was no difference in epigenotype diversity between affected and unaffected sites ($p = .993$), and as in our previous studies of S. alterniflora (Foust et al., 2016), h-MS-AFLP tended to be lower than h-AFLP (Table 2).

Hierarchical AMOVA failed to detect a significant effect of oil contamination on epigenetic differentiation, but among populations within site type explained 7% of the epigenetic variance (Table 3), and 38% of the pairwise comparisons between sites were significant (Table 4). The

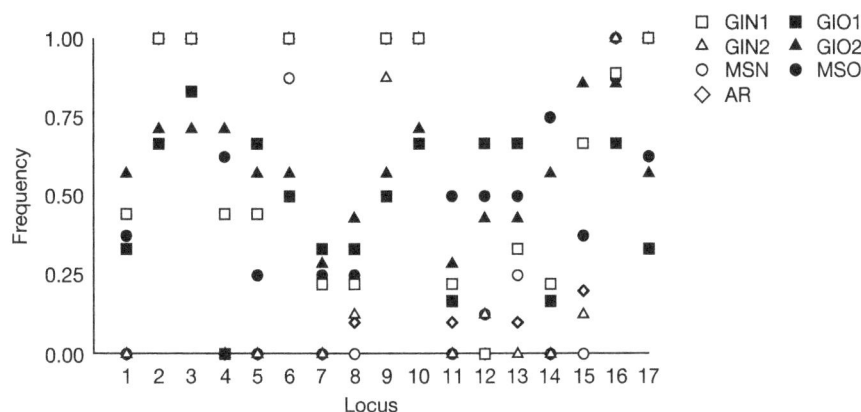

FIGURE 3 Frequencies of genetic loci significantly correlated to oil contamination across seven populations in locus-by-locus analysis. Contaminated sites are shown in closed shapes and uncontaminated sites in open shapes

TABLE 2 Mean AFLP haplotype and MS-AFLP epigenotype diversity (*h*) and percent polymorphic loci by site (%P), based on 71 AFLP and 39 MS-AFLP loci

	AFLP		MS-AFLP	
Population	*h*- (SE)	% P	*h*- (SE)	% P
Oil-contaminated				
GIO1	0.216 (0.031)	40.85	0.179 (0.037)	41.03
GIO2	0.246 (0.035)	42.25	0.185 (0.037)	43.59
MSO	0.216 (0.028)	50.70	0.161 (0.030)	41.03
Unaffected				
GIN1	0.246 (0.027)	57.75	0.226 (0.031)	66.67
GIN2	0.190 (0.022)	57.75	0.152 (0.031)	46.15
MSN	0.138 (0.020)	50.70	0.204 (0.037)	48.72
AR	0.103 (0.021)	28.17	0.132 (0.033)	33.33

lack of an effect of oil contamination on overall epigenetic variation was supported by perMANOVA (*F* = 0.373, *p* = .815) and redundancy analysis (*F* = 6.7269, *p* = .22). Locus-by-locus AMOVA revealed five loci were significantly differentiated between oil-contaminated and unaffected sites (Figure 4).

4 | DISCUSSION

Following the *Deepwater Horizon* oil spill in 2010, we sampled contaminated and uncontaminated populations of *S. alterniflora* along

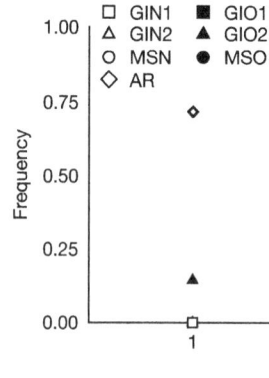

FIGURE 4 Frequencies of epigenetic loci significantly correlated to oil contamination across seven sites in locus-by-locus analysis. Contaminated sites are shown in closed shapes and uncontaminated sites in open shapes

the coast of the Gulf of Mexico in populations that had experienced heavy oiling and complete aboveground dieback. Despite reports of full recovery of aboveground biomass and stem density in heavily oiled populations after 18 months (Lin et al., 2016), our hierarchical AMOVA returned evidence of genetic differentiation among oil-contaminated and noncontaminated populations. However, we did not find evidence of decreased genetic diversity in contaminated populations, as nearly all individuals displayed a unique genotype in both contaminated and noncontaminated sites. These findings are consistent with other genetic surveys of *S. alterniflora* (Foust et al., 2016; Hughes & Lotterhos, 2014; Richards et al., 2004), which also show high levels of genetic variation. With our small sample size

TABLE 3 Summary of hierarchical AMOVA for AFLP and MS-AFLP data sets among site type (Φ_{RT}), among populations within site type (Φ_{PR}), and within populations (Φ_{PT}). Φ-statistics were calculated using 9,999 permutations

	Genetic			Epigenetic		
	Φ-statistics	% variation	df	Φ-statistics	% variation	df
Among site type	0.056*	6	1	0.017[NS]	1	1
Among populations within site type	0.168***	16	5	0.076***	7	5
Within subpopulations	0.215***	78	49	0.071**	92	49

df, degrees of freedom.
*$p \leq .05$, **$p \leq .01$, ***$p \leq .001$
[NS]nonsignificant following sequential Bonferroni correction.

TABLE 4 Pairwise Φ_{PT} comparisons of variation among study sites. Epigenetic comparisons are shown above the diagonal, genetic below

	Unaffected sites				Oil-contaminated sites		
	GIN1	GIN2	MSN	AR	GIO1	GIO2	MSO
GIN1		0.007	0.012	**0.124**	0.000	0.064	0.042
GIN2	**0.193**		0.013	0.205	0.024	0.039	0.133
MSN	**0.191**	0.150		0.145	0.018	0.032	0.077
AR	**0.189**	0.067	**0.180**		**0.166**	**0.160**	**0.208**
GIO1	0.123	**0.216**	**0.167**	0.225		0.000	0.137
GIO2	0.067	**0.261**	**0.247**	**0.319**	0.062		**0.159**
MSO	0.038	**0.197**	**0.190**	**0.217**	0.119	0.078	

Statistical significance after sequential Bonferroni correction denoted by bolded numbers.

(n < 10 at most sites), it is possible that we were unable to capture a change in genetic diversity among populations if one occurred in response to the oil spill.

In addition, we found no evidence of epigenetic differentiation over all loci between oil-contaminated and uncontaminated populations, but five loci showed epigenetic differentiation due to oil exposure in the locus-by-locus analysis. Further study is required to determine whether these loci are indicative of a regulatory response acting in concert with a few, but important epigenetic loci. However, redundancy analysis shows that overall patterns of methylation were not significantly correlated with oil exposure when controlling for the effects of genetic variation, which suggests that patterns of DNA methylation are explained almost entirely by genetic effects. Although we did expect to find epigenetic differentiation due to oil presence, it is possible either that oil did not induce any epigenetic changes between the population types, or that any existing epigenetic signature was too labile or too weak to be detected given the high epigenetic variation between individuals at our sites. Alternatively, our MS-AFLP may provide too few, anonymous markers to quantify epigenetic differentiation, and our small sample size may not have sufficient power to detect effects of rare epigenetic alleles or weak signatures of epigenetic change among the genetically differentiated populations. Many previous studies of epigenetic variation have taken advantage of low genetic diversity in natural systems to more clearly delineate population epigenetic effects (e.g., Gao, Geng, Li, Chen, & Yang, 2010; Richards et al., 2012). However, *S. alterniflora* is an outcrossing, wind-pollinated grass with extremely high genetic diversity (Foust et al., 2016; Hughes & Lotterhos, 2014; Richards et al., 2004). These high levels of genetic polymorphism make it more difficult to partition epigenetic structure due to increased statistical noise and genetic-dependent effects, particularly using anonymous genetic markers such as AFLP (but see e.g., Foust et al., 2016).

4.1 | Genetic and epigenetic response to pollution

Human-mediated environmental impacts have been well documented as potential evolutionary drivers of population differentiation. A classic example is the rapid phenotypic change experienced by the peppered moth as a result of coal pollution (Kettlewell, 1958), which was recently explained by the activity of transposable elements that alter its development (van't Hof et al., 2016). Several studies also describe molecular differentiation in marine organisms across the eastern coast of the United States in response to aquatic pollution (Chapman et al., 2011; Whitehead et al., 2012; Williams & Oleksiak, 2008). For example, populations of Atlantic killifish (*Fundulus heteroclitus*) in severely polluted habitats show broad genetic differentiation, including an allelic variant of cytochrome CYP1A (Williams & Oleksiak, 2008, 2011), which is correlated with changes in gill morphology. Populations of the related Gulf killifish (*F. grandis*) in the Gulf of Mexico also showed differential expression of CYP1A among affected and unaffected populations following the *Deepwater Horizon* oil spill (Whitehead et al., 2012). Together, these studies highlight the role of anthropogenic stress in selection, adaptation, and divergence (Hoffmann &

Sgrò, 2011; Lande, 1998). Despite previous literature suggesting that *S. alterniflora* is robust to heavy oil exposure, we found a signature of genetic differentiation between oil-exposed and unexposed populations. These results suggest at least some mortality in oil-exposed populations, consistent with findings of initial losses in live belowground biomass (Lin et al., 2016). By examining the genetic and epigenetic composition of marshes after the *DWH* oil spill, our study adds to the growing number of ecological and evolutionary genomics studies describing population-level response to pollution.

Populations in coastal habitats, and salt marshes in particular, have long been models for phenotypic differentiation across natural environmental gradients (Schmidt & Rand, 1999; Schmidt et al., 2008), and we expected to detect population-level differentiation of DNA methylation in response to oil contamination as well (Foust et al., 2016; Richards et al., 2012). The idea that epigenetic mechanisms can contribute to population differentiation as a source of heritable phenotypic variation has been challenged in recent literature (Laland et al., 2014; Wibowo et al., 2016). However, DNA methylation has been posited as a mechanism of phenotypic plasticity as well as a marker of stress response, and a number of studies have found a relationship between epigenetic variation and environment in support of this hypothesis (Herman & Sultan, 2016; Jablonka & Raz, 2009; Verhoeven, Jansen, et al., 2010; Verhoeven, Van Dijk, et al. 2010). Environmental stressors can induce variation in DNA methylation and in some cases, these environmentally induced methylation patterns can be inherited (Herrera & Bazaga, 2010, 2011; Verhoeven, Jansen, et al., 2010; Verhoeven, Van Dijk, et al. 2010), suggesting the potential for a signature of environmental response that is partially distinct from genetic variation.

Although we found high levels of epigenetic variation among individuals within and among populations, we failed to detect epigenetic differentiation in response to oil contamination. Our previous work showed a weak correlation between environmental conditions and epigenetic variation in *S. alterniflora* in a Georgia salt marsh (Foust et al., 2016). However, these data were collected from relatively protected habitat, and populations from this area are unlikely to have been exposed to a stress as severe as the *DWH* oil spill, which resulted in total aboveground dieback, and reduction by approximately 84%–95% of belowground biomass of the leading 5–10 meters of *S. alterniflora* in heavily oiled Gulf of Mexico marshes (Lin et al., 2016; Silliman et al., 2012). This impact may be far beyond what is normally experienced by *S. alterniflora* including natural disturbance events (Pennings & Bertness, 2001).

Epigenetic mechanisms of response, such as DNA methylation, are expected to be evolutionarily favorable when the periodicity of a stressor is short (Lachmann & Jablonka, 1996), such as cyclic patterns of rainfall, nutrient flows, and salinity that cause the zonation patterns observed among salt marsh plants (Pennings & Bertness, 2001). In contrast, the *Deepwater Horizon* oil spill may have acted as a single, discrete event that changed the makeup of the extensive genetic variation present in *S. alterniflora* rather than inducing a plastic or regulatory response that could be captured by assaying DNA methylation.

As studies of epigenetic variation in natural populations move away from quantifying the amount of standing genetic and epigenetic

variation in natural populations to describing the role of that variation and the relative contribution of genetic and epigenetic variation
to population differentiation, more precise sampling techniques and
analyses will be needed. In future studies, a reduced-representation
bisulfite sequencing approach would allow the direct comparison of
genetic and epigenetic data sets, and at a much finer scale, with substantially increased statistical power to detect epigenetic differences
between populations (van Gurp et al., 2016; Robertson & Richards,
2015; Schrey et al., 2013; Trucchi et al., 2016). In addition, sequencing-
based methods provide an increased ability to disentangle the relationship of methylation variation and gene function when fragments
overlap with the promoters or coding regions of genes. By increasing
the number of loci surveyed, future studies may better identify the
environmental conditions under which genetic or epigenetic variation
is associated with environmental cues (Robertson & Richards, 2015).

ACKNOWLEDGEMENTS

We thank the Office of Undergraduate Research at the University of
South Florida for financial support. We thank B DeLoach McCall and
the S Pennings' laboratory for help locating and accessing field sites.
We thank the Louisiana University Marine Consortium, especially A
Kolker, for advice on oiled site locations in Louisiana, Superintendent
J Malbrough for providing boat access, and Captain M Wike for navi-
gating the field sites and troubleshooting issues related to collections.
We also thank C Strobel at Aransas NWR, J Dingee at Brazoria NWR,
P Walther at Anahuac NWR, R Gosnell at Sabine NWR, T Augustine
at Grand Isle State Park, A Rupp at the Mississippi Coastal Preserves
Program, F Coleman and R Hughes at FSU Marine Laboratory for pro-
viding access to field sites.

REFERENCES

Ainouche, M. L., Fortune, P., Salmon, A., Parisod, C., Grandbastien, M.-A.,
    Fukunaga, K., & Misset, M.-T. (2009). Hybridization, polyploidy and inva-
    sion: Lessons from Spartina (Poaceae). Biological Invasions, 11, 1159–1173.

Ayres, D. R., Smith, D. L., Zaremba, K., Klohr, S., & Strong, D. R. (2004).
    Spread of exotic cordgrasses and hybrids (Spartina sp.) in the tidal
    marshes of San Francisco Bay, California USA. Biological Invasions, 6,
    221–231.

Bewick, A. J., Ji, L., Niederhuth, C. E., Willing, E.-M., Hofmeister, B. T., Shi,
    X., & Hartwig, B. (2016). On the origin and evolutionary consequences
    of gene body DNA methylation. Proceedings of the National Academy of
    Sciences, 113, 9111–9116.

Chapman, R. W., Mancia, A., Beal, M., Veloso, A., Rathburn, C., Blair, A.,
    & Sokolova, I. M. (2011). The transcriptomic responses of the eastern
    oyster, Crassostrea virginica, to environmental conditions. Molecular
    Ecology, 20, 1431–1449.

Clausen, J., Keck, D. D., & Hiesey, W. M. (1948). Experimental studies on the
    nature of species. III. Environmental responses of climatic races of Achillea.
    Washington, DC: Carnegie Institute. Carnegie Institute of Washington
    Publication 581.

Dlugosch, K. M., & Parker, I. (2008). Founding events in species invasions:
    Genetic variation, adaptive evolution, and the role of multiple introduc-
    tions. Molecular Ecology, 17, 431–449.

Dowen, R. H., Pelizzola, M., Schmitz, R. J., Lister, R., Dowen, J. M., Nery, J.

R., & Ecker, J. R. (2012). Widespread dynamic DNA methylation in re-
    sponse to biotic stress. Proceedings of the National Academy of Sciences,
    109, E2183–E2191.

Edwards, K. R., Travis, S. E., & Proffitt, C. E. (2005). Genetic effects of
    a large-scale Spartina alterniflora (smooth cordgrass) dieback and
    recovery in the northern Gulf of Mexico. Estuaries and Coasts, 28,
    204–214.

Evanno, G., Regnaut, S., & Goudet, J. (2005). Detecting the number of clus-
    ters of individuals using the software STRUCTURE: A simulation study.
    Molecular Ecology, 14, 2611–2620.

Falush, D., Stephens, M., & Pritchard, J. K. (2003). Inference of population
    structure using multilocus genotype data: Linked loci and correlated
    allele frequencies. Genetics, 164, 1567–1587.

Falush, D., Stephens, M., & Pritchard, J. K. (2007). Inference of population
    structure using multilocus genotype data: Dominant markers and null
    alleles. Molecular Ecology Notes, 7, 574–578.

Fitzpatrick, B. M. (2009). Power and sample size for nested analysis of mo-
    lecular variance. Molecular Ecology, 18, 3961–3966.

Foust, C., Preite, V., Schrey, A. W., Alvarez, M., Robertson, M., Verhoeven,
    K., & Richards, C. (2016). Genetic and epigenetic differences associated
    with environmental gradients in replicate populations of two salt marsh
    perennials. Molecular Ecology, 25, 1639–1652.

Fulneček, J., & Kovařík, A. (2014). How to interpret methylation sensitive
    amplified polymorphism (MSAP) profiles? BMC Genetics, 15(1), 1.

Gao, L., Geng, Y., Li, B., Chen, J., & Yang, J. (2010). Genome-wide DNA
    methylation alterations of Alternanthera philoxeroides in natural and
    manipulated habitats: Implications for epigenetic regulation of rapid re-
    sponses to environmental fluctuation and phenotypic variation. Plant,
    Cell & Environment, 33, 1820–1827.

Gugger, P. F., Fitz-Gibbon, S., Pellegrini, M., & Sork, V. L. (2016). Species-
    wide patterns of DNA methylation variation in Quercus lobata
    and its association with climate gradients. Molecular Ecology, 25,
    1665–1680.

van Gurp, T. P., Wagemaker, N. C., Wouters, B., Vergeer, P., Ouborg, J. N.,
    & Verhoeven, K. J. (2016). epiGBS: Reference-free reduced representa-
    tion bisulfite sequencing. Nature Methods, 13, 322–324.

Hamrick, J., & Allard, R. W. (1972). Microgeographical variation in allozyme
    frequencies in Avena barbata. Proceedings of the National Academy of
    Sciences, 69(8), 2100–2104.

Herman, J. J., & Sultan, S. E. (2016). DNA methylation mediates genetic vari-
    ation for adaptive transgenerational plasticity. Proceedings of the Royal
    Society B, 283, 20160988. https://doi.org/10.1098/rspb.2016.0988

Hermisson, J., & Pennings, P. S. (2005). Soft sweeps. Genetics, 169,
    2335–2352.

Herrera, C. M., & Bazaga, P. (2010). Epigenetic differentiation and rela-
    tionship to adaptive genetic divergence in discrete populations of the
    violet Viola cazorlensis. New Phytologist, 187, 867–876. https://doi.
    org/10.1111/j.1469-8137.2010.03298.x

Herrera, C. M., & Bazaga, P. (2011). Untangling individual variation in nat-
    ural populations: Ecological, genetic and epigenetic correlates of long-
    term inequality in herbivory. Molecular Ecology, 20, 1675–1688. https://
    doi.org/10.1111/j.1365-294X.2011.05026.x

van't Hof, A. E., Campagne, P., Rigden, D. J., Yung, C. J., Lingley, J., Quail,
    M. A., … Saccheri, I. J. (2016). The industrial melanism mutation in
    British peppered moths is a transposable element. Nature, 534(7605),
    102–105.

Hoffmann, A. A., & Sgrò, C. M. (2011). Climate change and evolutionary
    adaptation. Nature, 470(7335), 479–485.

Hubisz, M. J., Falush, D., Stephens, M., & Pritchard, J. K. (2009). Inferring
    weak population structure with the assistance of sample group infor-
    mation. Molecular ecology resources, 9, 1322–1332.

Hughes, A. R., & Lotterhos, K. E. (2014). Genotypic diversity at multiple
    spatial scales in the foundation marsh species, Spartina alterniflora.
    Marine Ecology Progress Series, 497, 105–117.

Jablonka, E., & Raz, G. (2009). Transgenerational epigenetic inheritance: Prevalence, mechanisms, and implications for the study of heredity and evolution. *The Quarterly review of biology*, *84*, 131–176.

Keller, T. E., Lasky, J. R., & Yi, S. V. (2016). The multivariate association between genome-wide DNA methylation and climate across the range of Arabidopsis thaliana. *Molecular Ecology*, *25*, 1823–1837.

Kettlewell, H. D. (1958). A survey of the frequencies of Biston betularia (L.) (Lep.) and its melanic forms in Great Britain. *Heredity*, *12*, 51–72.

Lachmann, M., & Jablonka, E. (1996). The inheritance of phenotypes: An adaptation to fluctuating environments. *Journal of theoretical biology*, *181*, 1–9.×

Laland, K., Uller, T., Feldman, M., Sterelny, K., Müller, G. B., Moczek, A., & Hoekstra, H. E. (2014). Does evolutionary theory need a rethink? *Nature*, *514*(7521), 161.

Lande, R. (1998). Anthropogenic, ecological and genetic factors in extinction and conservation. *Researches on population ecology*, *40*, 259–269.

Lin, Q., & Mendelssohn, I. A. (2012). Impacts and recovery of the Deepwater Horizon oil spill on vegetation structure and function of coastal salt marshes in the Northern Gulf of Mexico. *Environmental Science & Technology*, *46*, 3737–3743.

Lin, Q., Mendelssohn, I. A., Graham, S. A., Hou, A., Fleeger, J. W., & Deis, D. R. (2016). Response of salt marshes to oiling from the Deepwater Horizon spill: Implications for plant growth, soil surface-erosion, and shoreline stability. *Science of the Total Environment*, *557*, 369–377.

Lira-Medeiros, C. F., Parisod, C., Fernandes, R. A., Mata, C. S., Cardoso, M. A., & Ferreira, P. C. (2010). Epigenetic variation in mangrove plants occurring in contrasting natural environment. *PLoS ONE*, *5*, e10326. https://doi.org/10.1371/journal.pone.0010326

Medrano, M., Herrera, C. M., & Bazaga, P. (2014). Epigenetic variation predicts regional and local intraspecific functional diversity in a perennial herb. *Molecular Ecology*, *23*, 4926–4938.

Mirouze, M., Lieberman-Lazarovich, M., Aversano, R., Bucher, E., Nicolet, J., Reinders, J., & Paszkowski, J. (2012). Loss of DNA methylation affects the recombination landscape in *Arabidopsis*. *Proceedings of the National Academy of Sciences*, *109*, 5880–5885.

Nelson, M. F., & Anderson, N. O. (2013). How many marker loci are necessary? Analysis of dominant marker data sets using two popular population genetic algorithms. *Ecology and Evolution*, *3*, 3455–3470.

Nixon, Z., Zengel, S., Baker, M., Steinhoff, M., Fricano, G., Rouhani, S., & Michel, J. (2016). Shoreline oiling from the Deepwater Horizon oil spill. *Marine pollution bulletin*, *107*, 170–178.

Oksanen, J., Blanchet, F. G., Friendly, M., Kindt, R., Legendre, P., McGlinn, D., ... Wagner, H. (2017). *vegan: Community ecology package*. R package version 2.4-2. http:// CRAN.R-project.org/package=vegan.

Orr, H. A., & Betancourt, A. J. (2001). Haldane's sieve and adaptation from the standing genetic variation. *Genetics*, *157*, 875–884.

Paun, O., Bateman, R. M., Fay, M. F., Hedren, M., Civeyrel, L., & Chase, M. W. (2010). Stable epigenetic effects impact adaptation in allopolyploid orchids (*Dactylorhiza*: Orchidaceae). *Molecular Biology and Evolution*, *27*, 2465–2473. https://doi.org/10.1093/molbev/msq150

Peakall, R., & Smouse, P. E. (2012). GenAlEx 6.5: Genetic analysis in excel. Population genetic software for teaching and research—an update. *Bioinformatics*, *28*, 2537–2539.

Pennings, S. C., & Bertness, M. D. (2001). Salt marsh communities. In M. D. Bertness & S. D. Gaines (Ed.), *Marine community ecology* (pp.289–316). Sunderland, MA: Sinauer.

Pritchard, J. K., Stephens, M., & Donnelly, P. (2000). Inference of population structure using multilocus genotype data. *Genetics*, *155*, 945–959.

Rapp, R. A., & Wendel, J. F. (2005). Epigenetics and plant evolution. *New Phytologist*, *168*, 81–91. https://doi.org/10.1111/j.1469-8137.2005.01491.x

Reyna-Lopez, G., Simpson, J., & Ruiz-Herrera, J. (1997). Differences in DNA methylation patterns are detectable during the dimorphic transition of fungi by amplification of restriction polymorphisms. *Molecular and General Genetics MGG*, *253*, 703–710.

Richards, C. L., Hamrick, J., Donovan, L. A., & Mauricio, R. (2004). Unexpectedly high clonal diversity of two salt marsh perennials across a severe environmental gradient. *Ecology Letters*, *7*, 1155–1162.

Richards, C. L., Pennings, S. C., & Donovan, L. A. (2005). Habitat range and phenotypic variation in salt marsh plants. *Plant Ecology*, *176*, 263–273.

Richards, C. L., Schrey, A. W., & Pigliucci, M. (2012). Invasion of diverse habitats by few Japanese knotweed genotypes is correlated with epigenetic differentiation. *Ecology Letters*, *15*, 1016–1025.

Richards, C. L., Walls, R. L., Bailey, J. P., Parameswaran, R., George, T., & Pigliucci, M. (2008). Plasticity in salt tolerance traits allows for invasion of novel habitat by Japanese knotweed sl (*Fallopia japonica* and F.× bohemica, Polygonaceae). *American Journal of Botany*, *95*, 931–942.

Robertson, M., & Richards, C. (2015). Opportunities and challenges of next-generation sequencing applications in ecological epigenetics. *Molecular Ecology*, *24*, 3799–3801.

Salmon, A., Clotault, J., Jenczewski, E., Chable, V., & Manzanares-Dauleux, M. J. (2008). Brassica oleracea displays a high level of DNA methylation polymorphism. *Plant Science*, *174*, 61–70. https://doi.org/10.1016/j.plantsci.2007.09.012

Salzman, A. G. (1985). Habitat selection in a clonal plant. *Science*, *228*(4699), 603–604.

Schmidt, P. S., & Rand, D. M. (1999). Intertidal microhabitat and selection at MPI: Interlocus contrasts in the northern acorn barnacle, *Semibalanus balanoides*. *Evolution*, *53*, 135–146.

Schmidt, P. S., Serrão, E. A., Pearson, G. A., Riginos, C., Rawson, P. D., Hilbish, T. J., & Wethey, D. S. (2008). Ecological genetics in the North Atlantic: Environmental gradients and adaptation at specific loci. *Ecology*, *89*(sp11), 91–107.

Schrey, A. W., Alvarez, M., Foust, C. M., Kilvitis, H. J., Lee, J. D., Liebl, A. L., & Robertson, M. (2013). Ecological epigenetics: Beyond MS-AFLP. *Integrative and comparative biology*, *53*, 340–350.

Schulz, B., Eckstein, R. L., & Durka, W. (2014). Epigenetic variation reflects dynamic habitat conditions in a rare floodplain herb. *Molecular Ecology*, *23*, 3523–3537. https://doi.org/10.1111/mec.12835

Silliman, B. R., van de Koppel, J., McCoy, M. W., Diller, J., Kasozi, G. N., Earl, K., & Zimmerman, A. R. (2012). Degradation and resilience in Louisiana salt marshes after the BP–Deepwater Horizon oil spill. *Proceedings of the National Academy of Sciences*, *109*, 11234–11239.

Travis, S. E., Proffitt, C. E., & Ritland, K. (2004). Population structure and inbreeding vary with successional stage in created *Spartina alterniflora* marshes. *Ecological Applications*, *14*, 1189–1202.

Trucchi, E., Mazzarella, A. B., Gilfillan, G. D., Romero, M. L., Schönswetter, P., & Paun, O. (2016). BsRADseq: Screening DNA methylation in natural populations of non-model species. *Molecular Ecology*, *25*, 1697–1713.

Verhoeven, K. J., Jansen, J. J., van Dijk, P. J., & Biere, A. (2010). Stress-induced DNA methylation changes and their heritability in asexual dandelions. *New Phytologist*, *185*, 1108–1118. https://doi.org/10.1111/j.1469-8137.2009.03121.x

Verhoeven, K. J., Van Dijk, P. J., & Biere, A. (2010). Changes in genomic methylation patterns during the formation of triploid asexual dandelion lineages. *Molecular Ecology*, *19*, 315–324.

Verhoeven, K. J., Vonholdt, B. M., & Sork, V. L. (2016). Epigenetics in ecology and evolution: What we know and what we need to know. *Molecular Ecology*, *25*(8), 1631–1638.

Whitehead, A., Dubansky, B., Bodinier, C., Garcia, T. I., Miles, S., Pilley, C., & Walter, R. B. (2012). Genomic and physiological footprint of the Deepwater Horizon oil spill on resident marsh fishes. *Proceedings of the National Academy of Sciences*, *109*, 20298–20302.

Wibowo, A., Becker, C., Marconi, G., Durr, J., Price, J., Hagmann, J., ... Becker, J. (2016). Hyperosmotic stress memory in Arabidopsis is mediated by distinct epigenetically labile sites in the genome and is restricted in the male germline by DNA glycosylase activity. *Elife*, *5*, e13546.

Williams, L. M., & Oleksiak, M. F. (2008). Signatures of selection in natural populations adapted to chronic pollution. *BMC Evolutionary Biology, 8,* 282.

Williams, L. M., & Oleksiak, M. F. (2011). Evolutionary and functional analyses of cytochrome P4501A promoter polymorphisms in natural populations. *Molecular Ecology, 20,* 5236–5247.

Wilschut, R. A., Oplaat, C., Snoek, L. B., Kirschner, J., & Verhoeven, K. J. (2016). Natural epigenetic variation contributes to heritable flowering divergence in a widespread asexual dandelion lineage. *Molecular Ecology, 25,* 1759–1768.

Xie, H. J., Li, A. H., Liu, A. D., Dai, W. M., He, J. Y., Lin, S., … Qiang, S. (2015). ICE1 demethylation drives the range expansion of a plant invader through cold tolerance divergence. *Molecular Ecology, 24,* 835–850.

Zengel, S., Weaver, J., Pennings, S. C., Silliman, B., Deis, D. R., Montague, C. L., & Zimmerman, A. R. (2016). Five years of Deepwater Horizon oil spill effects on marsh periwinkles Littoraria irrorata. *Marine Ecology Progress Series.*, http://dx.doi.org.ezproxy.lib.usf.edu/10.3354/meps11827.

Zilberman, D., Gehring, M., Tran, R. K., Ballinger, T., & Henikoff, S. (2007). Genome-wide analysis of Arabidopsis thaliana DNA methylation uncovers an interdependence between methylation and transcription. *Nature Genetics, 39,* 61–69.

10

Elucidation of the genetic architecture of self-incompatibility in olive: Evolutionary consequences and perspectives for orchard management

Pierre Saumitou-Laprade[1,a] (ID) | Philippe Vernet[1,a] | Xavier Vekemans[1] | Sylvain Billiard[1] | Sophie Gallina[1] | Laila Essalouh[2] | Ali Mhaïs[2,3,4] | Abdelmajid Moukhli[3] | Ahmed El Bakkali[5] | Gianni Barcaccia[6] | Fiammetta Alagna[7,8] | Roberto Mariotti[8] | Nicolò G. M. Cultrera[8] | Saverio Pandolfi[8] | Martina Rossi[8] | Bouchaïb Khadari[2,9,a] | Luciana Baldoni[8,a]

[1]CNRS, UMR 8198 Evo-Eco-Paleo, Université de Lille - Sciences et Technologies, Villeneuve d'Ascq, France

[2]Montpellier SupAgro, UMR 1334 AGAP, Montpellier, France

[3]INRA, UR Amélioration des Plantes, Marrakech, Morocco

[4]Laboratoire AgroBiotechL02B005, Faculté des Sciences et Techniques Guéliz, University Cadi Ayyad, Marrakech, Morocco

[5]INRA, UR Amélioration des Plantes et Conservation des Ressources Phytogénétiques, Meknès, Morocco

[6]Laboratory of Genomics and Plant Breeding, DAFNAE - University of Padova, Legnaro, PD, Italy

[7]Research Unit for Table Grapes and Wine Growing in Mediterranean Environment, CREA, Turi, BA, Italy

[8]CNR, Institute of Biosciences and Bioresources, Perugia, Italy

[9]INRA/CBNMed, UMR 1334 Amélioration Génétique et Adaptation des Plantes (AGAP), Montpellier, France

Correspondence
Pierre Saumitou-Laprade, CNRS, UMR 8198 Evo-Eco-Paleo, Université de Lille - Sciences et Technologies, Villeneuve d'Ascq, France.
Email: pierre.saumitou@univ-lille1.fr

Funding information
This research was supported by the Project OLEA – Genomics and Breeding of Olive, funded by MiPAAF, Italy, D.M. 27011/7643/10, by the Project "BeFOre - Bioresources for Oliviculture," 2015–2019, H2020-MSCA-RISE-Marie Skłodowska-Curie Research and Innovation Staff Exchange, Grant Agreement N. 645595, by the French National Research Agency through the project "TRANS" (ANR-11-BSV7-013-03), by French-Moroccan scientific cooperation PRAD 14-03, and by Agropolis Fondation "OliveMed" N° 1202-066 through the "Investissements d'avenir" / Labex Agro ANR-10-Labex-0001-01., managed by French National Research Agency.

Abstract

The olive (*Olea europaea* L.) is a typical important perennial crop species for which the genetic determination and even functionality of self-incompatibility (SI) are still largely unresolved. It is still not known whether SI is under gametophytic or sporophytic genetic control, yet fruit production in orchards depends critically on successful ovule fertilization. We studied the genetic determination of SI in olive in light of recent discoveries in other genera of the Oleaceae family. Using intra- and interspecific stigma tests on 89 genotypes representative of species-wide olive diversity and the compatibility/incompatibility reactions of progeny plants from controlled crosses, we confirmed that *O. europaea* shares the same homomorphic diallelic self-incompatibility (DSI) system as the one recently identified in *Phillyrea angustifolia* and *Fraxinus ornus*. SI is sporophytic in olive. The incompatibility response differs between the two SI groups in terms of how far pollen tubes grow before growth is arrested within stigma tissues. As a consequence of this DSI system, the chance of cross-incompatibility between pairs of varieties in an orchard is high (50%) and fruit

[a]These four authors contributed equally to this work.

production may be limited by the availability of compatible pollen. The discovery of the DSI system in *O. europaea* will undoubtedly offer opportunities to optimize fruit production.

KEYWORDS

diallelic self-incompatibility system, homomorphic system, *Olea europaea* L., Oleaceae, olive diversity, plant mating systems, sporophytic genetic control, trans-generic conservation of SI functionality

1 | INTRODUCTION

Self-incompatibility (SI), a postpollination prezygotic mechanism preventing self-fertilization in simultaneous hermaphroditic individuals, is a common feature in flowering plants, occurring in around 40% of angiosperm species (Igic, Lande, & Kohn, 2008). Genetic determination of SI is highly variable, with a single locus or several loci (diallelic or multi-allelic) and gametophytic or sporophytic control of the pollen SI phenotype (Castric & Vekemans, 2004; De Nettancourt, 1977). Because distinct individuals can share identical SI genotypes, incompatible crosses are not limited to self-pollination (Bateman, 1952). By limiting compatible matings, SI can cause a direct decrease in seed production and can be an important demographic factor, a phenomenon known as the S-Allee effect (Leducq et al., 2010; Wagenius, Lonsdorf, & Neuhauser, 2007). This effect is especially important in populations with low genetic diversity (Byers & Meagher, 1992; Vekemans, Schierup, & Christiansen, 1998). Despite its widespread occurrence in angiosperms, the genetic basis of SI has been identified in a limited number of cases, and the underlying molecular mechanism has been shown in only a handful of plant families. These include the Brassicaceae (Kitashiba & Nasrallah, 2014; Tantikanjana, Rizvi, Nasrallah, & Nasrallah, 2009), Papaveraceae (Eaves et al., 2014), Solanaceae, Plantaginaceae, and Rosaceae (Iwano & Takayama, 2012; Sijacic et al., 2004; Williams, Wu, Li, Sun, & Kao, 2015).

Among plant species possessing a functional SI system, crop species are of particular importance because the SI system can interfere with plant production and breeding, representing a major obstacle for constant high yield (Sassa, 2016). A reduction in genetic diversity in commercial varieties may also potentially limit seed and fruit production in field conditions, with adverse economic consequences (Matsumoto, 2014). This issue has stimulated active crossing programs to assess allelic diversity at the SI locus in crop species showing functional SI, such as apple (Broothaerts, 2003), Japanese pear, sweet cherry, apricot (Sassa, 2016; Wünsch & Hormaza, 2004), cabbage (Ockendon, 1974), chicory (Gonthier et al., 2013), and sugarbeet (Larsen 1977). However, despite the obvious interest for breeders to use the SI system to their advantage as part of their breeding programs, proper understanding of the genetic factors and molecular mechanisms involved in SI is lacking for most species and generally technically difficult for breeding companies.

The mechanisms controlling SI are often conserved and shared among species belonging to a given plant family (Allen & Hiscock, 2008; Charlesworth, 1985; Weller, Donoghue, & Charlesworth, 1995). Hence, evolutionary approaches can help to uncover SI mechanisms in crop species based on knowledge of related species. Although SI has evolved independently many times within angiosperms, the rate of evolution of new SI systems is thought to be low, and the occurrence of distinct mechanisms of SI genetic determination within a given family should be rare (Igic et al., 2008).

The olive (*Olea europaea* subsp. *europaea*) is the iconic tree of the Mediterranean area, present in cultivated (var. *europaea*) and wild (var. *sylvestris*) forms (Green, 2002). Despite the economical, ecological, cultural, and social importance of this species, its mating system is still largely controversial and no consensus model has been accepted. The genetic determination and even functionality of SI are still largely controversial. In this species, even the most basic biological details of SI are unresolved and we do not know whether SI is under gametophytic (Ateyyeh, Stosser, & Qrunfleh, 2000) or sporophytic (Breton & Berville, 2012; Collani et al., 2012) genetic control. The number of genes involved in the olive SI system and their pattern of linkage and chromosomal location are also unknown.

Cultivars able to produce seed by selfing are thought to exist (Farinelli, Breton, Famiani, & Berville, 2015; Wu, Collins, & Sedgley, 2002), but this contention is rarely supported by molecular paternity tests (De la Rosa, James, & Tobutt, 2004; Díaz, Martín, Rallo, Barranco, & De la Rosa, 2006; Díaz, Martín, Rallo, & De la Rosa, 2007; Mookerjee, Guerin, Collins, Ford, & Sedgley, 2005; Seifi, Guerin, Kaiser, & Sedgley, 2012). In a recent study involving paternity assessment of seed progenies from the Koroneiki cultivar (Marchese, Marra, Costa, et al. 2016), it was shown that seeds produced on twigs protected from outcross pollen were derived from self-fertilization. In addition, open pollination resulted in 11% of the seeds being produced via selfing. Together, these results suggest that SI is indeed functional in this variety, although leaky. Pollination experiments have been performed in several studies to characterize SI at the phenotypic level and identify groups of compatibility among varieties. These studies scored compatibility either at the prezygotic stage, by cytological observations of pollen tube elongation in stigmas, or at the postzygotic stage, by measuring seed production (supported or not by paternity assessment) after application of pollen from donor plants (Breton et al., 2014; Cuevas & Polito, 1997). Contradictory conclusions were drawn in terms of classifying cultivars into SI groups, as well as in terms of the quantitative strength of the incompatibility reaction. Some of these discrepancies may be caused by pollen contamination, likely because

O. europaea produces a large quantity of pollen typical of most wind-pollinated species (Ferrara, Camposeo, Palasciano, & Godini, 2007).

Here, we studied the genetic determination of SI in olive in light of recent discoveries in other genera in the same family, Oleaceae. SI has been investigated in *Phillyrea angustifolia* L. and *Fraxinus ornus* L, two androdioecious species in which males and hermaphrodites co-occur in the same population. Both species share a homomorphic sporophytic diallelic SI (DSI) system (Saumitou-Laprade et al., 2010; Vernet et al., 2016). Self-incompatible hermaphroditic individuals belong to one of two homomorphic SI groups: Individuals of a given SI group can only sire seeds on hermaphrodites from the other group, and cross-pollination between individuals of the same group elicits an incompatibility response (Saumitou-Laprade et al., 2010). The DSI system has been conserved in both species, and cross-species pollination tests have demonstrated that the recognition specificities currently segregating in the two species are identical. *P. angustifolia* and *F. ornus* belong to two different subtribes within the Oleaceae (subtribe Oleinae for *P. angustifolia* and subtribe Fraxininae for *F. ornus*). Hence, it has been suggested that this DSI system originated ancestrally within Oleaceae (Vernet et al., 2016) and thus was present in the most recent common ancestor of these two subtribes about 40 million years ago (Mya) (Besnard, de Casas, Christin, & Vargas, 2009). Because *O. europaea* and *P. angustifolia* belong to the same subtribe (Oleinae), we hypothesize that they share the same SI system.

We applied experimental approaches developed for *P. angustifolia* and *F. ornus* (Saumitou-Laprade et al., 2010; Vernet et al., 2016) to characterize the SI system in *O. europaea*, both phenotypically and genetically. First, we performed controlled pollinations in a full diallel crossing scheme between hermaphroditic individuals used as pollen recipients and pollen donors (including self-pollination). The objective was to compare the pattern of the incompatibility reactions with those described in *P. angustifolia* and *F. ornus*. Second, we analyzed the pattern of segregation of incompatibility phenotypes among 91 offspring from one single intervarietal cross (De la Rosa et al., 2003). The results were in agreement with a genetic model consisting of a DSI system with two mutually incompatible groups of hermaphrodites. The validity of this genetic model was assessed by performing controlled pollinations with 89 genotypes representative of a significant portion of the olive diversity present in a worldwide collection against two pairs of tester genotypes, each pair being composed of two reciprocally compatible hermaphrodites phenotyped in the first diallel crossing experiment and each assigned to one of the two SI groups. Using pollen from hermaphroditic individuals of *P. angustifolia* and *F. ornus*, we also demonstrated that the same two allelic specificities are shared among the three genera, thereby confirming our hypothesis that they share the same DSI system. The results are presented in light of previous attempts to characterize the SI system in olive. It is worth mentioning that this study focused on the cultivated form of olive (var. *europaea*), analyzing varieties representative of domesticated Mediterranean olive diversity (El Bakkali et al., 2013; Haouane et al., 2011). Our results suggest new avenues for the development of olive orchard management practices to optimize fruit production.

2 | MATERIALS AND METHODS

2.1 | Plant material

To avoid any misclassification of varietal clones and to allow their authentication by means of voucher samples, DNA was extracted from each individual tree phenotyped for SI and was genotyped by assaying 15 different polymorphic microsatellite (SSR) marker loci. Hence, we identified each individual tree with a reference DNA sample code, a physical position in the orchard, a genotype reference number corresponding to a specific marker allele combination for the 15 SSR marker loci (Table S1), and an SI phenotype.

In 2013, six genotypes were chosen in Italian orchards and tested for cross-compatibility using stigma tests in a reciprocal diallel design (Table 1). Because testers had to be used as pollen recipients in future tests, the six genotypes were chosen among those represented by several trees in the experimental stations, at different sites under different agroecological conditions favoring different flowering times for a single genotype, and located as close as possible to laboratory facilities to ensure quick transfer of receptive flowers to the laboratory over the whole study period. From these six genotypes, four were selected to constitute the two pairs of testers used for screening varieties from the olive collections. Receptive flowers sampled from the four chosen tester genotypes were used to phenotype SI in 2013 and 2014. For phenotyping, 118 trees, corresponding to 89 genotypes, were selected from different *ex situ* collections. In particular, 64 trees were kept from the worldwide Olive World Germplasm Bank (OWGB) of INRA Marrakech, at the experimental orchard (Tessaout, Morocco), 45 from the Perugia collection ((43°04′54.4″N; 12°22′56.8E), Italy), and three from the CNR—Institute of Biosciences and Bioresources (CNR-IBBR) experimental garden (Perugia, Italy), and six were derived from the olive germplasm collection of the Conservatoire Botanique National Méditerranéen (CBNMed) (Porquerolles Island, France) (Table S1). These were used as pollen donors, to define the genetic architecture of SI and to maximize the genetic diversity of sampled *O. europaea* (Belaj et al. 2012; El Bakkali et al., 2013).

To verify segregation of the SI phenotype in progeny from an F1 cross, pollen was collected from 91 trees (hereafter called LEDA) growing at the CNR-IBBR Institute (Table S2) that are the progeny of a controlled cross between Leccino and Dolce Agogia varieties (referenced as Oit27 and Oit15). Their paternity was previously confirmed using RAPD, AFLP, SSR, and RFLP markers (De la Rosa et al., 2003).

2.2 | Genotyping of the sampled trees with microsatellite markers

To genotype sampled trees, total DNA was extracted from 100 mg of fresh leaf tissue by GeneElute Plant Genomic DNA Miniprep Kit (Sigma-Aldrich), following manufacturer's instructions, and then quantified by a Nanodrop spectrophotometer. Genotype identification was performed by analyzing 15 informative nuclear SSR markers (Baldoni et al., 2009; El Bakkali et al., 2013). PCR products were separated

TABLE 1 Results from self-pollination and reciprocal stigma tests performed in a diallel crossing design among six *Olea europaea* genotypes

| | SI group | DNA database reference | Pollen donor | | | | | |
| | | | G1 | | | G2 | | |
			Oit27	Oit26	Oit24	Oit15	Oit30	Oit28
Pollen recipient	G1	Oit27	SI	0	0	1	1	1
		Oit26	0	SI	0	1	1	1
		Oit24	0	0	SI	1	1	1
	G2	Oit15	1	1	1	SI	0	0
		Oit30	1	1	1	0	SI	0
		Oit28	1	1	1	0	0	SI

SI, self-incompatibility reaction detected, no or only short pollen tubes observed in stigmatic tissue after self-pollination; 0, incompatibility reaction, no or only short pollen tubes observed in stigmatic tissue (Figure 1, panel a and d); 1, compatibility reaction, pollen tubes were observed converging through the stigmatic tissue toward the style (see Figure 1 panel b and c). Two incompatible genotypes were assigned to the same incompatibility group (either G1 or G2); two compatible genotypes were assigned to different incompatibility groups. DNA database reference corresponds to voucher specimen accessible in referenced collections (see Table S1).

using an automatic capillary sequencer (ABI 3130 Genetic Analyzer, Applied Biosystems), and electropherograms were then investigated for allele composition across marker loci using GenMapper 3.7 software (Applied Biosystems).

To verify the genetic representation of the selected sample set, SSR data obtained on the 118 trees (Table S1) were compared to a collection of 342 genotypes: the 309 olive genotypes present in the OWGB collection (El Bakkali et al., 2013) together with the 33 genotypes sampled in Italian (27) and French (6) collections not present in the OWGB.

2.3 | Assessments of the compatibility/incompatibility reactions

2.3.1 | Incompatibility tests

To ensure that receptive stigmas were free of contaminant pollen, branches about 40–50 cm long and bearing several flower buds were bagged on tester recipients at least one week before flowers opened and stigmas became receptive (using two PBS3d/50 bags, an outer bag enclosing an inner bag, each of size 16 × 50 × 16 cm; PBS International, Scarborough, UK). A 10 × 25 cm PVC window allowed us to monitor flowers or treat them without opening the bags. For prezygotic stigma tests, twigs were collected when flowers were mature (i.e., when 5%–10% of flowers present on a twig were open). When performing postzygotic tests by controlled crosses and scoring of produced seeds, to prevent pollen contamination during pollination, the outer bag was removed and the inner bag was pierced with a needle to inject pollen with a spray gun, the needle hole was carefully taped immediately after spraying, and the outer bag was put back in place. To ensure continuous pollen availability, freshly collected pollen was stored at −80°C (Vernet et al., 2016) until it was applied to recipient stigmas; this procedure also allowed us to collect pollen on the latest flowering tree in 2013 for use in phenotyping on stigmas in 2014.

2.3.2 | Stigma test

We scored cross-compatibility following the protocol in Vernet et al. (2016). Under these conditions, stigmas treated with pollen were fixed 16 hr after pollination, then stained with aniline blue, and observed under a UV fluorescent microscope, which allowed us to distinguish pollen grains and pollen tubes from maternal tissues (Figure 1). When the pollen recipient and the pollen donor are compatible, several pollen tubes converge through the stigmatic tissue toward the style until the base of the stigma and entrance of the style (Figure 1 panels b and c). The absence of pollen tubes or the presence of only short pollen tubes growing within the stigma but never reaching the style was used as the criteria to score incompatibility (Figure 1 panels a and d). Given the risk of contamination, a single pollen tube growing in the stigmatic tissue was never considered a reliable criterion to determine compatibility. Three replicate flowers were pollinated for each cross.

2.3.3 | Interspecific stigma tests between *O. europaea*, *P. angustifolia*, and *F. ornus*

To (i) substantiate our conclusions on the occurrence of DSI in *O. europaea*, (ii) determine whether SI recognition specificities have remained stable in the hermaphrodite lineage of the *Olea* genus as its divergence from the lineage containing the androdioecious species *P. angustifolia*, and (iii) take advantage of recent knowledge about the genetic architecture of SI in *P. angustifolia* (Billiard et al., 2015), we performed interspecific stigma tests using *P. angustifolia* and *F. ornus* hermaphrodites assigned to the two SI groups in previous studies (called G1 and G2) (Saumitou-Laprade et al., 2010; Vernet et al., 2016). Stigma tests were performed using the two pairs of *O. europaea* testers (Oit27/Oit15 and Oit30/Oit26) as recipients, with frozen pollen from *P. angustifolia* (Pa-01C.02 [G1] and Pa-06G.15 [G2]) and *F. ornus* (Fo-A17 [G1] and Fo-G$_{1999}$-48 [G2]) belonging to the G1 and G2 SI groups.

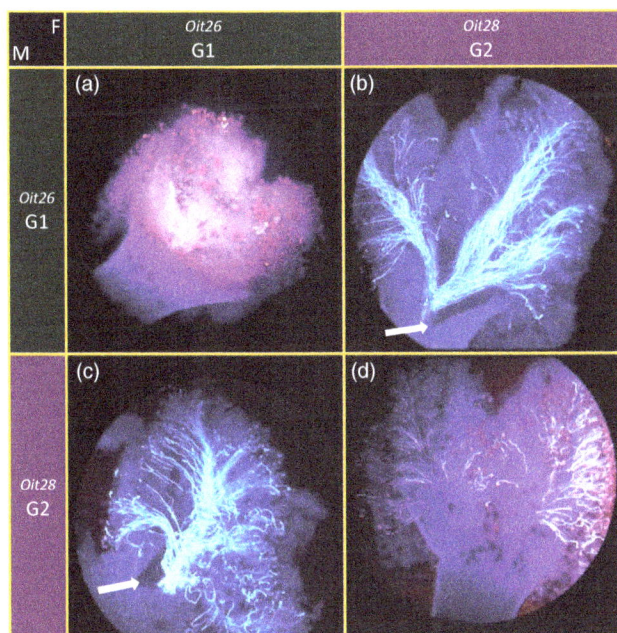

FIGURE 1   Stigma tests performed to assess self-incompatibility in *O. europaea*: examples of hermaphrodites Oit26 and Oit28. (a) The pollen of the hermaphrodite *Oit26* does not germinate on its own stigma demonstrating the self-sterility of this individual; (b) *Oit26* pollen germinates on hermaphrodite *Oit28* attesting to its viability; (c) the stigma from *Oit26* allows germination of *Oit28* pollen attesting to the stigma's functional receptivity when pollinated by compatible pollen; (d) the *Oit26* pollen does not germinate on its own stigma demonstrating the self-sterility of this individual. Arrows pinpoint the region corresponding to the base of the stigma and entrance of the style. M: genotype used as male pollen donor; F: genotype used as female recipient

We performed stigma tests by depositing pollen from one test sample on stigmas of a pair of cross-compatible testers (i.e., belonging to two different SI groups). Under the hypothesis that *O. europaea* exhibits DSI, we expected pollen from every sample to be compatible with one of the two tester lines (thereby confirming pollen viability) and incompatible with the other; indeed, cases in which pollen would be compatible with both tester lines would indicate either the presence of a third SI group (different from these represented by the tester recipients) or a nonfunctional SI genotype. Cases in which tested pollen was negative with both reference recipients were likely caused by either low pollen viability or low stigma receptivity. Hence, pollinations were repeated until compatibility was observed on at least one pollen recipient.

2.4 | Postzygotic validation of SI group assignment

To validate the compatibility versus incompatibility status assessed between pairs of genotypes, according to pollen tube behavior in the stigma tests, we carried out additional phenotypic assessments based on seed production after controlled pollination. We followed the protocol established and validated for *P. angustifolia* (Saumitou-Laprade et al., 2010). Each tested genotype was used as a pollen recipient and treated as follows: Two 40–50 cm long branches per

tree carrying numerous inflorescences were selected a week before the opening of the first flowers and carefully protected in two bags each. One inflorescence was pollinated with pollen collected from pollen donors belonging respectively to the G1 and G2 incompatibility groups (Oit27 and Oit15, respectively, see Results and Table 1). For each cross, pollination was repeated three times over a period of 8 days, beginning when the first flowers opened (between late May and mid-June 2014, depending on the flowering stage of the recipient). Isolation bags were removed several days after the end of flowering (on July 10) and replaced by net bags to prevent loss of fruit during ripening. Finally, fruits were collected and counted in mid-October, and to confirm paternity, genomic DNA was extracted from fruit embryos (Díaz et al., 2007) and from leaves of the parents. Parents and offspring were genotyped using 10 highly polymorphic microsatellite markers (Table S3) having high exclusion probability (El Bakkali et al., 2013). Paternity assignments were calculated with Cervus 3.0.3 (Table S4).

2.5 | Genetic assignment of the trees phenotyped for SI group and assessment of genetic diversity

To determine how our sampling represented the genetic diversity and the geographic structure of the Mediterranean olive tree, SSR alleles were scored in a single analysis (Tables S1 and S5) and combined with previous data obtained from the complete OWGB collection (El Bakkali et al., 2013).

The number of alleles per locus (Na), the observed (Ho) heterozygosity and expected (He) heterozygosity (Nei, 1987) were estimated using the Excel Microsatellite Toolkit v3.1 (Park, 2001). Principal coordinate analysis (PCoA), implemented in the DARWIN v.5.0.137 program (Perrier, Flori, & Bonnot, 2003), was carried out using a simple matching coefficient. To identify the genetic structure within the studied samples, in comparison with the Mediterranean olive germplasm, a model-based Bayesian clustering implemented in the program Structure ver. 2.2 (Pritchard, Stephens, & Donnelly, 2000) was used. Bayesian analysis was run under the admixture model for 1,000,000 generations after a burn-in period of 200,000, assuming correlation among allele frequencies. Analyses were run for values of K between one and six clusters with 10 iterations for each value. Validation of the most likely number of K clusters was performed using the ΔK statistics developed by Evanno, Regnaut, and Goudet (2005) with the R program, and the similarity index between 10 replicates for the same K clusters (H') was calculated using CLUMPP 1.1 (Greedy algorithm; (Jakobsson & Rosenberg, 2007)). For each selected K value, each accession was assigned to its respective cluster with a posterior membership coefficient ($Q > 0.8$).

We tested whether the allelic diversity observed in the 89 genotypes representing a subsample of the core collection was significantly lower than that of the overall OWGB collection using a Mann–Whitney U-test (p-value > .01 one-tailed test) after standardization of the dataset using the rarefaction method according to ADZE (Szpiech, Jakobsson, & Rosenberg, 2008).

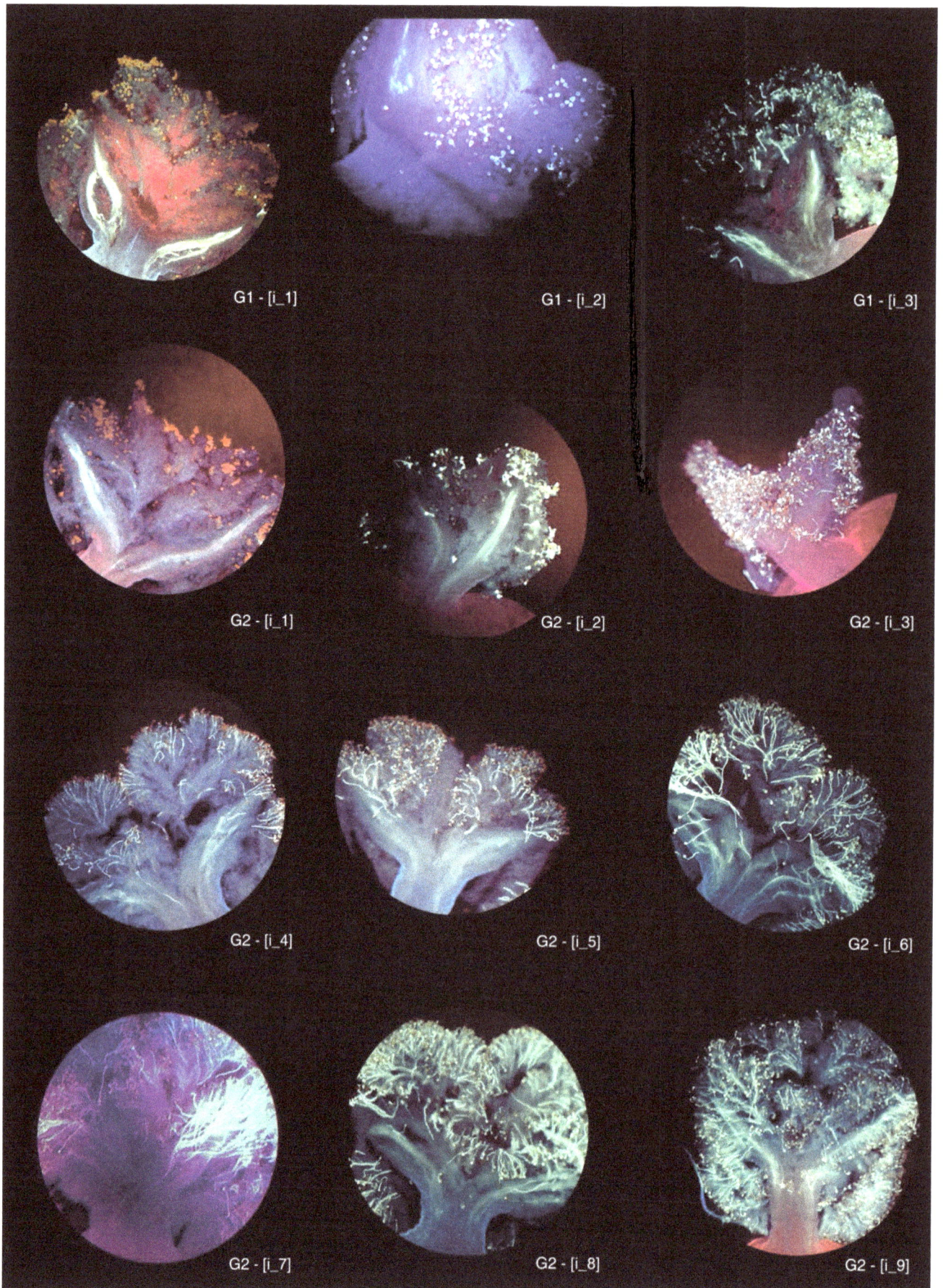

G1 - [i_1]

G1 - [i_2]

G1 - [i_3]

G2 - [i_1]

G2 - [i_2]

G2 - [i_3]

G2 - [i_4]

G2 - [i_5]

G2 - [i_6]

G2 - [i_7]

G2 - [i_8]

G2 - [i_9]

FIGURE 2   Classes of incompatibility phenotypes observed within self-incompatibility groups according to pollen (donor × recipient) interactions. On stigmas belonging to the G1 group, pollen tube length after growth was arrested was homogenous: from null to low (see the cases G1: [i_1] to [i_3]). On stigmas of G2 groups, pollen tube length after the arrest of growth varied widely among (donor/recipient) pairs: from null to high (see the cases G2: [i_1] to [i_9])

2.6 | Statistical analysis of pollen tube length scored in incompatible crosses

The specific length of pollen tubes within stigmatic tissue was measured for a given set of incompatible reactions (i.e., the growth that occurred prior to the arrest of further growth). Based on this growth, we defined nine discrete phenotypic classes, from i_1 to i_9 (Figure 2). Because an incompatible response scored in the highest phenotypic classes (i.e., longer pollen tubes, see [i_7], [i_8], and [i_9]) could be mistaken for a compatible response, we applied generalized linear model (GLM) analyses to the phenotypic data. A subset of 86 pollen donors was crossed with the four *O. europaea* testers involved in the stigma test described above (Table S6). For each pollen donor, we observed four crosses (two compatible and two incompatible), and for each cross, we photographed pollen tubes growing down stigmatic tissue and styles in three different flowers. The images were randomly labeled and observed four times independently, providing four reads for assignment to a phenotypic class (i.e., 12 independent scores for each cross).

First, we tested the effect of the SI group, replicate scoring, pollinator genotype, recipient genotype, and individual flowers on the SI phenotypic response (i.e., the length of pollen tube growth before growth arrest within stigmatic tissue in incompatible crosses, scored among nine phenotypic classes by the experimenter). We then used generalized linear mixed models (GLMM) on the categorized phenotype of the SI response to test the following: (i) whether the phenotype scoring based on digital images was repeatable, (ii) whether the phenotype was consistent among replicates of the same pollination test, (iii) whether the SI groups showed a different SI response, and (iv) whether the genotype of the recipient had an effect on the SI response. We considered the factors "flower read," "pollinator genotype," and "flower" as random effects, and "recipient genotype" and "SI group" as fixed effects. The SI response was the dependent variable and followed a Poisson distribution. To test whether a random or fixed factor had a significant effect, we performed a likelihood ratio test of nested models, using the package lme4 in R (Bates et al., 2014).

3 | RESULTS

3.1 | Phenotypic characterization of self-incompatibility in *O. europaea*

In 2013, six accessions (Oit27, Oit26, Oit24, Oit15, Oit30, and Oit28) corresponding to six different genotypes (Table S1) were used as both pollen recipients and pollen donors, in a reciprocal diallel design, for the stigma tests (Table 1). Self-fertilization was tested on the six genotypes, and no pollen tube successfully reached the style in any of the observed pistils, confirming strong SI reactions (Table 1). However, the length of pollen tubes within the stigmatic tissues varied between genotypes. Pollen tubes did not grow at all, or their growth stopped very early in the first layer of the stigma cells in the Oit26 genotype (Figure 1, panel a). In comparison, arrest of pollen tube growth did not occur until the pollen tubes had reached the deeper layers of the stigma cells, in the Oit28 genotype (Figure 1, panel d). However, even in this case, the pollen tubes stopped before reaching the transmitting tissue of the style.

For the intergenotype pollination tests, pollen tube/stigma interactions suggested the existence of SI reactions for each of the six different individual trees when crossed with specific partners (Table 1). An incompatibility phenotype similar to the Oit26 self-fertilization reaction (Figure 1, panel a: no or very short pollen tubes detected) was observed in the stigmatic tissues from Oit27, Oit26, and Oit24 when their pistils were pollinated by one another. The viability of their pollen and receptivity of their stigmas were verified in compatible crosses with Oit15, Oit30, and Oit28. A phenotype similar to the SI reaction observed with Oit28 (Figure 1, panel d: pollen tubes of variable length never reaching the style) was observed in the stigmas from Oit15, Oit30, and Oit28 when pollinated by one another. Here again, pollen viability and stigmatic receptivity of the same three individuals were checked in compatible crosses with Oit27, Oit26, and Oit24. We concluded that trees Oit27, Oit26, and Oit24 are incompatible with each other and belong to a single SI group, whereas trees Oit15, Oit30, and Oit28 belong to a different SI group. These results suggest that *O. europaea* individuals can be classified into at least two groups of SI, with incompatibility reactions between individuals belonging to the same group and compatible reactions between individuals belonging to different groups.

3.2 | The two *O. europaea* SI groups are functionally homologous to those of *P. angustifolia* and *F. ornus*

Nonambiguous and repeatable incompatibility phenotypes were observed when *P. angustifolia* and *F. ornus* G1 pollen was deposited on stigmas from Oit26 and Oit27 (Figure 3A,B, panel a), whereas compatibility phenotypes were scored on stigmas from Oit15 and Oit30 (Figure 3A,B, panel b). This demonstrated the capacity of trans-generic pollen to germinate and elicit both incompatible and compatible responses on *O. europaea* stigmas. Similarly, incompatibility phenotypes were scored on stigmas from Oit15 and Oit30 (Figure 3A,B, panel d), and compatibility phenotypes were observed with *P. angustifolia* and *F. ornus* G2 pollen on stigmas from Oit26 and Oit27 (Figure 3A,B, panel c). Therefore, we concluded that the SI system of *O. europaea* is functionally homologous to the DSI system previously reported for *P. angustifolia* and *F. ornus* (Saumitou-Laprade et al., 2010; Vernet et al., 2016). We assigned the Oit26 and Oit27 genotypes, and all their incompatible mates, to the G1 SI group, and the Oit15 and Oit30 genotypes, and all their incompatible mates, to the G2 SI group.

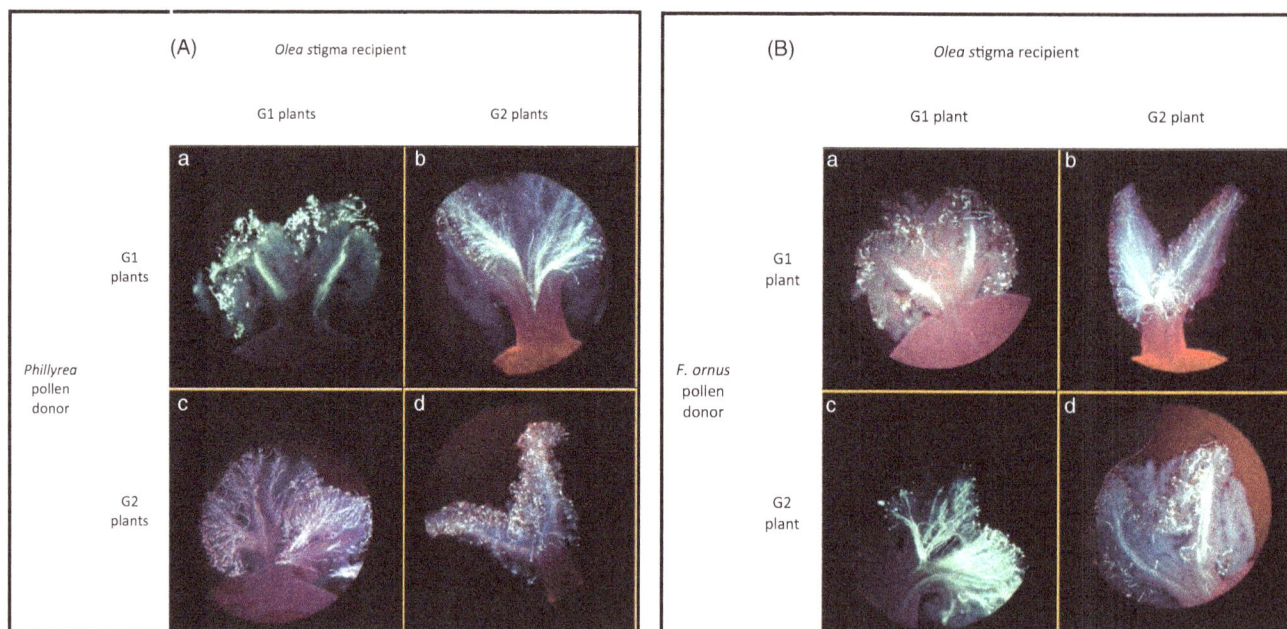

FIGURE 3 Trans-generic conservation of the self-incompatibility reaction between *Olea europaea* and two other Oleaceae species: (A) *Phillyrea angustifolia* and (B) *Fraxinus ornus*. In the photographs presented, stigma from *O. europaea* is pollinated by *P. angustifolia* and *F. ornus* pollen. (a) Incompatibility reaction between stigma of *Oit26* and G1 pollen; (b) compatibility reaction between stigma of *Oit15* and G1 pollen; (c) compatibility reaction between stigma of *Oit26* and G2 pollen; (d) incompatibility reaction between stigma of *Oit15* and G2 pollen

TABLE 2 Self-incompatibility phenotyping of the 91 LEDA F1 trees from the (Oit64 × Oit27) controlled cross

SI group	[G1] S1S2[a]	[G2] S1S1[a]	[Other] SxSy
	Incompatible with G1 and compatible with G2	Incompatible with G2 and compatible with G1	Compatible with G1 and G2
Total	41	50	0

Three types of behavior were scored. [G1], individual incompatible with G1 testers and compatible with G2 testers; [G2], individual incompatible with G2 testers and compatible with G1 testers; [Other], individual compatible with G1 and G2 testers and therefore belonging to a SI group different from G1 and G2 The S-locus segregates as a single locus with two alleles *S1* and *S2* (with *S2* dominant over *S1*) (Chi² test = 0.345, *df* = 1).
[a]Expected genotype deduced from genetic analyses in *P. angustifolia* (Billiard et al., 2015).

3.3 | Segregation of the self-incompatibility phenotypes in a controlled cross

Among the 91 LEDA full-sib trees that flowered in 2013 and/or in 2014, 41 were incompatible with G1 recipients and compatible with G2, indicating that they belong to the G1 compatibility group, and 50 were incompatible with G2 recipients and compatible with G1, indicating that they belong to G2. No offspring appeared compatible or incompatible with both groups of recipients in any of the tests (Table 2). The observed data agree with a genetic model assuming a 1:1 segregation of the two phenotypic groups (Chi² test statistic = 0.345, *df* = 1, ns).

3.4 | Two and only two self-incompatibility groups detected in *O. europaea*

The 118 sampled trees from the four germplasm collections (OWGB, Perugia collection, CNR-IBRR, and CBNMed) represented 89 distinct genotypes (20 genotypes were represented by more than one clonal

replicate, Table S1). We performed a total of 1,500 pollination tests, which allowed us to determine the SI phenotype of each individual tree with a mean of 2.6 replicates per tested genotype. All replicates were fully concordant (Fig. S1), demonstrating the robustness and reliability of the stigma tests performed. Among the 89 genotypes, 42 genotypes were incompatible with G1 recipients and compatible with G2, indicating that they belong to G1, and 47 genotypes were incompatible with G2 recipients and compatible with G1, revealing that they belong to G2. None of the genotypes were either compatible or incompatible with both groups of recipients, proving that two and only two SI groups were present in our extended sample (Table 3).

3.5 | Population genetic assessment of the sampling and representativeness of olive diversity

The samples we phenotyped for SI represented 89 distinct genotypes. We were highly conservative in our genotype identification: We grouped genotypes defined by a specific allele combination at 15 loci that differed by only one allele or two alleles into a single genotype

TABLE 3 Result of stigma tests performed with 89 *O. europaea* genotypes tested for compatibility and incompatibility with two pairs of pollen recipients used as testers

SI group	[G1]	[G2]	[Other]
	Incompatible with G1 and compatible with G2	Incompatible with G2 and compatible with G1	Compatible with G1 and G2
Total	42	47	0

Three types of behavior were scored (see Table 2 caption). The cultivars tested belong either to G1 or to G2, and none belong to a hypothetical third incompatibility group. In the sample tested, we detected only two incompatibility groups.

considering the possibility that their differences derived from somatic mutations that occurred within old clones (Haouane et al., 2011). Within the 89 genotypes, we detected 179 alleles over the 245 scored on the collection of 342 genotypes (Table S5-A). Hence, our sample captured 73% of the total allelic diversity observed in the collection of 342 genotypes. To check for the representativeness of olive diversity in our subsample of 89 genotypes, we compared allelic richness in the subsample with that of the collection of 342 genotypes after correction for difference in sample sizes based on the rarefaction method (see Table S5-B). Allelic richness of the two sample sets was not significantly different (Mann–Whitney U-test at $p \leq .01$ using one-tailed test; $U = 89$, P-value = .171; Table S5-B). Furthermore, we noted similar expected heterozygosity values (He = 0.745 in the 89 genotypes and He = 0.749 in the 342 genotypes collection; Table S5-A).

Most of the 89 genotypes phenotyped for SI were classified into one of the three western, central, and eastern Mediterranean clusters detected in previous studies (El Bakkali et al., 2013; Haouane et al., 2011), with a slight underrepresentation of the eastern gene pool (Table S5-C). This conclusion was further confirmed by their position on the first two axes of the principal coordinate analysis (PCoA, Fig. S3). Overall, despite the limited number of eastern olive trees, the 89 genotypes were distributed among the three Mediterranean gene pools, indicating they were a fair representation of domesticated olive diversity.

3.6 | The incompatibility response differs between the two self-incompatibility groups

When we analyzed variation in pollen tube lengths, we found no significant variation among replicate observations of the same flower and among flowers of a single individual when pollinated with incompatible pollen, indicating consistent incompatibility reactions. However, the SI group of the pollen recipient had a significant effect (p-value <.0001) on the distance that incompatible pollen tubes were able to grown within the stigma: Plants belonging to G1 showed an SI phenotype that fell in classes of low value (short pollen tubes), whereas plants belonging to G2 showed phenotypes that can fell in a wider panel of values (from short to long pollen tubes). The incompatibility reaction between G1 individuals seemed to occur almost immediately after pollen landed on the stigma as either pollen grains did not germinate or pollen tube growth stopped shortly after germination. In contrast, the incompatibility reaction between G2 individuals seems to occur later: Pollen grains germinated and pollen tubes grew into the stigmatic tissue before their growth was arrested.

Our analyses also revealed a significant effect of the recipient genotype on the score value within each SI group (p-value <.001 for both G1 and G2). Within G2, Oit30 showed a higher score than Oit15, and within G1, Oit27 showed a higher score than Oit26. This suggests consistent differences among genotypes in the timing of the SI response (early or late), whose functional significance remains to be determined.

3.7 | The SI assignment based on prezygotic stigma test validated by postzygotic genotyping

We verified at the postzygotic stage whether cases of incompatibility in which pollen tubes were able to germinate and grow substantial distances in the stigma (therefore the most ambiguous cases because of their relative similarity to compatible phenotypes) were cases in which fertilization was not achieved. The functional incompatibility of 10 different genotypes, belonging to SI group G2 and which scored in the highest phenotypic classes [i_7], [i_8], and [i_9] (Figure 2), was assessed at the postzygotic stage through progeny analysis (Table S4), as well as counts of the number of seeds produced (Table 4). All 99 progeny from crosses between putatively compatible mates (assigned based on the prezygotic stigma test), had genotypes compatible with both parents (Table S3). This confirms that the stigma test is reliable and suggests that our experimental design prevents pollen contamination. In contrast, the number of seeds collected following the 10 pollinations between parents belonging to the same SI group was extremely low (no seed produced in seven crosses and 2, 2, and 10 seeds in the three remaining crosses, respectively). In addition, none of the seeds harvested in these three crosses had a genotype that was consistent with its putative father. Again, these results confirm that the stigma test is a reliable procedure to predict which incompatibility group a plant belongs to, even in those cases in which some pollen tube growth occurs within the stigma. Interestingly, the few seeds obtained were all attributed to selfing.

4 | DISCUSSION

4.1 | Confirmation that the three genera *Olea*, *Phillyrea*, and *Fraxinus* share the same self-incompatibility system

The evolution of new SI systems in plants is thought to be a rare phenomenon, which is in agreement with the general observation that SI mechanisms are generally shared among species that exhibit SI within

TABLE 4 Number of seeds collected on G2 genotypes after controlled compatible and incompatible crosses performed in June 2014 and verified by paternity testing

[G2] Genotypes used as recipient	Pollen donors			
	[G1]: Oit27		[G2]: Oit15	
	Seeds produced	Paternity confirmed/tested	Seeds produced	Paternity confirmed/tested
LEDA_222	27	NA	0	–
LEDA_262	24	NA	0	–
LEDA_282	102	20/20	0	–
LEDA_301	98	20/20	0	–
Oit28	15	NA	2	0/2[a]
Oit03	30	10/10	0	–
Oit55	16	12/12	10	0/10[a]
Oit57	17	17/17	2	0/2[a]
Oit36	25	10/10	0	–
Oit22	40	10/10	0	–

[a]Selfing cannot be excluded with the 10 microsatellite markers used (see Tables S3 for genotyping results and S4 for estimation of exclusion probability based on markers and calculated using Cervus ver. 3.0.3.); NA, fruits not collected.

a given plant family (Allen & Hiscock, 2008; Charlesworth, 1985; Igic et al., 2008; Weller et al., 1995) and that losses of SI within a clade are much more common than gains (Igic, Bohs, & Kohn, 2006). As expected based on these arguments, we confirmed in *O. europaea* the occurrence of the same DSI discovered in *P. angustifolia* and *F. ornus* (Vernet et al., 2016), two androdioecious species that belong to two phylogenetic branches of the same family that diverged from each other more than 40 Mya (Besnard et al., 2009). While SI systems are often trans-generic, long-term stability of homomorphic DSI—that is the presence of only two alleles over a long time—is unexpected for two reasons. First, SI systems are susceptible to the rapid invasion of new incompatibility alleles, as a consequence of the strong frequency-dependent advantage of rare mating phenotypes (Wright, 1939). Gervais et al. (2011) showed that in a model where new alleles arise through self-compatible intermediates, selection for allelic diversification is inversely related to the number of segregating S-alleles, that is, more active diversification with a low number of alleles. Second, in hermaphroditic species, self-compatible mutants are expected to invade a homomorphic DSI population regardless of the extent of inbreeding depression (Charlesworth & Charlesworth, 1979). The stability of DSI was recently explained in the case of androdioecy with a theoretical model (Van de Paer, Saumitou-Laprade, Vernet, & Billiard, 2015), showing that androdioecy and DSI help maintain each other. DSI facilitates the maintenance of males (Billiard et al., 2015; Husse, Billiard, Lepart, Vernet, & Saumitou-Laprade, 2013; Pannell & Korbecka, 2010), and the full compatibility of males hinders the invasion of self-compatible mutants (Van de Paer et al., 2015).

The situation is quite different for *O. europaea*. The species belongs to the subgenus *Olea* which contains only hermaphrodite species and has diverged more than 30 Mya from the lineage containing androdioecious taxa such as *Osmanthus* and *Phillyrea* (Besnard et al., 2009). The evolutionary causes of the maintenance of DSI over 30 My remain

to be identified. Molecular characterization of the SI locus is a promising avenue of research to resolve issues related to the origin and maintenance of homomorphic DSI, because the simplest explanation is that the genetic architecture of the system does not allow the generation of additional SI phenotypes (e.g., a third SI allele). Molecular characterization will be facilitated by the trans-generic functionality of DSI that we observed among the *P. angustifolia*, *O. europaea*, and *F. ornus* species.

4.2 | Self-incompatibility in *O. europaea* is sporophytic

Our results are consistent with determination of SI on *O. europaea* by a single S-locus with only two alleles present in all cultivated forms of the species and demonstrate the sporophytic nature of this SI. First, the 1:1 proportion of the two parental SI groups in the controlled cross progeny excludes the possibility of gametophytic genetic control of self-incompatibility (GSI) (Bateman, 1952) in *O. europaea*. Second, with GSI, the incompatibility gene at the S-locus is expressed in the haploid pollen grains and interacts with the diploid tissue of the stigma. To be functional, a GSI system requires strict codominance between S-alleles in the pistil to avoid compatibility of heterozygous individuals with half of their own (self) pollen and a minimum of three alleles that define a minimum of three incompatibility groups (Hiscock & McInnis, 2003). In contrast, in the case of sporophytic genetic control of self-incompatibility (SSI), the incompatibility gene is expressed before meiosis in the diploid sporophytic tissue, and incompatibility arises with only two alleles, with a complete dominance of one allele over the other (see reviews by (Hiscock & McInnis, 2003; Billiard, Castric, & Vekemans, 2007). In our recent genetic study performed with *P. angustifolia* (Billiard et al., 2015), we showed a SI system governed by an S-locus with two alleles, *S2* and *S1* (with *S2* dominant over *S1*), which produced the two incompatibility groups G1 and G2 (Saumitou-Laprade et al., 2010), and with *S1S2* corresponding to G1 and *S1S1* to G2 (Billiard et al., 2015).

The gametophytic versus sporophytic nature of the SI system in *O. europaea* has been questioned for a long time in the literature, using indirect arguments, and several studies on the SI of olive cultivars have resulted in variable and conflicting results (for a review see Seifi, Guerin, Kaiser, & Sedgley, 2015). Features revealed by histological investigations, such as binucleate pollen and wet papillae stigma or a solid style and a large number of pollen grains germinating on the stigma surface, were reminiscent of species with GSI (De Nettancourt, 1997), whereas a dry papillae stigma was also reported in Oleaceae (Heslop-Harrison & Shivanna, 1977). Additional arguments based on the observation of pollen tube growth in incompatible crosses were in favor of GSI: The way pollen tubes halted in the proximal area of the style was interpreted as the intervention of programmed cell death, a frequent feature of GSI. Other arguments based on histochemical location of key enzyme activities involved in GSI were also reported in olive tree (Serrano & Olmedilla, 2012). Moreover, transcriptome analyses have been performed to screen for conserved transcripts typical of GSI in other plant species (e.g., S-ribonuclease transcripts such as in Solanaceae; (McClure, 2006)) or SSI (e.g., S-receptor kinase transcripts, such as in Brassicaceae; (Takasaki et al., 2000)). Transcripts similar to male and female SSI determinants of Brassicaceae were identified in olive (Alagna et al., 2016; Collani et al., 2010, 2012); however, there is no evidence of their functionality in SI reaction.

Here, we demonstrated that one of these numerous indirect arguments for assessing the gametophytic/sporophytic status of SI was wrong: For one SI group, the incompatibility reaction takes place at the stigma, whereas for the other SI group, the incompatibility reaction occurs later, sometimes at the entrance of the style. This feature may explain some of the past difficulty in identifying the gametophytic or sporophytic nature of the incompatibility system in *O. europaea*.

4.3 | Within-group incompatibility is stricter than within-individual self-incompatibility in *O. europaea*

One surprising observation from our experiments is the production of a small number of selfed seeds by G2 individuals following pollination with incompatible outcross pollen. This is the only indication of a partial breakdown of SI in the face of an otherwise very strong SI reaction. Why self-pollination seems to be promoted in the presence of incompatible outcross pollen remains to be determined. This feature is unexpected because in SI systems, the incompatibility phenotype of a pollen grain should only depend on the pollen parent genotype at the S-locus, which is shared among individuals from the same SI group. This result may indicate that, in olive or, at least, in some olive genotypes, the incompatibility reaction may be stronger with outcross pollen from the same group than with self-pollen. It is also possible that this observation results from the larger amount of self-pollen deposited on stigmas through autonomous self-pollination, compared with the outcross pollen transferred experimentally.

4.4 | *Olea europaea* is a true self-incompatible species in which some genotypes can produce seeds by selfing

All genotypes tested for SI in the present study were classified as self-incompatible according to the criteria of our stigma test, and all belong to one of the two SI groups identified in the species. These statements confirm that *O. europaea* is a true self-incompatible species. They are in agreement with conclusions of studies that tested SI in *O. europaea* at the postzygotic level, by measuring seed production after controlled crosses or open pollination, together with paternity analysis of the progeny (De la Rosa et al., 2004; Díaz et al., 2006; Marchese, Marra, Caruso, et al. 2016; Marchese, Marra, Costa, et al., 2016; Mookerjee et al., 2005). Just as in our study, many studies observed seeds produced by selfing either from controlled crosses with pollen from incompatible genotypes (see Oit28, Oit55, and Oit57 in Tables S3 and S4) or in controlled selfing (Farinelli et al., 2015) or open pollination (Marchese, Marra, Costa, et al., 2016). The self-incompatible status of a species does not exclude the possibility that the incompatibility reaction may be broken for self-pollen in some genotypes. The underlying mechanism allowing this remains to be studied. The occurrence of a low rate of selfing in individual plants with an active SI system is commonly reported and is referred to as pseudo-self-compatibility or leaky self-incompatibility. Leaky SI is generally thought to be a consequence of environmental factors interfering with the SI reaction or to the action of modifier genes (Busch & Schoen, 2008; Levin, 1996).

The leaky SI observed in olive has provided material for genetic mapping and sequencing (Marchese, Marra, Caruso, et al. 2016) and allows an opportunity to measure inbreeding depression. For example, in the wild relative *P. angustifolia*, 2% of 2,000 surveyed seedlings produced from controlled crosses were found to have been selfed (Billiard et al., 2015). Notably, none of these selfed seedlings ever flowered (unpublished results). In addition, leaky SI in olive might explain the gradient of results that have until now masked the real self-incompatibility system.

5 | CONCLUSION: ADDITIONAL EVOLUTIONARY APPLICATIONS

The level of interindividual incompatibility that we observed in our stigma test was very high: On average, half of the pairs of genetically distinct trees from the sampled collections were mutually incompatible. Similarly, most of the studies checking for compatibility within and among olive varieties, when using seed production and paternity analyses, detected numerous cases of cross-incompatibility (De la Rosa et al., 2003; Díaz et al., 2006; Mookerjee et al., 2005; Wu et al., 2002). In contrast, studies in orchards or crops of other domesticated species with SI, under either GSI (e.g., *Prunus, Malus, Pyrus, Amygdalus*) or SSI (e.g., *Brassica, Cichorium*), show high numbers of S-alleles and therefore high levels of cross-compatibility within or between cultivars (Dreesen et al., 2010; Ockendon, 1982; Wünsch & Hormaza, 2004). In the olive, the low number of elite varieties that co-occur in an orchard, together with the 50% chance of cross-incompatibility between pairs of varieties according to its DSI system, may limit fruit production. Limitation of the availability of compatible pollen, a phenomenon described as the S-Allee effect, occurs in wild populations of SI species with low S-allele diversity (Leducq et al., 2010; Wagenius et al., 2007). Small isolated populations or populations that have experienced a recent genetic bottleneck may have limited allelic diversity at the S-locus, leading to an increase in the probability of interindividual incompatibility, which in turn causes a reduction in seed production (Byers & Meagher, 1992; Vekemans et al., 1998).

The discovery of the DSI system in *O. europaea* will undoubtedly offer opportunities to optimize fruit production. First, it helps to understand the heretofore unexplained beneficial effect of ancestral practices that encourage the planting of a minimum number of varieties to ensure satisfactory olive production. Second, easy-to-use methods should be developed to determine the SI phenotype of each cultivated variety of olive to help guide the choice of varieties to be assembled in a given orchard, especially in nontraditional olive growing areas. Finally, ecological models can be developed to address the question of the optimal number of varieties to be introduced to ensure, effective pollination in an orchard, regardless of climate. Clearly, mono-varietal orchards must be avoided. In addition to the SI phenotype, the choice of varieties should take into account other important parameters such as flowering phenology, the direction of wind during the flowering period, and the relative positions of the different varieties within the orchard.

In the present study, we chose to present varieties through their reference genotype and not through their variety name, to assess the strict association between genotype and SI phenotype. Previous studies suggested possible discrepancies between varietal names and genotypes (El Bakkali et al., 2013; Haouane et al., 2011; Trujillo et al., 2014), and during our study, we observed different names associated with a single genotype (Table S1) as well as different genotypes associated with a single variety name; indeed, in more than 20% of cases, the genotypes associated with the same name were different in the Italian and OWGB collections (data not presented). Therefore, there is no strict association expected between variety name and SI phenotype. Therefore, each genotype of interest for olive producers needs to be assigned to one of the two SI groups. This will require characterizing these genotypes for their SI phenotype using the stigma test in rigorous conditions. Lastly, an effort should be devoted to identifying molecular markers with strong linkage with the S-locus to provide an easy-to-use diagnostic molecular assay for genotyping trees at the S-locus. We are confident that the evolutionary conservation of the functionality of the DSI among the *Olea*, *Phillyrea*, and *Fraxinus* genera will be an asset for accomplishing this task, through genomic and transcriptomic comparative analyses of the two groups within and among these three genera.

ACKNOWLEDGEMENTS

We warmly thank Drs. Vincent Castric, Isabelle De Cauwer, and Lynda Delph for scientific discussions and helpful comments on the manuscript. We thank Sylvia Lochon-Menseau for collecting pollen from the French germplasm collection in Porquerolles Island, and Sylvain Santoni, Pierre Mournet, and Ronan Rivallan for their technical support of molecular analysis in genotyping platform (UMR AGAP, Montpellier).

AUTHORS' CONTRIBUTIONS

All authors contributed significantly to the work presented in this manuscript. P.S.-L. and P.V. jointly designed and carried the sampling design and phenotyping strategies with B.K. and L.B.. P.S.-L. and P.V. performed phenotyping and crosses together with F.A., R.M., S.P., and M.R.. B.K. organized the collect of pollen in the OWGB collection in Marrakech (Morocco) and at the CBNMed in Porquerolles (France) together with A. Mh and A. Mo. L.B. organized the collect of pollen in the Perugia collection (Italy) and in the CNR-IBBR collections (including the LEDA F1 progeny). L. E. and N.G.M.C. performed DNA extraction and genotyping. B.K. and A.E.B. performed population genetic structure analyses. R.M. performed paternity analyses. S.B. performed statistical analyses of the variation in pollen tube lengths. S.G. created bioinformatics tools allowing management, comparison, and sharing among partners of the thousands of pictures produced for the SI phenotyping. G.B. provided expertise on olive SI and strongly contributed to initiate the project. X.V. provided expertise on the SI and population genetics analysis. P.S.-L., P.V., and X.V. wrote the paper.

REFERENCES

Alagna, F., Cirilli, M., Galla, G., Carbone, F., Daddiego, L., Facella, P., ... Perrotta, G. (2016). Transcript analysis and regulative events during flower development in olive (*Olea europaea* L.). *PLoS One*, *11*(4), 1–32.

Allen, A. M., & Hiscock, C. J. (2008). Evolution and phylogeny of self-incompatibility systems in angiosperms. In V. E. Franklin-Tong (Ed.), *Self-incompatibility in flowering plants – Evolution, diversity, and mechanisms*. Berlin Heidelberg: Springer-Verlag.

Ateyyeh, A. F., Stosser, R., & Qrunfleh, M. (2000). Reproductive biology of the olive (*Olea europaea* L.) cultivar'Nabali Baladi'. *Journal of Applied Botany*, *74*(5–6), 255–270.

Baldoni, L., Cultrera, N., Mariotti, R., Ricciolini, C., Arcioni, S., Vendramin, G., ... Testolin, R. (2009). A consensus list of microsatellite markers for olive genotyping. *Molecular Breeding*, *24*(3), 213–231.

Bateman, A. J. (1952). Self-incompatibility systems in angiosperms. *Heredity*, *6*(3).

Bates, D., Maechler, M., Bolker, B., Walker, S., Christensen, R. H. B., Singmann, H., ... LinkingTo Rcpp (2014). *Package 'lme4'*. Vienna: R Foundation for Statistical Computing.

Belaj, A., Dominguez-García, M. del C., Atienza, S. G., Urdíroz, N. M., De la Rosa, R., Satovic, Z., Martín, A., Kilian, A., Trujillo, I., & Valpuesta, V. (2012). Developing a core collection of olive (*Olea europaea* L.) based on molecular markers (DArTs, SSRs, SNPs) and agronomic traits. *Tree Genetics & Genomes*, *8*(2), 365–378.

Besnard, G., de Casas, R. R., Christin, P.-A., & Vargas, P. (2009). Phylogenetics of Olea (Oleaceae) based on plastid and nuclear ribosomal DNA sequences: Tertiary climatic shifts and lineage differentiation times. *Annals of Botany*, *104*(1), 143–160.

Billiard, S., Castric, V., & Vekemans, X. (2007). A general model to explore complex dominance patterns in plant sporophytic self-incompatibility systems. *Genetics*, *175*(3), 1351–1369.

Billiard, S., Husse, L., Lepercq, P., Godé, C., Bourceaux, A., Lepart, J., ... Saumitou-Laprade, P. (2015). Selfish male-determining element favors the transition from hermaphroditism to androdioecy. *Evolution*, 69, 683–693.

Breton, C. M., & Bervillé, A. (2012). New hypothesis elucidates self-incompatibility in the olive tree regarding S-alleles dominance relationships as in the sporophytic model. *Comptes Rendus Biologies*, 335(9), 563–572.

Breton, C. M., Farinelli, D., Shafiq, S., Heslop-Harrison, J. S., Sedgley, M., & Bervillé, A. J. (2014). The self-incompatibility mating system of the olive (*Olea europaea* L.) functions with dominance between S-alleles. *Tree Genetics & Genomes*, 10(4), 1055–1067.

Broothaerts, W. (2003). New findings in apple S-genotype analysis resolve previous confusion and request the re-numbering of some S-alleles. *Theoretical and Applied Genetics*, 106(4), 703–714.

Busch, J. W., & Schoen, D. J. (2008). The evolution of self-incompatibility when mates are limiting. *Trends in Plant Science*, 13(3), 128–136.

Byers, D. L., & Meagher, T. R. (1992). Mate availability in small populations of plant species with homomorphic sporophytic self-incompatibility. *Heredity*, 68(4), 353–359.

Castric, V., & Vekemans, X. (2004). Plant self-incompatibility in natural populations: A critical assessment of recent theoretical and empirical advances. *Molecular Ecology*, 13, 2873–2889.

Charlesworth, D. (1985). Distribution of dioecy and self-incompatibility in angiosperms. In P. J. Greenwood, P. H. Harvey, & M. Slatkin (Eds.), *Evolution—essays in honour of John Maynard Smith*. Cambridge: Cambridge University Press.

Charlesworth, D., & Charlesworth, B. (1979). The evolution and breakdown of S-allele systems. *Heredity*, 43, 41–55.

Collani, S., Alagna, F., Caceres, E. M., Galla, G., Ramina, A., Baldoni, L., ... Barcaccia, G. (2012). Self-incompatibility in olive: A new hypothesis on the S-locus genes controlling pollen-pistil interaction. *Acta Horticulturae*, 967, 133–140.

Collani, S., Moretto, F., Galla, G., Alagna, F., Baldoni, L., & Muleo, R. (2010). A new hypothesis on the mechanism of self-incompatibility occurring in olive (*Olea europaea* L.): Isolation, characterization and expression studies of SLG and SRK genes as candidates for a sporophytic self-incompatibility system. *Journal of Biotechnology*, 150, 502.

Cuevas, J., & Polito, V. S. (1997). Compatibility relationships in `Manzanillo' olive. *HortScience*, 32(6), 1056–1058.

De la Rosa, R., Angiolillo, A., Guerrero, C., Pellegrini, M., Rallo, L., Besnard, G., ... Baldoni, L. (2003). A first linkage map of olive (*Olea europaea* L.) cultivars using RAPD, AFLP, RFLP and SSR markers. *Theoretical and Applied Genetics*, 106(7), 1273–1282.

De la Rosa, R., James, C. M., & Tobutt, K. R. (2004). Using microsatellites for paternity testing in olive progenies. *HortScience*, 39(2), 351–354.

De Nettancourt, D. (1977). *Incompatibility in angiosperms*. Berlin: Springer-Verlag.

De Nettancourt, D. (1997). Incompatibility in angiosperms. *Sexual Plant Reproduction*, 10(4), 185–199.

Díaz, A., Martín, A., Rallo, P., Barranco, D., & De la Rosa, R. (2006). Self-incompatibility of 'Arbequina' and 'Picual' olive assessed by SSR markers. *Journal of the American Society for Horticultural Science*, 131(2), 250–255.

Díaz, A., Martín, A., Rallo, P., & De la Rosa, R. (2007). Cross-compatibility of the parents as the main factor for successful olive breeding crosses. *Journal of the American Society for Horticultural Science*, 132(6), 830–835.

Dreesen, R. S. G., Vanholme, B. T. M., Luyten, K., Van Wynsberghe, L., Fazio, G., Roldán-Ruiz, I., & Keulemans, J. (2010). Analysis of Malus S-RNase gene diversity based on a comparative study of old and modern apple cultivars and European wild apple. *Molecular Breeding*, 26(4), 693–709.

Eaves, D. J., Flores-Ortiz, C., Haque, T., Lin, Z., Teng, N., & Franklin-Tong, V. E. (2014). Self-incompatibility in Papaver: Advances in integrating the signalling network. *Biochemical Society Transactions*, 42(2), 370–376.

El Bakkali, A., Haouane, H., Moukhli, A., Costes, E., Van Damme, P., & Khadari, B. (2013). Construction of core collections suitable for association mapping to optimize use of Mediterranean olive (*Olea europaea* L.) genetic resources. *PLoS One*, 8(5), e61265.

Evanno, G., Regnaut, S., & Goudet, J. (2005). Detecting the number of clusters of individuals using the software STRUCTURE: A simulation study. *Molecular Ecology*, 14(8), 2611–2620.

Farinelli, D., Breton, C. M., Famiani, F., & Bervillé, A. (2015). Specific features in the olive self-incompatibility system: A method to decipher S-allele pairs based on fruit settings. *Scientia Horticulturae*, 181, 62–75.

Ferrara, G., Camposeo, S., Palasciano, M., & Godini, A. (2007). Production of total and stainable pollen grains in *Olea europaea* L. *Grana*, 46(2), 85–90.

Gervais, C. E., Castric, V., Ressayre, A., & Billiard, S. (2011). Origin and diversification dynamics of self-incompatibility haplotypes. *Genetics*, 188(3), 625–636.

Gonthier, L., Blassiau, C., Moerchen, M., Cadalen, T., Poiret, M., Hendriks, T., & Quillet, M.-C. (2013). High-density genetic maps for loci involved in nuclear male sterility (NMS1) and sporophytic self-incompatibility (S-locus) in chicory (*Cichorium intybus* L., Asteraceae). *Theoretical and Applied Genetics*, 126(8), 2103–2121.

Green, P. S. (2002). A revision of Olea L. (Oleaceae). *Kew Bulletin*, 57(1), 91–140.

Haouane, H., El Bakkali, A., Moukhli, A., Tollon, C., Santoni, S., Oukabli, A., ... Khadari, B. (2011). Genetic structure and core collection of the World Olive Germplasm Bank of Marrakech: Towards the optimised management and use of Mediterranean olive genetic resources. *Genetica*, 139(9), 1083–1094.

Heslop-Harrison, Y., & Shivanna, K. R. (1977). The receptive surface of the angiosperm stigma. *Annals of Botany*, 41(6), 1233–1258.

Hiscock, S. J., & McInnis, M. (2003). Pollen Recognition and rejection during the sporophytic self-incompatibility response: *Brassica* and beyond. *Trends in Plant Science*, 8(12), 606–613.

Husse, L., Billiard, S., Lepart, J., Vernet, P., & Saumitou-Laprade, P. (2013). A one-locus model of androdioecy with two homomorphic self-incompatibility groups: Expected vs. observed male frequencies. *Journal of Evolutionary Biology*, 26(6), 1269–1280.

Igic, B., Bohs, L., & Kohn, J. R. (2006). Ancient polymorphism reveals unidirectional breeding system shifts. *Proceedings of the National Academy of Sciences of the United States of America*, 103(5), 1359–1363.

Igic, B., Lande, R., & Kohn, J. R. (2008). Loss of self-incompatibility and its evolutionary consequences. *International Journal of Plant Sciences*, 169(1), 93–104.

Iwano, M., & Takayama, S. (2012). Self/non-self discrimination in angiosperm self-incompatibility. *Current Opinion in Plant Biology*, 15(1), 78–83.

Jakobsson, M., & Rosenberg, N. A. (2007). CLUMPP: A cluster matching and permutation program for dealing with label switching and multimodality in analysis of population structure. *Bioinformatics*, 23(14), 1801–1806.

Kitashiba, H., & Nasrallah, J. B. (2014). Self-incompatibility in Brassicaceae crops: Lessons for interspecific incompatibility. *Breeding Science*, 64(1), 23–37.

Larsen, K. (1977). Self-incompatibility in *Beta vulgaris* L. *Hereditas*, 85(2), 227–248.

Leducq, J.-B., Gosset, C. C., Poiret, M., Hendoux, F., Vekemans, X., & Billiard, S. (2010). An experimental study of the S-Allee effect in the self-incompatible plant Biscutella neustriaca. *Conservation Genetics*, 11(2), 497–508.

Levin, D. A. (1996). The evolutionary significance of pseudo-self-fertility. *American Naturalist*, 148(2), 321–332.

Marchese, A., Marra, F. P., Caruso, T., Mhelembe, K., Costa, F., Fretto, S., & Sargent, D. J. (2016). The first high-density sequence characterized SNP-based linkage map of olive (*Olea europaea* L. subsp. *europaea*) developed using genotyping by sequencing. *AJCS*, 10(6), 857–863.

Marchese, A., Marra, F. P., Costa, F., Quartararo, A., Fretto, S., & Caruso, T. (2016). An investigation of the self- and inter-incompatibility of the olive cultivars 'Arbequina' and 'Koroneiki' in the Mediterranean climate of Sicily. *Australian Journal of Crop Science, 10*(1), 88–93.

Matsumoto, S. (2014). Apple pollination biology for stable and novel fruit production: Search system for apple cultivar combination showing incompatibility, semicompatibility, and full-compatibility based on the S-RNase allele database. *International Journal of Agronomy, 2014*(9), 138271.

McClure, B. (2006). New views of S-RNase-based self-incompatibility. *Current Opinion in Plant Biology, 9*(6), 639–646.

Mookerjee, S., Guerin, J., Collins, G., Ford, C., & Sedgley, M. (2005). Paternity analysis using microsatellite markers to identify pollen donors in an olive grove. *Theoretical and Applied Genetics, 111*(6), 1174–1182.

Nei, M. (1987). *Molecular evolutionary genetics*. New York: Columbia University Press.

Ockendon, D. J. (1974). Distribution of self-incompatibility alleles and breeding structure of open-pollinated cultivars of Brussels sprouts. *Heredity, 33*(2), 159–171.

Ockendon, D. J. (1982). An S-allele survey of cabbage (Brassica oleracea var. capitata). *Euphytica, 31*(2), 325–331.

Pannell, J. R., & Korbecka, G. (2010). Mating-System Evolution: Rise of the Irresistible Males. *Current Biology, 20*(11), R482–R484.

Park, S.D.E. (2001). Trypanotolerance in West African cattle and the population genetic effects of selection., Dublin.

Perrier, X., Flori, A., & Bonnot, F. (2003). Data analysis methods. In P. Hamon, M. Seguin, X. Perrier, & J.-C. Glazmann (Eds.), *Genetic diversity of cultivated tropical plants*. Plymouth, UK: Enfield Science Publishers.

Pritchard, J. K., Stephens, M., & Donnelly, P. (2000). Inference of Population Structure Using Multilocus Genotype Data. *Genetics, 155*(2), 945–959.

Sassa, H. (2016). Molecular mechanism of the S-RNase-based gametophytic self-incompatibility in fruit trees of Rosaceae. *Breeding Science, 66*(1), 116.

Saumitou-Laprade, P., Vernet, P., Vassiliadis, C., Hoareau, Y., de Magny, G., Dommée, B., & Lepart, J. (2010). A self-incompatibility system explains high male frequencies in an androdioecious plant. *Science, 327*(5973), 1648–1650.

Seifi, E., Guerin, J., Kaiser, B., & Sedgley, M. (2012). Sexual compatibility of the olive cultivar" Kalamata" assessed by paternity analysis. *Spanish Journal of Agricultural Research, 3*, 731–740.

Seifi, E., Guerin, J., Kaiser, B., & Sedgley, M. (2015). Flowering and fruit set in olive: A review. *Journal of Plant Physiology, 5*(2), 1263–1272.

Serrano, I., & Olmedilla, A. (2012). Histochemical location of key enzyme activities involved in receptivity and self-incompatibility in the olive tree (*Olea europaea* L.). *Plant Science, 197*, 40–49.

Sijacic, P., Wang, X., Skirpan, A. L., Wang, Y. Y., Dowd, P. E., McCubbin, A. G., … Kao, T.-H. (2004). Identification of the pollen determinant of S-RNase-mediated self-incompatibility. *Nature, 429*(6989), 302–305.

Szpiech, Z. A., Jakobsson, M., & Rosenberg, N. A. (2008). ADZE: A rarefaction approach for counting alleles private to combinations of populations. *Bioinformatics, 24*(21), 2498–2504.

Takasaki, T., Hatakeyama, K., Suzuki, G., Watanabe, M., Isogai, A., & Hinata, K. (2000). The S receptor kinase determines self-incompatibility in Brassica stigma. *Nature, 403*(6772), 913–916.

Tantikanjana, T., Rizvi, N., Nasrallah, M. E., & Nasrallah, J. B. (2009). A Dual Role for the S-Locus Receptor Kinase in Self-Incompatibility and Pistil Development Revealed by an Arabidopsis rdr6 Mutation. *Plant Cell, 21*(9), 2642–2654.

Trujillo, I., Ojeda, M. A., Urdiroz, N. M., Potter, D., Barranco, D., Rallo, L., & Diez, C. M. (2014). Identification of the Worldwide Olive Germplasm Bank of Córdoba (Spain) using SSR and morphological markers. *Tree Genetics & Genomes, 10*(1), 141–155.

Van de Paer, C., Saumitou-Laprade, P., Vernet, P., & Billiard, S. (2015). The joint evolution and maintenance of self-incompatibility with gynodioecy or androdioecy. *Journal of Theoretical Biology, 371*, 90–101.

Vekemans, X., Schierup, M. H., & Christiansen, F. B. (1998). Mate availability and fecundity selection in multi-allelic self-incompatibility systems in plants. *Evolution, 52*, 19–29.

Vernet, P., Lepercq, P., Billiard, S., Bourceaux, A., Lepart, J., Dommée, B., & Saumitou-Laprade P. (2016). Evidence for the long-term maintenance of a rare self-incompatibility system in Oleaceae. *New Phytologist, 210*, 1408–1417.

Wagenius, S., Lonsdorf, E., & Neuhauser, C. (2007). Patch Aging and the S-Allee Effect: Breeding System Effects on the Demographic Response of Plants to Habitat Fragmentation. *The American Naturalist, 169*(3), 383–397.

Weller, S. G., Donoghue, M. J., & Charlesworth, D. (1995). The evolution of self-incompatibility in flowering plants: A phylogenetic approach. *Experimental and Molecular Approaches to Plant Biosystematics. St. Louis, Mo.: Missouri Botanical Garden ((Monographs in Systematic Botany, 53*, 355–382.

Williams, J. S., Wu, L., Li, S., Sun, P., & Kao, T.-H. (2015). Insight into S-RNase-based self-incompatibility in Petunia: Recent findings and future directions. *Frontiers in Plant Science, 6*, 41.

Wright, S. I. (1939). The distribution of self-sterility alleles in populations. *Genetics, 24*, 538–552.

Wu, S. B., Collins, G., & Sedgley, M. (2002). Sexual compatibility within and between olive cultivars. *Journal of Hortical Sciences and Biotechnology, 77*(6), 665–673.

Wünsch, A., & Hormaza, J. I. (2004). Genetic and molecular analysis in Cristobalina sweet cherry, a spontaneous self-compatible mutant. *Sexual Plant Reproduction, 17*(4), 203–210.

11

Plasticity in gene transcription explains the differential performance of two invasive fish species

Kyle W. Wellband[1] (iD) | Daniel D. Heath[1,2]

[1]Great Lakes Institute for Environmental Research, University of Windsor, Windsor, ON, Canada

[2]Department of Biological Sciences, University of Windsor, Windsor, ON, Canada

Correspondence
Daniel D. Heath, Great Lakes Institute for Environmental Research, University of Windsor, Windsor, ON, Canada.
Email: dheath@uwindsor.ca

Funding information
NSERC; Canadian Aquatic Invasive Species Network II

Abstract

Phenotypic plasticity buffers organisms from environmental change and is hypothesized to aid the initial establishment of nonindigenous species in novel environments and postestablishment range expansion. The genetic mechanisms that underpin phenotypically plastic traits are generally poorly characterized; however, there is strong evidence that modulation of gene transcription is an important component of these responses. Here, we use RNA sequencing to examine the transcriptional basis of temperature tolerance for round and tubenose goby, two nonindigenous fish species that differ dramatically in the extent of their Great Lakes invasions despite similar invasion dates. We used generalized linear models of read count data to compare gene transcription responses of organisms exposed to increased and decreased water temperature from those at ambient conditions. We identify greater response in the magnitude of transcriptional changes for the more successful round goby compared with the less successful tubenose goby. Round goby transcriptional responses reflect alteration of biological function consistent with adaptive responses to maintain or regain homeostatic function in other species. In contrast, tubenose goby transcription patterns indicate a response to stressful conditions, but the pattern of change in biological functions does not match those expected for a return to homeostatic status. Transcriptional plasticity plays an important role in the acute thermal tolerance for these species; however, the impaired response to stress we demonstrate in the tubenose goby may contribute to their limited invasion success relative to the round goby. Transcriptional profiling allows the simultaneous assessment of the magnitude of transcriptional response as well as the biological functions involved in the response to environmental stress and is thus a valuable approach for evaluating invasion potential.

KEYWORDS
biological invasions, gene expression, nonindigenous species, phenotypic plasticity, round goby, tubenose goby

1 | INTRODUCTION

In recent decades, there has been renewed interest in phenotypic plasticity as a mechanism that facilitates species persistence in novel and changing environments (Ghalambor, McKay, Carroll, & Reznick, 2007). Phenotypic plasticity is defined as the ability of organisms with identical genotypes to alter a specific aspect of their phenotype, either transiently or permanently, in response to environmental factors

(West-Eberhard, 2003). Traditionally regarded as a source of unpredictable phenotypic variance (e.g., Wright, 1931), plasticity was believed to retard evolution by natural selection by obscuring adaptive genetic variation from selective pressures. However, the ability to alter phenotype in an environmentally dependent manner may be advantageous for organisms experiencing variable environments if the phenotypic changes provide a fitness advantage (Schlichting & Smith, 2002). Not surprisingly, both empirical and theoretical considerations of plasticity have demonstrated conditions where plasticity is adaptive (provides a fitness advantage; Price, Qvarnstrom, & Irwin, 2003), demonstrated plasticity's role in facilitating genetic adaptation through genetic accommodation (West-Eberhard, 2003) and distinguished between plasticity that is adaptive (beneficial for an organism's fitness but not a product of selection) and plasticity that is an adaptation (beneficial for an organism's fitness and has been shaped by natural selection; Gotthard & Nylin, 1995). Plasticity that improves an organism's fitness is clearly an important trait for organisms experiencing environmental challenges such as those experienced when organisms colonize novel environments.

Biological invasions expose organisms to novel environments and provide an excellent opportunity to study the role of adaptive plasticity in population establishment, persistence, and expansion. Blackburn et al. (2011) developed a conceptual model to describe the invasion process as a series of barriers and stages that a species must pass through to be classified as invasive. Thus, a highly successful invasive species is not just one that survives and establishes in a non-native region but one that expands its range throughout the non-native region (Blackburn et al., 2011). Plasticity certainly plays a role in the survival of nonindigenous species during the "transport" and "establishment" stages of an introduction when environmental changes will be rapid and before evolutionary responses can occur; however, plasticity may also be critically important for the postestablishment range expansion that characterizes highly successful invasions. Species may rapidly evolve elevated plasticity to produce an optimal, yet responsive, phenotype during the range expansion phases of an invasion (Lande, 2015). This rapid increase in plasticity is then followed by assimilation of these traits by selection on standing genetic variation and relaxed selection for plasticity as populations stabilize (Lande, 2015). The role of plasticity in providing fitness advantages to organisms experiencing novel environments has generated interest in whether successful invaders are more plastic than unsuccessful invaders; however, support for the hypothesis that invaders are more plastic than noninvaders is inconsistent (Davidson, Jennions, & Nicotra, 2011; Godoy, Valladares, & Castro-Díez, 2011; Palacio-López & Gianoli, 2011). Phenotypic plasticity is expected to change through the stages of an invasion and the inconsistent support for plasticity as an important mechanism driving invasion success is likely a result of the varied amount of time since invasion for species included in these studies (Lande, 2015). As a result, direct tests of the hypothesis that more successful invaders have greater plasticity must compare species with similar invasion timing and histories.

There is a growing body of the literature implicating gene expression variation as a mechanism that facilitates plastic phenotypic responses to environmental change (Aubin-Horth & Renn, 2009; Schlichting & Smith, 2002). Gene expression is a phenotype that responds to environmental cues and is the mechanistic basis for different phenotypes expressed by different types of cells, tissues, and organisms (Wray et al., 2003). Gene transcription, the initial step in gene expression, has shown the capacity to evolve both changes in constitutive expression (Whitehead & Crawford, 2006) and altered responses to environmental cues (Aykanat, Thrower, & Heath, 2011). As a key regulator of the physiological status of organisms, there has been an increased focus on the role of gene transcription as a mechanism underlying plastic traits in wild populations; examples include salinity tolerance (Lockwood & Somero, 2011; Whitehead, Roach, Zhang, & Galvez, 2012), immune function (Stutz, Schmerer, Coates, & Bolnick, 2015), long-term thermal acclimation (Dayan, Crawford, & Oleksiak, 2015), and acute thermal tolerance (Fangue, Hofmeister, & Schulte, 2006; Quinn, McGowan, Cooper, Koop, & Davidson, 2011). Increased thermal tolerance has been linked to invasion success (Bates et al., 2013). Widespread transcriptional changes in response to both acute exposure and long-term acclimation to thermal stress have been documented in a diverse array of taxa including plants, yeast, invertebrates, fish, and mammals (Logan & Somero, 2011; Smith & Kruglyak, 2008; Sonna, Fujita, Gaffin, & Lilly, 2002; Sørensen, Nielsen, Kruhøffer, Justesen, & Loeschcke, 2005; Swindell, Huebner, & Weber, 2007) indicating that transcriptional plasticity plays an important and evolutionary conserved role in both short- and long-term responses to altered temperature (López-Maury, Marguerat, & Bähler, 2008). Given the important role of transcriptional plasticity in mediating physiological changes associated with thermal stress, the question arises: Do successful invasive species exhibit higher transcriptional plasticity in response to thermal stress? Indeed there is some evidence that transcriptional plasticity may be a feature of successful biological invasions as an increased capacity for transcriptional response to temperature exposure has also been observed in a highly successful marine invader *Mytilus galloprovincialis* compared to its native conger *Mytilus trossulus* on the west coast of North America (Lockwood, Sanders, & Somero, 2010).

Understanding attributes that make invaders successful is a critical aspect of the management of invasive species (Kolar & Lodge, 2001). Ideally, experiments testing the importance of invasive traits should compare congeners exhibiting a successful and failed invasion in the same environment (Kolar & Lodge, 2001); however, this presents the logistical challenge of studying organisms that do not exist (failed invader). In this study, we take advantage of a nearly analogous instance of a highly successful invasion (as determined by extent of range expansion) and a less successful invasion between two phylogenetically and invasion history paired species in the Laurentian Great Lakes of North America to test the hypothesis that more successful invasive species are more transcriptionally plastic than less successful invasive species.

Round goby (*Neogobius melanostomus*, Pallas) and tubenose goby (*Proterorhinus semilunaris*, Heckel) are two species of fish from the family Gobiidae that possess overlapping geographic ranges and habitat in their native Ponto-Caspian region of Eastern Europe. These species were both first detected in North America in the St. Clair River in 1990 (Jude, Reider, & Smith, 1992), presumably introduced via ballast water carried by cargo ships originating from the Black Sea (Brown & Stepien, 2009). Since introduction, round goby have spread throughout the entire Great Lakes basin and reached high population densities in many areas, while tubenose goby have mostly remained geographically restricted to the Huron–Erie corridor near the site of initial introduction and occur at low population densities (Figure 1). There is limited information about factors that may have differentially restricted range expansion for these species. Round goby have small home ranges (~5 m^2; Ray & Corkum, 2001) and typically do not disperse more than 500 m on their own (Lynch & Mensinger, 2012; Wolfe & Marsden, 1998). Similar information is unavailable for tubenose goby in the Great Lakes; however, it is difficult to imagine that the dispersal attributes described above would provide round goby with an advantage that would explain the differential range expansion and impact. The presence of both species in Lake Superior (Figure 1) suggests that differences in secondary transport due to shipping vectors within the Great Lakes are unlikely to explain the differential range expansion. Tubenose goby are slightly smaller on average than round goby (maximum total length in the Great Lakes: TNG ~ 130 mm, RG ~ 180 mm; Fuller, Benson, et al. 2017; Fuller, Nico, et al. 2017), but this does not appear to result in large differences in fecundity (MacInnis & Corkum, 2000b; Valová, Konečná, Janáč, & Jurajda, 2015).

Differences in phenotypic plasticity may explain the difference in invasion performance of round and tubenose goby. Round goby exhibit greater dietary plasticity compared to tubenose goby (Pettitt-Wade, Wellband, Heath, & Fisk, 2015). Thermal performance curves suggest that round goby has a broad thermal tolerance (Lee & Johnson, 2005). While similar curves are unavailable for tubenose goby, they have similar standard and resting metabolic rates at near optimum temperatures (O'Neil, 2013; Xin, 2016) but reduced performance at temperature extremes. Tubenose goby have a decreased upper critical thermal limit (31.9°C) compared with round goby (33.4°C; Xin, 2016) and exhibit higher standard metabolic rates at elevated temperatures (O'Neil, 2013) that may indicate a narrower range of temperature tolerance

than round goby. In addition to the difference in performance at elevated temperatures, the expansion and impact of invasive fish species in the Great Lakes are also typically limited by cold temperature tolerance (Kolar & Lodge, 2002); however, specific critical limits are unavailable for these species.

Changes in gene transcription underpin many adaptive responses to acute and long-term temperature exposure (e.g., Logan & Somero, 2011). To investigate the genetic mechanisms that underlie apparent differences in thermal tolerance, we use RNA sequencing (RNAseq) to characterize the liver transcriptomes of round and tubenose goby in response to acute exposure to increased and decreased temperatures. Liver tissue is a key regulator of a fish's metabolic processes and is known to play an important role in molecular reprogramming of metabolism in response to acute stressors (Wiseman et al., 2007). We predict that (i) the round goby will show generally higher transcriptional plasticity (more genes responding and at higher magnitudes of transcriptional change) across the liver transcriptome and (ii) the observed transcriptional variation will have greater functional relevance for maintaining homeostatic function in the round goby relative to the tubenose goby. Transcriptional profiling has enormous potential for applications in conservation biology (e.g., He et al., 2015; Miller et al., 2011) and a characterization of the evolutionary processes driving variation in transcription in invasive species may extend that utility to invasion biology.

2 | METHODS

2.1 | Sample collection and experimental design

Round and tubenose gobies were collected in the first week of October 2014 from the Detroit River using a 10-m beach seine net. Although we did not directly age the fish, they ranged in size from 48 to 69 mm total length, indicating that most were age-1 with possibly some age-2 for the larger round goby, although they are typically absent in samples by October (MacInnis & Corkum, 2000a). No individuals were reproductively mature as determined by the absence of developed gonads during tissue dissection, all fish appeared healthy and no fish died during the experimental procedures. Gobies were immediately transferred to the aquatics facility at the Great Lakes Institute for Environmental Research in aerated coolers where they were immediately placed into one of three different water temperature tanks (five fish per tank). Each temperature treatment consisted of paired 10-L tanks (one for round goby and one for tubenose goby) connected to a recirculation system that aerated the water and controlled water temperature. The three temperature conditions were the following: (i) control: ambient water conditions in the aquatics facility (18°C) that was drawn from the Detroit River immediately upstream from the sampling site (<100 m) and reflects the temperature both species were exposed to prior to sampling, (ii) high-temperature challenge: increasing the water temperature 2°C per hour from ambient to 25°C, and (iii) low-temperature challenge: decreasing the water temperature 3°C per hour from ambient to 5°C. Temperatures were chosen to represent a range of temperatures potentially experienced

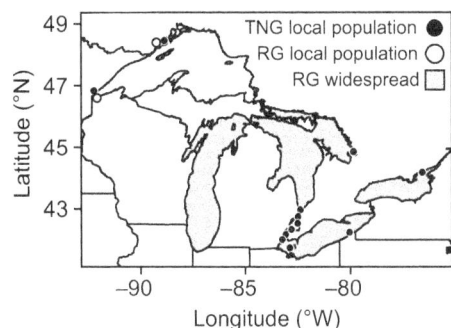

FIGURE 1 Map of the Laurentian Great Lakes contrasting the postinvasion dispersal and distribution of round and tubenose gobies. Round goby are widespread throughout lakes Michigan, Huron, Erie, and Ontario with local populations in Lake Superior (open circles). Local established populations of tubenose goby indicated by black circles. Distribution data from U.S. Geological Survey (2016)

during range expansion from the St. Clair River throughout the extent of the North American range expansion of round goby but less extreme than known critical thermal limits for these species (round goby: 33.4°C and tubenose goby 31.9°C, Xin, 2016). Once the treatment temperature was reached, fish were held in these conditions for 24 hr after which they were humanely euthanized in an overdose solution of tricaine methylsulfonate (200 mg/L MS-222, Finquel, Argent Laboratories, Redmond, WA). All fish (five per treatment, per species) were weighed and measured and liver tissue was immediately dissected, preserved in a high salt solution (700 g/L ammonium sulfate, 25 mM sodium citrate, 20 mM ethylenediaminetetraacetic acid, pH 5.2), and stored at −20°C.

2.2 | RNA sequencing and de novo transcriptome assembly

RNA was extracted from liver tissue using TRIzol® reagent (Life Technologies, Mississauga, ON) following the manufacturer's protocol. RNA was dissolved in sterile water and treated with TURBO™ DNase (Life Technologies, Mississauga, ON) to remove genomic DNA contamination. RNA quality was assessed using the Eukaryotic RNA 6000 Nano assay on a 2100 Bioanalyzer (Agilent, Mississauga, ON). Only samples with an RIN > 7 and a 28S:18S rRNA ratio >1.0 were used to prepare sequencing libraries. RNA sequencing libraries (one library per fish, three fish per treatment per species; total of 18 samples or libraries) were prepared and sequenced at the McGill University and Genome Quebec Innovation Centre (McGill University, Montreal, QC) using the TruSeq stranded mRNA library protocol and 100-bp paired-end sequencing in two lanes of an Illumina HiSeq 2000 sequencer (Illumina Inc., San Diego, CA).

Raw reads were pooled by species and de novo transcriptome assemblies were created for each species of goby using Trinity v3.0.3 (Grabherr et al., 2011). De novo assemblies were created using the default parameters and included a quality-filtering step using default Trimmomatic v0.32 (Bolger, Lohse, & Usadel, 2014) and in-silico normalization methods as implemented in Trinity. Raw reads for each sample were then individually quality filtered using Trimmomatic v0.32. Cleaned reads were multimapped to the reference transcriptome generated by Trinity for that species using Bowtie2 (Langmead & Salzberg, 2012) to report all valid mappings using the"−a" method. Further details of the specific parameters used for each software program are available in the Appendix S1 in the form of a custom unix shell script used to perform quality trimming and read mapping. Aligned reads for all samples of each species were processed using the program Corset v1.0.1 (Davidson & Oshlack, 2014), which uses information from the shared multimapping of sequence reads to hierarchically cluster the transcript contigs produced by de novo assembly into "genes" while using information about the treatment groups of individuals to split grouping of contigs when the relative expression difference between the contigs is not constant across treatments groups. Thus, Corset simultaneously clusters gene fragments generated during de novo assembly while separating paralogous genes and finally enumerates read counts for each of these genes (Davidson & Oshlack, 2014). This method performs as well or better than other current methods for clustering transcripts generated during de novo assembly (Davidson & Oshlack, 2014). To focus on biologically relevant transcriptional changes and avoid statistical issues for genes with low numbers of counts, we removed genes that did not meet a minimum expression level of at least one count per million reads in at least three samples (within one treatment) prior to analysis. To assess the consistency of our data and visually validate the use of three biological replicates per treatment, we conducted principal component analysis on centered and scaled count data as implemented in the "ade4" v1.7-4 package (Dray & Dufour, 2007) in R v3.1.3 (R Core Team 2016) for each species individually and then the two species combined for putative orthologous genes.

To test the hypothesis that round goby have an increased capacity for transcriptional response, we conducted two sets of complimentary analyses. The first set of analyses focused on the quantification of the ability of gobies to alter transcriptome-wide gene expression in response to environmental perturbation (temperature treatments). The second set of analyses focused on the function of responding genes, and whether genes with plastic responses to environmental perturbations represented relevant and coordinated biological functions for dealing with the temperature stress or random transcriptional changes lacking directed biological function.

2.2.1 | Transcriptome-wide plasticity

We used univariate generalized linear models (GLM) to identify differentially expressed genes in response to each temperature challenge for each species of goby separately. Negative binomial GLMs were implemented using the "edgeR" v3.8.6 package (Robinson, McCarthy, & Smyth, 2010) in R v3.1.3 (R Core Team 2016) using a false discovery rate of 0.05 to correct p-values for multiple comparisons (Benjamini & Hochberg, 1995). Briefly, the "edgeR" approach normalizes count data using trimmed mean of M-values (Robinson & Oshlack, 2010) that accounts for differences in library size among individuals. Negative binomial models are then fitted to the normalized count data for individuals, gene by gene, using gene-specific dispersion parameters estimated from the data using an empirical Bayes approach (McCarthy, Chen, & Smyth, 2012). Statistical significance of model terms is then tested using a likelihood ratio test. Genes identified as being differentially expressed in response to temperature represent gene transcription that is responding plastically to environmental cues.

To assess differences between round and tubenose goby for transcriptome-wide scope (magnitude of transcriptional change) for response, we first compared the distribution of Log_2 fold changes in transcription response to temperature challenges for all genes irrespective of statistical significance. We tested for differences in the rank order of fold change between species for upregulated (positive Log_2 fold change) and downregulated (negative Log_2 fold change) genes separately in each treatment using nonparametric Wilcoxon rank-sum tests in R v3.1.3 (R Core Team 2016). This analysis provides an estimate of transcriptional variability not explicitly influenced by temperature. We then considered the specific difference between species in the scope of transcriptional response for genes that were identified as statistically significantly responding to temperature challenge. For this analysis, we considered only Log_2 fold changes from the genes that were identified as being significantly differentially expressed individually by each species in the GLMs above. Nonparametric Wilcoxon rank-sum tests were again used to compare the rank order of fold change between species for upregulated and downregulated genes separately in each treatment.

To further facilitate comparison of gene transcription variation between species and allow combining the species-specific datasets, we identified putative orthologous genes using reciprocal best blast hits for round goby and tubenose goby transcripts using the blastn algorithm from BLAST+ v2.19 (Camacho et al., 2009). We retained valid putative orthologs only where both transcripts were each other's best matches. While this is a simple approach to identifying gene orthologs, it has been shown to outperform many more sophisticated algorithms (Altenhoff & Dessimoz, 2009). We recognize the need for further phylogenetic assessment to verify our putative gene pairs are in fact orthologs and not extra-paralogs and so we refer to our orthologs throughout as "putative" to reinforce their preliminary designation. We used the putative orthologous gene information to analyze paired comparisons of species-specific Log_2 fold changes to temperature in each challenge (Log_2 fold change from species-specific one-way GLMs above). We included only orthologous genes identified as statistically significantly responding to temperature challenge based on the two-factor GLMs. Here, we analyzed the paired comparison of Log_2 fold changes between the two species of goby for upregulated and downregulated genes separately in each treatment with Wilcoxon signed-rank tests, a nonparametric analog of a paired t-test.

We then combined the raw gene transcription count data from both species for genes that were putatively orthologous and tested for species differences in transcription at the shared expressed genes using two-factor GLMs for each temperature challenge. The two-factor negative binomial GLMs were implemented in "edgeR," with gene-specific dispersion parameters estimated as described above, using the following model:

$$X_{ijk} = T_i + S_j + I_{ij} + e_{ijk} \qquad (1)$$

where T_i represents the effect of temperature treatment (control versus treatment), S_j represents the effect of species, I_{ij} the species × temperature interaction, and e_{ijk} the residual error. Genes exhibiting a species-by-treatment interaction could reflect transcriptional response capacity possessed or utilized by one species but not the other and may thus be the basis of differential invasion success. Additionally, maintenance of biological function may be more transcriptionally demanding and the scope for response may be limited due to higher levels of constitutive transcription for genes in one species. To assess this, we identified orthologous genes that were statistically significantly differentially transcribed between species based on the likelihood ratio test for the species term from the two-factor GLMs. We then used the Log_2 fold change associated with statistically significant genes to assess the magnitude that one species over-transcribed a gene relative to the other. In this context, positive fold changes indicated genes consistently transcribed higher by tubenose goby irrespective of temperature treatment and negative fold changes indicated genes consistently transcribed

higher by round goby. Wilcoxon rank-sum tests were used to test for a difference between round and tubenose goby in the magnitude of over transcription between the two species. For this analysis, we only considered genes significantly differently transcribed between species and not exhibiting an interaction effect.

2.3 | Plasticity in gene function

The second set of analyses investigated differences in regulation of gene function between round and tubenose goby. We annotated our sequences with Gene Ontology (GO; Ashburner et al., 2000) information using Blast2GO v3.1 (Conesa et al., 2005). Briefly, transcript sequences were compared for sequence homology to records in the nonredundant (nr) protein database of the National Center for Biotechnology Information (http://www.ncbi.nlm.nih.gov) using the blastx algorithm from BLAST+ v2.19 (Camacho et al., 2009) with an e-value cutoff of 0.001. Goby transcripts were then associated with GO terms based on the GO annotations for the transcripts' top BLAST hits using the GO association database from 15 September 2015 (The Gene Ontology Consortium, 2015). To account for transcript length biases in the ability to detect differential expression from RNAseq data, we tested for over-representation of GO categories present in our contrasts of interest using the "goseq" v1.18 package (Young, Wakefield, Smyth, & Oshlack, 2010) in R v3.1.3 (R Core Team 2016). Specifically, we tested for functional enrichment (over-representation) for all GO categories represented by a minimum of five annotated genes. We tested up- and downregulation of biological processes to increased or decreased temperature relative to all genes with annotation for each species separately. We corrected for multiple comparisons using a false discovery rate of 0.05 (Benjamini & Hochberg, 1995). Additionally, we identified the genes that exhibited the strongest response to temperature challenge for each species (top 5% of fold increase or decrease in transcription in each temperature treatment). We tested for functional enrichment of GO biological processes represented by those genes in the same manner as above to discover the most plastic functions in each species that might be important for explaining the difference in performance between them.

3 | RESULTS

3.1 | RNA sequencing and de novo transcriptome assembly

We generated 214.9 million 100-bp paired-end reads for round goby and 214.2 million 100-bp paired-end reads for tubenose goby with an even distribution of data among samples (Table S1). The Trinity assembly software reconstructed 213,329 transcript clusters for round goby and 188,405 transcript clusters for tubenose goby. Quality filtering of individual sample read sets using Trimmomatic retained 93%–95% of read pairs (Table S1). Of these, a large proportion of high-quality read pairs (91%–94%) were mapped to the respective species de novo transcript reference (Table S1). Corset transcript clustering reduced the number of unique "genes," or transcript clusters, to 63,231 for round goby and

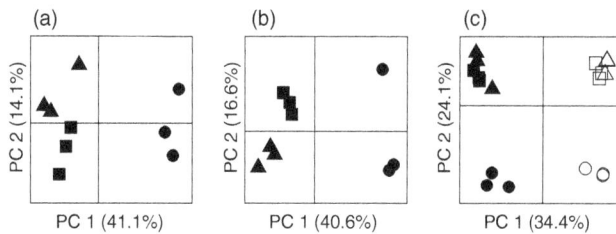

FIGURE 2 Principle component bi-plots of the first two principle components derived from gene transcription count data between samples for all genes for round goby (a), tubenose goby (b), and putative orthologous genes for both species combined (c) from three acute temperature treatments: control−18°C (squares), cold treatment−5°C (circles), and warm treatment−25°C (triangles). Round goby are represented by the solid symbols and tubenose goby by the open symbols in panel c

57,468 for tubenose goby, and of these, 26,215 genes for round goby and 23,648 genes for tubenose goby were retained following filtering for minimum expression level (>1 count per million reads, e.g., approximately 20–25 reads across at least three fish). Principal component (PC) bi-plots of the two largest PCs indicate good consistency among samples from each treatment (Figure 2). The first PC axis for both species describes approximately 40% of the transcriptional variation and is driven by the

difference in expression of the cold treatment and likely reflects the magnitude of temperature change for the cold treatment relative to the warm treatment. The second PC axis for both species explains approximately 15% of the transcriptional variation and generally separates the warm treatment from the control treatment (Figure 2), although it does capture some within-group variation especially for the cold treatment tubenose goby (Figure 2b). This within-group variation is unlikely to be due to age differences and all fish appeared to be in good condition prior to experimentation; however, it could reflect a sex difference, as we were unable to obtain sex information for these fish. The PCA combining round and tubenose goby for the putative orthologous genes identified similar patterns; however, species differences appear to explain as much or more of variance in transcription than the temperature challenge (Figure 2c).

3.2 | Transcriptome-wide plasticity

To first characterize transcriptome-wide patterns of plasticity, we identified differentially expressed genes using univariate GLMs for each species and temperature treatment. Results from the individual species GLMs indicate that only a minority of genes in both species responded plastically to temperature challenge (high temperature: ~2%; low temperature: ~22%; Table 1). The patterns of differential

TABLE 1 Gene transcriptional response of all genes and for paired putative orthologous genes from round and tubenose goby exposed to cold and hot temperature challenges (N: number of genes in category for RG: round goby or TNG: tubenose goby, mean (SD): average (standard deviation) of Log_2 fold change in response to temperature challenge, Wilcoxon W: W statistic for Wilcoxon test, p value: p-value for Wilcoxon test)

	RG		TNG		Wilcoxon	
	N	Mean (SD)	N	Mean (SD)	W	p value
All genes						
Increased temperature	26,215	0.423 (0.58)	23,648	0.417 (0.46)	2.96×10^8	$<2.2 \times 10^{-16}$
Decreased temperature	26,215	0.771 (0.82)	23,648	0.726 (0.77)	3.20×10^8	9.6×10^{-11}
Differentially expressed genes						
Increased temperature						
Upregulated	308	2.55 (1.50)	225	2.29 (1.32)	3.85×10^4	.029
Downregulated	334	−2.83 (1.56)	199	−2.01 (1.24)	4.64×10^4	1.6×10^{-14}
Not DE	25,573		23,224			
Decreased temperature						
Upregulated	2,922	1.84 (1.09)	2,806	1.83 (1.04)	4.02×10^6	.21
Downregulated	2,941	−1.80 (0.99)	2,264	−1.67 (0.91)	3.68×10^6	1.1×10^{-10}
Not DE	20,352		18,578			
Orthologous genes						
Increased temperature						
Upregulated	345	1.11 (0.90)	345	0.75 (0.81)	3.9×10^4	4.6×10^{-7}
Downregulated	338	−0.98 (0.99)	338	−1.01 (0.49)	2.1×10^4	2.1×10^{-5}
Not DE	10,481		10,481			
Decreased temperature						
Upregulated	2,313	0.99 (0.77)	2,313	1.00 (0.78)	1.4×10^6	.70
Downregulated	2,418	−1.01 (0.67)	2,418	−0.93 (0.60)	1.59×10^6	6.9×10^{-5}
Not DE	6,433		6,433			

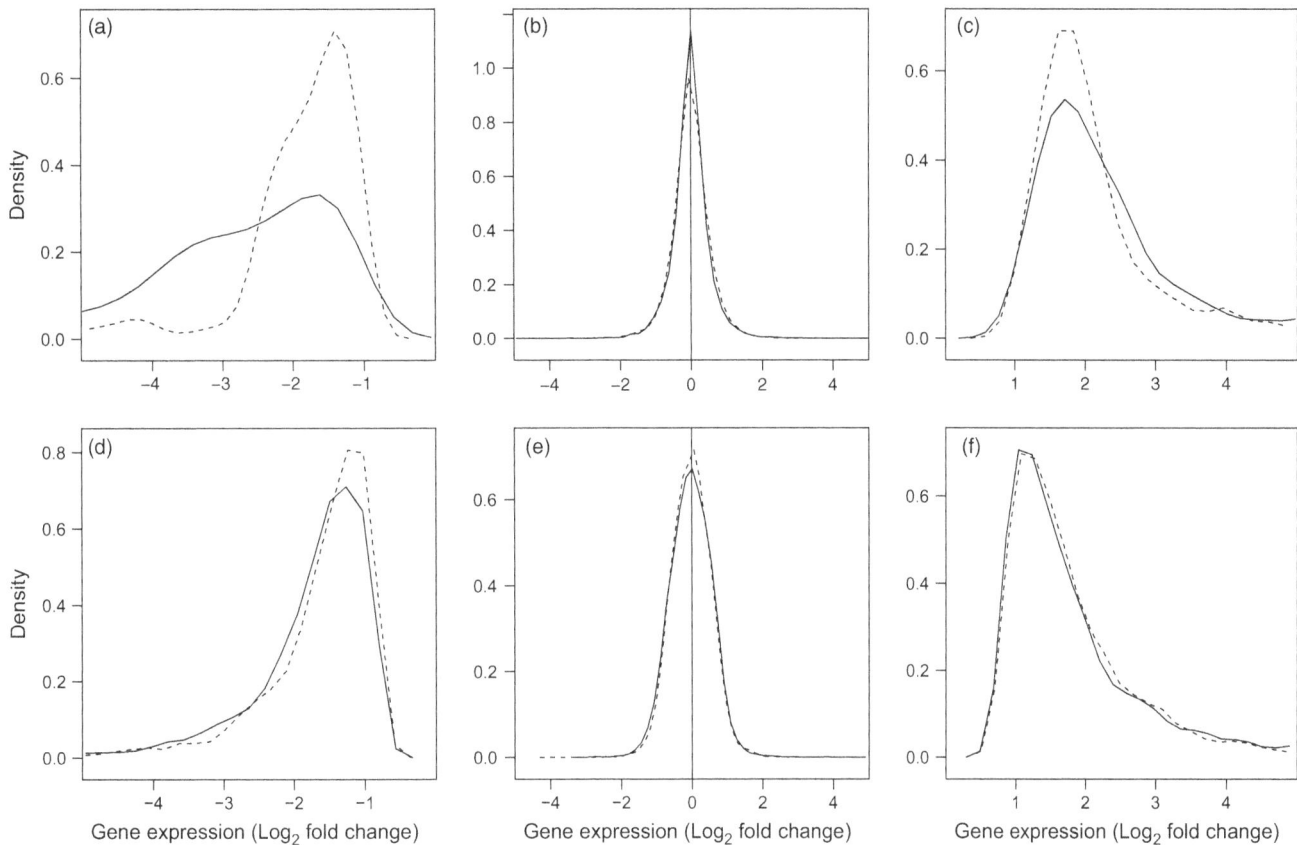

FIGURE 3 Differences between round and tubenose goby in the distribution of Log_2 fold changes of gene transcription in response to increased temperature challenge (a–c) and decreased temperature challenge (d–f). Lines represent the relative density (amount) of genes corresponding to the fold change indicated on the x-axis for round goby (solid lines) and tubenose goby (dashed lines). Panels present genes with statistically significant downregulation of transcription (a, d), no transcriptional plasticity (b, e), and statistically significant upregulation of transcription (c, f) as determined for each species using negative binomial generalized linear models (FDR < 0.05, see Section 2). The generally higher density of genes for tubenose goby at lower magnitude fold changes indicates reduced scope for transcriptional plasticity. The shift of the distribution between species is statistically significant for comparisons a, c, and d based on Wilcoxon rank-sum tests (Table 1)

transcription in terms of the proportions of differentially expressed genes are similar between the two species (Table 1). In contrast, Log_2 fold changes were on average greater in magnitude for round goby in all comparisons except for genes upregulated in response to cold, where there was no significant difference (Table 1; Figure 3). This indicates that round goby have an increased scope for transcriptional plasticity compared with tubenose goby. When considering only the putative orthologous genes, the pattern remains the same, except for genes downregulated in response to high temperature where the pattern of greater average fold change is higher for tubenose goby (Table 1; Figure 4).

The two-factor GLMs with species and temperature as factors identified 76 (0.7%) gene orthologs with a significant species-by-temperature interaction effect in the high-temperature treatment and 823 (7.3%) gene orthologs in the cold temperature treatment. Functional annotation was available for 44 gene orthologs demonstrating a significant interaction in the high-temperature treatment and 560 gene orthologs in the cold temperature treatment. The only biological process significantly over-represented by any of these responses was present in response to cold temperature challenge and

was for genes involved in steroid hormone-mediated signaling (GO: 0043401, 11 differentially expressed genes, 35 total genes with this GO annotation, FDR = 0.0097, Fig. S1). These genes, and the other genes demonstrating an interaction between species and temperature challenge (Table S2), may represent the transcriptomic basis of the differential performance of these species and are candidates for further study.

Of the 10,265 putative orthologs not exhibiting an interaction effect between species in either treatment, 6,782 (66.1%) of them are significantly differently transcribed between the two species. These represent 3,346 genes (49.3%) transcribed at a higher level in tubenose goby (mean Log_2 fold difference: 1.23) and 3,441 genes (50.7%) transcribed at a higher level in round goby (mean Log_2 fold difference: 1.08). There is a significant difference in the magnitude of differential transcription between goby species (W = 6.04×10^6, p = 1.8×10^{-15}). The genes that tubenose goby over-transcribes relative to round goby are over-transcribed to a greater degree than the genes that round goby over-transcribes relative to tubenose goby (Figure 5). This difference corresponds to tubenose goby having, on average, 11% higher transcription of orthologous genes compared to round goby. This

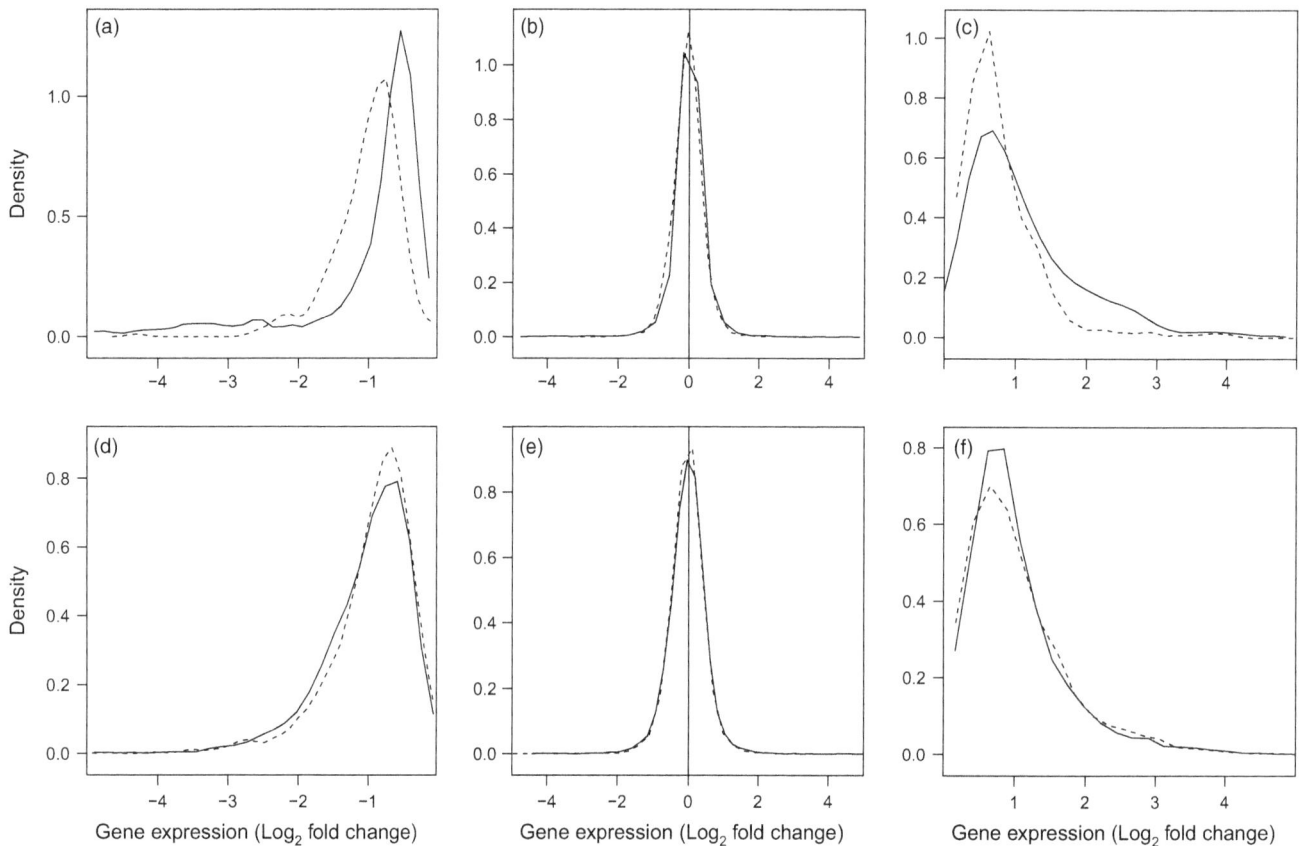

FIGURE 4 Differences between round and tubenose goby in the distribution of Log_2 fold changes of transcription for identified putative orthologous genes in response to increased temperature challenge (a–c) and decreased temperature challenge (d–f). Lines represent the relative density (amount) of genes corresponding to the fold change indicated on the x-axis for round goby (solid lines) and tubenose goby (dashed lines). Panels present genes with statistically significant downregulation of transcription (a, d), no transcriptional plasticity (b, e), and statistically significant upregulation of transcription (c, f) as determined for each species using negative binomial generalized linear models (FDR < 0.05, see Section 2). The generally higher density of genes for tubenose goby at lower magnitude fold changes indicates reduced scope for transcriptional plasticity. The shift of the distribution between species is statistically significant for comparisons a, c, and d based on Wilcoxon rank-sum tests (Table 1)

pattern of higher average transcription in tubenose goby is largely driven by differences in constitutive expression of genes not responding plastically to temperature challenge (Table 2), although there is a significant difference in the magnitude of transcription between species for genes upregulated in response to decreased temperature.

3.3 | Plasticity in gene function

The second set of analyses investigated biological function associated with transcriptional changes in response to temperature challenge. Functional annotation was possible for 10,777 genes in round goby and 10,695 genes in tubenose goby. We characterized biological process categories in the Gene Ontology framework that were over-represented by genes either up- or downregulated in response to increased and decreased temperature for each species separately.

Round goby did not exhibit over-representation of upregulated transcription for any biological processes in response to increased temperature but did exhibit over-representation of downregulation for a variety of biological processes (N = 89), most of which were related to cell cycle, DNA replication, and cell division (Figure 6, Table S3). The round goby also exhibited over-representation of downregulated genes involved in the repression of ubiquitin-mediated proteolysis, which should result in the upregulation of this function. In contrast, tubenose goby exhibited over-representation of upregulated transcription of five biological processes, all involved in humoral immunity and activation of the immune response. Tubenose goby exhibited over-representation of downregulated transcription of biological processes (N = 7) mostly involved in rRNA and tRNA metabolic processes and tRNA activation (Figure 6, Table S3) suggesting a general reduction in gene translational activity in response to increased temperature.

In response to decreased temperature, round goby exhibited over-representation of many upregulated biological processes (N = 81), including carboxylic acid metabolic processes typical of phospholipid membrane alterations, transport of basic amino acids (arginine and lysine), and biosynthesis of carbohydrates typical of antifreeze functions, negative regulation of apoptosis, and proteosomal activity characteristic of targeted degradation or turnover of proteins (Figure 6, Table S3). Tubenose goby also exhibited over-representation of many

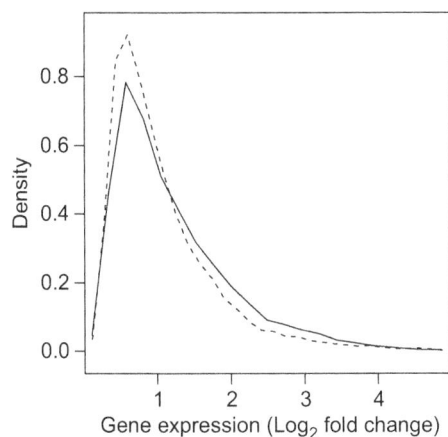

FIGURE 5 Distribution of Log_2 fold changes of transcription for putative orthologous genes differentially transcribed (FDR < 0.05) between round and tubenose goby. Lines represent the relative density (amount) of genes corresponding to the magnitude of fold change indicated on the x-axis for orthologous genes one species over-transcribes relative to the other. Genes transcribed higher in round goby are represented by the solid lines and genes transcribed higher in tubenose goby are represented by the dashed lines. Tubenose goby over-transcribes genes to a greater magnitude than round goby based on a Wilcoxon rank-sum test (p < .0001)

upregulated biological processes (N = 57) in response to decreased temperature, but with very different functional implications. The majority of upregulated processes were response to stimulus processes indicative of detection of stimulus, cell signaling cascades, regulation of gene expression, and immune system processes (Figure 6, Table S3). Neither species of goby exhibited any over-representation of downregulated biological processes in response to reduced temperature, after correction for multiple tests. Interestingly, round and tubenose goby shared 14 biological processes that were over-represented by genes upregulated in response to decreased temperature (Figure 6, Table S3). All of these processes were for response to stimulus suggesting that these species were both able to detect the changes in their environment and produce signaling cascades to direct biological

functions as a result. The lack of many other processes regulated by tubenose goby could suggest either they lack specific mechanisms to deal with the stress they experienced or that there may be a difference in the timing of the onset of the response.

To characterize the most plastic biological functions for each species in response to temperature challenge, we identified genes with the largest Log_2 fold changes (top 5%) within the significantly up- and downregulated genes separately in each temperature treatment (Table 3). Significantly over-represented biological processes represented by these highly plastic genes were only evident for upregulated genes in response to the cold temperature treatment for both species. Round goby demonstrated over-representation of 28 biological processes, whereas tubenose goby only demonstrated over-representation of five biological processes (Table S4). Two processes were shared between both species relating to alcohol and polyol biosynthesis that may be related to antifreeze capacity and cold tolerance. Round goby exhibited extreme plasticity for additional processes related to oxygen binding and carbohydrate metabolism, while tubenose goby exhibited plasticity for ceramide metabolic process potentially related to signaling cellular stress.

4 | DISCUSSION

We demonstrated liver tissue transcriptional differences between round and tubenose gobies in response to acute temperature challenges that may contribute to the dramatic differences in the geographic extent of invasion of these two species. Round goby possessed a greater scope for transcriptional response to altered temperature compared with tubenose goby. The two species exhibited a similar number of genes with significantly altered transcriptional state; however, the transcriptional changes by tubenose goby failed to represent the same biological processes altered by round goby. Furthermore, the functions of the genes that responded to the challenges in round goby, but did not in tubenose goby, were consistent with adaptive responses to maintain or regain homeostasis following rapid changes in temperature. The capacity for transcriptional plasticity to environmental

TABLE 2 Magnitude of Log_2 fold difference between round and tubenose gobies for genes plastically responding to increased or decreased temperature and those not responding to temperature (N: number of genes in category higher for RG: round goby or TNG: tubenose goby, mean (SD): average (standard deviation) of Log_2 fold increase over the other species, Wilcoxon W: W statistic for Wilcoxon rank-sum test for rank order of RG versus TNG for that category of genes, p value: p-value for Wilcoxon rank-sum test)

	RG		TNG		Wilcoxon	
	N	Mean (SD)	N	Mean (SD)	W	p value
Increased temperature						
Upregulated	51	1.33 (0.90)	92	1.37 (0.87)	2.25×10^3	.712
Downregulated	95	1.02 (0.85)	43	1.16 (0.77)	1.78×10^3	.232
Decreased temperature						
Upregulated	639	1.01 (0.75)	538	1.18 (0.78)	1.43×10^5	8.28×10^{-7}
Downregulated	693	1.02 (0.67)	700	1.13 (0.71)	2.18×10^5	.001
No temperature response						
No difference	1,806	1.11 (0.83)	1,825	1.27 (0.94)	1.48×10^6	1.43×10^{-7}

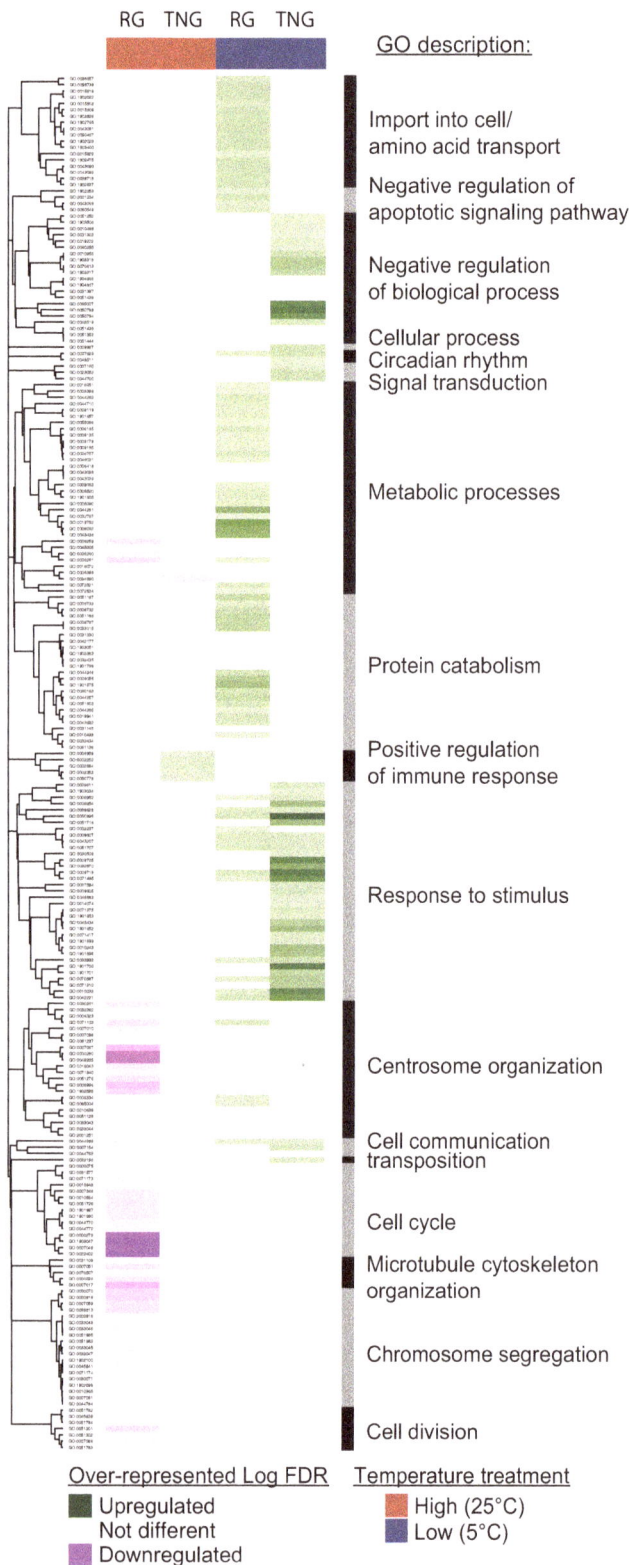

FIGURE 6 Heatmap of gene ontology (GO) biological process categories over-represented by genes either upregulated (green) or downregulated (purple) in round goby (RG) and tubenose goby (TNG) liver tissue in response to two acute thermal challenges. GO biological process over-representation tests were performed using the "goseq" v1.18 package in R v3.1.3 (R Core Team 2016; Young et al., 2010). Statistically significant processes after false discovery rate correction (Benjamini & Hochberg, 1995) were grouped across species and treatment and clustered based on semantic similarity criterion of Schlicker, Domingues, Rahnenführer, and Lengauer (2006) as implemented in the "GOSemSim" v1.99.4 package (Yu et al., 2010) in R and "complete" hierarchical clustering as implemented in the "hclust" function in R. Full GO over-representation results are available in Table S3

broader range of temperatures than tubenose goby. This result is consistent with round goby having a higher thermal limit than tubenose goby (Xin, 2016). Given the more dramatic differences we observed in transcriptional response to cold treatment between species and the role of cold tolerance in determining invasion success in the Great Lakes (Kolar & Lodge, 2002), we suggest further investigation into the thermal performance curves for tubenose goby and determination of lower thermal limits for these species would be worthwhile. Broad thermal tolerance has been previously associated with higher invasion success (Bates et al., 2013), and our transcriptional results suggest that capacity for transcriptional response is a potential mechanism that explains the differential invasion success between goby species in our study.

Reduced scope of gene transcription response to specific environmental challenges (in our case, temperature) implies a reduced capacity to acclimate to a broad range of environments and may have limited the range expansion of tubenose goby. Indeed, Antarctic fishes that have evolved in very stable environments have completely lost a heat shock response (for a review, see: Logan & Buckley, 2015). Reduced transcriptional capacity to respond to heat stress has also been documented for fish species that only have a moderate temperature tolerance range (*Hypomesus transpacificus*, Komoroske, Connon, Jeffries, & Fangue, 2015) compared to the transcriptional responses of fish species that are known to tolerate a broader range of temperatures (e.g., *Gillichthys miribilis*, Logan & Somero, 2011). The evolution of plasticity is thought to be constrained by the relative cost of having a plastic phenotype compared with exhibiting a canalized phenotype (Agrawal, 2001). It is possible that tubenose goby have experienced a greater cost to being transcriptionally plastic in its native range than round goby that resulted in the evolution of a reduced transcriptional response to acute thermal challenge; however, we cannot rule out genetic drift as a mechanism explaining the difference either (Whitehead, 2012). Alternatively, increased transcriptional response may not always be indicative of tolerance; for example, if a stressor is mild, a highly tolerant species may not respond transcriptionally at all, and there are examples of pollutant tolerant fish that have evolved a muted transcriptional response to pollution exposure (Whitehead, Triant, Champlin, & Nacci, 2010). In our case, the combination of the species-level performance (invasion range expansion and impact) and physiological differences (thermal limits and metabolic rates) makes it unlikely that tubenose goby were able

stressors has potential as an important predictor of the physiological tolerances of organisms (López-Maury et al., 2008; Whitehead, 2012). Physiological tolerances ultimately define species' distributions, capacity for range expansion, and, therefore, potential for invasion success.

The response of round goby to thermal stress suggests that it can transcriptionally respond to maintain biological function over a

TABLE 3 Magnitudes of most plastic gene transcription (top 5% of Log$_2$ fold change) for round goby (RG) and tubenose goby (TNG) in response to acute temperature challenge. N = number of genes in top 5% of fold change, R = range of Log$_2$ fold changes for genes

	RG		TNG	
	N	R	N	R
Increased temperature				
Upregulated	6	4.1–8.2	4	3.9–10.1
Downregulated	8	5.2–8.1	6	2.6–8.3
Decreased temperature				
Upregulated	67	3.1–8.1	60	3.1–9.5
Downregulated	56	3.1–7.8	50	2.8–7.4

to maintain homeostasis despite a reduced transcriptional responses to temperature challenge.

In addition to increased capacity for transcriptional plasticity, the transcriptional changes exhibited by round goby are more consistent with adaptive responses to thermal challenge than those observed in the tubenose goby. Round goby altered biological processes that are characteristic of acute responses to temperature reported in other species with broad thermal tolerance (e.g., ubiquitin-dependent protein degradation and negative regulation of apoptosis; Logan & Somero, 2011) and are believed to help organisms survive and recover from acute stress events (Wiseman et al., 2007). In contrast, tubenose goby responded to the challenge by altering a similar number of genes; however, with the exception of innate immune response to tissue damage, tubenose goby did not respond with the same biological processes as round goby. This highlights an important difference between adaptive and maladaptive phenotypic plasticity. That is, phenotypic plasticity is only beneficial for an organism when it alters phenotype (partially or fully) in the direction of a peak on a fitness landscape (increases fitness; Ghalambor et al., 2007). If plasticity alters a phenotype in a direction other than toward a fitness peak, as it does for tubenose goby where a similar number of transcriptional changes as round goby do not represent a similar functional response, these plastic changes may result in no or even negative fitness consequences for the organism. Variation in the timing of transcriptional response to a stressor (e.g., Whitehead et al., 2012) could explain the observed difference between species; however, delayed induction of biological responses by tubenose goby would likely also be maladaptive, especially if it resulted in delayed compensatory responses that are necessary for short-term survival.

The reduced scope of transcriptional response of tubenose goby suggests either that it lacked the biological mechanisms to respond to acute thermal stress or that tubenose goby found the handling procedures stressful and thus suffered reduced capacity to respond to the heat stress. While we could have conducted a laboratory acclimation experiment to isolate temperature as the sole factor driving transcriptional changes in our gobies, temperature is not the only environmental stressor encountered by these organisms. We provide a comparison of transcriptional response to temperature stressors that reflects the

organisms' ecological context while controlling for prior environmental exposure by sampling these organisms from the same habitat at the same time. Presumably, sensitivity to the synergistic effects of multiple stressors expressed as a reduction in a potential aquatic invader's transcriptional capacity would not be adaptive for the invading species. Our use of three biological replicates has the potential to result in inflated variance estimates that inhibit our ability to detect more subtle differential expression; thus, our list of differentially transcribed genes should be considered conservative. Despite this limitation, we have characterized hundreds to thousands of differentially transcribed genes in each treatment (Table 1) and our treatments are well separated in multivariate space suggesting within-group error is not a limiting factor (Figure 2). The proportions of differentially responding genes we report are comparable to other studies of acute thermal stress (Logan & Somero, 2011; Quinn et al., 2011) suggesting that despite the lack of laboratory acclimation, we still captured important biological responses in an ecological context.

The process of invasion or range expansion often results in genetic founder effects and bottlenecks (Dlugosch & Parker, 2008) and the resulting reductions in genetic diversity have potential consequences for adaptive capacity. Phenotypic plasticity, when adaptive, is widely believed to help buffer species from the selective forces of novel environments (Ghalambor et al., 2007; Lande, 2015); however, plasticity itself can evolve. The evolution of increased plasticity is expected to be favored early in the process of invasion, while selection in the invaded range is expected to eventually reduce plasticity (Lande, 2015). One of the key issues regarding empirical assessment of the role of plasticity in invasions is controlling for the time since invasion (Lande, 2015). The goby species presented here have similar invasion histories (both first detected in St. Clair River in 1990, Jude et al., 1992) and have similar ages at maturity (females at age 1; round goby: MacInnis & Corkum, 2000b; tubenose goby: Valová et al., 2015) indicating that a similar number of generations since invasion have occurred for both species. It is therefore unlikely that tubenose goby has had enough time to evolve a loss of plasticity in North America, while the round goby has not. Alternatively, the stochastic processes associated with founder effects may have prevented tubenose gobies bearing the full range of plastic phenotypes in the native range from becoming established in the first place. There is no evidence that tubenose goby have experienced greater founder or bottleneck effects during their North American invasion than round goby (Stepien & Tumeo, 2006) making differences in genetic diversity an unlikely explanation for the observed differences in transcriptional plasticity.

The lower transcriptional plasticity we found in the tubenose goby may reflect source population characteristics if selection pressures among assemblages of tubenose goby in their native range resulted in local adaptation, while the round goby in their native range are one broadly tolerant species. Round goby is known to exhibit broad environmental tolerance to other abiotic stressors, including salinity (Karsiotis, Pierce, Brown, & Stepien, 2012) and contaminants (McCallum et al., 2014). While less is known about the specific physiological tolerances of tubenose goby, the two species are found in similar habitats in both their native (Kottelat & Freyhof, 2007) and invaded ranges (Jude & DeBoe, 1996) suggesting they have evolved under similar conditions for at least the past several thousand years. The phylogeny of tubenose goby in the northern Black Sea is represented by multiple divergent lineages (Neilson & Stepien, 2009; Sorokin, Medvedev, Vasil'ev, & Vasil'eva, 2011) only one of which has invaded North America (Neilson & Stepien, 2009). In contrast, round goby from this same region form one monophyletic group (Brown & Stepien, 2008).

There has been a tendency for invasion biologists to treat organisms as static entities and ignore the role of plasticity and evolution in determining invasion risk (Whitney & Gabler, 2008). Plasticity may confer invasion success by either increasing fitness in both unfavorable and favorable environments (Richards, Bossdorf, Muth, Gurevitch, & Pigliucci, 2006). Broad thermal tolerance should increase fitness in unfavorable environments and has been associated with range expansions (Bates et al., 2013). The role of transcriptional plasticity in determining thermal tolerance suggests that assessment of transcriptional profiles under thermal stress may be a valuable tool to assess invasion risk. Our results demonstrate the power of using measures of transcriptional variation to detect meaningful biological responses to thermal stress in an ecological context that would be directly relevant to a species' ability to survive, uptake transport, and establishment in a novel environment. Comparative genomics has enormous potential to identify the mechanistic basis of variable acclimation capacity among groups of organisms (Whitehead, 2012). We have used a comparative approach to further demonstrate that differences in transcriptional response to acute temperature challenge may underlie the difference in invasion success between our two study species. Conservation biologists have embraced the use of transcriptomic profiles to identify and select more plastic source populations to maximize the success of species reintroductions (He, Johansson, & Heath, 2016). Managing invasive species is simply applying this approach in reverse, where managers would want to prioritize prevention of transport and establishment of the most plastic invaders. Assessing transcriptional plasticity in response to acute stressors, such as temperature, combined with knowledge of the relationship between transcription and physiology (e.g., high transcriptional response is beneficial for thermal acclimation but may be maladaptive for pollution tolerance) would provide managers with objective measures of the plastic capacity of potential invasive species. Such data are critical for effective invasion risk assessment and the incorporation of quantitative approaches into invasion risk assessment will change how invasive species are managed and their impacts minimized.

ACKNOWLEDGEMENTS

We would like to thank Stacey MacDonald, Felicia Vincelli, Lida Nguyen-Dang, and Meghan Donovan for assistance collecting samples and the McGill University and Genome Quebec Innovation Centre for providing sequencing services. We thank two anonymous reviewers and the Associate Editor for their constructive comments that improved this manuscript. We would like to acknowledge NSERC and the Canadian Aquatic Invasive Species Network II for funding that supported this project.

DATA ACCESSIBILITY

Raw sequencing data for both species are available at the NCBI Sequence Read Archive under project accession numbers SRP075124 and SRP075141. Scripts used to process raw data, assemble the transcriptome, and generate the count data file as well as the count data file and R scripts used to perform the differential expression analysis are available on Dryad: https://doi.org/10.5061/dryad.408ht.

REFERENCES

Agrawal, A. A. (2001). Phenotypic plasticity in the interactions and evolution of species. *Science, 294,* 321–326.

Altenhoff, A. M., & Dessimoz, C. (2009). Phylogenetic and functional assessment of orthologs inference projects and methods. *PLoS Computational Biology, 5,* e1000262.

Ashburner, M., Ball, C. A., Blake, J. A., Botstein, D., Butler, H., Cherry, J. M., … Sherlock, G. (2000). Gene ontology: Tool for the unification of biology. *Nature Genetics, 25,* 25–29.

Aubin-Horth, N., & Renn, S. C. P. (2009). Genomic reaction norms: Using integrative biology to understand molecular mechanisms of phenotypic plasticity. *Molecular Ecology, 18,* 3763–3780.

Aykanat, T., Thrower, F. P., & Heath, D. D. (2011). Rapid evolution of osmoregulatory function by modification of gene transcription in steelhead trout. *Genetica, 139,* 233–242.

Bates, A. E., McKelvie, C. M., Sorte, C. J. B., Morley, S. A., Jones, N. A. R., Mondon, J. A., … Quinn, G. (2013). Geographical range, heat tolerance and invasion success in aquatic species. *Proceedings of the Royal Society B: Biological Sciences, 280,* 1958.

Benjamini, Y., & Hochberg, Y. (1995). Controlling the false discovery rate: A practical and powerful approach to multiple testing. *Journal of the Royal Statistical Society B, 57,* 289–300.

Blackburn, T. M., Pyšek, P., Bacher, S., Carlton, J. T., Duncan, R. P., Jarošík, V., … Richardson, D. M. (2011). A proposed unified framework for biological invasions. *Trends in Ecology & Evolution, 26,* 333–339.

Bolger, A. M., Lohse, M., & Usadel, B. (2014). Trimmomatic: A flexible trimmer for Illumina sequence data. *Bioinformatics, 30,* 2114–2120.

Brown, J. E., & Stepien, C. A. (2008). Ancient divisions, recent expansions: Phylogeography and population genetics of the round goby *Apollonia melanostoma. Molecular Ecology, 17,* 2598–2615.

Brown, J. E., & Stepien, C. A. (2009). Invasion genetics of the Eurasian round goby in North America: Tracing sources and spread patterns. *Molecular Ecology, 18,* 64–79.

Camacho, C., Coulouris, G., Avagyan, V., Ma, N., Papadopoulos, J., Bealer, K., & Madden, T. L. (2009). BLAST+: Architecture and applications. *BMC Bioinformatics, 10,* 421.

Conesa, A., Gotz, S., Garcia-Gomez, J. M., Terol, J., Talon, M., & Robles, M. (2005). Blast2GO: A universal tool for annotation, visualization and analysis in functional genomics research. *Bioinformatics, 21,* 3674–3676.

Davidson, A. M., Jennions, M., & Nicotra, A. B. (2011). Do invasive species show higher phenotypic plasticity than native species and if so, is it adaptive? A meta-analysis. *Ecology Letters, 14,* 419–431.

Davidson, N. M., & Oshlack, A. (2014). Corset: Enabling differential gene expression analysis for. *Genome Biology, 15,* 410.

Dayan, D. I., Crawford, D. L., & Oleksiak, M. F. (2015). Phenotypic plasticity in gene expression contributes to divergence of locally adapted populations of *Fundulus heteroclitus. Molecular Ecology, 24,* 3345–3359.

Dlugosch, K. M., & Parker, I. (2008). Founding events in species invasions: Genetic variation, adaptive evolution, and the role of multiple introductions. *Molecular Ecology, 17,* 431–449.

Dray, S., & Dufour, A. B. (2007). The ade4 Package: Implementing the duality diagram for ecologists. *Journal of Statistical Software, 22,* 1–20.

Fangue, N. A., Hofmeister, M., & Schulte, P. M. (2006). Intraspecific variation in thermal tolerance and heat shock protein gene expression in common killifish, *Fundulus heteroclitus*. *Journal of Experimental Biology, 209*, 2859–2872.

Fuller, P., Benson, A., Maynard, E., Neilson, M., Larson, J., & Fusaro, A. (2017). *Neogobius melanostomus*. USGS Nonindigenous Aquatic Species Database, Gainesville, FL. Retrieved from https://nas.er.usgs.gov/queries/FactSheet.aspx?speciesID=713 Revision Date: 1/7/2016.

Fuller, P., Nico, L., Maynard, E., Neilson, M., Larson, J., Makled, T. H., & Fusaro, A. (2017). *Proterorhinus semilunaris*. USGS Nonindigenous Aquatic Species Database, Gainesville, FL. Retrieved from https://nas.er.usgs.gov/queries/FactSheet.aspx?SpeciesID=714 Revision Date: 9/21/2015.

Ghalambor, C. K., McKay, J. K., Carroll, S. P., & Reznick, D. N. (2007). Adaptive versus non-adaptive phenotypic plasticity and the potential for contemporary adaptation in new environments. *Functional Ecology, 21*, 394–407.

Godoy, O., Valladares, F., & Castro-Díez, P. (2011). Multispecies comparison reveals that invasive and native plants differ in their traits but not in their plasticity. *Functional Ecology, 25*, 1248–1259.

Gotthard, K., & Nylin, S. (1995). Adaptive plasticity and plasticity as an adaptation: A selective review of plasticity in animal morphology and life history. *Oikos, 74*, 3–17.

Grabherr, M. G., Haas, B. J., Yassour, M., Levin, J. Z., Thompson, D. A., Amit, I., ... Regev, A. (2011). Full-length transcriptome assembly from RNA-Seq data without a reference genome. *Nature Biotechnology, 29*, 644–652.

He, X., Johansson, M. L., & Heath, D. D. (2016). Role of genomics and transcriptomics in selection of reintroduction source populations. *Conservation Biology, 30*, 1010–1018.

He, X., Wilson, C. C., Wellband, K. W., Houde, A. L. S., Neff, B. D., & Heath, D. D. (2015). Transcriptional profiling of two Atlantic salmon strains: Implications for reintroduction into Lake Ontario. *Conservation Genetics, 16*, 277–287.

Jude, D. J., & DeBoe, S. F. (1996). Possible impact of gobies and other introduced species on habitat restoration efforts. *Canadian Journal of Fisheries and Aquatic Sciences, 53*, 136–141.

Jude, D. J., Reider, R. H., & Smith, G. R. (1992). Establishment of Gobiidae in the Great Lakes Basin. *Canadian Journal of Fisheries and Aquatic Sciences, 49*, 416–421.

Karsiotis, S. I., Pierce, L. R., Brown, J. E., & Stepien, C. A. (2012). Salinity tolerance of the invasive round goby: Experimental implications for seawater ballast exchange and spread to North American estuaries. *Journal of Great Lakes Research, 38*, 121–128.

Kolar, C. S., & Lodge, D. M. (2001). Progress in invasion biology: Predicting invaders. *Trends in Ecology and Evolution, 16*, 199–204.

Kolar, C. S., & Lodge, D. M. (2002). Ecological predictions and risk assessment for alien fishes in North America. *Science, 298*, 1233–1236.

Komoroske, L. M., Connon, R. E., Jeffries, K. M., & Fangue, N. A. (2015). Linking transcriptional responses to organismal tolerance reveals mechanisms of thermal sensitivity in a mesothermal endangered fish. *Molecular Ecology, 24*, 4960–4981.

Kottelat, M., & Freyhof, J. (2007). *Handbook of European freshwater fishes*. Cornol, Switzerland: Kottelat and Freyhof, Berlin, Germany.

Lande, R. (2015). Evolution of phenotypic plasticity in colonizing species. *Molecular Ecology, 24*, 2038–2045.

Langmead, B., & Salzberg, S. L. (2012). Fast gapped-read alignment with Bowtie 2. *Nature Methods, 9*, 357–359.

Lee, V. A., & Johnson, T. B. (2005). Development of a bioenergetics model for the round goby (*Neogobius melanostomus*). *Journal of Great Lakes Research, 31*, 125–134.

Lockwood, B. L., Sanders, J. G., & Somero, G. N. (2010). Transcriptomic responses to heat stress in invasive and native blue mussels (genus Mytilus): Molecular correlates of invasive success. *Journal of Experimental Biology, 213*, 3548–3558.

Lockwood, B. L., & Somero, G. N. (2011). Transcriptomic responses to salinity stress in invasive and native blue mussels (genus Mytilus). *Molecular Ecology, 20*, 517–529.

Logan, C. A., & Buckley, B. A. (2015). Transcriptomic responses to environmental temperature in eurythermal and stenothermal fishes. *Journal of Experimental Biology, 218*, 1915–1924.

Logan, C. A., & Somero, G. N. (2011). Effects of thermal acclimation on transcriptional responses to acute heat stress in the eurythermal fish *Gillichthys mirabilis* (Cooper). *AJP: Regulatory, Integrative and Comparative Physiology, 300*, R1373–R1383.

López-Maury, L., Marguerat, S., & Bähler, J. (2008). Tuning gene expression to changing environments: From rapid responses to evolutionary adaptation. *Nature Reviews Genetics, 9*, 583–593.

Lynch, M. P., & Mensinger, A. F. (2012). Seasonal abundance and movement of the invasive round goby (*Neogobius melanostomus*) on rocky substrate in the Duluth-Superior Harbor of Lake Superior. *Ecology of Freshwater Fish, 21*, 64–74.

MacInnis, A. J., & Corkum, L. D. (2000a). Age and growth of round goby *Neogobius melanostomus* in the Upper Detroit River. *Transactions of the American Fisheries Society, 129*, 852–858.

MacInnis, A. J., & Corkum, L. D. (2000b). Fecundity and reproductive season of the round goby *Negobius melanosomus* in the upper Detroit river. *Transactions of the American Fisheries Society, 129*, 136–144.

McCallum, E. S., Charney, R. E., Marenette, J. R., Young, J. A. M., Koops, M. A., Earn, D. J. D., ... Balshine, S. (2014). Persistence of an invasive fish (*Neogobius melanostomus*) in a contaminated ecosystem. *Biological Invasions, 16*, 2449–2461.

McCarthy, D. J., Chen, Y., & Smyth, G. K. (2012). Differential expression analysis of multifactor RNA-Seq experiments with respect to biological variation. *Nucleic Acids Research, 40*, 4288–4297.

Miller, K. M., Li, S., Kaukinen, K. H., Ginther, N., Hammill, E., Curtis, J. M. R., ... Farrell, A. P. (2011). Genomic signatures predict migration and spawning failure in wild Canadian Salmon. *Science, 331*, 214–217.

Neilson, M. E., & Stepien, C. A. (2009). Evolution and phylogeography of the tubenose goby genus Proterorhinus (Gobiidae: Teleostei): Evidence for new cryptic species. *Biological Journal of the Linnean Society, 96*, 664–684.

O'Neil, J. (2013). *Determination of standard and field metabolic rates in two Great Lakes invading fish species: Round goby (Neogobius melanostomus) and tubenose goby (Proterorhinus semilunaris)*. M.Sc. Thesis, GLIER, University of Windsor, Windsor, ON.

Palacio-López, K., & Gianoli, E. (2011). Invasive plants do not display greater phenotypic plasticity than their native or non-invasive counterparts: A meta-analysis. *Oikos, 120*, 1393–1401.

Pettitt-Wade, H., Wellband, K. W., Heath, D. D., & Fisk, A. T. (2015). Niche plasticity in invasive fishes in the Great Lakes. *Biological Invasions, 17*, 2565–2580.

Price, T. D., Qvarnstrom, A., & Irwin, D. E. (2003). The role of phenotypic plasticity in driving genetic evolution. *Proceedings of the Royal Society B: Biological Sciences, 270*, 1433–1440.

Quinn, N. L., McGowan, C. R., Cooper, G. A., Koop, B. F., & Davidson, W. S. (2011). Identification of genes associated with heat tolerance in Arctic charr exposed to acute thermal stress. *Physiological Genomics, 43*, 685–696.

R Core Team. (2016). *R: A language and environment for statistical computing*. R Foundation for Statistical Computing, Vienna, Austria. URL: https://www.R-project.org/

Ray, W. J., & Corkum, L. D. (2001). Habitat and site affinity of the round goby. *Journal of Great Lakes Research, 27*, 329–334.

Richards, C. L., Bossdorf, O., Muth, N. Z., Gurevitch, J., & Pigliucci, M. (2006). Jack of all trades, master of some? On the role of phenotypic plasticity in plant invasions. *Ecology Letters, 9*, 981–993.

Robinson, M. D., McCarthy, D. J., & Smyth, G. K. (2010). edgeR: A bioconductor package for differential expression analysis of digital gene expression data. *Bioinformatics, 26*, 139–140.

Robinson, M. D., & Oshlack, A. (2010). A scaling normalization method for differential expression analysis of RNA-seq data. *Genome Biology, 11*, R25.

Schlichting, C. D., & Smith, H. (2002). Phenotypic plasticity: Linking molecular mechanisms with evolutionary outcomes. *Evolutionary Ecology*, *16*, 189–211.

Schlicker, A., Domingues, F. S., Rahnenführer, J., & Lengauer, T. (2006). A new measure for functional similarity of gene products based on Gene Ontology. *BMC Bioinformatics*, *7*, 302.

Smith, E. N., & Kruglyak, L. (2008). Gene-environment interaction in yeast gene expression. *PLoS Biology*, *6*, e83.

Sonna, L. A., Fujita, J., Gaffin, S. L., & Lilly, C. M. (2002). Invited review: Effects of heat and cold stress on mammalian gene expression. *Journal of Applied Physiology*, *92*, 1725–1742.

Sørensen, J. G., Nielsen, M. M., Kruhøffer, M., Justesen, J., & Loeschcke, V. (2005). Full genome gene expression analysis of the heat stress response in *Drosophila melanogaster*. *Cell Stress and Chaperones*, *10*, 312–328.

Sorokin, P. A., Medvedev, D. A., Vasil'ev, V. P. & Vasil'eva, E. D. (2011). Further studies of mitochondrial genome variability in Ponto-Caspian proterorhinus species (actinopterygii: Perciformes: Gobiidae) and their taxonomic implications. *Acta Ichthyologica et Piscatoria*, *41*, 95–104.

Stepien, C. A., & Tumeo, M. A. (2006). Invasion genetics of Ponto-Caspian gobies in the Great Lakes: A "Cryptic" species, absence of founder effects, and comparative risk analysis. *Biological Invasions*, *8*, 61–78.

Stutz, W. E., Schmerer, M., Coates, J. L., & Bolnick, D. I. (2015). Among-lake reciprocal transplants induce convergent expression of immune genes in threespine stickleback. *Molecular Ecology*, *24*, 4629–4646.

Swindell, W. R., Huebner, M., & Weber, A. P. (2007). Plastic and adaptive gene expression patterns associated with temperature stress in *Arabidopsis thaliana*. *Heredity*, *99*, 143–150.

The Gene Ontology Consortium (2015). Gene ontology consortium: Going forward. *Nucleic Acids Research*, *43*, D1049–D1056.

Valová, Z., Konečná, M., Janáč, M., & Jurajda, P. (2015). Population and reproductive characteristics of a non-native western tubenose goby (*Proterorhinus semilunaris*) population unaffected by gobiid competitors. *Aquatic Invasions*, *10*, 57–68.

West-Eberhard, M. (2003). *Developmental plasticity and evolution*. New York: Oxford University Press.

Whitehead, A. (2012). Comparative genomics in ecological physiology: Toward a more nuanced understanding of acclimation and adaptation. *Journal of Experimental Biology*, *215*, 884–891.

Whitehead, A., & Crawford, D. L. (2006). Neutral and adaptive variation in gene expression. *Proceedings of the National Academy of Sciences*, *103*, 5425–5430.

Whitehead, A., Roach, J. L., Zhang, S., & Galvez, F. (2012). Salinity- and population-dependent genome regulatory response during osmotic acclimation in the killifish (*Fundulus heteroclitus*) gill. *Journal of Experimental Biology*, *215*, 1293–1305.

Whitehead, A., Triant, D. A., Champlin, D., & Nacci, D. (2010). Comparative transcriptomics implicates mechanisms of evolved pollution tolerance in a killifish population. *Molecular Ecology*, *19*, 5186–5203.

Whitney, K. D., & Gabler, C. A. (2008). Rapid evolution in introduced species, "invasive traits" and recipient communities: Challenges for predicting invasive potential. *Diversity and Distributions*, *14*, 569–580.

Wiseman, S., Osachoff, H., Bassett, E., Malhotra, J., Bruno, J., VanAggelen, G., … Vijayan, M. M. (2007). Gene expression pattern in the liver during recovery from an acute stressor in rainbow trout. *Comparative Biochemistry and Physiology—Part D: Genomics and Proteomics*, *2*, 234–244.

Wolfe, K. R., & Marsden, E. J. (1998). Tagging methods for the round goby (*Neogobius melanostomus*). *Journal of Great Lakes Research*, *24*, 731–735.

Wray, G. A., Hahn, M. W., Abouheif, E., Balhoff, J. P., Pizer, M., Rockman, M. V., & Romano, L. A. (2003). The evolution of transcriptional regulation in eukaryotes. *Molecular Biology and Evolution*, *20*, 1377–1419.

Wright, S. (1931). Evolution in Mendelian populations. *Genetics*, *16*, 97–159.

Xin, S. (2016). *Comparison of physiological performance characteristics of two Great Lakes invasive fish species: Round Goby (Neogobius melanostomus) and Tubenose Goby (Proterorhinus semilunaris)*. M.Sc. Thesis, GLIER, University of Windsor, Windsor, ON.

Young, M. D., Wakefield, M. J., Smyth, G. K., & Oshlack, A. (2010). Gene ontology analysis for RNA-seq: Accounting for selection bias. *Genome Biology*, *11*, R14.

Yu, G., Li, F., Qin, Y., Bo, X., Wu, Y., & Wang, S. (2010). GOSemSim: An R package for measuring semantic similarity among GO terms and gene products. *Bioinformatics*, *26*, 976–978.

12

Population genetic structure and connectivity of deep-sea stony corals (Order Scleractinia) in the New Zealand region: Implications for the conservation and management of vulnerable marine ecosystems

Cong Zeng[1,2,3] (ID) | Ashley A. Rowden[3] | Malcolm R. Clark[3] | Jonathan P. A. Gardner[2]

[1]College of Animal Science and Technology, Hunan Agricultural University, Changsha, China

[2]School of Biological Sciences, Victoria University of Wellington, Wellington, New Zealand

[3]National Institute for Water and Atmospheric Research, Kilbirnie, Wellington, New Zealand

Correspondence
Cong Zeng, College of Animal Science and Technology, Hunan Agricultural University, Changsha, China.
Email: congzeng@live.cn

Funding information
Startup Foundation for Advanced Talents, Hunan Agricultural University; New Zealand Ministry of Business, Innovation and Employment, Predicting the occurrence of vulnerable marine ecosystems for planning spatial management in the South Pacific region, Grant/Award Number: CO1X1229; NIWA under the Marine Biological Resources programme

Abstract

Deep-sea stony corals, which can be fragile, long-lived, late to mature and habitat-forming, are defined as vulnerable marine ecosystem indicator taxa. Under United Nations resolutions, these corals require protection from human disturbance such as fishing. To better understand the vulnerability of stony corals (*Goniocorella dumosa*, *Madrepora oculata*, *Solenosmilia variabilis*) to disturbance within the New Zealand region and to guide marine protected area design, genetic structure and connectivity were determined using microsatellite loci and DNA sequencing. Analyses compared population genetic differentiation between two biogeographic provinces, amongst three subregions (north–central–south) and amongst geomorphic features. Extensive population genetic differentiation was revealed by microsatellite variation, whilst DNA sequencing revealed very little differentiation. For *G. dumosa*, genetic differentiation existed amongst regions and geomorphic features, but not between provinces. For *M. oculata*, only a north–central–south regional structure was observed. For *S. variabilis*, genetic differentiation was observed between provinces, amongst regions and amongst geomorphic features. Populations on the Kermadec Ridge were genetically different from Chatham Rise populations for all three species. A significant isolation-by-depth pattern was observed for both marker types in *G. dumosa* and also in *ITS* of *M. oculata*. An isolation-by-distance pattern was revealed for microsatellite variation in *S. variabilis*. Medium to high levels of self-recruitment were detected in all geomorphic populations, and rates and routes of genetic connectivity were species-specific. These patterns of population genetic structure and connectivity at a range of spatial scales indicate that flexible spatial management approaches are required for the conservation of deep-sea corals around New Zealand.

KEYWORDS
deep-sea conservation, gene flow, genetic connectivity, marine protected areas, Scleractinia

1 | INTRODUCTION

The deep sea is the largest habitat on Earth, but knowledge about this biome is limited because of the logistical constraints and costs of sampling such a large and inaccessible area (Ramirez-Llodra et al., 2011). Whilst deep-sea communities and their ecological structures and functions are still being described, they face ongoing or increasing threats from anthropogenic activities such as fishing, mining, dumping, pollution and climate change (Ramirez-Llodra et al., 2011). To reduce the impact of human activities in the deep sea, vulnerable marine ecosystems (VMEs) have been selected as a protection priority under United Nations General Assembly resolutions (61/105 and 59/25). VMEs are habitats and ecosystems characterized by uniqueness or rarity of their species, their significant ecological function, that are easily disturbed by anthropogenic activities and that may exhibit slow or even no recovery from disturbance (FAO, 2009). Based on the FAO characteristics, Parker, Penney, and Clark (2009) identified sponges, anemones, soft corals, sea fans, sea pens, stony corals, black corals, hydrocorals, sea lilies and armless sea stars as VME indicator taxa. The presence of these taxa can be used to identify VMEs that may be vulnerable to impacts from fishing activities in the South Pacific Ocean and guide management measures designed to protect them (Penny, Parker, & Brown, 2009).

Deep-sea stony corals (Order Scleractinia) may form complex three-dimensional structures that are classified as VMEs because they provide habitats and refuges for many other species (e.g., Bongiorni et al., 2010). Amongst these habitat-forming corals, the three most common species in the New Zealand region are *Goniocorella dumosa*, *Madrepora oculata* and *Solenosmilia variabilis*, and these three corals are commonly found on seamount features (Tracey, Rowden, Mackay, & Compton, 2011). Seamounts (including knolls and hills) are the focus of several deep-sea fisheries in New Zealand waters, and both *G. dumosa* and *S. variabilis* have been recorded in large quantities as bycatch from seamount fisheries (Anderson & Clark, 2003). In addition to seamounts, *G. dumosa* is also associated with phosphorite nodules on soft sediments of the Chatham Rise to the east of New Zealand (Kudrass & Rad, 1984), which suggests that *G. dumosa* may be impacted by any future mining of this resource. Unfortunately, the mode of larval development of these three corals is still poorly understood, and information about their patterns of larval movement and gene flow (connectivity) is sparse (e.g., Addamo, Reimer, Taviani, Freiwald, & Machordom, 2012; Baco et al., 2016; Miller & Gunasekera, 2017).

An understanding of connectivity is important for management decisions related to the conservation of corals because if limited gene exchange exists amongst populations, then loss of areas of reef may be detrimental to the overall genetic diversity of the species. In this article, we describe species-specific patterns of genetic connectivity amongst populations of three deep-sea corals from the New Zealand region. Newly developed microsatellite markers were genotyped, and a nuclear gene (internally transcribed spacer, *ITS2*) and a mitochondrial region (the control region, *D-loop*) were sequenced for the three VME indicator species, *G. dumosa*, *M. oculata* and *S. variabilis*. We employed a province, region and geomorphic feature hierarchical

framework to test the hypotheses that (i) the water mass characteristics of two biogeographic provinces (based on Watling, Guinotte, Clark, & Smith, 2013) influence the north–south population distributions, which in turn will be reflected in a province-scale pattern of genetic structure; (ii) current flows from north and south of New Zealand that meet along the Chatham Rise to the east and that form the Subtropical Front will act as conduits and barriers to larval dispersal, which will be reflected in a north–central–south regional-scale pattern of genetic structure, and higher genetic diversity in the central putative mixing region on the Chatham Rise; and (iii) hydrodynamic conditions (e.g., current flows, eddies, turbulent mixing) associated with particular topographic features (such as slopes, seamounts, plateaux, rises, ridges, troughs, basins) influence the dispersal of larvae, potentially restricting dispersal amongst these features, and thereby generating genetic structure amongst populations of corals found on these features (Figure 1). Tests were also conducted to examine the influence of geographic distance and depth on genetic population structure. Post hoc analyses were conducted on data to examine patterns of genetic population structure and potential barriers to gene flow independent of the hypothesis testing framework and to identify potential larval migration patterns. Assessments of effective population size were also made. Results of these analyses are interpreted in terms of their significance for management of VME indicator species around New Zealand and the possible distribution of offshore marine protected areas.

2 | MATERIALS AND METHODS

2.1 | Samples

The total number of specimens was 78 for *M. oculata*, 134 for *G. dumosa* and 208 for *S. variabilis*. Samples were mainly sourced from the NIWA Invertebrate Collection, which archives specimens obtained from a range of sampling expeditions since the 1950s. Samples of *G. dumosa*, *M. oculata* and *S. variabilis* were collected from depth ranges of 198–1,270 m, 236–1,537 m and 322–1,805 m, respectively. Most specimens that were used in this study were preserved in ethanol, and the rest were dry-preserved. The majority of specimens were from seamount and slope habitats (198–1,805 m water depth) (Figure 1). Preliminary testing revealed that most specimens more than 5 years old did not provide sufficient quantity and/or quality of DNA, meaning that temporal variability amongst our samples was low, especially in the context of species that are long-lived. Specimens within 1° of latitude and longitude were grouped for haplotypic distribution charts, and the groups were further assigned into different populations at scales that reflect the various environmental features that could influence connectivity amongst populations (i.e., province, region, geomorphic feature). Due to patchy sampling efforts and the different distributions of the species, samples for all species were not available from all locations. To achieve a balance between the validation of results and extracting maximum information content from the specimens, the minimum sample size was set at four for population analysis.

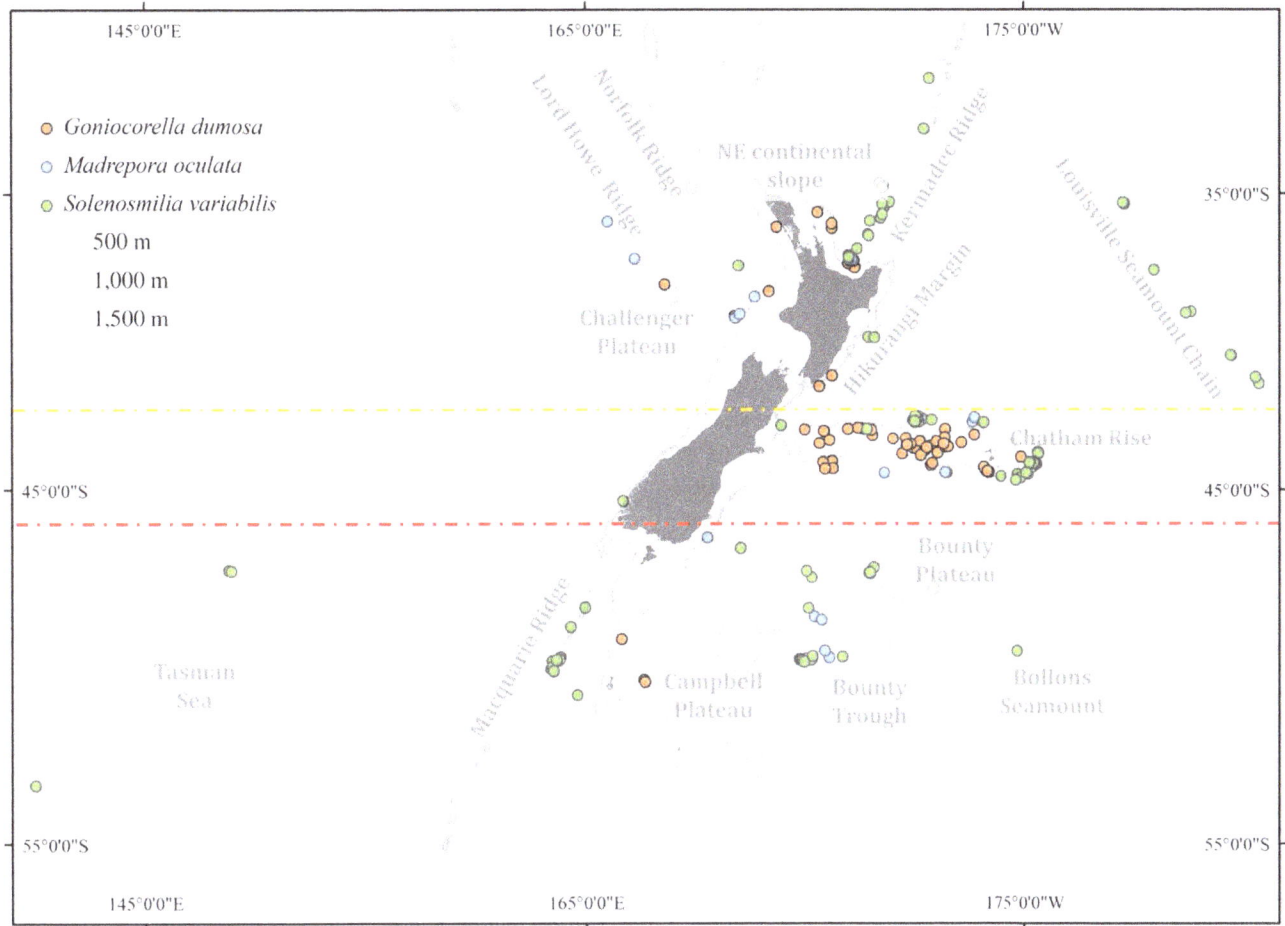

FIGURE 1 Map showing the distribution of samples amongst lower bathyal biogeographic provinces (yellow dashed line is the boundary between the northern and southern provinces), regions (yellow and red dashed lines indicate the boundaries for the north-central-south regions), and geomorphic features (named features) used for the analysis of genetic population structure for three species of deep-sea stony corals

2.2 | Molecular methods

All methods follow Zeng (2016). DNA sequencing of one mitochondrial region (*D-loop*) and one nuclear region (*ITS*) was employed to estimate genetic connectivity and population genetic differentiation (Table S1). *ITS* data were obtained for all three species, whereas *D-loop* data were obtained for two species (the *D-loop* is absent in *M. oculata*). In total, 27 microsatellite loci were assayed for *G. dumosa* and *S. variabilis*, and 11 microsatellite loci were assayed for *M. oculata*. Of these loci, most were developed for this study (Zeng, 2016), but others were obtained from colleagues (Karen Miller, AIMS, Australia, for *S. variabilis* and Sophie Arnoud-Haond, IFREMER, France, for *M. oculata*).

2.3 | Data analysis

2.3.1 | D-loop and ITS variation

Multiple sequences were aligned using the plugin ClustalW Alignment (Gap Open Cost = 100, Gap Extend Cost = 10), and then all alignments were viewed by eye using Geneious v7 (Biomatters Ltd, New Zealand). For all populations with two or more individuals, the intraspecific genetic diversity was evaluated by computing the number of haplotypes, the number of polymorphic sites, haplotypic diversity (h) and nucleotide diversity (π).

Analyses of molecular variance (AMOVA) between locations (between north and south biogeographic provinces, or amongst northern, central and southern regions, or amongst geomorphic features) were tested using Arlequin (Excoffier & Lischer, 2010). Pairwise comparisons of population differentiation and significance values were estimated after 1,000 permutations. Between-location or within-location Φ_{ST} statistics (based on sequence divergence) were calculated to test for genetic differentiation amongst populations. If significant differentiation amongst populations was detected, the location of the genetic discontinuity was identified using BARRIER v2.2 (Manni, Guerard, & Heyer, 2004).

To visualize spatial patterns of genetic variation, specimens were colour-coded according to haplotype and their geographic coordinates of collection were plotted using ArcGIS (ESRI, USA). The Mantel test (Mantel, 1967) was employed to test for isolation by depth by comparing the matrix of Φ_{ST} values to the matrix of depth (m) values and to test for isolation by distance by comparing the matrix of Φ_{ST} values to the matrix of shortest actual distances (km) between pairs of sites (GenAlEx v6.5, Peakall & Smouse, 2012).

2.3.2 | Microsatellite variation

Micro-Checker v2.2.1 with default settings was used to identify stuttering, large allele dropout, and null alleles (van Oosterhout, Hutchinson, Wills, & Shipley, 2004). Loci that were putatively neutral or under selection were identified using LOSITAN with default settings (Antao, Lopes, Lopes, Beja-Pereira, & Luikart, 2008). Loci under selection and loci with null alleles present at >10% frequency (Oosterhout score) were removed, and a second reduced data set (neutral loci) was created. Both the full and the reduced data sets were tested, where appropriate, for population genetic differentiation because non-neutral loci may often be informative about population genetic structure and therefore useful for management purposes (e.g., Gagnaire et al., 2015; Wei, Wood, & Gardner, 2013). Allelic frequencies, number of alleles, departures from Hardy–Weinberg equilibrium (HWE), linkage disequilibrium (LD) and observed and expected heterozygosities were estimated using GenAlEx. Arlequin was used to estimate unbiased estimator of Wright's F statistic (F_{ST}), and hierarchical F_{ST} analyses (AMOVA) were conducted in GenAlEx. If significant differentiation amongst populations was detected, the location of the genetic discontinuity, isolation by depth and isolation by distance were estimated as described for the *D-loop* and *ITS* markers. Multilocus matches were employed to detect the asexual/colonial relationship between individuals (GenAlEx).

Discriminant analysis of principle components (DAPC) (Jombart, Devillard, & Balloux, 2010) was used to identify population structure amongst the microsatellite multilocus genotypes of all individuals per species. Analysis was implemented in the R package "adegenet" (Jombart, 2008). The optimal number of genetic clusters (K) was chosen when Bayesian information criterion (BIC) values were the lowest. Scatter plots of microsatellite genotypes in relation to discriminant functions were created in "adegenet". STRUCTURE (v2.3.4), with a 5×10^6 burn-in period and 5×10^6 MCMC runs after burn-in, was employed to infer the number of distinct genetic groups based on Bayesian assignment analysis (Falush, Stephens, & Pritchard, 2007). The models were run in three iterations for K (number of distinct genetic clusters) values to evaluate likelihood scores and consistency amongst runs. The values of K were set from 1 to the number of geomorphic feature populations, that is, 4, 4 and 8 for *G. dumosa*, *M. oculata* and *S. variabilis*, respectively. The optimal number for K was chosen as the maximum number of clusters which yielded likelihoods lower than those observed at lower values of K. This number was computed in the online program STRUCTURE HARVESTER (Earl & von Holdt, 2012).

Where AMOVA results indicated significant genetic structure at the three different spatial scales, we used assignment tests implemented in GeneClass2 (Piry et al., 2004) to generate estimates of contemporary dispersal and to identify first-generation migrants (e.g., Wei et al., 2013). Although this assignment methodology assumes HWE, simulations suggest that small heterozygote deficits have little effect on test performance (Cornuet, Piry, Luikart, Estoup, & Solignac, 1999). This approach does not require that the true population of origin has been sampled and it is therefore likely to be a relatively accurate method (Berry, Tocher, & Sarre, 2004).

Effective population size (*Ne*) was estimated using the software NeEstimator v2 (Do, Waples, Peel, Macbeth, & Tillett, 2014). We estimated contemporary effective population size (i.e., the *Ne* estimate applies to the time period encompassed by sampling) using multilocus data in the single-sample bias-corrected method based on linkage disequilibrium (Waples & Do, 2010). Both the full locus and the neutral only locus data sets were analysed. $P_{critical}$ values were set according to the formula $P_{critical} >1/2S$, where S = number of individuals per population per bioprovince or region or geomorphic feature. Parametric confidence intervals were estimated, and following Waples and Do (2010) negative values of *Ne* were set to infinity.

3 | RESULTS

There was considerable variability in the number of DNA samples obtained per species and per geomorphic feature, region and province due to our ability to extract high-quality DNA, based on the age and state of sample preservation, as well as the spatial differences in sampling effort.

3.1 | Population diversity and structure based on mitochondrial and nuclear DNA sequence variation

3.1.1 | Population genetic diversity

For *G. dumosa*, *ITS* haplotypic diversity was greater in the southern than in the northern province, but nucleotide diversity in the northern was greater than in the southern province. *D-loop* haplotypic and nucleotide diversity values in the southern were greater than in the northern province (Figure 2a; Table S2). However, given the small sample sizes in the south, these values must be interpreted with care. For *M. oculata*, *ITS* haplotypic and nucleotide diversity values of the northern were greater than the southern province, and values for both decreased from the north via the central to the south region (Figure 2b; Table S2). For *S. variabilis*, both the haplotypic and nucleotide diversity values of *ITS* and *D-loop* of the southern were greater than those of the northern province, and values for both decreased from the south via the central to the north region (Figure 2c; Table S2). Common haplotypes of both *ITS* and *D-loop* were found in most of the sampled areas for all three species (Fig. S1).

3.1.2 | Population genetic structure

For *G. dumosa*, AMOVA for both *ITS* and *D-loop* did not reveal any genetic structure amongst geomorphic populations (*p* > .05 in all three cases). Comparable analyses were not possible for the between northern/southern provinces or for the amongst north-central-south regions because of very low sample sizes (*n* < 4) in the south. For *M. oculata*, AMOVA of *ITS* variation revealed significant differentiation amongst the three regions (north, central, south; *p* < .05). In contrast, for *S. variabilis*, *ITS* and *D-loop* variation showed no evidence of structure at any spatial scale.

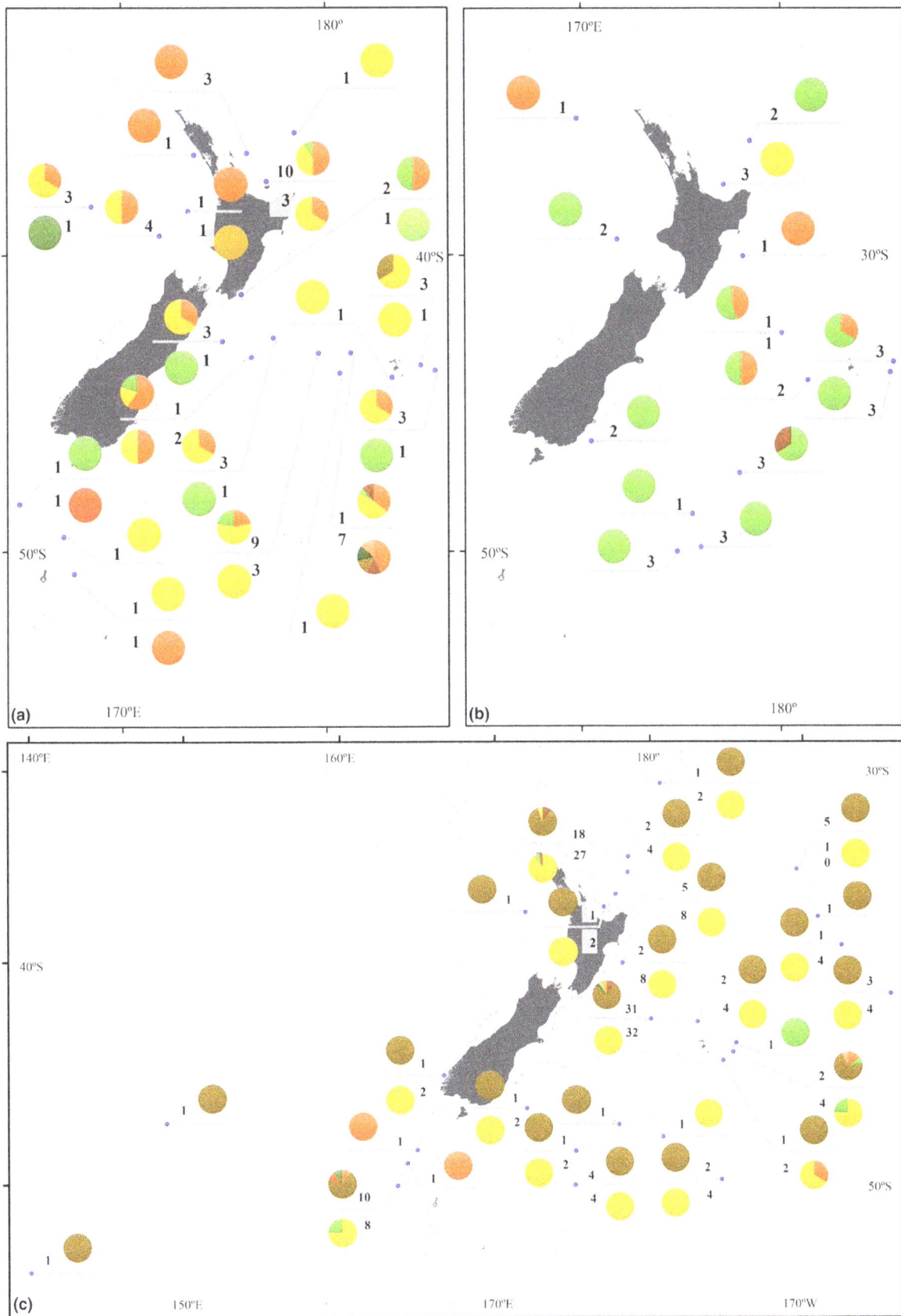

FIGURE 2 Map of haplotype distributions for *Goniocorella dumosa* (a), *Madrepora oculata* (b) and *Solenosmilia variabilis* (c) in the New Zealand region. Pie charts for *ITS* (above line) and *D-loop* (below line) indicate haplotypic composition of each location, and numbers indicate total number of sequences from each location

The Φ_{ST} values for *ITS* and *D-loop* were variable but relatively low for all three species. In *G. dumosa*, no significant Φ_{ST} values were observed for either marker. For *M. oculata*, a significant pairwise value between the Kermadec Ridge and Chatham Rise populations was observed for *ITS*. For *S. variabilis*, significant pairwise population values of Φ_{ST} for *ITS* were observed between the Macquarie Ridge and Louisville Seamount Chain populations, and between the Macquarie Ridge and Kermadec Ridge populations (Table S3). No significant Φ_{ST} values were observed for *D-loop*.

There was no statistically significant isolation by distance ($p > .05$) for any of the species, but *ITS* sequence variation exhibited a significant pattern of isolation by depth for *M. oculata* and *G. dumosa* ($p < .05$).

3.2 | Population diversity and structure based on microsatellite variation

Micro-Checker detected null alleles at 4, 5 and 5 loci, and LOSITAN revealed that 11, 1 and 8 loci were subject to putative selection in *G. dumosa*, *M. oculata* and *S. variabilis*, respectively. The reduced data sets (putatively neutral loci only) included eight loci for *G. dumosa*, six for *M. oculata* and 12 for *S. variabilis* (Table S4). Where appropriate, analyses were conducted on the species-specific full (all loci) and the reduced microsatellite data sets. Because the results were concordant, we report here the results for the full data set but highlight major differences between the two data sets.

3.2.1 | Genotype identity test

All genotyped loci were utilized to screen out any genetically identical individuals (clones or fragments of individuals). For *S. variabilis*, one pair of specimens shared the same multilocus genotype. One specimen was excluded from analysis. In *G. dumosa*, no specimens with the same multilocus genotype were detected amongst 108 individuals. In

M. oculata, three pairs of specimens were identical across all assayed loci, and duplicates were therefore excluded.

3.2.2 | Population genetic diversity

For *G. dumosa*, the number of alleles per locus varied from 7 to 27 (average = 8.20). Observed heterozygosity (H_O) ranged from 0.500 to 1.000 per locus, and most H_O values were greater in all populations than expected (Table S5a). For *M. oculata*, the number of alleles per locus ranged from 5 to 19 (average = 9.96). H_O values ranged from 0.273 to 0.944 per locus, and unlike the other two coral species, were lower in all populations than expected (Table S5b). For *S. variabilis*, the number of alleles varied from 7 to 31 (average = 14.42), and H_O ranged from 0.211 to 1.000 (Table S5c). In all three species, many loci exhibited significant departures from HWE (Table S5).

Pairwise values of F_{ST} for all species and for both data sets were rarely greater than 0.1 (Table S6). For *G. dumosa*, only one significant result (of six tests per data set) was observed (Chatham Rise–Kermadec Ridge), no significant results were observed *M. oculata*, or for *S. variabilis*, 12 of 28 (all loci) and 14 of 28 (neutral loci) tests were significant (Table S6).

3.2.3 | Population genetic structure

For all three species, there was very strong concordance in AMOVA results for the reduced and for the all loci data sets, indicating that the presence or absence of a signal of genetic differentiation is conserved across all surveyed loci (Tables 1–3). The AMOVA of genetic structure for *G. dumosa* detected large-scale structure amongst north, central, south regions and amongst the geomorphic features, but not between northern and southern provinces. The only AMOVA evidence of genetic structure in *M. oculata* was amongst the north, central and south regions. The AMOVA for *S. variabilis* detected large-scale structure between northern and southern provinces, amongst north,

TABLE 1 AMOVA results of microsatellites for *Goniocorella dumosa* at three different spatial scales (in bold)

Source of variation	All loci			Neutral loci only		
	df	ss	Var. comp.	df	ss	Var. comp.
Between provinces	1	9.373	0.000	1	3.629	0.000
Amongst Individuals	106	1,058.678	2.390**	106	422.144	0.769**
Within Individuals	108	562.500	5.208**	108	264.000	2.444**
Total	215	1,630.551	7.598	215	689.773	3.213
Amongst regions	2	27.947	0.088*	2	10.180	0.025*
Amongst Individuals	105	1,040.104	2.349**	105	415.593	0.757**
Within Individuals	108	562.500	5.208**	108	264.000	2.444**
Total	215	1,630.551	7.645	215	689.773	3.226
Amongst geomorphic features	3	0.196	0.196**	3	16.653	0.063**
Amongst Individuals	96	2.318	2.318**	96	377.757	0.740**
Within Individuals	100	5.240	5.240**	100	245.500	2.455**
Total	199	7.754	7.754	199	639.910	3.258

Significant values *$p < .05$, and **$p < .01$.

Source of variation	All loci			Neutral loci only		
	df	ss	Var. comp.	df	ss	Var. comp.
Between provinces	1	6.420	0.013	1	3.234	0.004
Amongst Individuals	91	507.634	1.926**	91	269.846	0.934**
Within Individuals	93	160.500	1.726**	93	102	1.097**
Total	185	674.554	3.665	185	375.081	2.035
Amongst regions	2	13.474	0.022*	2	7.581	0.016*
Amongst Individuals	90	500.580	1.918**	90	265.5	0.927**
Within Individuals	93	160.500	1.726**	93	102	1.097**
Total	185	674.554	3.666	185	375.081	2.039
Amongst geomorphic features	3	18.763	0.017	3	11.377	0.029
Amongst Individuals	55	324.898	1.886**	55	174.724	0.893**
Within Individuals	59	126.000	2.136**	59	82.000	1.390**
Total	117	469.661	4.038	117	268.102	2.312

TABLE 2 AMOVA results of microsatellites for *Madrepora oculata* at three different spatial scales (in bold)

Significant values *p < .05, and **p < .01.

Source of variation	All			Neutral only		
	df	ss	Var. comp.	df	ss	Var. comp.
Between provinces	1	16.070	0.038*	1	11.039	0.044**
Amongst Individuals	206	2,322.236	2.207**	206	1,112.833	0.700**
Within Individuals	208	1,426.500	6.858**	208	832.500	4.002**
Total	415	3,764.805	9.103	415	1,956.373	4.747
Amongst regions	2	52.773	0.115**	2	32.454	0.083**
Amongst Individuals	205	2,285.532	2.145**	205	1,091.419	0.661**
Within Individuals	208	1,426.500	6.858**	208	832.500	4.002**
Total	415	3,764.805	9.119	415	1,956.373	4.746
Amongst geomorphic features	7	151.751	0.259**	7	80.076	0.150**
Amongst Individuals	192	2,080.067	1.971**	192	989.109	0.570**
Within Individuals	200	1,378.500	6.893**	200	802.500	4.013**
Total	399	3,610.318	9.122	399	1,871.685	4.732

TABLE 3 AMOVA results of microsatellites for *Solenosmilia variabilis* at three different spatial scales (in bold)

Significant values *p < .05, and **p < .01.

central, south regions and amongst the geomorphic features. Notably, a north–central–south regional differentiation was found in all three species for both data sets (Tables 1–3).

For all three species, tests of differentiation (F_{ST} values) amongst populations from the geomorphic features for both the reduced and all loci data sets showed similar patterns (Table S6). In *G. dumosa*, the Kermadec Ridge population was significantly different from the Chatham Rise population (Table S6a). In *M. oculata*, no significant F_{ST} values were observed for pairwise testing amongst all (Campbell Plateau, Challenger Plateau, Chatham Rise and Kermadec Ridge) populations (data not shown). For *S. variabilis*, pairwise population F_{ST} values revealed no significant genetic differences between the Bounty Trough population and all other populations, and the Kermadec Ridge and Chatham Rise populations (Table S6b).

Tests for microsatellite genetic isolation by distance and by depth identified only two significant results—in *S. variabilis* for distance and in *G. dumosa* for depth.

3.3 | Post hoc analyses

3.3.1 | Geographic differentiation

In *G. dumosa*, the DAPC (Figure 3a) and the STRUCTURE HARVESTER plot (Fig. S3) provided evidence of more than one population, but the number of putative genetic clusters was inconsistent between Structure (K = 2) and DAPC (K = 3, Northeast Slope, Kermadec Ridge and all other populations). In *M. oculata*, DAPC revealed that the Campbell Plateau and Chatham Rise populations were genetically

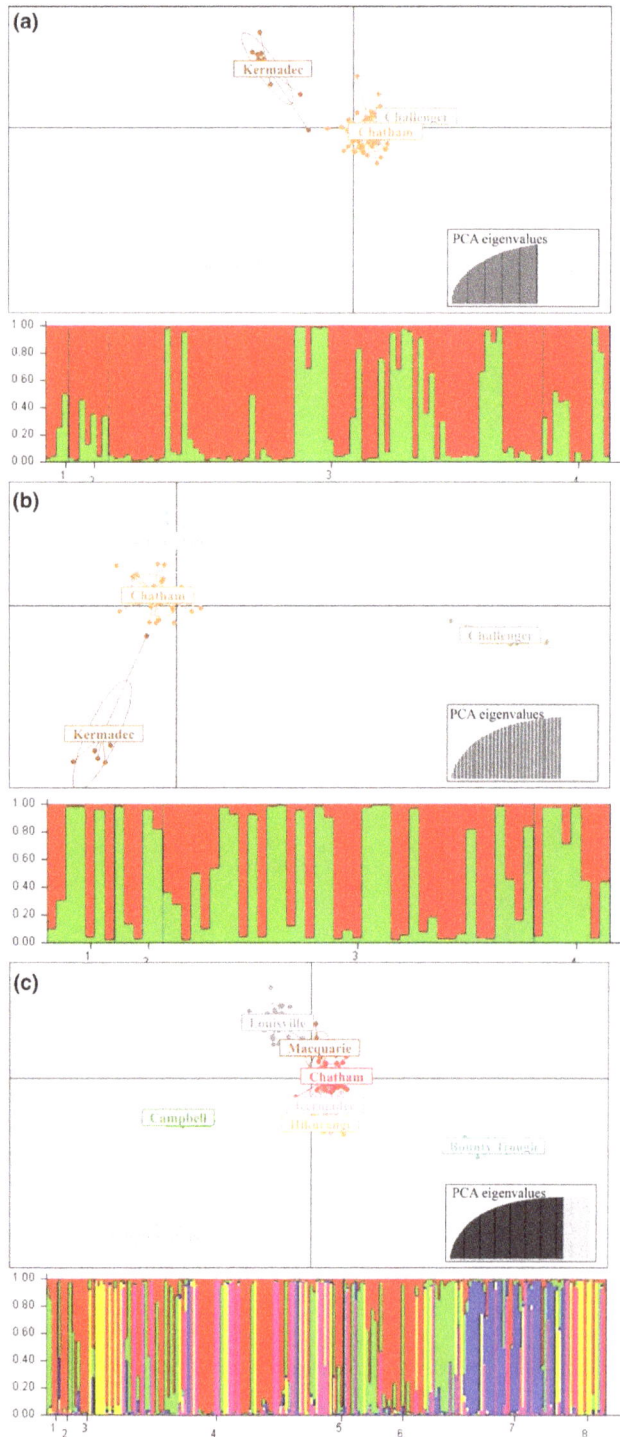

FIGURE 3 (a) DAPC scatter plot (Above) and posterior estimates from STRUCTURE (K = 2) (Below) of *Goniocorella dumosa* based on variation at all microsatellite loci. X-axis—1. Northeast continental slope, 2. Challenger Plateau, 3. Chatham Rise, 4. Macquarie Ridge. (b) DAPC scatter plot (Above) and posterior estimates from STRUCTURE (K = 2) (Below) of *Madrepora oculata* based on variation at all microsatellite loci. X-axis—1. Campbell Plateau, 2. Challenger Plateau, 3. Chatham Rise, 4. Kermadec Ridge. (c) DAPC scatter plot (Above) and posterior estimates from STRUCTURE (K = 5) (Below) of *Solenosmilia variabilis* based on variation at all microsatellite loci. X-axis—1. Bounty Plateau, 2. Bounty Trough, 3. Campbell Plateau, 4. Chatham Rise, 5. Hikurangi Margin, 6. Kermadec Ridge, 7. Louisville Seamount Chain, 8. Macquarie Ridge

3.3.2 | Barriers to gene flow

It is likely that the BARRIER results for analysis of microsatellite variation better reflect contemporary patterns of gene flow, whereas results for DNA sequence data better reflect historical (phylogeographic) patterns of gene flow.

For *G. dumosa*, based on microsatellite variation, BARRIER predicted that barriers to gene flow may exist for the Challenger Plateau and Chatham Rise populations (Figure 4a). In *M. oculata*, because there were no significant pairwise F_{ST} values calculated from the microsatellite data sets, the barriers to gene flow were predicted from the Φ_{ST} values of the *ITS* sequence, which indicated that the Kermadec Ridge (to the north) and Chatham Rise populations are isolated (Figure 4b). There was generally very good agreement for the location of barriers to gene flow amongst populations of *S. variabilis* based on Φ_{ST} values of the *ITS* sequence (Figure 4c) and F_{ST} values of the full microsatellite data set (Figure 4d). In both instances, four barriers to gene flow were predicted, isolating the far north (Kermadec Ridge), the far east (Louisville Seamount Chain), the east (Chatham Rise), the south central (Campbell Plateau, Bounty Plateau, Bounty Trough) and the far south (Macquarie Ridge) populations.

3.3.3 | *Ne* estimation

Estimates of *Ne* determined from the software NeEstimator indicated that in most cases, contemporary effective population size was very small and in the range of ~20 to 60 individuals (Table S7). Estimates of *Ne* based on variation in the all loci data set were usually slightly higher than those based on the neutral loci data set. The *Ne* estimates for all three species at all three spatial hierarchies were very low.

3.3.4 | Assignment accuracy and estimates of first-generation migrants

For *G. dumosa*, assignment rate success ranged between 51.9% (neutral data set, regional spatial scale) and 69.4% (all loci, regional spatial scale) (Table S8a, c). At both spatial scales, the assignment success rates were higher for all loci than for the neutral loci. First-generation migrants were observed amongst all regions, but were more numerous for the north and central than for the south region (Table S8b).

similar to each other (Figure 3b). The structure complots of the posteriors from the DAPC confirm the group separations within all three species (Fig. S2), as visualized by the DAPC plots (Figure 3). The STRUCTURE HARVESTER plot indicated the presence of three groups (Fig. S3). For *S. variabilis*, the DAPC differentiated five genetic clusters, Campbell Plateau, Bounty Plateau, Bounty Trough, Louisville Seamount Chain and all other populations (Figure 3c). The best grouping cluster number was equally *K* = 3 or *K* = 5, based on Δ*K* (Fig. S3). Overall, the DAPC was able to resolve more within-species groups than was STRUCTURE.

First-generation migrants were observed between most, but not all, pairs of populations on geomorphic features (Table 8Sb). Most first-generation migrants were associated with the Chatham Rise (Table 8Sb). The percentage of first-generation migrants ranged from 30.6% (all loci, regional spatial scale) to 48.2% (neutral loci, regional spatial scale).

For *M. oculata*, assignment success rates at the regional scale were low (all loci = 34.2%; neutral loci = 39.5%). Individuals sampled from the central region were assigned at nearly equal rates to the north, central and south regions, whilst individuals sampled from the north or south regions exhibited higher rates of assignment success (Table S8e). First-generation migrants were identified in all cases except one (from south to north, all loci) and occurred at high rates (65.7% of all individuals for the all loci data set, 60.5% for the neutral data set) (Table S8f).

For *S. variabilis*, assignment success rates at the bioprovince scale were high (67.8% for all loci, 63.9% for the neutral loci), but were lower at the regional scale (44.7% and 48.6%, respectively) and at the geomorphic features scale (42.0% and 42.5%, respectively) (Table S8g, i, k). Estimates of first-generation migrants were lowest at the bioprovince scale (32.2% and 36.1%, respectively), highest at the regional scale (55.3% and 51.4%, respectively) and intermediate at the geomorphic features scale (42.0% and 42.5%, respectively). First-generation migrants were identified between all pairs of sites for the bioprovince and the regional analyses (Table S8h, j). At the geomorphic features scale, most first-generation migrants were associated with the Chatham Rise (Table S8l); this was also the case for the central region in the regional-scale analysis (Table S8j).

4 | DISCUSSION

Three habitat-forming deep-sea corals, *G. dumosa*, *M. oculata* and *S. variabilis*, were examined for genetic connectivity and gene flow amongst populations in the New Zealand region for the purpose of contributing to effective spatial management planning options to protect VMEs. We employed an a priori hierarchical and spatially explicit approach to hypothesis testing for population genetic structure, estimated contemporary and historical effective population sizes and identified barriers to gene flow as well as source and sink populations, all of which may contribute significantly to management decisions with conservation outcomes in the deep sea.

4.1 | Influence of sample sizes and marker selection on detection of population genetic structure

Small sample sizes and an absence of fine spatial scale sampling are widely acknowledged as inevitable constraints in genetic studies of deep-sea organisms (reviewed by Baco et al., 2016). Although limited sampling made it difficult to reveal small-scale population genetic structure, we develop here an approach that provides a broad spatial-scale view of the connectivity of three coral species and that reveals sufficient information about patterns of genetic structure to be useful to support management decisions.

Preliminary data screening identified a number of outlier loci within the microsatellite data set, a well-known problem for taxa such as deep-sea corals. It is unknown for sure why outlier loci occur at such high frequencies, but it may be for several different reasons, including null alleles, coloniality, inbreeding, selection, the Wahlund effect and poor state of sample preservation (Baco et al., 2016; Becheler et al., 2015; Miller & Gunasekera, 2017; Quattrini, Baums, Shank, Morrison, & Cordes, 2015). We always tested our DNA to ensure that we worked with the best quality possible, so poor quality DNA (e.g., arising from age of sample or preservation in formalin) is unlikely to be an explanation. To address this problem, our analytical approach has been to test both the full locus and the neutral locus data sets, given that both may be informative. This is particularly important from the perspective of including non-neutral loci in analyses that are (usually) designed to meet neutral expectations. Importantly, there was very little difference in the results of all analyses between the full (all loci) and the neutral (reduced number of loci) data sets. This strongly suggests that the presence/absence of a signal of genetic differentiation is conserved across all surveyed loci. That is, addition or removal of the outlier loci is not critical to the interpretation of the results in the context of management options for future protection of New Zealand's vulnerable marine ecosystems and their associated taxa.

As our hierarchical spatial testing increased in definition, so our sample sizes decreased. Thus, our analyses have small sample sizes for most geomorphic features but larger sample sizes for the regional and biogeographic province levels of testing. All estimates for populations on geomorphic features should be interpreted with care and are best viewed in combination with other analyses. However, population groups (provinces and regions) and populations on some geomorphic features (such as the Chatham Rise) had large enough sample sizes to reveal patterns of genetic differentiation and connectivity. Overall, our analyses are probably best viewed as a conservative interpretation of patterns of population genetic differentiation.

Use of the software package STRUCTURE (Falush et al., 2007) to identify the number of distinct genetics groups (i.e., identification of *K*) is now widespread in population genetics analyses (e.g., Ruiz-Ramos, Saunders, Fisher, & Baums, 2015). STRUCTURE is a parametric analysis and is known to perform best when sample sizes are equal, when null alleles are absent and when populations are in HWE (Puechmaille, 2016; Putman & Carbone, 2014; Wang, 2017). Our analyses using STRUCTURE with no a priori groupings did not provide reliable results, most probably because of large differences in population/region sample sizes and the absence of HWE in several samples. Absence of HWE amongst deep-sea corals, in particular the presence of significant heterozygote deficiencies, is routinely reported (Baco, Clark, & Shank, 2006; Le Goff & Rogers, 2002; Lunden, McNicholl, Sears, Morrison, & Cordes, 2014; Miller & Gunasekera, 2017). One interpretation of multilocus heterozygote deficiencies is that recruitment in many populations is locally derived with infrequent long-distance dispersal events. Such an interpretation is consistent with our results for the three coral species we examined (see below). Because of the uncertainty associated with the STRUCTURE results, we have

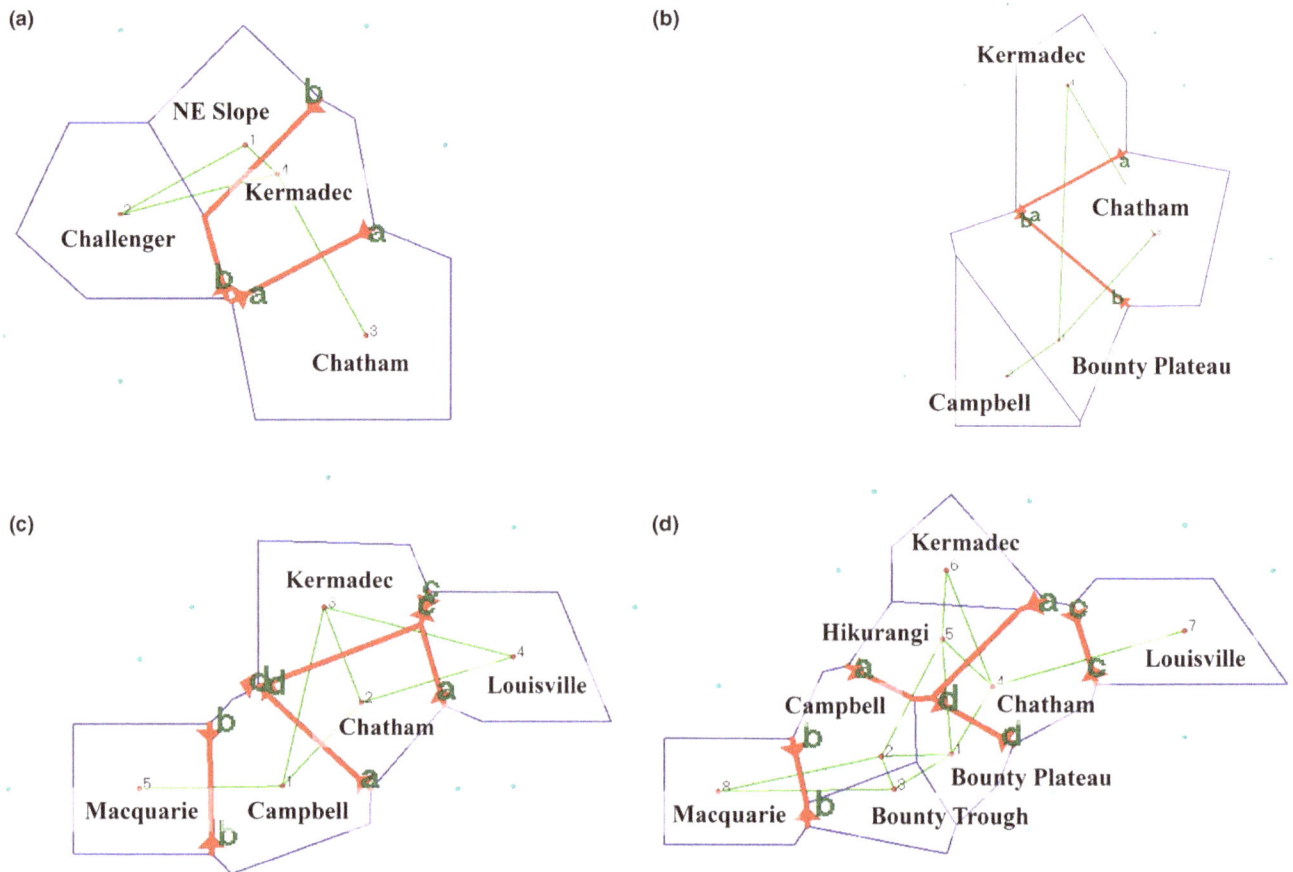

FIGURE 4   Barriers to gene flow indicated by software Barriers based on the full microsatellite data set for *Goniocorella dumosa* (a), ITS data for *Madrepora oculata* (b), ITS data for *Solenosmilia variabilis* (c) and the full microsatellite data set for Solenosmilia variabilis (d)

focussed on other analyses for which there are greater robustness and more confirmation from other approaches.

Miller, Williams, Rowden, Knowles, and Dunshea (2010) noted that low genetic diversity of nuclear and mitochondrial DNA sequences limits the utility of these markers in population genetics studies of *S. variabilis* and *M. oculata*. In the present study, genetic diversities of *ITS* and *D-loop* regions were low, and neither marker provided evidence of population structure in *G. dumosa* or *S. variabilis*, but *ITS* in *M. oculata* revealed a north-central-south regional pattern of differentiation. There were no unique haplotypes in any of the geomorphic populations and only four significant Φ_{ST} values were observed between pairs of populations on geomorphic features (Fig. S1). These findings for the DNA sequences indicate that insufficient variation exists for these markers types to be informative at the scale of our work. In contrast, microsatellite variation provided much greater resolution of genetic structure, connectivity and isolation. Where significant results exist for both marker types for the same species, they are in agreement (e.g., location of putative barriers to gene flow and estimation of exchange of migrants). In the present study, to obtain a comprehensive understanding of the population genetic structure of the three corals, both DNA sequence and microsatellite marker types

are considered, and robust results are highlighted when both marker results are in agreement.

4.2 | Effective population sizes

Estimation of effective population size (*Ne*) depends on a signal that is a function of 1/*Ne*. Such methods are therefore most powerful with small populations when the signal is strong, but have difficulty distinguishing large populations from infinite ones when the signal is small (Waples & Do, 2010). Our estimates of *Ne* were typically in the range of 20–60 individuals, meaning that at all spatial scales all three corals have very small effective contemporary population sizes (Waples & Do, 2010). This may occur because these species are typically late (old) to reach sexual maturity with the result that few adult (mature) colonies exist in an area and therefore few colonies contribute to reproductive success. Alternatively, given the topographic complexity of the seafloor habitat in general, it is possible that certain locations are favoured in terms of reproductive success for corals because of local topography and/or its interaction with local currents. Thus, whichever individual is in possession of a "good" spot may be able to contribute disproportionally more to future generations that any number of other individuals that occupy poorer quality locations. From a management or protection perspective what is of considerable concern here is that all species show such small effective population sizes. This result is consistent with recently published estimates of low Ne values for *S. variabilis* from the Australian EEZ (Miller & Gunasekera, 2017). This implies that any (further) damage to the corals could have a serious negative effect on the regional population.

Historical events will likely have altered the ancestral and contemporary effective sizes of populations (Palstra, O'Connell, & Ruzzante, 2007). Trying to quantify ancestral values of Ne is problematical, in particular for deep-sea taxa when sample sizes are small and spatial coverage is limited in its extent (Baco et al., 2016). The force altering contemporary Ne values is expected to be destruction from fishing activity, given that no other impact is known to have such a potentially broad spatial effect in the region. However, the extent of damage to corals caused by bottom trawling is not quantified, although qualitative estimates indicate that such damage can be extensive, in particular in localized areas around fishing hotspots, such as the seamounts where corals are predominantly found (Althaus et al., 2009; Clark & Rowden, 2009; Parker et al., 2009). Other factors, including environmental factors such as depth-dependent water temperature, pH, calcite and aragonite saturation concentrations, are also likely to contribute both historically and contemporaneously (Henry et al., 2014; Miller, Rowden, Williams, & Häussermann, 2011; Tracey et al., 2013), but are thought to be of lesser importance than fishing pressure in modern times. However, the most recent study of possible impacts of fishing activity on deep-sea corals (Miller & Gunasekera, 2017) reported that genetic diversity of *S. variabilis* and the solitary cup coral *Desmophyllum dianthus* on fished and unfished seamounts was similar. This question clearly requires further investigation.

4.3 | Patterns of genetic connectivity

Populations of *M. oculata* exhibited regional structure, populations of *G. dumosa* were differentiated at the regional and geomorphic features scale, and populations of *S. variabilis* exhibited structure between provinces, amongst regions and also amongst geomorphic features. In other comparable studies, little or no evidence of province-scale structure has been observed amongst populations of crustaceans and sponges in the New Zealand region (Bors, Rowden, Maas, Clark, & Shank, 2012; Zeng, 2016). Thus, we conclude that there is little support for the hypothesis that water mass characteristics associated with biogeographic provinces within the New Zealand region result in genetic population structure of *G. dumosa* and *M. oculata*. However, a north-central-south regional pattern of structure was observed in another deep-sea scleractinian, the cup coral, *D. dianthus* based on *ITS* variation (Miller et al., 2010), and also in the deep-sea sponge *Poecillastra laminaris* based on *COI*, *Cytb* and microsatellite variation (Zeng, 2016). In combination, these findings provide evidence for the hypothesis that currents/fronts associated with the Chatham Rise act as a barrier to gene flow for several species across multiple phyla. Significant genetic differentiation has already been reported at the geomorphic population level in noncoral species between, for example,

the Challenger Plateau and the Chatham Rise (Bors et al., 2012; Knox et al., 2012). These results, combined with those for the deep-sea corals, support the hypothesis that oceanic dynamics contribute to the formation of fine-scale population structure amongst populations on different geomorphic features in the New Zealand region.

4.4 | Oceanographic dynamics

We hypothesized that currents associated with the Subtropical Front would influence north-central-south regional scale population connectivity, which was indeed observed for all three corals. Surface and intermediate currents flow from northwest New Zealand in a southerly direction to the Chatham Rise following the eastern coastline of the North Island. Currents also flow from the southeast of New Zealand northward along the east coast of the South Island towards the Chatham Rise. The result is that currents from the warmer north and the cooler south meet and mix along the Chatham Rise, forming the Subtropical Front, before heading east into the Pacific Ocean (Chiswell, Bostock, Sutton, & Williams, 2015). These two contrasting sets of currents, their associated eddies and the Subtropical Front may therefore impede larval migration between the south and north regions and provide an explanation for the increased genetic diversity values observed amongst populations on the Chatham Rise (central region) for all three corals and also for some sponges (Zeng, 2016).

At the geomorphic features level, our analyses provided specific evidence for a lack of genetic connectivity between populations on the Kermadec Ridge and the Chatham Rise (significant F_{ST} and clear split in DAPC plots). No previous study has investigated connectivity between these two areas, but significant genetic subdivision between populations of *D. dianthus* was detected between the Auckland Island slope (far south) and seamounts on the northern Chatham Rise (Miller et al., 2010). This differentiation is consistent with patterns of flow at depth, with water from the far south flowing north until it reaches the southern edge of the Chatham Rise, before it is deflected eastward (Chiswell et al., 2015). In other words, migrants from the Auckland Island slope are not expected to reach the northern edge of the Chatham Rise, which will result in some degree of genetic differentiation, which is indeed what is observed. For *S. variabilis*, the most apparent difference between populations on geomorphic features was that between the Louisville Seamount Chain and all other populations. The general lack of current flow from the Louisville Seamount Chain towards most topographic features elsewhere in the New Zealand region (Chiswell & Rickard, 2006) is likely to be the reason that populations of *S. variabilis* on this feature are differentiated genetically from those in the rest of the region.

4.5 | Depth and related environmental gradients

Depth is an important factor contributing to patterns of genetic connectivity for some benthic fauna, particularly for taxa that are distributed across a wide depth range (Brandão, Sauer, & Schön, 2010; Miller et al., 2010; O'Hara, England, Gunasekera, & Naughton, 2014; Ruiz-Ramos et al., 2015). In the present study, there was no evidence for isolation by depth (across the depth range 322–1,805 m) for *S. variabilis*. Populations of *S. variabilis* from Australian waters were also not differentiated by depth, but populations were sampled within a more restricted depth range (1,000–1,400 m) (Miller & Gunasekera, 2017). For populations of both *G. dumosa* (from 300 to 600 m) and *M. oculata* (from 700 to 1,200 m), at least one marker revealed the existence of a pattern of isolation by depth. As neither *G. dumosa* nor *M. oculata* populations were genetically isolated by distance, this implies that depth and related environmental gradients may play an important role in the larval migration of these species. Within the New Zealand region, Miller et al. (2011) reported that the stony cup coral *D. dianthus* from different depth strata (<600 m, 1,000–1,500 m, >1,500 m) are strongly differentiated based on DNA sequence variation. Unfortunately, the limited sample sizes of the present study precluded detailed analysis to determine at what depth(s) genetic differentiation for *G. dumosa* and *M. oculata* populations takes place (i.e., the response may just as easily be a gradient as an abrupt break).

4.6 | Life history and dispersal strategies

Differences in reproductive strategies may contribute to differences in patterns of gene flow and genetic structure observed across taxa in the deep sea (Hilário et al., 2015). All three species showed evidence of self-recruitment (i.e., high self assignment rates) at larger spatial scales which doubtless have contributed to the patterns of genetic structure detected in all three species, as well as the limited connectivity that exists amongst regions. Interestingly, the migration estimates amongst populations on different geomorphic features for the three corals may be related to maximum oocyte size. *M. oculata* has the largest mean oocyte size (2~3 times larger than other two species) (Table S9) and was the only species for which significant differentiation amongst populations on geomorphic features was not observed (it also had the lowest assignment success rates, something that is consistent with higher levels of gene flow), whereas *G. dumosa* and *S. variabilis* with a smaller mean oocyte diameter exhibited less connectivity (i.e., significant structure existed amongst populations on geomorphic features and rates of assignment success were lower). Given its similarity to *S. variabilis* in terms of morphology and reproduction, the larvae of *G. dumosa* are likely to share the same characteristics. Based on this assumption, the larger oocyte size, providing more nutritional support, may promote a longer larval life and greater dispersal potential for *M. oculata*. Thus, life-history characteristics may vary even between closely related deep-sea corals (Burgess & Babcock, 2005) and therefore can influence patterns of population connectivity, as recently reported by Miller and Gunasekera (2017) for *S. variabilis* (limited dispersal) and *D. dianthus* (widespread dispersal).

In the present study, none of the three coral species showed strong evidence of asexual reproduction as assessed using multilocus microsatellite genotypes. Whilst there was no direct evidence to indicate asexual reproduction in any of the three species, the smaller effective population size with heterozygote deficits at most loci of *S. variabilis* suggests that this species has a higher asexual or inbreeding rate in populations on geomorphic features than the other two species. Miller

and Gunasekera (2017) predicted that clonal reproduction may account for as much as 76% of recruitment in Australian populations of *S. variabilis*, and if applicable in the New Zealand context, then this high asexual recruitment rate would explain the self-recruitment result in this study. Asexual reproduction has also been observed in *G. dumosa*, but the proportion of clonal reproduction in recruitment is unknown, whilst for *M. oculata*, there is no evidence for asexual reproduction (Burgess & Babcock, 2005). However, it has been reported that mixed clonal/sexual reproduction is nearly indistinguishable from strict sexual reproduction as long as the proportion of clonal reproduction is not large (Balloux, Lehmann, & De Meeûs, 2003). Therefore, asexual reproduction would not be the main explanation for the genetic structure and connectivity patterns of three corals observed in this study.

4.7 | Conservation and management implications

Seventeen seamount closure areas and 17 Benthic Protection Areas (BPAs) were established in 2001 and 2007, respectively, to protect benthic fauna (including VMEs such as coral reefs) from bottom trawl fisheries throughout the New Zealand EEZ (Fig. S4). These closed areas were not designed with input from population connectivity studies beyond the general principle that protected areas should be large and distributed amongst different environments throughout the EEZ. Understanding connectivity amongst areas is fundamental to designing an effective and new protected areas network or for modifying the existing distribution of protected areas to create a network (Hilário et al., 2015; Palumbi, 2003). The maintenance of genetic diversity and the protection of genetically distinct populations should also be considered as an integral component of protected area design (Bors et al., 2012).

The results of the present study for *G. dumosa*, *M. oculata* and *S. variabilis* (as proxies for similar VME indicator species) provide new information useful for assessing the effectiveness of existing protection measures, and the design of future protected area networks in the EEZ. The populations of the three species all showed evidence of pronounced self-recruitment within the scale of the geomorphic features examined. This finding suggests that additional protected areas will be required to maintain genetic diversity (through self-recruitment) of populations at features that do not already receive some protection. For *S. variabilis*, populations on all the geomorphic features examined were to some extent source populations for other populations, but those from the Kermadec Ridge and the Bounty Trough seem to be particularly important. Of particular significance as a migrant source for *G. dumosa* and *S. variabilis* populations was the Kermadec Ridge, with the NE Slope (*G. dumosa*), the Bounty Trough (*S. variabilis*) and Louisville Seamount Chain (*S. variabilis*) populations also being important for genetic connectivity.

Currently, most geomorphic features examined contain one or more BPAs or seamount closure areas. The exceptions are the Hikurangi Margin, Bounty Trough and Macquarie Ridge, as well as the Lord Howe Rise and the Louisville Seamount Chain which are outside the New Zealand EEZ. The Hikurangi Margin is an area subject to bottom trawling (Clark & O'Driscoll, 2003), and populations of VME indicator taxa, such as reef-forming corals, are therefore at risk from disturbance. The

Bounty Trough is a geomorphic feature where deep-water drilling for hydrocarbons may occur in the future (Wood & Davy, 2008), making populations of VME indicator taxa that exist there also at risk from anthropogenic activities. Although there are no BPAs on the Macquarie Ridge, two seamount closures to the west and east of the ridge, and a BPA to the south, may provide some protection for this southern source of genetic connectivity. However, this will depend on whether they provide suitable habitat for the corals. Source populations of corals on the Kermadec Ridge are mainly protected from bottom trawling by the Kermadec and Tectonic Reach BPAs and will likely receive additional protection should a proposed Kermadec Ocean Sanctuary be established in the near future (http://www.mfe.govt.nz/marine/kermadec-ocean-sanctuary). However, outside of the sanctuary area corals and other VME taxa are still vulnerable to disturbance from future seabed mining of sulphide deposits in the region (Boschen et al., 2016). While there are two BPAs on the Chatham Rise, they do not cover the depth range of the two deeper-distributed species, in particular *S. variabilis*, nor do they provide much protection along the axis of the Rise, including some areas where *G. dumosa* forms dense thickets in an area of interest for phosphorite nodule mining. Considering the relative importance of the Chatham Rise as a sink population and as a hot spot area with the highest genetic diversity but relative lower effective population sizes, additional protected areas should be considered here for the future management of human activities on this feature.

The results of the present study, and those of previous genetic connectivity studies for various species around New Zealand (Bors et al., 2012; Boschen, Rowden, Clark, & Gardner, 2015; Dueñas et al., 2016; Knox et al., 2012; Miller et al., 2010) provide genetic connectivity information at different spatial scales. Not surprisingly, their connectivity patterns vary across the range of taxa with differing ecological characteristics. These variable patterns demonstrate the need for a flexible spatial management system that can be periodically adjusted to accommodate increased understanding about the connectivity of a range of deep-sea benthic taxa at a variety of spatial scales.

ACKNOWLEDGEMENTS

This work was funded by the New Zealand Ministry of Business, Innovation and Employment as part of the NIWA-led project "Predicting the occurrence of vulnerable marine ecosystems for planning spatial management in the South Pacific region" (CO1X1229). The research was also benefitted from additional funding provided by NIWA under the Marine Biological Resources programme (Objective 1: Discovery and definition of the marine biota of New Zealand). The payment of open access for this manuscript was supported by Startup Foundation for Advanced Scholars, Hunan Agricultural University. We are grateful to Sadie Mills and Kareen Schnabel (NIWA Invertebrate Collection) for their diligent assistance with sample location and processing. We also thank Di Tracey (NIWA) for specimen identification, Karen Miller (AIMS, Australia) for the use of unpublished microsatellite primers for *Solenosmila variabilis* and Sophie Arnoud-Haond (IFREMER, France) for the use of unpublished microsatellite primers for *Madrepora oculata*. Further acknowledgements for the funding of specific sample collection voyages are provided in the Supplementary Materials section.

REFERENCES

Addamo, A. M., Reimer, J. D., Taviani, M., Freiwald, A., & Machordom, A. (2012). *Desmophyllum dianthus* (Esper, 1794) in the Scleractinian phylogeny and its intraspecific diversity. *PLoS ONE, 7*(11), e50215.

Althaus, F., Williams, A., Schlacher, T. A., Kloser, R. J., Green, M. A., Barker, B. A., ... Schlacher-Hoenlinger, M. A. (2009). Impacts of bottom trawling on deep-coral ecosystems of seamounts are long-lasting. *Marine Ecology Progress Series, 397,* 279–294.

Anderson, O. F., & Clark, M. R. (2003). Analysis of bycatch in the fishery for orange roughy, *Hoplostethus atlanticus*, on the South Tasman Rise. *Marine and Freshwater Research, 54*(5), 643–652.

Antao, T., Lopes, A., Lopes, R. J., Beja-Pereira, A., & Luikart, G. (2008). LOSITAN: A workbench to detect molecular adaptation based on a F_{ST}-outlier method. *BMC Bioinformatics, 9*(1), 1–5.

Baco, A. R., Clark, A. M., & Shank, T. M. (2006). Six microsatellite loci from the deep-sea coral *Corallium lauuense* (Octocorallia: Coralliidae) from the islands and seamounts of the Hawaiian archipelago. *Molecular Ecology Notes, 6,* 147–149.

Baco, A. R., Etter, R. J., Ribeiro, P. A., von der Heyden, S., Beerli, P., & Kinlan, B. P. (2016). A Synthesis of genetic connectivity in deep-sea fauna and implications for marine reserve design. *Molecular Ecology, 25*(4), 3276–3298.

Balloux, F., Lehmann, L., & De Meeûs, T. (2003). The population genetics of clonal and partially clonal diploids. *Genetics, 164*(4), 1635–1644.

Becheler, R., Cassone, A. L., Noël, P., Mouchel, O., Morrison, C. L., & Arnaud-Haond, S. (2015). Low incidence of clonality in cold water corals revealed through the novel use of a standardized protocol adapted to deep sea sampling. *Deep Sea Research Part II: Topical Studies in Oceanography*. https://doi.org/10.1016/j.dsr2.2015.11.013

Berry, O., Tocher, M. D., & Sarre, S. D. (2004). Can assignment tests measure dispersal? *Molecular Ecology, 13,* 551–561.

Bongiorni, L., Mea, M., Gambi, C., Pusceddu, A., Taviani, M., & Danovaro, R. (2010). Deep-water scleractinian corals promote higher biodiversity in deep-sea meiofaunal assemblages along continental margins. *Biological Conservation, 143*(7), 1687–1700.

Bors, E. K., Rowden, A. A., Maas, E. W., Clark, M. R., & Shank, T. M. (2012). Patterns of deep-sea genetic connectivity in the New Zealand region: Implications for management of benthic ecosystems. *PLoS ONE, 7*(11), e49474.

Boschen, R. E., Collins, P. C., Tunnicliffe, V., Carlsson, J., Gardner, J. P. A., Lowe, J., ... Swaddling, A. (2016). A primer for use of genetic tools in selecting and testing the suitability of set-aside sites protected from deep-sea seafloor massive sulfide mining activities. *Ocean and Coastal Management, 122,* 37–48.

Boschen, R. E., Rowden, A. A., Clark, M. R., & Gardner, J. P. A. (2015). Limitations in the use of archived vent mussel samples to assess genetic connectivity among seafloor massive sulfide deposits: A case study with implications for environmental management. *Frontiers in Marine Science, 2,* 1–14.

Brandão, S. N., Sauer, J., & Schön, I. (2010). Circumantarctic distribution in Southern Ocean benthos? A genetic test using the genus *Macroscapha* (Crustacea, Ostracoda) as a model. *Molecular Phylogenetics and Evolution, 55*(3), 1055–1069.

Burgess, S., & Babcock, R. (2005). Reproductive ecology of three reef-forming, deep-sea corals in the New Zealand region. In A. Freiwald, & J. M. Roberts (Eds.), *Cold-water corals and ecosystems* (pp. 701–713). Berlin: Springer, Berlin Heidelberg.

Chiswell, S. M., Bostock, H. C., Sutton, P. J., & Williams, M. J. (2015). Physical oceanography of the deep seas around New Zealand: A review. *New Zealand Journal of Marine and Freshwater Research, 49,* 1–32.

Chiswell, S. M., & Rickard, G. J. (2006). Comparison of model and observational ocean circulation climatologies for the New Zealand region. *Journal of Geophysical Research: Oceans, 111*(C10), 264–270.

Clark, M. R., & O'Driscoll, R. (2003). Deepwater fisheries and aspects of their impact on seamount habitat in New Zealand. *Journal of Northwest Atlantic Fishery Science, 31*, 441–458.

Clark, M. R., & Rowden, A. A. (2009). Effect of deepwater trawling on the macro-invertebrate assemblages of seamounts on the Chatham Rise, New Zealand. *Deep-Sea Research Part I: Oceanographic Research Papers, 56*, 1540–1554.

Cornuet, J., Piry, S., Luikart, G., Estoup, A., & Solignac, M. (1999). New methods employing multilocus genotypes to select or exclude populations as origins of individuals. *Genetics, 153*, 1989–2000.

Do, C., Waples, R. S., Peel, D., Macbeth, G. M., & Tillett, B. J. (2014). NeEstimator v2: re-implementation of software for the estimation of contemporary effective population size (Ne) from genetic data. *Ular Ecology Resources, 14*(1), 209.

Dueñas, L. F., Tracey, D. M., Crawford, A. J., Wilke, T., Alderslade, P., & Sánchez, J. A. (2016). The Antarctic Circumpolar Current as a diversification trigger for deep-sea octocorals. *BMC Evolutionary Biology, 16*(1), 2.

Earl, D. A., & von Holdt, B. M. (2012). Structure Harvester: A website and program for visualizing STRUCTURE output and implementing the Evanno method. *Conservation Genetics Resources, 4*(2), 359–361.

Excoffier, L., & Lischer, H. E. L. (2010). Arlequin suite ver 3.5: A new series of programs to perform population genetics analyses under Linux and Windows. *Molecular Ecology Resources, 10*(3), 564–567.

Falush, D., Stephens, M., & Pritchard, J. K. (2007). Inference of population structure using multilocus genotype data: Dominant markers and null alleles. *Molecular Ecology Notes, 7*(4), 574–578.

FAO. (2009). International guidelines for the management of deep-sea fisheries in the high seas. Roma, (Italia). Retrieved from http://www.fao.org/fishery/topic/166308/en

Gagnaire, P. A., Broquet, T., Aurelle, D., Viard, F., Souissi, A., Bonhomme, F., ... Bierne, N. (2015). Using neutral, selected, and hitchhiker loci to assess connectivity of marine populations in the genomic era. *Evolutionary Applications, 8*(8), 769–786.

Henry, L. A., Vad, J., Findlay, H. S., Murillo, J., Milligan, R., & Roberts, J. M. (2014). Environmental variability and biodiversity of megabenthos on the Hebrides Terrace Seamount (Northeast Atlantic). *Scientific Reports, 4*, 5589.

Hilário, A., Metaxas, A., Gaudron, S. M., Howell, K. L., Mercier, A., Mestre, N. C., ... Young, C. (2015). Estimating dispersal distance in the deep sea: Challenges and applications to marine reserves. *Frontiers in Marine Science, 2*, 1–14.

Jombart, T. (2008). Adegenet: A R package for the multivariate analysis of genetic markers. *Bioinformatics, 24*(11), 1403–1405.

Jombart, T., Devillard, S., & Balloux, F. (2010). Discriminant analysis of principal components: A new method for the analysis of genetically structured populations. *BMC Genetics, 11*(1), 94.

Knox, M. A., Hogg, I. D., Pilditch, C. A., Lörz, A. N., Hebert, P. D. N., & Steinke, D. (2012). Mitochondrial DNA (*COI*) analyses reveal that amphipod diversity is associated with environmental heterogeneity in deep-sea habitats. *Molecular Ecology, 21*(19), 4885–4897.

Kudrass, H. R., & Rad, U. Von (1984). Underwater television and photography observations, side-scan sonar and acoustic reflectivity measurements of phosphorite-rich areas on the Chatham Rise (New Zealand). *Geology Journal, 65*, 69–89.

Le Goff, M. C., & Rogers, A. D. (2002). Characterization of 10 microsatellite loci for the deep-sea coral *Lophelia pertusa* (Linnaeus 1758). *Molecular Ecology Notes, 2*, 164–166.

Lunden, J. J., McNicholl, C. G., Sears, C. R., Morrison, C. L., & Cordes, E. (2014). Acute survivorship of the deep-sea coral *Lophelia pertusa* from the Gulf of Mexico under acidification, warming, and deoxygenation. *Frontiers in Marine Science, 1*, 74.

Manni, F., Guerard, E., & Heyer, E. (2004). Geographic patterns of (genetic, morphologic, linguistic) variation: How barriers can be detected by using Monmonier's algorithm. *Human Biology, 76*(2), 173–190.

Mantel, N. (1967). The detection of disease clustering and a generalized regression approach. *Cancer Research, 27*(1), 209–220.

Miller, K. J., & Gunasekera, R. M. (2017). A comparison of genetic connectivity in two deep sea corals to examine whether seamounts are isolated islands or stepping stones for dispersal. *Scientific Reports, 7*, 46103.

Miller, K. J., Rowden, A. A., Williams, A., & Häussermann, V. (2011). Out of their depth? Isolated deep populations of the cosmopolitan coral *Desmophyllum dianthus* may be highly vulnerable to environmental change. *PLoS ONE, 6*(5), e19004.

Miller, K. J., Williams, A., Rowden, A. A., Knowles, C., & Dunshea, G. (2010). Conflicting estimates of connectivity among deep-sea coral populations. *Marine Ecology, 31*, 144–157.

O'Hara, T. D., England, P. R., Gunasekera, R. M., & Naughton, K. M. (2014). Limited phylogeographic structure for five bathyal ophiuroids at continental scales. *Deep-Sea Research Part I: Oceanographic Research Papers, 84*, 18–28.

van Oosterhout, C., Hutchinson, W. F., Wills, D. P. M., & Shipley, P. (2004). MICRO-CHECKER: Software for identifying and correcting genotyping errors in microsatellite data. *Molecular Ecology Notes, 4*(3), 535–538.

Palstra, F. P., O'Connell, M. F., & Ruzzante, D. E. (2007). Population structure and gene flow reversals in Atlantic salmon (*Salmo salar*) over contemporary and long-term temporal scales: Effects of population size and life history. *Molecular Ecology, 16*(21), 4504–4522.

Palumbi, S. R. (2003). Population genetics, demographic connectivity, and the design of marine reserves. *Ecological Applications, 13*, 146–158.

Parker, S. J., Penney, A. J., & Clark, M. R. (2009). Detection criteria for managing trawl impacts on vulnerable marine ecosystems in high seas fisheries of the South Pacific Ocean. *Marine Ecology Progress Series, 397*, 309–317.

Peakall, R., & Smouse, P. E. (2012). GenALEx 6.5: Genetic analysis in Excel. Population genetic software for teaching and research-an update. *Bioinformatics, 28*(19), 2537–2539.

Penny, A., Parker, S. J., & Brown, J. H. (2009). Protection measures implemented by New Zealand for vulnerable marine ecosystems in the South Pacific Ocean. *Marine Ecology Progress Series, 397*, 341–354.

Piry, S., Alapetite, A., Cornuet, J. M., Paetkau, D., Baudouin, L., & Estoup, A. (2004). GeneClass2: A software for genetic assignment and first generation migrant detection. *Journal of Heredity, 95*, 536–539.

Puechmaille, S. J. (2016). The program structure does not reliably recover the correct population structure when sampling is uneven: Subsampling and new estimators alleviate the problem. *Molecular Ecology Resources, 16*, 608–627.

Putman, A. I., & Carbone, I. (2014). Challenges in analysis and interpretation of microsatellite data for population genetic studies. *Ecology and Evolution, 4*, 4399–4428.

Quattrini, A. M., Baums, I. B., Shank, T. M., Morrison, C. L., & Cordes, E. E. (2015). Testing the depth-differentiation hypothesis in a deepwater octocoral. *Proceedings Biological Sciences / The Royal Society, 282*(1807), 20150008.

Ramirez-Llodra, E., Tyler, P. A., Baker, M. C., Bergstad, O. A., Clark, M. R., Escobar, E., ... van Dover, C. L. (2011). Man and the last great wilderness: Human impact on the deep sea. *PLoS ONE, 6*(8), e22588.

Ruiz-Ramos, D. V., Saunders, M., Fisher, C. R., & Baums, I. B. (2015). Home bodies and wanderers: Sympatric lineages of the deep-sea black coral *Leiopathes glaberrima*. *PLoS ONE, 10*(10), e0138989.

Tracey, D., Bostock, H., Currie, K., Mikaloff-Fletcher, S., Williams, M., Hadfield, M., ... Cummings, V. (2013). The potential impact of ocean acidification on deep-sea corals and fisheries habitat in New Zealand waters. *New Zealand Aquatic Environment and Biodiversity Report, 117*, 105.

Tracey, D. M., Rowden, A. A., Mackay, K. A., & Compton, T. (2011). Habitat-forming cold-water corals show affinity for seamounts in the New

Zealand region. *Marine Ecology Progress*, *430*(8), 1–22.

Wang, J. (2017). The computer programme STRUCTURE for assigning individuals to populations: easy to use but easier to misuse. Molecular Ecology, https://doi.10.1111/1755-0998.12650

Waples, R. S., & Do, C. (2010). Linkage disequilibrium estimates of contemporary *Ne* using highly variable genetic markers: A largely untapped resource for applied conservation and evolution. *Evolutionary Applications*, *3*(3), 244–262.

Watling, L., Guinotte, J., Clark, M. R., & Smith, C. R. (2013). A proposed biogeography of the deep ocean floor. *Progress in Oceanography*, *111*, 91–112.

Wei, K. J., Wood, A. R., & Gardner, J. P. A. (2013). Population genetic variation in the New Zealand greenshell mussel: Locus-dependent conflicting signals of weak structure and high gene flow balanced against pronounced structure and high self-recruitment. *Marine Biology*, *160*, 931–949.

Wood, R., & Davy, B. (2008). New Zealand's UNCLOS Project - Defining the continental margin: A summary and way forward. *New Zealand Science Review*, *64*, 3–4.

Zeng, C. (2016). Patterns of genetic connectivity in deep-sea vulnerable marine ecosystems and implications for conservation. Unpublished doctoral dissertation thesis, Victoria University of Wellington, New Zealand. http://hdl.handle.net/10063/5551

13

Geographic extent of introgression in *Sebastes mentella* and its effect on genetic population structure

Atal Saha[1] (iD) | Torild Johansen[1] | Rasmus Hedeholm[2] | Einar E. Nielsen[3] | Jon-Ivar Westgaard[1] | Lorenz Hauser[4] | Benjamin Planque[1,5] | Steven X. Cadrin[6] | Jesper Boje[2,3]

[1]Tromsø Department, Institute of Marine Research, Tromsø, Norway

[2]Greenland Institute of Natural Resources, Nuuk, Greenland

[3]DTU Aqua – National Institute of Aquatic Resources, Charlottenlund, Denmark

[4]School of Aquatic and Fishery Sciences, University of Washington, Seattle, WA, USA

[5]Hjort Centre for Marine Ecosystem Dynamics, Bergen, Norway

[6]School for Marine Science and Technology, University of Massachusetts Darmouth, Fairhaven, MA, USA

Correspondence
Atal Saha and Torild Johansen, Tromsø Department, Institute of Marine Research, Tromsø, Norway.
Emails: atal.saha@imr.no; torildj@imr.no

Funding information
Greenland Institute of Natural Resources, Norwegian Institute of Marine Research, Research Council of Norway, Grant/Award Number: NFR-196691.

Abstract

Genetic population structure is often used to identify management units in exploited species, but the extent of genetic differentiation may be inflated by geographic variation in the level of hybridization between species. We identify the genetic population structure of *Sebastes mentella* and investigate possible introgression within the genus by analyzing 13 microsatellites in 2,562 redfish specimens sampled throughout the North Atlantic. The data support an historical divergence between the "shallow" and "deep" groups, beyond the Irminger Sea where they were described previously. A third group, "slope," has an extended distribution on the East Greenland Shelf, in addition to earlier findings on the Icelandic slope. Furthermore, *S. mentella* from the Northeast Arctic and Northwest Atlantic waters are genetically different populations. In both areas, interspecific introgression may influence allele frequency differences among populations. Evidence of introgression was found for almost all the identified *Sebastes* gene pools, but to a much lower extent than suggested earlier. Greenland waters appear to be a sympatric zone for many of the genetically independent *Sebastes* groups. This study illustrates that the identified groups maintain their genetic integrity in this region despite introgression.

KEYWORDS
gene flow, hybrid zone, incipient speciation, oceanic, redfish

1 | INTRODUCTION

Identification of genetic heterogeneity and its application to define fishery management units are important for the sustainable utilization of living marine resources (Shaklee & Bentzen, 1998). Significant population genetic structure caused by a diverse array of factors has been described for many marine species (Gagnaire et al., 2015; Hauser & Carvalho, 2008; Salmenkova, 2011) despite the apparent lack of physical barriers to migration in the marine environment. Nevertheless, marine species usually display low genetic differentiation, indicating that some gene flow exists between apparently isolated groups. On the other hand, distinct species or populations may mate in a particular marine habitat while maintaining reproductive barrier in surrounding regions and thereby form so-called hybrid zone (e.g., Nielsen, Hansen, Ruzzante, Meldrup, & Gronkjaer, 2003; Roques, Sevigny, & Bernatchez, 2001). Introgressive hybridization (i.e., introgression), where hybrids back-cross with one of their parental genotypes (Baskett & Gomulkiewicz, 2011), may also influence allele frequency differences between populations causing intraspecific diversification (Artamonova et al., 2013; Roques et al., 2001). The effects of introgression on the genetic population structure of closely related marine fish species remains largely unexplored.

In the North Atlantic, redfishes (genus: *Sebastes*) are represented by four species: *Sebastes mentella* Travin 1951 (beaked redfish), *Sebastes norvegicus* Ascanius 1772 (golden redfish, previously called *S. marinus*), *Sebastes fasciatus* Storer 1854 (acadian redfish) and *Sebastes viviparus* Krøyer 1845 (Norwegian redfish). *Sebastes mentella* is the most economically important species of the genus. It displays a high degree of genetic population diversity across its distributional range. In the Irminger Sea, two distinct groups, "shallow (=shallow pelagic)" (50–550 m) and "deep (=deep pelagic)" (550–800 m), have been reported (Cadrin et al., 2010, 2011; Pampoulie & Daníelsdóttir, 2008; Stefansson et al., 2009). These two groups have been discriminated on the basis of morphological characteristics; the "deep" group has brighter red color, stouter appearance and larger size at sexual maturity (Magnusson & Magnusson, 1995). They have also been reported to show different rates of parasite infestation (Magnusson & Magnusson, 1995). Microsatellite DNA and morphological analyses suggested that these are two incipient species maintained by an ecological isolation mechanism, although evidence for hybridization was found (Stefánsson et al., 2009). The historical divergence between the "shallow" and "deep" groups was further supported by DNA analysis of the mitochondrial control region, microsatellites and the gene coding for the visual-pigment rhodopsin (Shum, Pampoulie, Kristinsson, & Mariani, 2015; Shum, Pampoulie, Sacchi, & Mariani, 2014). Around Iceland, an additional "slope" component of *S. mentella* has been described (e.g., Cadrin et al., 2010). However, the genetic connectivity among these groups and their geographical distribution in the North Atlantic is not well understood.

Sebastes species coexist in different marine areas and overlap in their depth range, normally between 100 and 400 m (Barsukov, Litvinenko, & Serebryakov, 1984). For instance, *S. mentella* co-occurs with *S. norvegicus* and *S. fasciatus* in the Northwest Atlantic (Roques et al., 2001), *S. norvegicus* in Greenland waters (ICES 2001), and *S. norvegicus* and *S. viviparus* in Icelandic and Norwegian waters (ICES 2001). Studies across the North Atlantic have indicated hybridization of *S. mentella* with *S. norvegicus* (Pampoulie & Daníelsdóttir, 2008), *S. viviparus* (Artamonova et al., 2013) and *S. fasciatus* (Roques et al., 2001). However, except for the evidence of hybridization in the Northwest Atlantic (Roques et al., 2001), the geographic extent of hybridization has not been well studied, which partly relates to the uncertainty about *Sebastes* mating grounds. The waters off East Greenland are assumed to be the main nursery area for *S. mentella* and *S. norvegicus* juveniles, which are believed to be extruded along the Reykjanes Ridge (Anderson, 1984; Magnusson & Johannesson, 1995). Morphological species identification remains uncertain (particularly for fish <25 cm) in these geographical regions, possibly because of unrecognized hybridization (Johansen, 2003). Hybridization has also been suspected to affect estimates of genetic population structure within species (Artamonova et al., 2013). Specifically, introgression with *S. viviparus* has been indicated as one of the main reasons for the apparent differentiation between the "deep" and "shallow" groups of *S. mentella* (Artamonova et al., 2013). The role of hybridization on the genetic population structure is therefore still uncertain.

Despite the importance of Greenland waters as habitat and fisheries area for *Sebastes* species, the genetic population structure of *S. mentella* in this region has not yet been described in sufficient detail for fisheries management. For instance, the "slope" group in the East Greenland and Iceland waters are assessed separately by the International Council for the Exploration of the Sea (ICES 2015a) although it is known that the stock identity is uncertain. ICES perceives this as an interim status until the stock structure of *S. mentella* on the East Greenland slope is better understood. The objectives of this study were: I) to examine the genetic structure of *S. mentella* in Greenland, Iceland, and Irminger Sea waters, II) to investigate the extent of introgression among co-occurring *Sebastes* species/gene pools, and III) to assess whether introgression may influence apparent species and population structure in the region.

2 | MATERIALS AND METHODS

2.1 | Sampling

In total, 35 samples (collections of fish) consisting of 2,562 redfish specimens were included in the study (Table 1). Nineteen *S. mentella* samples were collected during 2011 and 2012 from Greenland waters by commercial fishing vessels and research surveys. All the fish were caught by trawl. Samples were from different seasons (spring and fall) and life stages (juveniles and adults). Norwegian samples were collected from the shelf and pelagic waters in different seasons of the years (Table 1). *Sebastes mentella* reference samples were included from Canadian, Icelandic (representing the Icelandic shelf component) waters, and Irminger Sea from the EU REDFISH project (1998–2001). The Irminger Sea reference samples were characterized as "deep" and "shallow" *S. mentella* based on both morphological characters and sampling depths as described by Magnusson and Magnusson (1995). We included reference samples of *S. norvegicus*, *S. viviparus*, and *S. fasciatus* from Greenland, Iceland, Norwegian, and Canadian waters (Table 1, Figure 1) to study introgression with *S. mentella*. Species identification was conducted on board the ships based on morphological characters (e.g., body size, beak size, eye diameter, direction of spines in the pre-operculum) as suggested by Barsukov et al. (1984). The adults and juveniles were categorized by length (Barsukov et al., 1984) [for instance, in *S. mentella*, adults ≥29 cm and juveniles = (4–28) cm].

2.2 | Microsatellite genotyping

DNA was isolated from ethanol-preserved gill tissue using the E-Z 96 Tissue DNA kit according to the manufacturer's instructions (Omega Bio-Tek, Inc, Norcross, GA, USA). We analyzed a total of 13 microsatellite loci: Seb09, Seb25, Seb31, Seb33, Seb45, Smen05 (Roques, Pallotta, Sevigny, & Bernatchez, 1999), Sal1, Sal3, Sal4 (Miller, Schulze, & Withler, 2000), Smen10 (Stefánsson et al., 2009), and Spi4, Spi6, Spi10 (Gomez-Uchida, Hoffman, Ardren, & Banks, 2003), arranged in three multiplexes (Table S1). PCR was performed in 2 μl volume comprising 1× Qiagen Multiplex Master Mix (Qiagen, Hilden,

TABLE 1 Details of the different *Sebastes* spp. samples analyzed. The sex ratio is given as % female and life stages as % adult (see text). Juvenile *Sebastes* (Seb) were collected from both East and West Greenland. The *Sebastes mentella* samples (M) from Greenland waters were collected both on surveys (R) and by commercial (C) vessels over 2 years. Q refers to sampling zones around Greenland (cf. Figure 1). ID = sample ID, N = sample sizes, and NA = data not available. Samples 5, 6, and 7 represent reference samples of *S. mentella* included from the EU REDFISH project

Species	ID	Code	Location	Lat/Long (mean)	Time	N	Depth (m)	Avg. length (cm)	Female (%)	Adult (%)
S. mentella	1	M-Nor 1	Northeast Arctic	72.18/10.25	October, 2006	91	340	34	56	100
	2	M-Nor 2		66.93/8.13	2006, 2009[a]	155	445	34	59	100
	3	M-Nor 3		67/8.09	March, 2009	76	508	38	83	100
	4	M-Nor 4		69.38/15.14	November, 2011	91	575	38	60	100
	5	M-Oc	Irminger Sea	60.41/−39.01	1995, 2001[a]	80	240	37	36	100
	6	M-Deep		62.55/−27.01	2001	73	845	43	35	100
	7	M-ICL	Icelandic Shelf	63.28/−26.17	October, 2001	59	651	37	50	92
	8	MC11Q2Q3	East Greenland	64.24/−35.14	March, 2011	137	372	36	56	64
	9	MC11Q2		64.25/−35.16	March, 2011	108	367	37	58	100
	10	MR11Q1Q2		65.95/−33.16	August, 2011	80	348	31	48	31
	11	MR11Q3		63.97/−36.3	August, 2011	69	462	32	38	71
	12	MR11Q5		62.2/−40.65	August, 2011	49	430	31	45	61
	13	MR11Q5Q6		61.15/−41.66	August, 2011	48	452	30	42	56
	14	MU11		NA	November, 2011	26	NA	36	64	76
	15	M1C12Q2		64.57/−35.08	May, 2012	49	375	34	45	99
	16	M2C12Q2		64.52/−35.18	February, 2012	83	375	37	NA	100
	17	M3C12Q2		64.40/−35.23	April, 2012	96	375	35	34	100
	18	MC12Q3		64.33/−35.33	February, 2012	45	375	37	NA	100
	19	MSeb12WGL	West Greenland	68.39/−58.36	June, 2012	91	325	13	NA	NA
	20	MR12Q2_6	East Greenland	59.60/−43.85	June, 2012	93	227	32	48	28
	21	MSeb12Q2		65.46/−30.39	August, 2012	92	561	12	NA	NA
	22	MR12Q6		61.04/−41.65	August, 2012	28	437	33	NA	98
S. mentella	23	MR12WGL	West Greenland	68.09/−56.79	June, 2012	89	478	25	NA	33
	24	MR12Q3	East Greenland	64.28/−37.39	August, 2012	89	580	36	47	69
	25	MR12Q4		64.15/−36.81	August, 2012	81	471	29	60	84
	26	MR12Q5		62.19/−40.67	August, 2012	80	448	31	50	89
	27	M-FC	Flemish Cap	48/−45.16	July, 2001	95	495	28	55	35

(Continues)

TABLE 1 (Continued)

Species	ID	Code	Lat/Long (mean)	Location	Time	N	Depth (m)	Avg. length (cm)	Female (%)	Adult (%)
Sebastes norvegicus	28	Nor-Nor	69.19/15.08	Northeast Arctic	October, 2001	41	258	39	15	100
	29	Nor-GL A	64.26/-35.15	East Greenland	February, 2011	108	365	38	58	87
	30	Nor-GL B	62.2/-40.65	West Greenland	August, 2011	70	239	28	44	29
	31	Nor-WGL	69.28/-53.1	West Greenland	June, 2012	49	521	53	65	100
	32	Nor-Giant	61/-29	East Greenland	August, 1996	17	704	79	82	100
Sebastes viviparus	33	VV-Ice	64.1/-13.47	Iceland	March, 2001	53	174	18	47	45
	34	VV-Nor	70.5/20.52	Northeast Arctic	1992, 2001[a]	26	136	20	NA	100
Sebastes fasciatus	35	Fasc	45.79/-47.16	Flemish Cap	October, 2001	45	294	21	62	36

[a]Samples collected in 2 years. No temporal genetic differences between samples observed.

Germany), 0.1–1.0 μM primer, and 15–25 ng DNA. The 5′ end on the forward primer was labeled with a fluorescent dye by the manufacturer (Applied Biosystems, Foster City, CA, USA). The GeneAmp 9700 (Applied Biosystems) thermal cycler was used for the amplification, with a PCR profile consisting of an initial denaturation step of 95°C for 15 min followed by 25 cycles of 95°C for 30 s, 56°C for 90 s, and 72°C for 60 s, ending with 60°C for 45 min. The PCR products were size-separated and genotyped using an ABI 3130 XL automated sequencer (Applied Biosystems) and the GeneMapper 4.0 software (Applied Biosystems). Micro-Checker (Van Oosterhout, Hutchinson, Wills, & Shipley, 2004) was used to detect possible scoring errors or null alleles in the data. Three of the loci (Spi4, Spi6, and Seb31) were not successfully genotyped for some of the samples of S. norvegicus and S. viviparus, likely due to poor DNA quality. Eventually those loci were omitted from the analyses of introgression.

2.3 | Descriptive statistics

We used FSTAT(Goudet, 1995) to calculate the number of alleles and gene diversities for different loci. Allelic richness was estimated for a minimum sample size of 17 individuals in ADZE 1.0 (Szpiech, Jakobsson, & Rosenberg, 2008). Pairwise F_{ST} values (Weir & Cockerham, 1984) were estimated in Arlequin 3.5 (Excoffier & Lischer, 2010). Deviations from linkage (LD) and Hardy–Weinberg equilibria were tested in Genepop 4.2 (Rousset, 2008) by Fisher's exact test with the implemented Markov chain Monte Carlo (MCMC) method (dememorization 10,000, batches 1,000, iterations 10,000). Hierarchical analysis of molecular variance (AMOVA) was performed in Arlequin for the S. mentella data with three-group configuration after Bayesian clustering (as described below). False discovery rate control (FDR, Benjamini & Yekutieli, 2001) was applied to avoid type I error, while not losing much power, when multiple comparisons were involved. Genetic diversities, deviations from Hardy–Weinberg and linkage equilibria were also estimated within detected clusters (as described below) to verify pure and admixed fish. For these estimations, comparable sample sizes of fish were selected randomly to represent the pure and admixed clusters.

2.4 | Population cluster and individual admixture analyses

Population cluster analysis was applied for the 27 samples of S. mentella (cf. Table 1; ID: 1–27, N = 2,153) genotyped with 13 microsatellites to detect genetic structuring within the species. To estimate the magnitude of introgression, individual admixture analysis was performed for the total dataset consisting of all 35 samples (N = 2,562) genotyped for ten loci (see rationale above). A Bayesian approach, as implemented in STRUCTURE (Pritchard, Stephens, & Donnelly, 2000), was used for clustering of genotypes and estimation of individual admixture proportion in an effort to identify possible hybrids (burn-in = 350,000, MCMC = 500,000, replication = 10 for each K). The method clusters individuals to minimize Hardy–Weinberg and gametic phase disequilibria between loci within groups. STRUCTURE

(a)

(b)

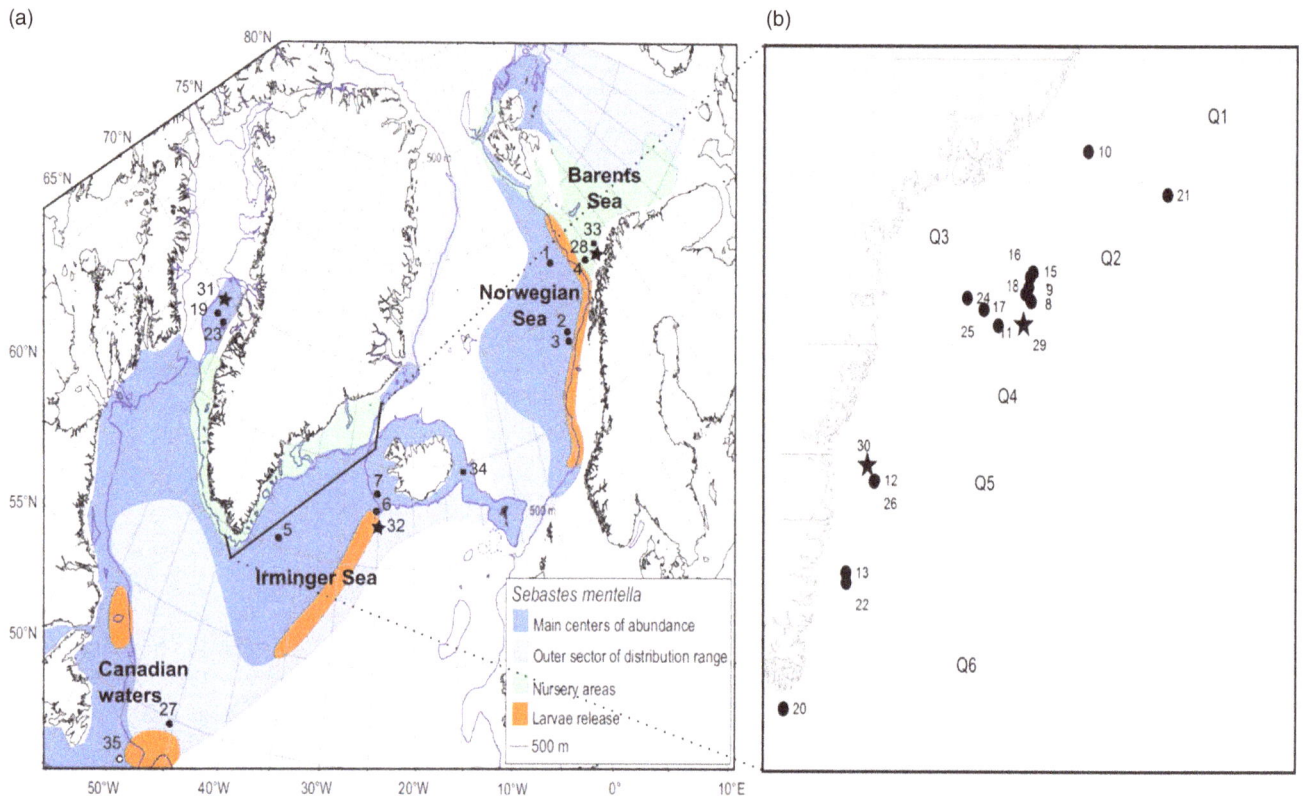

FIGURE 1 (a) Sampling locations of *Sebastes* spp. across the North Atlantic. For *Sebastes mentella*, the species distribution range, and larval release and nursery areas are presented (source: Planque et al. 2013). *S. mentella, Sebastes norvegicus, Sebastes viviparus,* and *Sebastes fasciatus* samples are indicated by numbers 1–27(●), 28–32 (★), 33–34 (■), and 35(○), respectively. (b) Samples from East Greenland waters are illustrated. Samples 5, 6, and 7 represent reference samples of *S. mentella* included from the EU REDFISH project (see Table 1)

may predict fewer than the actual number of clusters in a dataset with hierarchical structure (Kalinowski, 2010), so we ran the program using all samples followed by a cluster by cluster (hierarchical) approach. The Evanno method (Evanno, Regnaut, & Goudet, 2005) as implemented in STRUCTURE HARVESTER (Earl & vonHoldt, 2012) was used to estimate the number of clusters in each dataset.

The incidence of hybridization was further quantified by additional STRUCTURE analyses in clusters that indicated hybridization. To estimate the individual admixture proportions (Q = genome ancestry fraction) in STRUCTURE, an admixture model was used. Furthermore, a correlated allele frequency model was applied as the differentiation among the clusters was low (as suggested by Nielsen et al., 2003).

2.5 | Identification of hybrids

Hybridization of *Sebastes* in waters close to Greenland (Greenland, Irminger Sea, and Iceland) was compared with hybridization in the Northeast Arctic (Norwegian), and Northwest Atlantic waters (Flemish cap) following the approach of Nielsen et al. (2003). An individual was initially identified as possible hybrid if at least 10% of its genome (Q) originated from other groups (see Randi, 2008). The most pure parental individuals ($Q \geq 0.90$) were used as base population for generating *in silico* "pure" individuals using the program HYBRIDLAB (Nielsen, Bach, & Kotlicki, 2006). Simple mechanical mixing was

simulated by generating pure parental genotypes (i.e., randomly drawing alleles from the allele frequency distribution of the "pure population") equal to the numbers of individuals of the estimated clusters. A hybrid swarm (i.e., random mating between species/populations) was generated by random drawing of alleles from the observed allele frequency distribution of the combined species/populations. Differences between these simulated and empirical individual admixture proportions were tested with a Kolmogorov–Smirnov two-sample (K-S) test. Potential simulation bias caused by differences in individual admixture proportions between the observed and simulated pure individuals was tested by comparing individual admixture proportions of 20 randomly chosen "pure" individuals of each cluster (i.e., total 40) with 40 "simulated pure" individuals. As *S. norvegicus* could not be sampled from the Northwest Atlantic (Flemish Cap), Norwegian *S. norvegicus* was used as baseline sample for identification of hybrids. Hence, hybrids were investigated only in *S. mentella* and *S. fasciatus* samples for the Northwest Atlantic. The same approach was used for identification of hybrids with *S. viviparus* in Greenland waters as the species has not been identified west of Iceland (Johansen, 2003).

2.6 | Isolation with migration

The isolation-with-migration (IM) model as implemented in IMa2 (Hey, 2009) was applied to estimate introgression and population demographic parameters for the cluster pairs from Greenland waters. The analyses were conducted only for clusters in which hybridization had been indicated through earlier Bayesian clustering analyses (i.e., within *S. mentella* clusters, and across *S. mentella* and *S. norvegicus* clusters). A total of 45 fish within each cluster were randomly selected from the relevant samples. Rather than running all of the clusters together, pairwise cluster comparisons were chosen to reduce computational time. The IM model assumes random mating, free recombination among loci, and no recombination within loci. The MCMC-based method of the program uses sampling of gene genealogies to estimate posterior probability densities of demographic parameters scaled by mutation rate per generation per year (μ). A stepwise mutation (SM) model was applied for the ten microsatellite loci.

A few preliminary analyses were performed with wider priors. The estimated effective sample size values of parameter t (time since divergence), swapping rates, autocorrelation value, trend-line plots were evaluated. After a burn-in of 500,000 steps, 2,000,000 more steps were run to save 20,000 genealogies for each pair of clusters. To ensure convergence for the estimation, 150 chains were used for all of runs with high heating schemes ("−ha0.975 −hb0.75") so that the updates for the chains are accepted at higher rates. Introgression/bidirectional migration rates (m1 and m2) were estimated as effective number of migrants per generation ($2NM = 4N_e\mu^*M/\mu/2$, independent of mutation rate), population size parameters ($\Theta = 4N_e\mu$, N_e = effective population size), and time since divergence in generations ($t\mu$) without converting for a given mutation rate, and generation time.

3 | RESULTS

3.1 | Descriptive statistics

The number of alleles per locus ranged from 13 to 69, and mean allelic richness ($N = 17$) varied between 3.51 and 20.49. Gene diversities per locus within samples were between 0.14 and 0.969 (mean = 0.77). Significant heterozygote deficits were observed for all loci (most frequently in Greenland waters, Table S2). Nineteen of 35 samples deviated significantly from Hardy–Weinberg expectation before FDR control, and 18 of them remained significant after FDR (Table S2). For the 35 *Sebastes* samples, 162 of 1,575 pairwise tests (10.29%) showed significant deviations from linkage equilibrium. LD was found for fish from all four species (between one and ten significant pairwise comparisons between loci). However, only 21 LD tests remained significant after FDR control. For the 27 *S. mentella* samples, 212 of 2106 tests for linkage equilibrium (approx. 10%) deviated significantly from expected values. After FDR, 57 of those tests remained significant without any spatial pattern. No evidence for null alleles or short allele dominance was detected by Micro-Checker.

3.2 | Pattern of genetic differentiation

For the *S. mentella* population structure analysis (27 samples), an initial investigation identified 138 redfish as *S. norvegicus*, which were removed from the subsequent analyses. As no differentiation was observed between temporal samples from two different *S. mentella* locations (Table 1: sample ID 2 and 5), samples were pooled. The subsequent STRUCTURE analyses of the 27 *S. mentella* samples suggested three clusters (Figure 2 and Figure S1). In many cases, individuals from one sample assigned to different clusters. The clusters were sorted according to the occurrence of the reference samples as "shallow (sample 5)/deep (sample 6)/slope (sample 7)." The "shallow" cluster was the largest, including 80% of the "shallow" sampling group fish (M-Oc, Figure 2) plus individuals from all other *S. mentella* samples (depth = 375 to 575 m). The "deep" cluster consisted of 89% fish from the "deep" sampling group (depth = 830–850 m), fish from East and West Greenland waters (no depth record) in addition to individuals from the Icelandic Shelf ($N = 40$, 612–784 m), Northwest Atlantic ($N = 26$, 450–711 m), Irminger Sea "shallow" (M-Oc: $N = 7$, 451–735 m), and Northeast Arctic ($N = 6$, 508–575 m) waters. In the "slope" cluster, most individuals (89%) were from East Greenland waters sampled on the shelf (227–350 m). Approximately 29% ($N = 17$) of the "Icelandic Shelf" sample clustered with the "slope" (Figure 2). "Shallow" and "deep" clusters from Greenland waters were distributed in both East and West Greenland, whereas the "slope" cluster was only found in the east. The juvenile sample from East Greenland (MSeb12Q2, depth = 561 m) consisted mainly of the "deep" group fish, with a few individuals from the "slope" group. The two juvenile samples from West Greenland (MSeb12WGL and MR12WGL) were dominated by the "deep" group fish, with few individuals from the "shallow" group. No major seasonal shifts in pattern of the distribution of clusters were observed. However, spring samples of adult fish from commercial vessels were found to be dominated by fish from the "slope" cluster.

When the total dataset (i.e., 35 samples) was considered, six clusters were identified by STRUCTURE (Figure 3 and Figure S2). Three of the clusters identified within the *S. mentella* samples were consistent with the results described above. Two clusters were detected within the *S. norvegicus* samples, while *S. fasciatus* and *S. viviparus* samples were placed into a single cluster. However, with the application of hierarchical cluster approach, three clusters were detected within the *S. norvegicus* samples and *S. fasciatus* and *S. viviparus* samples were separated into two distinct clusters. For the *S. norvegicus* samples, three clusters were designated as "Norvegicus-A," "Norvegicus-B," and "Giants" (described in Saha et al., 2016). Mechanical mixing, as consistent with the observed deviations from the Hardy–Weinberg equilibrium, was therefore apparent in multiple samples. As no introgression was associated with the "Giants" cluster, the cluster was not used in the downstream analyses. Evidence of hybridization was observed for almost all possible cluster pairs (Figure 3, see details below).

The three-group configuration ("shallow–deep–slope") of *S. mentella* samples predicted by Bayesian approaches was supported by the AMOVA (F_{CT} = 0.03, P = .000). A much lower, but still significant proportion of the genetic variance could be ascribed to differentiation among samples within groups (F_{SC} = 0.002, P = .000). The *S. mentella* "shallow" group from Northwest Atlantic (FC) and Northeast Arctic (Nor) were significantly differentiated from each other, and from the Irminger Sea "shallow" group plus "shallow" group samples (except

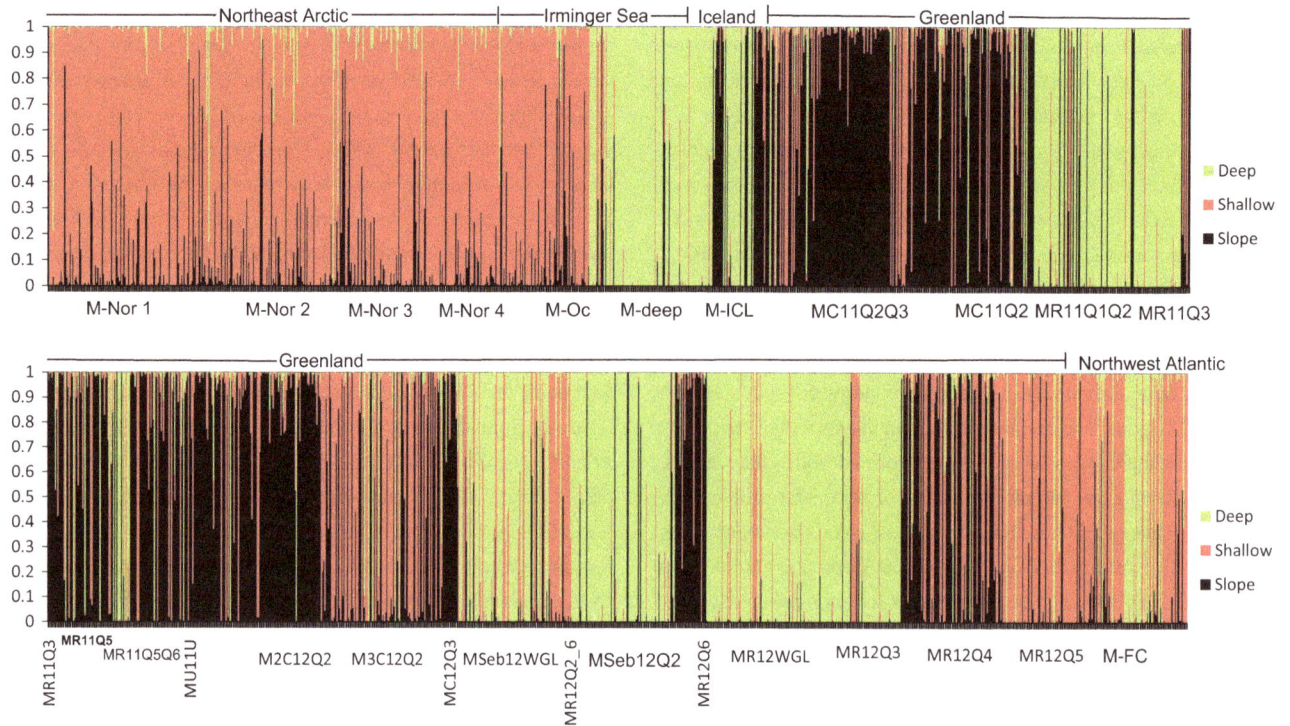

FIGURE 2 Column charts illustrating individual cluster ancestries obtained from the STRUCTURE analysis (Pritchard et al., 2000) using 13 microsatellites. Three clusters were predicted for the 27 *Sebastes mentella* samples (cf. Table 1). Sample names are provided at the *X*-axis, while genome ancestry fractions (*Q*) are given at the *Y*-axis

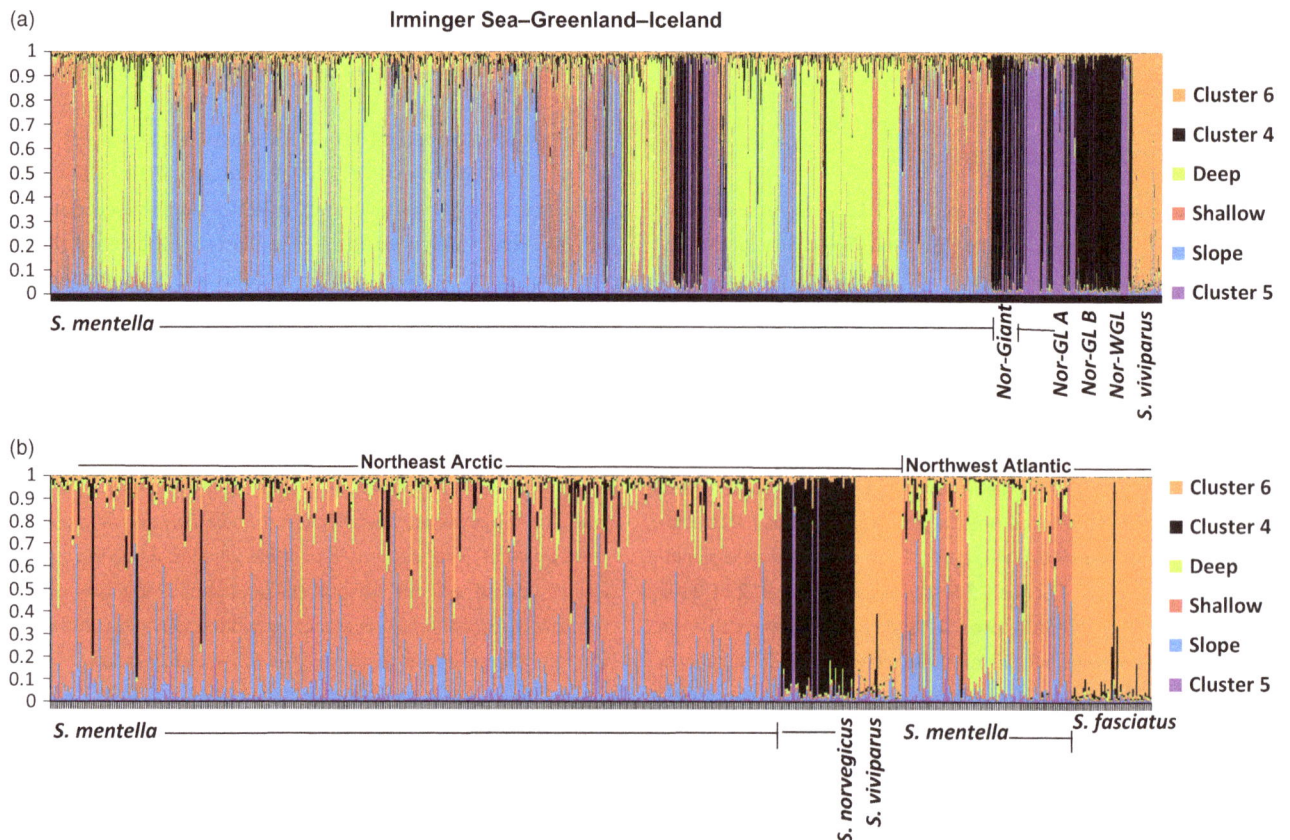

FIGURE 3 Six genetic clusters (as indicated by different colors) predicted by STRUCTURE for the *Sebastes* spp. samples. Two clusters within the *Sebastes norvegicus* samples are shown in black and purple, and the samples for *Sebastes viviparus* and *Sebastes fasciatus* were clustered together (in orange). Samples have been sorted as per the geographic locations: (a) Irminger Sea–Greenland–Iceland, (b) Northeast Arctic, and Northwest Atlantic

two) from Greenland waters (Table 3). In many of the cases, the "deep" groups identified from the Northwest Atlantic, Irminger Sea, Iceland, and Greenland waters were significantly differentiated from one another (Table 4). Finally, the "slope" group identified from the Northeast Arctic waters was significantly differentiated from Greenland waters (Table 5).

3.3 | Genetic diversity, Hardy–Weinberg and linkage equilibria within clusters

Genetic diversities, Hardy–Weinberg and linkage equilibria were estimated within the clusters with designated pure and hybrid ($Q > 10\%$ from other genomes) individuals identified through the admixture analyses. Clusters were formed with comparable numbers of fish (Table S3) to obtain similar statistical power. The cluster of pure *S. fasciatus* (Table S3: Fasciatus) showed significant heterozygote

deficiency, only before FDR control. All the admixed clusters except for *S. mentella* and *S. fasciatus*, and the "deep" and "shallow" groups of *S. mentella* were out of Hardy–Weinberg equilibrium. However, none of them remained significant except the admixed cluster of *S. mentella* and *S. viviparus* ("Shallow" × "Viviparus") after FDR. The admixed clusters had higher gene diversities than one or both of the parental clusters (except "deep–shallow"). A total of 59 (6.39%) tests for deviations from linkage equilibrium within clusters were significant, with more deviations observed in the admixed clusters (eight in Mentella-Fasciatus, six in Norvegicus-A-B and five in "deep–shallow").

3.4 | Introgressive hybridization in the genus *Sebastes*

STRUCTURE revealed admixed individuals for all geographically co-occurring clusters. Within Greenland waters, we found the highest rate of hybridization among the three *S. mentella* clusters. There were also admixed individuals between *S. mentella* and *S. norvegicus*, and *S. mentella* and *S. viviparus*. Patterns of individual admixture proportions were intermediate between a scenario of mechanical mixing and that of a hybrid swarm (see Figures S3–S7). No significant simulation bias was observed (Figure S17). Interspecific hybridization appeared to be most prevalent in the Northeast Arctic waters. In these waters, admixed individuals were observed both for *S. mentella*–*S. norvegicus* and *S. mentella*–*S. viviparus* (Figures S8 and S9). No significant admixture between *S. norvegicus* and *S. viviparus* was detected (Figure S10). In the Northwest Atlantic, hybrid genotypes were observed for *S. mentella*–*S. norvegicus*, *S. mentella*–*S. fasciatus*, and *S. norvegicus*–*S. fasciatus* (Figures S11–S16), but none of the cases conformed to a hybrid swarm scenario.

In Greenland waters, evidence of significant introgression was observed in all cluster pairs studied using the IM model (Figure 4). The highest introgression, which was also the most asymmetric, was observed between the clusters of "shallow" and "deep" *S. mentella*. Hybridization between the "shallow–slope" pair was just higher than for the "deep–slope" pair. In contrast, symmetric patterns

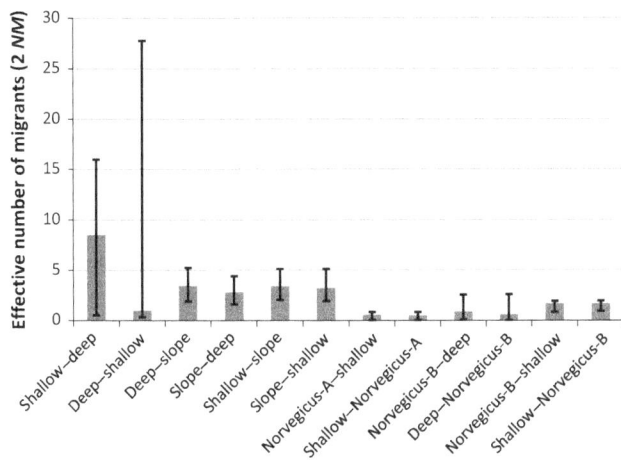

FIGURE 4 Bidirectional introgression rates for different clusters of *Sebastes* in Greenland waters. The extents of introgression are expressed as effective number of migrants per generation (2*NM*). For the values with highest posterior probabilities, 95% confidence intervals are provided with bars

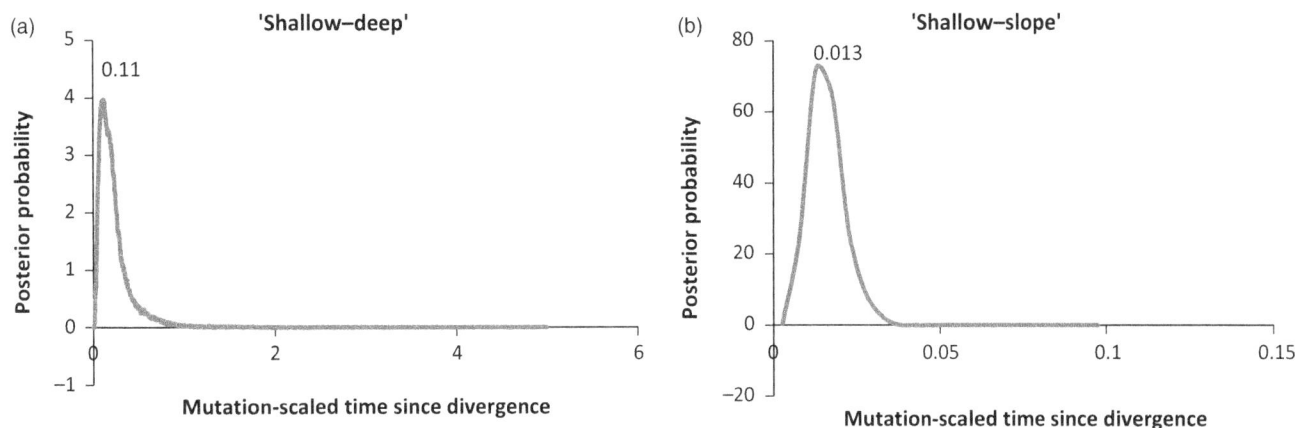

FIGURE 5 Posterior probability plots of divergence times for the clusters of *S. mentella* complex in Greenland waters. The estimated mutation-scaled time since divergences in generations (tμ) are presented at the apexes of the distributions

of introgression of much lower levels were estimated between Norvegicus-A-"shallow" Mentella, Norvegicus-B-"shallow" Mentella, and Norvegicus-B-"deep" Mentella.

The coalescent-based analysis suggested that the divergence between "shallow" and "deep" groups was more ancient (Figure 5a) than that between the "shallow" and "slope" groups (Figure 5b). The estimates of the population size parameter (Θ) were comparable for the three S. mentella clusters (Figure 6).

4 | DISCUSSION

Using an unprecedented large number of target and reference samples from different years, seasons, and life stages, the present investigation provided improved population delineation on a larger geographic scale for the S. mentella complex. Our primary results illustrate the presence of three distinct genetic groups within the species across its distribution range. The "shallow" group had a wider distribution than the "deep" and "slope" groups. The occurrence of the "slope" group on the East Greenland Shelf was supported by adult fish collected in both spring and fall, as well as by juveniles. In contrast, the "slope" sample from the Icelandic Shelf was found to be a mixture of fish belonging to both "deep" and "slope" groups. Connectivity in terms of effective number of migrants between the "shallow–slope" pair was found to be greater than connectivity between the "deep–slope" pair. It was also evident that S. mentella from the Northeast Arctic and Northwest Atlantic waters are genetically differentiated. Low, but statistically significant, evidence of introgression was observed among the clusters.

4.1 | Genetic population structure of *Sebastes mentella*

The three groups of S. mentella observed in this study are in agreement with earlier findings (for review, see Cadrin et al., 2010). The observation that the "shallow" group includes fish from all 27

geographical samples reveals the widest geographical distribution of this genetic group. Likewise, fish from inside and outside the Irminger Sea were assigned to the "deep" group. These observations are congruent with earlier findings (Shum et al., 2015). The F_{ST} estimate between these two distinct genetic groups in the present investigation (F_{ST} = 0.03) was higher than that estimated by Stefansson et al. (2009, F_{ST} = 0.009) but almost identical to that of Shum et al. (2014, F_{ST} = 0.031). Estimates of the demographic history by the IM model revealed that the divergence between these two groups was more ancient ($t\mu$ = 0.11) than that between "Giants" and "Norvegicus-B" ($t\mu$ = 0.06), supporting a deep evolutionary divergence between "shallow" and "deep" groups as suggested by Stefánsson et al. (2009), Shum et al. (2014, 2015).

The "slope" group was mainly observed in East Greenland in catches from the commercial fleets. These samples consisted of adult fish caught during the larval extrusion period (spring) and of juveniles collected in autumn. The finding of juvenile fish is not surprising, because Greenland waters are suggested to be nursery grounds for all S. mentella and S. norvegicus groups of the region (Anderson, 1984; Magnusson & Johannesson, 1995). However, the presence of adult "slope" individuals suggests that East Greenland waters may also act as an important area of distribution for the adults of this group, which has not previously been recognized.

The finding that 29% of the individuals from the Icelandic Shelf sample were assigned to the "slope" group, but the remaining individuals clustered with "deep," indicates sympatric occurrence of different groups on the Icelandic Shelf. Previously, it has been indicated that the Icelandic Shelf was the main distributional area for the "slope" group (Cadrin et al., 2010). However, the results presented herein suggest that this distribution extends further into East Greenland waters. At present, the "slope" group S. mentella on the Greenland and Iceland Shelf waters are assessed separately by ICES. In light of the results presented here, this should be reconsidered as stock dynamics appear to be linked across continental slopes, and the effect of fishing intensity on either shelf could affect the entire stock unit. A similar scenario has also been described for Atlantic cod where the East Greenland and Iceland slopes are inhabited by the same populations (Therkildsen et al., 2013), and for Greenland halibut in the entire region that is considered as a single unit (ICES 2015b).

The genetic differentiation observed for the "deep–slope" pair was the largest among the three S. mentella genetic groups (Table 2). Congruent with this observation, the F_{ST} estimates suggested the closest connectivity for the "shallow–slope" pair, which was further

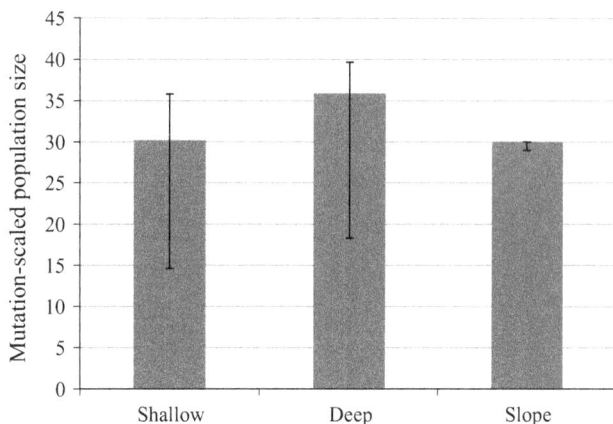

FIGURE 6 Estimates of the mutation-scaled population size parameters (Θ) for the three groups of *Sebastes mentella* from Greenland waters. For the values with highest posterior probabilities, 95% confidence intervals are provided with bars

TABLE 2 Pairwise F_{ST} values (Weir & Cockerham 1984) among the three groups of *Sebastes mentella* predicted by STRUCTURE (cf. Figure 2)

	N	Shallow	Slope
Shallow	786	–	
Slope	634	0.025	
Deep	595	0.030	0.037

The values are significant at P = .000. N = Sample size.

TABLE 3 Pairwise F_{ST} values (Weir & Cockerham 1984) within *Sebastes mentella* "shallow" group (cf. Figure 2)

	N	Greenland commercial					Greenland research			
		C11Q2Q3	C12Q2Q3	FC	Nor	M-Oc	R11Q3Q5Q6	R12Q2Q4	R12Q3Q5Q6	R12WGL
C11Q2Q3	33	–								
C12Q2Q3	66	0.001								
FC	60	**0.004**	0.003							
Nor	377	0.001	0.001	**0.004***						
M-Oc (shallow)	64	−0.001	**0.002**	**0.005***	0.002					
R11Q3Q5Q6	19	−0.003	−0.001	0.001	−0.001	0.000				
R12Q2Q4	39	−0.003	−0.002	0.001	−0.003	−0.003	−0.006			
R12Q3Q5Q6	75	**0.004**	**0.005***	**0.007***	**0.005***	**0.004***	0.003	0.001		
R12WGL	12	0.002	0.007	0.006	0.004	0.003	0.001	−0.003	0.004	
Seb12WGL	27	0.001	−0.001	0.003	0.003	**0.004**	−0.002	−0.002	0.004	0.006

N = Sample size. Closely located small samples from Greenland waters were pooled. For the sample codes, Nor = Samples 1–4, M-Oc = Sample 5 (reference sample "shallow"), FC = Sample 27, Seb12WGL = Sample 19, and R12WGL = Sample 23 (cf. Table 1).

The values in bold are significant at $P = .05$, while asterisks indicate significance after false discovery rate control.

supported by the estimates of gene flow (Figure 4) and time of divergence (Figure 5b). In contrast, allozyme studies by Johansen (2003) and Daníelsdóttir et al. (2008) indicated closer relationship for "slope-deep" pair. The discrepancy might be associated with the types of marker applied, because markers can be of varied mode of inheritance, function, and statistical properties. Moreover, the allozyme studies considered each locus separately, rendering low power for estimating connectivity among the genetic groups.

Based on the results from the hierarchical variance analysis (AMOVA), there was evidence of statistically significant genetic variance among samples within groups. This result was also supported by the many significant pairwise F_{ST} estimates between samples within groups. Most importantly, the identification of significant differentiation between the Northeast Arctic sample and other samples within the "shallow" group (Table 3) suggests the existence of isolated genetic components in the region. The elaborate sampling facilitated the testing of temporal stability (i.e., samples from 2006, 2009, and 2011) in the occurrence and genetic composition of the Northeast Arctic component. This component was previously identified by Roques, Sevigny, and Bernatchez (2002) and by Johansen (2003), but not by Stefansson et al. (2009). An isolated genetic component of *S. mentella* in the Northeast Arctic is also supported by the finding of a separate larval extrusion and nursery ground in this region (e.g., Cadrin et al., 2010). The finding that *S. mentella* from the Northwest Atlantic were

TABLE 4 Pairwise F_{ST} values (Weir & Cockerham 1984) within *Sebastes mentella* "deep" group (cf. Figure 2)

	N	Greenland commercial					Greenland research				
		C11Q2Q3	C12Q2Q3	M-Deep	FC	M-ICL	R11Q1Q2	R11Q3Q5Q6	R12Q3Q5Q6	R12WGL	Seb12Q2
C11Q2Q3	9	–									
C12Q2Q3	9	0.000									
M-Deep	65	0.005	−0.001								
FC	26	**0.011**	0.004	**0.006**							
M-ICL	40	0.002	0.000	**0.003**	0.002						
R11Q1Q2	65	0.008	0.005	**0.006***	0.003	−0.001					
R11Q3Q5Q6	61	0.009	0.004	**0.005***	0.003	−0.001	0.000				
R12Q3Q5Q6	81	0.006	−0.001	0.001	0.001	−0.001	0.000	0.000			
R12WGL	74	0.003	−0.003	**0.002**	**0.004**	−0.001	0.001	0.002	−0.001		
Seb12Q2	80	0.008	0.000	**0.002**	**0.004**	−0.001	0.000	0.002	−0.001	0.000	
Seb12WGL	54	0.004	−0.002	0.001	0.000	0.001	0.001	**0.003**	−0.001	0.000	0.000

N = Sample size. Closely located small samples from Greenland waters were pooled. For the sample codes, FC = Sample 27, M-Deep = Sample 6 (reference sample "deep"), M-ICL = Sample 7 (reference sample "Icelandic slope"), Seb12WGL = Sample 19, Seb12Q2 = Sample 21, and R12WGL = Sample 23 (cf. Table 1).

Significant at $P = .05$ in bold. Asterisks indicate significance after false discovery rate control.

significantly differentiated from other *S. mentella* samples (Tables 3 and 4) supports the idea of a "Western stock" as proposed by Cadrin et al. (2010). The highest substructuring was found within the "deep" group (Tables 3–5), indicating higher habitat segregation for the group in comparison with "shallow" and "slope." This observation is consistent with Shum et al. (2015) who hypothesized based on their mtDNA data that female "deep" *S. mentella* may exhibit some degree of philopatry.

Both the "deep" and "slope" groups were mainly concentrated in the central North Atlantic. Nevertheless, a significant number of fish from the Northwest Atlantic assigned to "deep" group, and some fish from Northeast Arctic ascribed to the "slope" group. The "shallow" group was distributed across the North Atlantic with evidence of variance among samples within the group. Such trans-Atlantic genetic structuring have been indicated for the entire *S. mentella* complex (Cadrin et al., 2010), but is genetically confirmed only for the "shallow" group (Shum et al., 2015), which could be associated with the less extensive sampling in earlier studies. However, trans-Atlantic patterns of genetic population structure have been observed in other species with comparable genetic structuring such as Atlantic cod (Bradbury et al., 2013; Hemmer-Hansen et al., 2013). It may be hypothesized that the apparent genetic pattern is possibly associated with separate glacial refugia and trans-Atlantic gene flow driven by the warm interglacial periods, an explanation in line with Shum et al. (2015), Bradbury et al. (2013), and Hemmer-Hansen et al. (2013).

4.2 | Introgression within the genus

Strong evidence of introgression among different clusters of *Sebastes* was found, in terms of deviations from Hardy–Weinberg expectations and linkage equilibrium, which was also consistent with the downstream Bayesian admixture and IM analyses. Although deviations from Hardy–Weinberg expectations and linkage equilibrium only provide

qualified support for introgression, an admixed group of individuals should show more extensive deviations than a group of pure individuals (Scribner, 1993), as seen when comparing clusters with pure and admixed individuals (Table S3). The reason why the apparently non-admixed *S. fasciatus* cluster (Figure 3) still displayed such deviations could be due to the inclusion of undetected hybrids with other clusters not included in the baseline such as *S. fasciatus–S. norvegicus*. This was supported by the detection of admixed individuals in the simulation including Norwegian *S. norvegicus* as baseline. In contrast, a few apparently admixed clusters did not display such deviations, which could be associated with the low power of those tests (i.e., as suggested by Nielsen et al., 2003). Another indication of introgression is the elevated levels of polymorphism observed, which is expected for mixed gene pools (e.g., Roques et al., 2001).

The extent of introgression was greater among the three groups of *S. mentella* than among other clusters (Figure 4 and Figures S3–S16). However, the magnitude observed in our study was less than those reported for *S. mentella–S. norvegicus* (Pampoulie & Daníelsdóttir, 2008), or for *S. mentella–S. viviparus* (Artamonova et al., 2013 cf. Figure S7). A low level of introgression within the redfish from these regions has been reported by Saha et al. (2016) and Schmidt (2005). The discrepancies with other studies might be associated with differences in statistical power among different studies. Incomplete or relaxed reproductive barriers among closely related species may provide opportunities for hybridization when in sympatry (Barton & Hewitt, 1989). Speciation within the genus *Sebastes* is a recent event (Briggs, 1995), and thus, these species may have relaxed reproductive barriers allowing hybridization to some extent (e.g., Roques et al., 2001). Nevertheless, *Sebastes* are ovoviviparous and display particular mating behavior during copulation (Helvey, 1982; Kendall, 1991) which implies that large-scale introgression may require more than simple sympatric existence, as evident in the Gulf of St. Lawrence (Roques et al., 2001).

TABLE 5 Pairwise F_{ST} values (Weir & Cockerham 1984) within *Sebastes mentella* "slope" group (cf. Figure 2)

	N	Greenland commercial					Greenland research			
		C11Q2Q3	C12Q2Q3	M-ICL	Nor	M-Oc	R11Q1Q2	R11Q3Q5Q6	R12Q2Q4	R12Q3Q5Q6
C11Q2Q3	219	–								
C12Q2Q3	155	0.000								
M-ICL	17	0.000	0.000							
Nor	27	**0.009***	**0.005**	0.004						
M-Oc (shallow)	9	0.001	−0.004	−0.003	−0.011					
R11Q1Q2	11	0.001	0.001	0.005	**0.018***	−0.001				
R11Q3Q5Q6	85	0.000	0.001	0.004	**0.009***	0.001	0.000			
R12Q2Q4	51	0.000	0.001	0.004	**0.009***	−0.001	0.001	0.000		
R12Q3Q5Q6	37	0.002	0.003	0.004	**0.009***	−0.001	0.005	0.003	0.003	
Seb12Q2	8	0.003	0.005	0.010	**0.016**	0.006	0.003	0.002	−0.009	0.002

N = Sample size. Closely located small samples from Greenland waters were pooled. For the sample codes, Nor = Samples 1–4, M-Oc = Sample 5 (reference sample "shallow"), M-ICL = Sample 7 (reference sample "Icelandic slope"), and Seb12Q2 = Sample 21 (cf. Table 1).
Significant at P = .05 in bold. Asterisks indicate significance after false discovery rate control.

The pattern of introgression between "shallow" and "deep" groups in Greenland waters appeared asymmetric (Figure 4). Similar observations have been made in other studies of *Sebastes*, and the roles of selection and differential abundance have been considered (Roques et al., 2001; Seeb, 1998). Considering the estimates of population size parameters in Greenland waters (Figure 6), more introgression would be expected from the "deep" group toward the "shallow," because smaller populations are more likely to be introgressed by more abundant populations over time (Arnold, Hamrick, & Bennett, 1993). However, the opposite was observed. The estimates of the population size were comparable with overlapping confidence intervals. Therefore, other factors than population size, such as selection, may be more likely as drivers of the asymmetric pattern of introgression. The spatial and temporal overlap between the groups during mating may also not be reflected by the population sizes *per se*.

The greatest extent of interspecific hybridization was observed in the Northeast Arctic waters (Figures S8 and S9). The results imply that *S. mentella* ("shallow") in the Northeast Arctic waters hybridize more frequently with both *S. viviparus* and *S. norvegicus* ("Norvegicus-B") than in other areas. One plausible reason for this might be the differential abundance of species. Recent surveys conducted in the Northeast Arctic waters have indicated a much greater abundance of *S. mentella* than *S. norvegicus* (Drevetnyak, Nedreaas, & Planque, 2011; ICES 2015c). Data on the *S. viviparus* stock size are sparse. But, the observed introgression in *S. mentella* might be one of the factors influencing allele frequency differences in the samples from the Northeast Arctic, an explanation in agreement with the Northwest component of this species (Roques et al., 2001).

Heterozygote deficiencies were apparent in some loci for several samples, consistent with observations using some of these loci in other studies (e.g., Pampoulie & Daníelsdóttir, 2008; Roques et al., 2001; Schmidt, 2005). No null alleles or short allele dominances were detected. Furthermore, sampling was not always in the mating area or during the mating period, and morphological identification of these species is uncertain (Barsukov et al., 1984). Therefore, mechanical mixing was considered as likely explanation of the observed heterozygote deficiencies.

Although the present investigation provides evidence of introgression between the identified *Sebastes* gene pools in sympatry, it may be difficult in some cases to distinguish the signal of introgression from noise. STRUCTURE has a tendency to misclassify "pure" individuals as "hybrids" when low differentiation is observed between the gene pools (Bohling, Adams, & Waits, 2013). As discussed earlier, introgression was supported by other genetic analyses such as IM results and the magnitude of introgression observed in our study is less than those reported in other investigations (e.g., Artamonova et al., 2013; Stefánsson et al., 2009). It is evident that hybridization was not of similar magnitude throughout the North Atlantic. For instance, interspecific hybridization appeared to be larger in Norwegian waters, whereas very low magnitude of introgression was observed in other areas (e.g., Figures S6, S7, S10 and S12–S14). Inclusion of more loci might have provided more power for the hybridization analyses. However, the number of loci used in our hybridization analyses was larger than that used by Roques et al. (2001) in a study of genus *Sebastes* and Nielsen et al. (2003) on cod.

The generally wide confidence intervals for our demographic estimates may reflect that the data do not allow precise estimation of the demographic history of these species and populations. Only one pair of clusters was considered at a time, thereby ignoring the role of other contemporary clusters (i.e., "ghost populations"). This approach may limit the applicability of the IM method for providing realistic estimates of introgression and demographic parameters. However, no severe bias has been observed in other comparable studies (Chan et al., 2013; Jacobsen & Omland, 2012; Won & Hey, 2005), so possible bias in our findings is believed to be minor and have little impact on our main conclusions.

The SM model is only option for IM analyses of microsatellite data, but microsatellites probably do not follow this model. Strasburg and Rieseberg (2009) analyzed sequence data and compared infinite sites and HKY substitution models for IM analyses, which may be similar between SM and infinite site model for microsatellite data. They report limited bias in IM results, with the exception of ancestral population size (which is not the objective of our study). Moreover, if the mutation model is violated, it is violated in all analyses. We are interested in relative magnitude of gene flow and time since divergence, and even if absolute values may be affected by such violations of assumptions, the relative comparison is likely to produce valid results.

Lack of knowledge on stock identity and scientific discrepancy in perception of stock entities of *S. mentella* especially in the Irminger Sea have caused failure of an efficient management of the stocks for the latest decade. The poor knowledge of biological parameters for the species and of stock identity have resulted in uncertain advice over the past (ICES 2015a). This lack of robustness in advice has caused the main client, the North East Atlantic Fisheries Commission (NEAFC), to fail reaching consensus on common management for the pelagic stocks of *S. mentella*, and catches have consequently exceeded biological advice by more than three times due to autonomous quota setting by each fishing nation (ICES 2015a). The further clarification of stock identity for *S. mentella* in the present study is anticipated to lead to a more robust advice which again likely will ensure a sustainable fishery by a common fishery management.

5 | CONCLUSIONS

We provide the first genetic investigation of *S. mentella* throughout its range, including Greenland waters. The identification of all three *S. mentella* groups in Greenland waters supports the interpretation of this region as a nursery area and population mixing zone. Fish from the "shallow" group were identified across the North Atlantic, but divided into three populations: Northeast Arctic, Irminger Sea/ Greenland, and Northwest Atlantic. Therefore, we clarify the genetic stock identity of *S. mentella* from the Northeast Arctic, which was previously disputed. The "deep" group *S. mentella* were identified in the Irminger Sea and Greenland waters, but some were also found in the Northwest Atlantic. The "slope" fish from the Icelandic Shelf and Greenland waters were not genetically differentiated, which suggests genetic connectivity of "slope" fish in East Greenland–Iceland Shelves. In Greenland waters, the "slope" group was the main target for the commercial fleets in spring. Although genetic heterogeneity was evident, low-to-moderate extent of gene flow was observed across the North Atlantic, implying incomplete reproductive isolation for *Sebastes* clusters possibly due to their close evolutionary relationship. The genetic differentiation between *S. mentella* groups were less than that between the other *Sebastes* clusters. Our findings mostly support the existing management practice of *S. mentella* throughout the North Atlantic (Cadrin et al., 2010, 2011), which strengthen the population genetic basis of stock boundaries. The consideration of all stock identities for these species into the management practices will help to manage sustainable utilization of this important fishery resource.

ACKNOWLEDGEMENTS

The project was funded by the Greenland Institute of Natural Resources, Norwegian Institute of Marine Research and by the Research Council of Norway (NFR-196691, SNIPFISK). We thank R/V Pamiut and commercial vessels (Atlantic Star and others) for collecting the samples. The reference samples were collected from the EU REDFISH project (QLK5-CT1999-01222). We thank Tula Skarstein and Tanja Hanebrekke for their technical assistances. We thank Prof. Svein Erik Fevolden, the three anonymous reviewers, and associate editor for the constructive comments to improve the manuscript.

LITERATURE CITED

Anderson, J. T. (1984). Early life history of redfish (*Sebastes* spp.) on Flemish Cap. *Canadian Journal of Fisheries and Aquatic Sciences, 41*, 1106–1116.

Arnold, M. L., Hamrick, J. L., & Bennett, B. D. (1993). Interspecific pollen competition and reproductive isolation in *Iris*. *Journal of Heredity, 84*, 13–16.

Artamonova, V. S., Makhrov, A. A., Karabanov, D. P., Rolskiy, A. Y., Bakay, Y. I., & Popov, V. I. (2013). Hybridization of beaked redfish (*Sebastes mentella*) with small redfish (*Sebastes viviparus*) and diversification of redfish (Actinopterygii: Scorpaeniformes) in the Irminger Sea. *Journal of Natural History, 47*, 1791–1801.

Barsukov, V. V., Litvinenko, N. I., & Serebryakov, V. P. (1984). *Manual for identification of redfish of the North Atlantic and adjacent areas*. Kalingrad, USSR: AtlantNIRO.

Barton, N. H., & Hewitt, G. M. (1989). Adaptation, speciation and hybrid zones. *Nature, 341*, 497–503.

Baskett, M. L., & Gomulkiewicz, R. (2011). Introgressive hybridization as a mechanism for species rescue. *Theoretical Ecology, 4*, 223–239.

Benjamini, Y., & Yekutieli, D. (2001). The control of the false discovery rate in multiple testing under dependency. *Annals of Statistics, 29*, 1165–1188.

Bohling, J. H., Adams, J. R., & Waits, L. P. (2013). Evaluating the ability of Bayesian clustering methods to detect hybridization and introgression using an empirical red wolf data set. *Molecular Ecology, 22*, 74–86.

Bradbury, I. R., Hubert, S., Higgins, B., Bowman, S., Borza, T., Paterson, I. G., ... Bentzen, P. (2013). Genomic islands of divergence and their consequences for the resolution of spatial structure in an exploited marine fish. *Evolutionary Applications, 6*, 450–461.

Briggs, J. C. (1995). *Global biogeography*. New York, NY: Elsevier.

Cadrin, S. X., Bernreuther, M., Danielsdottir, A. K., Hjorleifsson, E., Johansen, T., Kerr, L., ... Stransky, C. (2010). Population structure of beaked redfish, *Sebastes mentella*: Evidence of divergence associated with different habitats. *ICES Journal of Marine Science, 67*, 1617–1630.

Cadrin, S. X., Mariani, S., Pampoulie, C., Bernreuther, M., Danielsdottir, A. K., Johanssen, T., ... Stransky, C. (2011). Counter-comment on: Cadrin et al. (2010) "Population structure of beaked redfish, *Sebastes mentella*: Evidence of divergence associated with different habitats. ICES Journal of Marine Science, 67: 1617–1630." *ICES Journal of Marine Science 68*, 2016–2018.

Chan, Y.-C., Roos, C., Inoue-Murayama, M., Inoue, E., Shih, C.-C., Pei, K. J.-C., & Vigilant, L. (2013). Inferring the evolutionary histories of divergences in *Hylobates* and *Nomascus* gibbons through multilocus sequence data. *BMC Evolutionary Biology, 13*, 82.

Daníelsdóttir, A. K., Gíslason, D., Kristinsson, K., Stefánsson, M., Johansen, T., & Pampoulie, C. (2008). Population structure of deep-sea and oceanic phenotypes of deepwater redfish in the Irminger Sea and Icelandic continental slope: Are they cryptic species? *Transactions of the American Fisheries Society, 137*, 1723–1740.

Drevetnyak, K. V., Nedreaas, K. H., & Planque, B. (2011). Redfish. In T. Jakobsen & V. K. Ozhigin (Eds.), *The Barents Sea ecosystem, resources, management* (pp. 292–305). Trondheim, Norway: Tapir Academy Press.

Earl, D., & vonHoldt, B. (2012). STRUCTURE HARVESTER: A website and program for visualizing STRUCTURE output and implementing the Evanno method. *Conservation Genetics Resources, 4*, 359–361.

Evanno, G., Regnaut, S., & Goudet, J. (2005). Detecting the number of clusters of individuals using the software structure: A simulation study. *Molecular Ecology, 14*, 2611–2620.

Excoffier, L., & Lischer, H. E. L. (2010). Arlequin suite ver 3.5: A new series of programs to perform population genetics analyses under Linux and Windows. *Molecular Ecology Resources, 10*, 564–567.

Gagnaire, P.-A., Broquet, T., Aurelle, D., Viard, F., Souissi, A., Bonhomme, F., ... Bierne, N. (2015). Using neutral, selected, and hitchhiker loci to assess connectivity of marine populations in the genomic era. *Evolutionary Applications, 8*, 769–786.

Gomez-Uchida, D., Hoffman, E. A., Ardren, W. R., & Banks, M. A. (2003). Microsatellite markers for the heavily exploited canary (*Sebastes pinniger*) and other rockfish species. *Molecular Ecology Notes, 3*, 387–389.

Goudet, J. (1995). FSTAT (version 1.2): A computer program to calculate F-statistics. *Journal of Heredity, 86*, 485–486.

Hauser, L., & Carvalho, G. R. (2008). Paradigm shifts in marine fisheries genetics: Ugly hypotheses slain by beautiful facts. *Fish and Fisheries, 9*, 333–362.

Helvey, M. (1982). First observations of courtship behavior in rockfish, genus *Sebastes*. *Copeia, 1982*, 763–770.

Hemmer-Hansen, J., Nielsen, E. E., Therkildsen, N. O., Taylor, M. I., Ogden, R., Geffen, A. J., ... Carvalho, G. R. (2013). A genomic island linked to ecotype divergence in Atlantic cod. *Molecular Ecology, 22*, 2653–2667.

Hey, J. (2009). Isolation with migration models for more than two populations. *Molecular Biology and Evolution, 27*, 905–920.

ICES (2001). *Planning group on redfish stocks*. ICES CM 2001/D:04 Ref ACFM, Bergen: ICES.

ICES (2015a). Beaked redfish (*Sebastes mentella*) in subareas V, XII, and XIV (Iceland and Faroes grounds, north of Azores, east of Greenland) and NAFO subareas 1 + 2 (deep pelagic stock > 500 m). *Report of the ICES Advisory Committee 2015. ICES Advice, 2015. Book 2. Section 2.3.4a.* ICES Headquarters: Copenhagen, Denmark.

ICES (2015b). 2.3.11 Greenland halibut (Reinhardtius hippoglossoides) in Subareas V, VI, XII, and XIV (Iceland and Faroes grounds, West of Scotland, North of Azores, East of Greenland). Report of the ICES Advisory Committee 2015. ICES Advice, 2015. Book 2. Section 2.3.11.

ICES Headquarters, Copenhagen, Denmark.

ICES (2015c). *Report of the arctic fisheries working group (AFWG), 23–29 April 2015*, pp. 639. ICES CM 2015/ACOM:05, Hamburg, Germany: ICES.

Jacobsen, F., & Omland, K. E. (2012). Extensive introgressive hybridization within the northern oriole group (genus *Icterus*) revealed by three-species isolation with migration analysis. *Ecology and Evolution, 2,* 2413–2429.

Johansen, T. (2003). *Genetic study of genus Sebastes (redfish) in the North Atlantic with emphasis on the stock complex in the Irminger Sea.* Bergen: University of Bergen.

Kalinowski, S. T. (2010). The computer program STRUCTURE does not reliably identify the main genetic clusters within species: Simulations and implications for human population structure. *Heredity, 106,* 625–632.

Kendall, A. Jr (1991). Systematics and identification of larvae and juveniles of the genus *Sebastes*. In G. Boehlert & J. Yamada (Eds.), *Rockfishes of the genus Sebastes: Their reproduction and early life history* (pp. 173–190). Dordrecht, the Netherlands: Springer.

Magnusson, J. V., & Johannesson, G. (1995). Distribution and abundance of 0-group redfish in the Irminger Sea and at the East-Greenland in 1970-94 and its relation to *Sebastes marinus* abundance index from Icelandic groundfish survey. In: ICES (Eds.), *CM 1995/G39* (pp. 1–21). Reykjavik, Iceland: International Council for Exploration of the Sea.

Magnusson, J., & Magnusson, J. V. (1995). Oceanic redfish (*Sebastes mentella*) in the Irminger Sea and adjacent waters. *Scientia Marina, 59,* 241–254.

Miller, K. M., Schulze, A. D., & Withler, R. E. (2000). Characterization of microsatellite loci in *Sebastes alutus* and their conservation in congeneric rockfish species. *Molecular Ecology, 9,* 240–242.

Nielsen, E. E., Bach, L. A., & Kotlicki, P. (2006). Hybridlab (version 1.0): A program for generating simulated hybrids from population samples. *Molecular Ecology Notes, 6,* 971–973.

Nielsen, E. E., Hansen, M. M., Ruzzante, D. E., Meldrup, D., & Gronkjaer, P. (2003). Evidence of a hybrid-zone in Atlantic cod (*Gadus morhua*) in the Baltic and the Danish Belt Sea revealed by individual admixture analysis. *Molecular Ecology, 12,* 1497–1508.

Pampoulie, C., & Daníelsdóttir, A. K. (2008). Resolving species identification problems in the genus *Sebastes* using nuclear genetic markers. *Fisheries Research, 93,* 54–63.

Planque, B., Kristinsson, K., Astakhov, A., Bernreuther, M., Bethke, E., Drevetnyak, K., ... Stransky, C. (2013). Monitoring beaked redfish (*Sebastes mentella*) in the North Atlantic, current challenges and future prospects. *Aquatic Living Resources, 26,* 293–306.

Pritchard, J. K., Stephens, M., & Donnelly, P. (2000). Inference of population structure using multilocus genotype data. *Genetics, 155,* 945–959.

Randi, E. (2008). Detecting hybridization between wild species and their domesticated relatives. *Molecular Ecology, 17,* 285–293.

Roques, S., Pallotta, D., Sevigny, J.-M., & Bernatchez, L. (1999). Isolation and characterization of polymorphic microsatellite markers in the North Atlantic redfish (Teleostei: Scorpaenidae, genus *Sebastes*). *Molecular Ecology, 8,* 685–702.

Roques, S., Sevigny, J.-M., & Bernatchez, L. (2001). Evidence for broadscale introgressive hybridization between two redfish (genus *Sebastes*) in the North-west Atlantic: A rare marine example. *Molecular Ecology, 10,* 149–165.

Roques, S., Sevigny, J. M., & Bernatchez, L. (2002). Genetic structure of deep-water redfish, *Sebastes mentella*, populations across the North Atlantic. *Marine Biology, 140,* 297–307.

Rousset, F. (2008). GENEPOP'007: A complete re-implementation of the genepop software for Windows and Linux. *Molecular Ecology Resources, 8,* 103–106.

Salmenkova, E. A. (2011). New view on the population genetic structure of marine fish. *Russian Journal of Genetics, 47,* 1279–1287.

Saha, A., Hauser, L., Planque, B., Fevolden, S.-E., Hedeholm, R., Boje, J., & Johansen, T. (2016). Cryptic *Sebastes norvegicus* species in Greenland waters revealed by microsatellites. In preparation.

Schmidt, C. (2005). *Molecular genetic studies of species and population structure of North Atlantic redfish (genus Sebastes; Cuvier 1829).* Hamburg: University of Hamburg.

Scribner, K. T. (1993). Hybrid zone dynamics are influenced by genotype-specific variation in life-history traits: Experimental evidence from hybridizing *Gambusia* species. *Evolution, 47,* 632–646.

Seeb, L. (1998). Gene flow and introgression within and among three species of rockfishes, *Sebastes auriculatus, S. caurinus,* and *S. maliger. Journal of Heredity, 89,* 393–403.

Shaklee, J. B., & Bentzen, P. (1998). Genetic identification of stocks of marine fish and shellfish. *Bulletin of Marine Science, 62,* 589–621.

Shum, P., Pampoulie, C., Kristinsson, K., & Mariani, S. (2015). Three-dimensional post-glacial expansion and diversification of an exploited oceanic fish. *Molecular Ecology, 24,* 3652–3667.

Shum, P., Pampoulie, C., Sacchi, C., & Mariani, S. (2014). Divergence by depth in an oceanic fish. *PeerJ, 2,* e525.

Stefansson, M. O., Reinert, J., Sigurdsson, P., Kristinsson, K., Nedreaas, K., & Pampoulie, C. (2009). Depth as a potential driver of genetic structure of *Sebastes mentella* across the North Atlantic Ocean. *ICES Journal of Marine Science, 66,* 680–690.

Stefánsson, M. Ö., Sigurdsson, T., Pampoulie, C., Daníelsdóttir, A. K., Thorgilsson, B., Ragnarsdóttir, A., ... Bernatchez, L. (2009). Pleistocene genetic legacy suggests incipient species of *Sebastes mentella* in the Irminger Sea. *Heredity, 102,* 514–524.

Strasburg, J. L., & Rieseberg, L. H. (2009). How robust are "Isolation with Migration" analyses to violations of the IM model? A simulation study. *Molecular Biology and Evolution, 27,* 297–310.

Szpiech, Z. A., Jakobsson, M., & Rosenberg, N. A. (2008). ADZE: A rarefaction approach for counting alleles private to combinations of populations. *Bioinformatics, 24,* 2498–2504.

Therkildsen, N. O., Hemmer-Hansen, J., Hedeholm, R. B., Wisz, M. S., Pampoulie, C., Meldrup, D., ... Nielsen, E. E. (2013). Spatiotemporal SNP analysis reveals pronounced biocomplexity at the northern range margin of Atlantic cod *Gadus morhua. Evolutionary Applications, 6,* 690–705.

Van Oosterhout, C., Hutchinson, W. F., Wills, D. P. M., & Shipley, P. (2004). micro-checker: Software for identifying and correcting genotyping errors in microsatellite data. *Molecular Ecology Notes, 4,* 535–538.

Weir, B. S., & Cockerham, C. C. (1984). Estimating F-statistics for the analysis of population structure. *Evolution, 36,* 1358–1370.

Won, Y.-J., & Hey, J. (2005). Divergence population genetics of chimpanzees. *Molecular Biology and Evolution, 22,* 297–307.

14

Genetic diversity and structure of *Lolium perenne* ssp. *multiflorum* in California vineyards and orchards indicate potential for spread of herbicide resistance via gene flow

Elizabeth Karn ⓘD | Marie Jasieniuk

University of California Davis, Department of Plant Sciences, Davis, CA, USA

Correspondence
Elizabeth Karn, University of California Davis, Department of Plant Sciences, Davis, CA, USA.
Email: evkarn@ucdavis.edu

Funding information
USDA-NIFA-AFRI, Award Number: 2015-67013-22949

Abstract

Management of agroecosystems with herbicides imposes strong selection pressures on weedy plants leading to the evolution of resistance against those herbicides. Resistance to glyphosate in populations of *Lolium perenne* L. ssp. *multiflorum* is increasingly common in California, USA, causing economic losses and the loss of effective management tools. To gain insights into the recent evolution of glyphosate resistance in *L. perenne* in perennial cropping systems of northwest California and to inform management, we investigated the frequency of glyphosate resistance and the genetic diversity and structure of 14 populations. The sampled populations contained frequencies of resistant plants ranging from 10% to 89%. Analyses of neutral genetic variation using microsatellite markers indicated very high genetic diversity within all populations regardless of resistance frequency. Genetic variation was distributed predominantly among individuals within populations rather than among populations or sampled counties, as would be expected for a wide-ranging outcrossing weed species. Bayesian clustering analysis provided evidence of population structuring with extensive admixture between two genetic clusters or gene pools. High genetic diversity and admixture, and low differentiation between populations, strongly suggest the potential for spread of resistance through gene flow and the need for management that limits seed and pollen dispersal in *L. perenne*.

KEYWORDS
agricultural weed, glyphosate, glyphosate resistance, herbicide, Italian ryegrass, *Lolium perenne* ssp. *multiflorum*, microsatellite markers

1 | INTRODUCTION

Weedy plants pose a major problem to agricultural production causing significant crop losses worldwide and economic damages estimated to total $33 billion annually in the United States (Oerke, 2006; Pimentel, Zuniga, & Morrison, 2005). Weeds are an ongoing challenge for farmers as weed control practices exert strong selection for the evolution of weed adaptations that render the management practices less

effective over time (Barrett, 1983; Owen, Michael, Renton, Steadman, & Powles, 2011; Powles & Yu, 2010). One of the best examples of this process is the evolution of resistance to herbicides. In weed populations containing phenotypic variation for susceptibility to an herbicide, those individuals with an inherited ability to survive and reproduce following an herbicide application are favored and resistance increases in the population over time (Delye, Jasieniuk, & Le Corre, 2013; Neve, Vila-Aiub, & Roux, 2009). To date, over 470 cases of resistance in 250 species have been documented to a wide variety of herbicides worldwide (Heap, 2016).

Whether or not a weed population is able to adapt in response to management practices depends on whether that population contains the necessary genetic variation (Jasieniuk, Brule-Babel, & Morrison, 1996; Sakai et al., 2001). Population size, standing genetic variation, selection, and gene flow with other populations all play a role in the spatial distribution of evolved adaptive traits (Delye, Jasieniuk, et al., 2013; Lawton-Rauh, 2008). For studies of adaptation in agricultural weeds, strong selection pressures on weed populations such as tillage or herbicide application are usually known and population sizes are often large (Neve et al., 2009). Population sizes of common weeds vary across an agricultural landscape with some areas containing heavy infestations, allowing for high genetic diversity within a species across a region through the accumulation of mutations over time. In self-pollinating weeds, populations may be genetically uniform as individuals within populations often share nearly identical highly homozygous genotypes because of repeated inbreeding, but populations are likely to differ genetically (Ward & Jasieniuk, 2009). In contrast, obligately outcrossing weeds are expected to contain high genetic diversity within populations but low genetic differentiation among populations. The amount and distribution of phenotypic and genetic variation within weed populations influence the potential for adaptation in agricultural landscapes, which are variable in both space and time as a result of habitat fragmentation due to diverse crops and associated crop and weed management practices. Ultimately, the adaptation of weed populations to a variable environment across an agricultural landscape may lead to population structuring in both selfing and outcrossing weeds.

Genetic diversity in weed populations is required for weed adaptation, but is also impacted by it as a result of strong positive selection and population bottlenecks (Neve et al., 2009). Successful herbicide applications kill 95%–99% of individuals in susceptible weed populations. This substantial reduction in population size may mean that the alleles of only a small fraction of individuals are passed on to the next generation, potentially causing some alleles to be lost by genetic drift. Alternatively, strong selection will favor selectively advantageous alleles, if present in the population, and reduce population genetic diversity. For instance, individuals which survive herbicide treatment due to heritable mechanisms will pass on their resistance-conferring alleles to their progeny, and resistance will increase in frequency in the population over time. As the frequency of resistant individuals increases in a population, further herbicide applications will become less effective in reducing population size, leading to restoration of populations to their original size but with decreased genetic diversity. Strong selection for resistance may also be associated with a selective sweep at causative loci which not only results in the loss of susceptible alleles at the adaptive locus but also any alleles at loci in gametic disequilibrium with it (Maynard-Smith & Haig, 1974; Menchari, Delye, & Le Corre, 2007). In summary, weed populations with a high frequency of resistant individuals are expected to contain lower genetic diversity than populations with a low frequency of resistant individuals both due to population bottlenecks while an herbicide is still

effective in controlling the weed and due to selection as resistance to the herbicide evolves.

Lolium perenne ssp. *multiflorum* (Italian ryegrass) is an annual grass weed that causes economic losses in annual and perennial cropping systems worldwide (Preston, Wakelin, Dolman, Bostamam, & Boutsalis, 2009). *L. perenne* has an obligately outcrossing, self-incompatible mating system with wind-mediated pollen movement (Fearon, Hayward, & Lawrence, 1983). Populations of *L. perenne* ssp. *multiflorum* and the closely related *L. perenne* ssp. *rigidum* have repeatedly evolved resistance to several herbicides from different classes (Heap, 2016; Owen, Martinez, & Powles, 2014; Preston et al., 2009). The ability of *L. perenne* to rapidly evolve resistance to herbicides has been attributed to high genetic diversity within populations resulting from large population sizes and a self-incompatible outcrossing mating system (Balfourier, Charmet, & Ravel, 1998; Busi & Powles, 2009). However, while the genetic diversity of cultivated and wild accessions of *L. perenne* have been reasonably well characterized (e.g., Brazauskas, Lenk, Pedersen, Stender, & Lübberstedt, 2011; Kubik, Sawkins, Meyer, & Gaut, 2001; McGrath, Hodkinson, & Barth, 2007; Wang, Dobrowolski, Cogan, Forster, & Smith, 2009), the genetic variation and structure of weedy populations in agricultural settings (crop fields, orchards, vineyards) have not been examined, to our knowledge, despite the unique demographic processes and selective pressures in agricultural systems that are likely to shape genetic diversity in weeds. To date, studies of weedy *L. perenne* have largely focused on characterizing herbicide resistance phenotypes and resistance levels (e.g., Jasieniuk et al., 2008; Busi & Powles, 2009, Busi, Neve, & Powles, 2013; Liu, Hulting, & Mallory-Smith, 2016) and determining the underlying physiological and genetic mechanisms of resistance (e.g., Avila-Garcia, Sanchez-Olguin, Hulting, & Mallory-Smith, 2012; Gaines et al., 2014; Ge et al., 2012; Mahmood, Mathiessen, Kristensen, & Kudsk, 2016; Yu, Abdallah, Han, Owen, & Powles, 2009). In California, a population of *L. perenne* was identified with resistance to glyphosate in 1998 (Simarmata, Kaufmann, & Penner, 2003), and glyphosate resistance was later found to have spread in perennial cropping systems of the Central Valley of California (Jasieniuk et al., 2008). In 2013, populations of *L. perenne* suspected of containing individuals resistant to glyphosate were identified in Sonoma County and Lake County in northwestern California, outside of the Central Valley, after 2 years of failed control with glyphosate.

It has been hypothesized that gene flow may spread herbicide resistance among weed populations within an agricultural landscape to a greater degree than novel mutations as rates of gene flow are generally believed to be higher than rates of mutation (Jasieniuk et al., 1996). Herbicide resistance alleles may be present in populations prior to the onset of selection pressure by an herbicide (Delye, Deulvot, et al., 2013), and may spread by gene flow even before the trait is selectively advantageous. Evidence for the spread of herbicide resistance among populations by seed dispersal has been shown in several highly self-pollinating weed species, based on patterns of molecular marker and phenotypic variation (Okada et al., 2013, 2014; Osuna, Okada, Ahmad, Fischer, & Jasieniuk, 2011). Interestingly, however, neutral genetic and phenotypic variation in *Ipomoea purpurea*, a weed species with a mixed mating system, provided support for independent origins of resistance in multiple geographic locations (Kuester, Chang, & Baucom, 2015). In outcrossing weeds, analyses of neutral genetic variation revealed low population differentiation, and possible spread of resistance through local gene flow (Delye, Clement, Pernin, Chauvel, & Le Corre, 2010) but independent origins through novel mutations (Menchari et al., 2007).

The goal of the current study was to characterize genetic variation of northwestern California *L. perenne* populations where herbicide resistance evolution is very recent and likely ongoing. We examined the frequency of glyphosate-resistant plants in populations across the landscape along with microsatellite marker variation to address the following questions: (i) do populations of outcrossing weeds contain high genetic diversity and is this diversity reduced in populations with a high frequency of glyphosate-resistant individuals, (ii) is there evidence of genetic structuring and differentiation among populations of this widespread weed across an agricultural landscape, and (iii) is there potential for spread of resistance alleles across the landscape through gene flow?

2 | MATERIALS AND METHODS

2.1 | Population sampling

To determine whether glyphosate-resistant individuals are present in *L. perenne* ssp. *multiflorum* populations in northwest California, we sampled 13 orchards and vineyards in 2013 from Sonoma County and Lake County (Table 1) in the general regions where growers had reported difficulty controlling plants with glyphosate to farm advisors and in surrounding areas where populations may be experiencing gene flow with resistant plants. One population identified as resistant to glyphosate from Butte County was also sampled to serve as a comparison with an area which had evolved resistance greater than 10 years ago (Jasieniuk et al., 2008; Simarmata et al., 2003). Within

each population, young leaf tissue and panicles with mature seed were collected from each of 30–40 individuals at least one meter apart from one another while walking randomly selected tree or vine rows. Leaf tissue was transported to the laboratory for DNA extraction. Seed panicles were stored in paper envelopes for 3 months to allow seeds to after-ripen and overcome dormancy before planting and testing plants for resistance to glyphosate.

2.2 | Phenotyping plant response to glyphosate

Eight seeds from each sampled plant were germinated on moistened filter paper in petri dishes at 20°C and a 12-hr photoperiod. Germinated seedlings were transplanted into 8×8 cm square pots filled with UC soil mix (sand, compost, and peat in 1:1:1 ratio with 1.8 kg/m^3 dolomite) with two seedlings per pot and grown in the glasshouse at 27/15°C with ambient light conditions. At the tillering stage, individual plants were divided into genetically identical clones following the method described by Boutsalis (2001) and grown in the glasshouse to the 2–3 leaf stage. One clone of a genotype was treated with water, which served as a control. The second clone was treated with glyphosate (Roundup PowerMax, Monsanto, St. Louis, MO) at the rate of acid equivalent 1,681 g/ha, which is twice the recommended (label) field rate for the control of annual *Lolium perenne* plants under six inches tall. All treatments were applied in an enclosed cabinet track sprayer equipped with an 8002E nozzle (TeeJet, Spraying Systems Co., Wheaton, IL) delivering 200 L/ha. Three weeks after glyphosate treatment, we scored each plant as alive or dead and characterized the percentage of resistant plants in each population by the percentage of plants surviving glyphosate treatment of the total number of plants treated. Plants from a previously characterized susceptible reference seed collection (Jasieniuk et al., 2008) were included during each herbicide application to confirm herbicide activity.

TABLE 1 *Lolium perenne* ssp. *multiflorum* populations sampled for this study

Pop ID	Cropping system	County	Latitude (N)	Longitude (W)	N_S	N_G	N_P	%R
1	Orchard	Butte	39.80	−121.98	32	23	128	73.8
2	Vineyard	Sonoma	38.23	−122.52	30	28	128	9.7
3	Vineyard	Sonoma	38.24	−122.42	37	29	171	22.5
4	Vineyard	Sonoma	38.24	−122.36	33	18	212	29.1
6	Vineyard	Sonoma	38.359	−122.502	34	31	123	22.7
7	Vineyard	Sonoma	38.214	−122.457	33	32	65	32.1
8	Vineyard	Sonoma	38.587	−122.829	31	31	176	26.6
9	Vineyard	Sonoma	38.662	−122.825	33	32	150	31.6
10	Vineyard	Sonoma	38.673	−122.811	32	31	55	35.2
11	Vineyard	Sonoma	38.761	−122.976	41	41	166	40.6
12	Vineyard	Lake	38.989	−122.821	20	19	91	85.1
13	Orchard	Lake	38.997	−122.834	36	36	186	89.0
14	Orchard	Lake	38.996	−122.84	31	31	153	87.9
15	Orchard	Lake	39.086	−122.943	30	30	145	20.6

N_S, number of individuals sampled for leaf tissue and seeds from each population; N_G, number of individuals genotyped; N_P, number of progeny phenotyped for response to glyphosate; % R, percentage of individuals surviving treatment with glyphosate at 1681 g a.e. ha^{-1}.

2.3 | Genotyping plants using microsatellite markers

DNA was extracted from collected leaf tissue following the CTAB method (Doyle & Doyle, 1987), then quantified, and diluted to 25 ng/μl. We genotyped individuals at 12 polymorphic loci using 11 microsatellite primer pairs (Table 2), which included b1b1, b1b3, b3b1, b3b8, b3c5, b1a8, b4d3, b5d12 (Lauvergeat, Barre, Bonnet, & Ghesquiere, 2005), pr3 (Kubik et al., 2001), and 14C9, 44A7 (King et al., 2008). Primer pair 14C9 amplified two loci, called 14C9-1 and 14C9-2. Two sets of alleles, with one or two alleles in each of the two regions amplified by this primer pair, were observed in all individuals genotyped at these loci. For all primer pairs, forward primers labeled with either 6-Fam or Hex (Integrated DNA Technologies, Coralville, IA) were used in PCRs consisting of 25 ng DNA template, 1 × Qiagen PCR buffer (Valencia, CA), 0.25 mM additional $MgCl_2$, 0.4 μM forward and reverse primers, 0.125 mM DNTPs, and 0.5 units Taq polymerase. The PCR program consisted of an initial denaturing period of 3 min at 94°C, followed by 30 cycles of 1 min at 94°C, 1 min at x°C, 2 min at 72°C, and a final extension of 10 min at 72°C, where x is a primer-dependent annealing temperature (Table 2). PCR products were multiplexed into six pairs of PCR product and separated using an ABI 3100 Genetic Analyzer (Applied Biosystems, Foster City, CA) with GENESCAN 400HD as an internal size standard (Applied Biosystems). Fragments were sized with GeneMapper 3.7 (Applied Biosystems). Genotypes were also inspected manually.

2.4 | Microsatellite marker error rate

To assess genotyping errors, we tested for null alleles or genotyping stutter using Microchecker version 2.2.3 (Van Oosterhout, Hutchinson, Wills, & Shipley, 2004). Because null alleles were detected for several loci, FreeNa software (Chapuis & Estoup, 2007) was used to estimate the frequency of null alleles and their effect on F_{ST} estimates using the standard F_{ST} estimation method and the unbiased adjusted "excluding null alleles" F_{ST} (ENA) method with 1,000 bootstrap replicates. We tested standard and adjusted F_{ST} values for differences using Student's t-test.

2.5 | Genetic diversity and structure

To estimate the allelic diversity of each locus, we calculated the total number of alleles detected at each locus (N_A) using F_{STAT} software

TABLE 2 Characteristics and sources of 12 microsatellite loci used to genotype *Lolium perenne* ssp. *multiflorum*. Microsatellite markers were selected based on polymorphism and consistent amplification of alleles

Marker		Forward and Reverse Primers	Ta	Repeat Motif	Source
b1b1	f	CAGGTCCAGCGCTAGTGTTA	57	(CT)4(CA)2N	Lauvergeat et al. (2005)
	r	GAGGTGTGGTGCTGGGATAG		136(AC)7	
b1b3	f	AGGTGTCCTGTTGCTTTGGA	57	(TG)7	Lauvergeat et al. (2005)
	r	TTTACCCCCAGGGATCAAAT			
b3b1	f	TTTCCCTGGGATAGCGTTAG	57	(TG)10	Lauvergeat et al. (2005)
	r	TTAGCATAAACAGATGAAGCATAAC			
b3b8	f	TGTCATGTCCGCTGTCTACG	57	(CA)10	Lauvergeat et al. (2005)
	r	GAGAGTGGGCGATCATCTTC			
b3c5	f	TGTCATGTTCAGAAAGTGCG	55	(GT)8	Lauvergeat et al. (2005)
	r	TGTCCACATAAATGCACCTCA			
b1a8	f	GACTTTCAGGCATCGGTCAT	57	(TG)7	Lauvergeat et al. (2005)
	r	CCCAGCTCCATTCTTAATGC			
b4d3	f	ATTGATGGTGCCACTCCTCT	53	(CA)7	Lauvergeat et al. (2005)
	r	ATGGACAAAGCAGGGGTTC			
b5d12	f	GAATCCTCGATGTGGGCTAC	53	(GT)5	Lauvergeat et al. (2005)
	r	TAAAACGGAACCACCCATTC			
pr3	f	GTATAGTACCCATTCCGT	53	(CA)22	Kubik et al. (2001)
	r	GCCGCCCTGCCATGCTG			
14C9-1	f	AATGATGGCACGGAGCAATCG	50	(CT)22	King et al. (2008)
	r	CTGTAATTCCAGGTCACTACC			
14C9-2	f	AATGATGGCACGGAGCAATCG	50	CT	King et al. (2008)
	r	CTGTAATTCCAGGTCACTACC			
44A7	f	CACGTAGAAGCCACACTTTAC	50	(CT)60	King et al. (2008)
	r	GTCACATTCCATTCACTTCCG			

Ta, annealing temperature.
Primer pair 14C9 amplified two independent microsatellite loci (see Results section).

version 2.9.3.2 (Goudet, 1995). Wright's inbreeding coefficient (F_{IS}) was estimated for each of the 14 populations and averaged over populations for each locus, and Wright's fixation index (F_{ST}) was calculated over the 14 populations by locus, also in F_{STAT}. Statistical significances of F_{IS} and F_{ST} were determined with 1,000 permutations. Observed heterozygosity (H_O) and total gene diversity or expected heterozygosity (H_E) at each locus were calculated using Genepop on the Web software version 4.2 (option 5, suboption 1; Raymond & Rousset, 1995; Rousset, 2008).

To estimate genetic diversity within each population across all 12 loci, we calculated the average number of alleles detected per locus (N_A), mean allelic richness (A_R) defined as the number of alleles rarified to the smallest population sample size, and Wright's inbreeding coefficient (F_{IS}) in F_{STAT}. Statistical significance of F_{IS} was determined with 1,000 permutations. We also calculated H_O and H_E for each population in Genepop on the Web (option 5, suboption 1). We did not estimate population differentiation based on the statistic R_{ST} because stepwise mutations at microsatellite loci likely contribute relatively little to genetic differentiation between recently founded populations, such as those of weedy *Lolium perenne* (Kalinowski, 2002).

Departures from Hardy–Weinberg proportions per locus and per population were tested using the Hardy–Weinberg exact test with default Markov chain parameters in Genepop on the Web (option 1, suboption 3). To test for linkage disequilibrium between each pair of loci in each population, we performed a pairwise log likelihood ratio test for disequilibrium in Genepop on the Web (option 2, suboption 1) with default Markov chain parameters. To detect recent changes in effective population sizes or population bottlenecks, we performed a one-tailed Wilcoxon sign-rank test for heterozygote excess compared to that expected under a drift–mutation model using Bottleneck software version 1.2.02 (Cornuet & Luikart, 1997; Piry, Luikart, & Cornuet, 1999). A sequential Bonferroni correction was applied to adjust significance levels for multiple comparisons (Rice, 1989).

To investigate the spatial structuring of genetic variation among populations and counties, we performed multiple distance-based analyses using GenAlEx 6.5 (Peakall & Smouse, 2006, 2012). First, we calculated a matrix of genetic distances, based on Nei's D (Nei, 1978), between each pair of populations with GenAlEx software interpolating for missing data, and then used the pairwise genetic distances in the following analyses. We performed a principal coordinate analysis (PCA) using the standardized covariance method in GenAlEx to assess whether populations located within the same county and/or geographically near each other were also genetically similar and thus may share an evolutionary history. To test for isolation by distance (IBD) and determine whether genetic structuring correlates with geographic structuring between populations, we conducted a Mantel test (Mantel, 1967) in GenAlEx with 1,000 permutations between the matrix of pairwise genetic distances described above and a matrix of pairwise geographic distances calculated using the GPS coordinates (latitude and longitude) of each population. We also performed a hierarchical analysis of molecular variance (AMOVA; Excoffier, Smouse, & Quattro, 1992) in GenAlEx, which examined the distribution of genetic variation at five hierarchical levels: among counties within the total

sample, among populations within those counties, among populations in the total sample, among individuals within populations, and among individuals within the total sample. The AMOVA was performed with 1,000 permutations.

To assess population structure and determine the degree of admixture among populations, we used a model-based Bayesian clustering algorithm in STRUCTURE software version 2.3.4 (Pritchard, Stephens, & Donnelly, 2000). STRUCTURE infers genetic clusters or populations based on the multilocus genotypes of all individuals, independent of sampled location, by probabilistically assigning individuals to a cluster or jointly to multiple clusters if their genotypes indicate admixture between clusters, while simultaneously maximizing Hardy–Weinberg equilibrium and minimizing linkage disequilibrium within those clusters. STRUCTURE analysis was performed with a burn-in period of 1,000,000 iterations followed by 1,000,000 iterations for the number of genetic clusters (K) ranging from $K = 1$ to $K = 12$. Five independent runs at each value of K were performed using the population admixture model for potentially interbreeding populations and correlated allele frequencies. Likelihood values of ln P(D) were assessed for each run. The most likely value of K was inferred using the ΔK method (Evanno, Regnaut, & Goudet, 2005) in STRUCTURE Harvester online (Earl & von Holdt, 2012). Each individual's probability of assignment to each cluster (q), also interpreted as the proportion of an individual's genome that originated in each cluster (Pritchard et al., 2000), was visualized for all individuals using Distruct software version 1.1 (Rosenberg, 2004). To examine substructuring within genetic clusters, the multilocus genotypes of individuals with $q > 0.6$ to a cluster were analyzed independently, as suggested by Evanno et al. (2005), using the same parameters as above.

3 | RESULTS

3.1 | Plant response to glyphosate

Within sampled populations of *L. perenne* ssp. *multiflorum*, resistance to glyphosate, estimated as the percentage of individuals surviving glyphosate treatment per population, varied from 9.7% to 89.0% (Figure 1, Table 1). The Butte County population sampled from the area where glyphosate resistance was first reported in California (Simarmata et al., 2003) contained 73.8% resistant individuals. In Lake County, three populations (populations 12, 13, and 14) from an area where growers reported possible resistance contained 85%–89% resistant individuals, while a population (population 15) bordering the area contained 21% resistant individuals. In Sonoma County, populations show a gradient of survivorship ranging from 9.7% survivorship in the southern end to 40.6% in the northern end of the county (Table 1).

3.2 | Genetic diversity and structure

3.2.1 | Genetic diversity of microsatellite loci

Information on null alleles for each locus can be found in the Data S1 section. We detected 259 distinct alleles in 412 individuals of *L.*

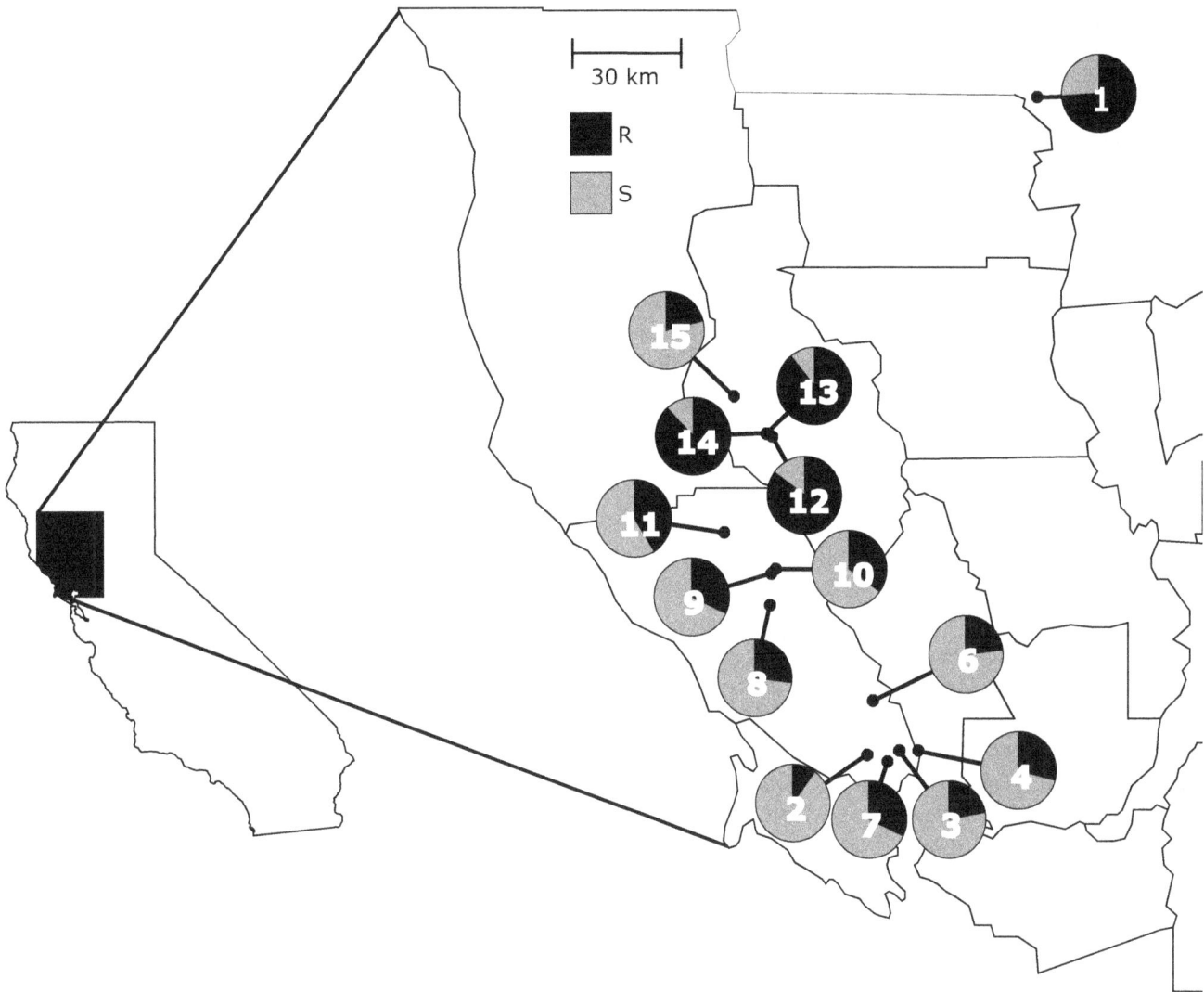

FIGURE 1 Geographic distribution of the populations sampled for this study in northwest California. Circles indicate the proportion of glyphosate-resistant (black) and glyphosate-susceptible (gray) individuals in each population, based on glasshouse screening of plants grown from field-collected seeds. Numbers are the population IDs (see Table 1)

TABLE 3 Genetic diversity detected at 12 microsatellite loci in 412 individuals of *Lolium perenne* spp. *multiflorum*

Marker	N_A	Allele sizes (bp)	% missing	H_E	H_O	F_{IS}	F_{ST}
b1b1	17	290–328	5.3%	0.861	0.382	0.541**	0.049*
b1b3	18	207–246	2.7%	0.705	0.392	0.435**	0.034*
b3b1	25	253–321	10.4%	0.878	0.461	0.476**	0.015
b3b8	26	292–335	2.9%	0.899	0.578	0.339**	0.020*
b3c5	16	113–146	4.6%	0.794	0.441	0.395**	0.088*
b1a8	26	237–303	1.2%	0.874	0.754	0.142**	0.015
b4d3	13	286–320	13.6%	0.788	0.388	0.492**	0.042
b5d12	8	102–116	0.7%	0.678	0.598	0.068*	0.051*
pr3	19	122–170	12.6%	0.786	0.352	0.538**	0.013
14C9-1	22	208–256	0.5%	0.867	0.640	0.263**	0.006
14C9-2	13	353–393	12.9%	0.825	0.147	0.824**	0.011
44A7	56	131–278	0.0%	0.826	0.385	0.593**	0.015*

N_A, total number of alleles detected; allele sizes, range of PCR product sizes (bp); % missing, % missing data at each locus; H_E, expected heterozygosity; H_O, observed heterozygosity; F_{IS}, Wright's inbreeding coefficient; F_{ST}, Wright's fixation index were averaged over 14 populations sampled in California. *p* values: *p < .01, **p < .001.

perenne ssp. *multiflorum* across 12 microsatellite loci. The total number of alleles detected per locus ranged from 8 to 56 (Table 3), and all loci were polymorphic in each population. Observed heterozygosity (H_O) ranged from 0.147 to 0.754, while expected heterozygosity (H_E) varied from 0.678 to 0.899. All loci revealed a reduction in observed heterozygosity compared to expected heterozygosity. Correspondingly, values of the inbreeding coefficient, F_{IS}, were statistically significant for all loci and ranged from 0.068 to 0.824. Because a reduction in heterozygosity compared to that expected under Hardy–Weinberg conditions was observed across all loci regardless of whether null alleles were detected, it is unlikely that null alleles are the major cause of heterozygosity deficits. Per locus estimates of F_{ST} ranged from 0.006 to 0.088, indicating little genetic differentiation among populations at each locus (Table 3). Among 923 pairwise comparisons of loci and populations, linkage disequilibrium was detected in only eight locus-by-population test combinations following Bonferroni correction (Table S2). Linkage disequilibrium was never detected between the same pair of loci twice, consistent with independently segregating loci. One microsatellite primer pair, 14C9, was found to amplify two separate loci (14C9-1 and 14C9-2) with nonoverlapping allele sizes (Table 2, Table 3). Both loci were scored and treated as separate microsatellite loci in data analyses. At locus 44A7, 56 alleles were identified ranging in size from 131 to 278 base pairs (Table 3). Despite the large range of allele sizes, no division of alleles into separate size classes that might indicate amplification of multiple loci was detected, and all individuals contained either one or two alleles as expected for diploids genotyped at a single microsatellite locus.

3.2.2 | Genetic diversity of populations

Within populations, the average number of alleles detected per locus (N_A) was high ranging from 8.7 to 12.1. Correspondingly, allelic

richness (A_R) ranged from 7.7 to 9.2 with an average of 8.4 over all populations (Table 4). Expected heterozygosity (H_E) ranged from 0.74 to 0.81 among populations, whereas observed heterozygosity (H_O) ranged from 0.40 to 0.52, indicating a heterozygote deficiency relative to Hardy–Weinberg expectations in all populations. Accordingly, values of the inbreeding coefficient F_{IS} for populations were high ranging from 0.374 to 0.475 (Table 4). Population bottlenecks as indicated by heterozygote excess relative to expectations under a drift–mutation model were detected in seven populations (Table 4).

Three populations (1, 12, and 14) containing a high frequency of resistant individuals (% R > 70%) all show a lower than average allelic richness (A_R = 7.7, 8.1, and 7.7, respectively) (Table 4), which might indicate that populations with a high frequency of resistant individuals have lower genetic diversity. Bottlenecks were detected in these three populations (Table 4). However, the population (population 13) containing the highest frequency of glyphosate-resistant individuals (89% R) (Table 1) also had the highest number of alleles detected per locus (N_A = 12.1) and the highest allelic richness (A_R = 9.2), and no bottleneck was detected. Correspondingly, there is no significant correlation between frequency of resistant plants within populations and allelic richness (Spearman's rank coefficient ρ = −0.22, p = .449). If population 13 is removed from the analysis, the correlation is stronger but still not significant (Spearman's rank coefficient ρ = −0.525, p = .065).

3.3 | Population structure

A principal coordinate analysis (PCA) of genetic distances between populations revealed differentiation among populations with grouping of some populations by geographic origin (Figure 2). Most populations from the southern end of Sonoma County (2, 3, and 4) group tightly together, while populations from the central and northern parts of Sonoma County (6, 8, 9, 10, and 11) along with population 13 from

Pop ID	N_G	N_A	A_R	H_E	H_O	F_{IS}	B
1	23	8.75	7.68	0.771	0.454	0.431**	0.0003*
2	28	10.92	8.72	0.776	0.482	0.396**	0.0052
3	29	11.08	8.81	0.788	0.431	0.467**	0.0212
4	18	9.42	8.72	0.805	0.517	0.384**	0.0052
6	31	11.75	8.87	0.800	0.441	0.462**	0.0017*
7	32	11.25	8.62	0.792	0.427	0.475**	0.0052
8	31	10.75	8.38	0.782	0.501	0.374**	0.0023*
9	32	11.25	8.54	0.786	0.490	0.391**	0.0008*
10	31	11.00	8.82	0.774	0.443	0.442**	0.0017*
11	41	10.92	7.58	0.738	0.396	0.474**	0.0052
12	19	8.67	8.18	0.797	0.498	0.400**	0.0012*
13	36	12.08	9.22	0.809	0.459	0.445**	0.0261
14	31	9.58	7.71	0.757	0.470	0.395**	0.0006*
15	30	9.58	7.82	0.794	0.430	0.473**	0.0319

TABLE 4 Genetic diversity within populations of *L. perenne* ssp. *multiflorum* based on variation at 12 microsatellite loci

N_G, number of individuals genotyped; N_A, average number of alleles detected per locus; A_R, mean allelic richness; H_E, expected heterozygosity; H_O, observed heterozygosity; F_{IS}, Wright's inbreeding coefficient, and B, the p value of Wilcoxon sign-rank test for genetic bottleneck. p values: *p < .05 following Bonferroni correction. **p < .001.

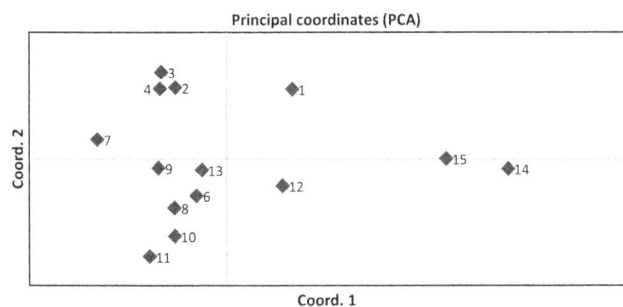

FIGURE 2 Principal coordinate analysis (PCoA) of pairwise genetic distances between populations. The first two axes explain 32.2% and 19.1% of genetic variation

Lake County group together. Populations located in Lake County (12, 13, 14, and 15) are more genetically distant from each other and do not group tightly together. The percentage of variation explained by the first two axes are 32.2% and 19.1%. A Mantel test revealed a weak, nonsignificant correlation between genetic and geographic distances (slope = 0.0105, R^2 = 0.14, p = .468), indicating that the genetic differentiation observed among populations is not related solely to geographic isolation.

As population clustering by county was not explained by geographic distance, an AMOVA was conducted to determine how much of the genetic variance could be attributed to county or population differences. The AMOVA revealed low, but significant, genetic differentiation among counties (F_{RT} = 0.018, p = .001), among populations within counties (F_{SR} = 0.018, p = .001), and among populations within the total sample (F_{ST} = 0.036, p = .001) (Table 5). Genetic differences among individuals within populations (F_{IS} = 0.466, p = .001) and among individuals within the total sample (F_{IT} = 0.485, p = .001) are high. Because of large genetic differences among individuals, 45.6% of the genetic variation is distributed among individuals within populations and 47.4% among individuals in the total sample, with 1.8% of genetic variation distributed between counties, 1.8% among populations within counties, and 3.5% among populations in the total sample, based on the F-statistic for the corresponding measure.

STRUCTURE analysis (Pritchard et al., 2000) was used to further examine genetic structuring. STRUCTURE revealed increasing values of ln P(D) with increasing K values ranging from 1 to 12 with no clear maximum likelihood (Figure 3a). ΔK (Evanno et al., 2005) clearly showed the highest value at K = 2, indicating two genetic clusters (Figure 3b). The proportion of the genome, as represented by the 12 microsatellite loci, that assigns to each cluster, q, was calculated for each individual. Individuals assigning to cluster 1 with q > 0.7 were comprised of some individuals from Sonoma County and Butte County and a few individuals from Lake County, while most individuals from Lake County and some from Sonoma County and Butte County assigned to cluster 2 (Figure 4a). While individuals from Sonoma County and Butte County assigned to both genetic clusters, individuals from Lake County assigned highly to cluster 2. All populations contained some individuals that assigned partially to each cluster (q < 0.6) indicating admixture between genetic clusters (Figure 4a).

There is no apparent pattern of genetic structuring based on whether the individuals originated from field populations that were predominantly resistant or susceptible to glyphosate (Table 1). The majority of individuals from populations 12, 13, and 14, where the frequency of glyphosate resistance was 85%–89%, assigned to cluster 2 with q > 0.7, but most individuals from population 15 where resistance frequency was 21% also assigned to the same cluster. Of the individuals from population 1 genotyped, 48% and 52% assigned to clusters 1 and 2, respectively, but 77% of individuals phenotyped were resistant to glyphosate, indicating that glyphosate-resistant individuals likely assign to both clusters.

To examine patterns of hierarchical population structure, individuals assigning to each cluster with q > 0.6 were separated and analyzed independently. STRUCTURE analysis of each cluster revealed evidence of subclustering, with K = 3 within cluster 1 and K = 4 within cluster 2 (Figure 3c–f). Among individuals that assigned to cluster 1, most individuals assigned highly to one of the three subclusters (Figure 4b). However, there is little apparent geographic substructuring of genetic variation, with the exception that most individuals from Butte County assigned to subcluster 1. Among individuals that assigned to cluster 2, most individuals were admixed, assigning to multiple subclusters (Figure 4c).

4 | DISCUSSION

4.1 | High genetic variation observed in weedy *L. perenne* regardless of resistance frequency

Our analyses indicate very high genetic diversity within populations of *L. perenne* ssp. *multiflorum* as expected based on the biology of this widespread obligately outcrossing species. However, given the large number of detected alleles and the allele frequencies observed, a higher level of heterozygosity was expected than was observed across all populations (Table 4). While some of this reduction in heterozygosity may be due to null alleles that do not amplify during PCR creating

TABLE 5 Analysis of molecular variance (AMOVA) results showing F-statistics for codominant allelic data for genetic variation distributed among five hierarchical levels

Effect	F-statistic	F	p	% of total genetic variation
Among counties in total sample	F_{RT}	0.018	.001	1.8
Among populations in counties	F_{SR}	0.018	.001	1.8
Among populations in total sample	F_{ST}	0.036	.001	3.5
Among individuals in populations	F_{IS}	0.466	.001	45.6
Among individuals in total sample	F_{IT}	0.485	.001	47.4

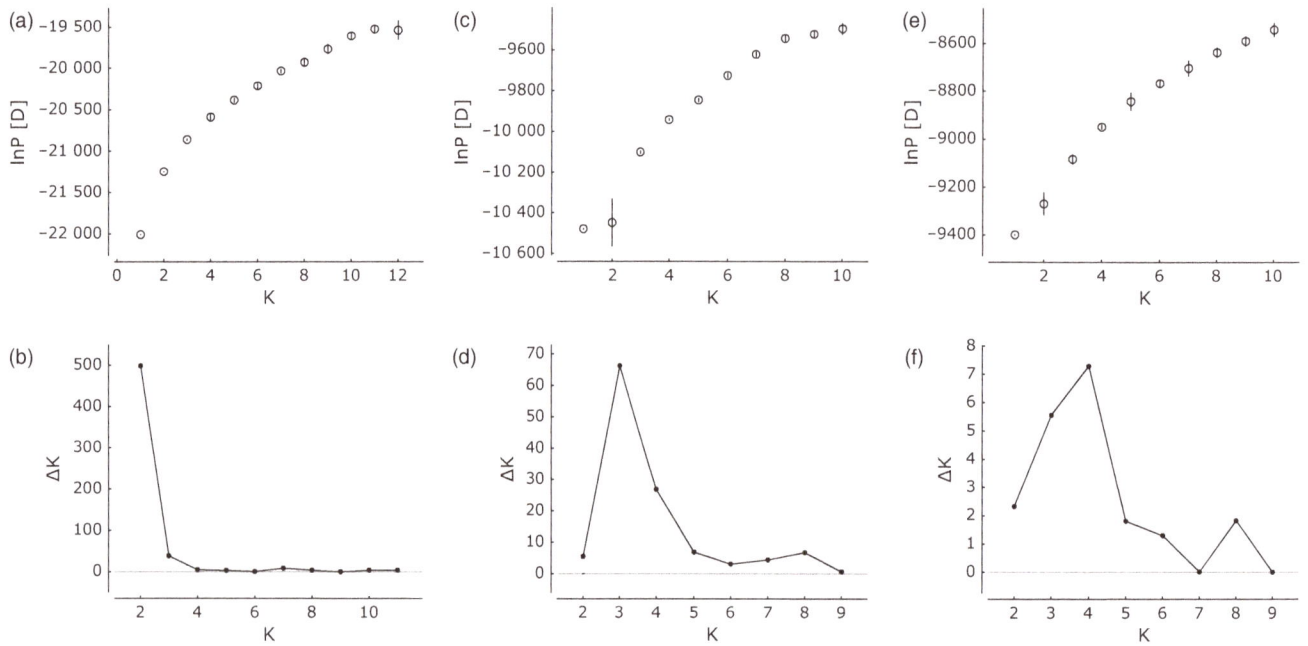

FIGURE 3 Bayesian clustering analysis (STRUCTURE, Pritchard et al., 2000) of *Lolium perenne* plots of (a, c, e) the log likelihood ln P[D] for five runs at each value of K, and (b, d, f) the second order of change in ln P[D], ΔK, as a function of the number of clusters or gene pools, K, from the analysis of all samples (a, b) and subclustering analysis of individuals assigning with $q > 0.6$ to cluster 1 (c, d) and to cluster 2 (e, f) within the global analysis

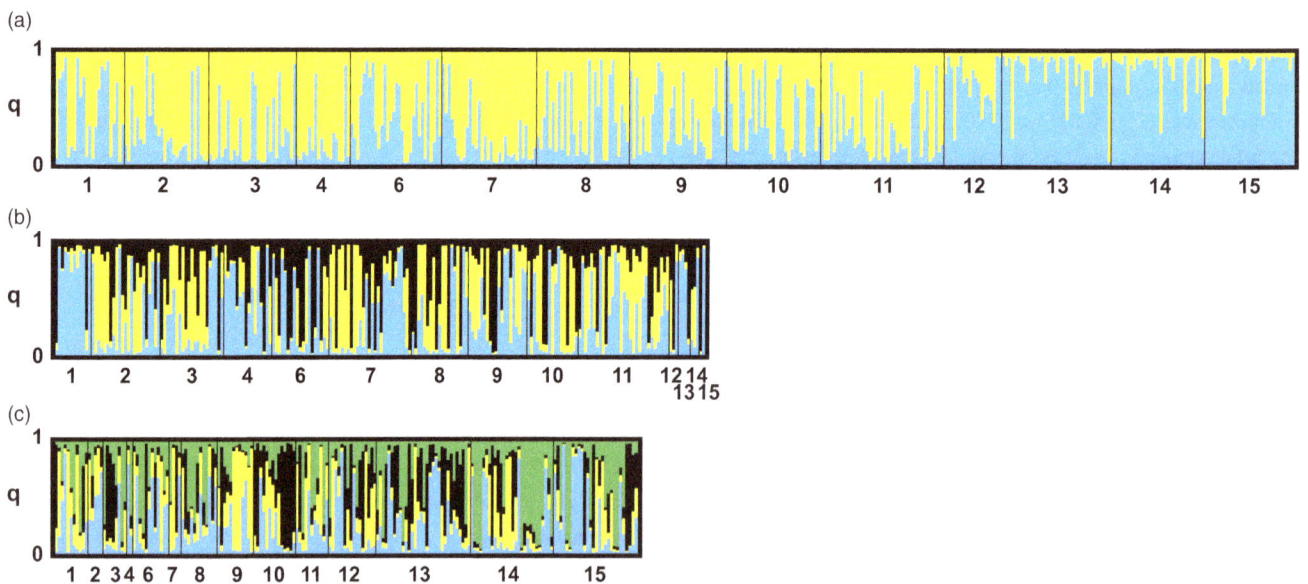

FIGURE 4 Assignment of 412 individuals of *L. perenne* ssp. *multiflorum* to the genetic clusters inferred by Bayesian clustering analysis (STRUCTURE). Each vertical bar corresponds to a distinct individual and its probability of assignment, q, to each cluster. (a) $K = 2$, the most likely number of genetic clusters for the global data set, (b) $K = 3$, the most likely number of subclusters among individuals assigning with $q > 0.6$ to cluster 1, and (c) $K = 4$, the most likely number of subclusters among individuals assigning with $q > 0.6$ to cluster 2

the false appearance of homozygous individuals, this would result in reduced observed heterozygosity only in loci with a high frequency of null alleles. However, a reduction in observed heterozygosity is present across all loci, leading to the conclusion that there is a biological cause for the lower than expected heterozygosity (Table 3). This could be due to a very recent or ongoing population bottleneck in

L. perenne populations due to control by glyphosate or other weed management practices. Seven of the 14 sampled populations show evidence of a past bottleneck. These seven populations may be located in fields with more intensive weed management and thus have undergone a stronger bottleneck. In populations with no detected bottleneck, extensive gene flow with other populations could have restored genetic diversity and erased the genetic signature of a past population bottleneck. Bottlenecks are only detectable if they are either very strong or very recent (Luikart & Cornuet, 1998), although the bottleneck imposed by intense selection by glyphosate and the evolution of glyphosate resistance likely has been recent, probably occurring within the last 20 years, that is 20 generations, based on when resistance to glyphosate first was identified in *Lolium* in the United States and worldwide (Powles, Lorraine-Colwill, Dellow, & Preston, 1998; Simarmata et al., 2003). Although resistance alleles may have been present at low frequencies previous to 20 years ago, they did not rise to frequencies high enough to be problematic until glyphosate use in agriculture intensified, following the introduction of inexpensive generic formulations of the herbicide and widespread adoption of reduced- or no-tillage cropping systems and glyphosate-resistant transgenic crops.

Populations which have undergone recent population bottlenecks and strong selective sweeps for adaptive traits may be expected to contain lower genetic diversity than populations, which have not yet adapted to the selection pressure. Populations with a high frequency of resistant individuals would be expected to contain lower genetic diversity than those with a lower frequency of resistant individuals. However, *L. perenne* populations with a high frequency of resistant plants (>70%) still contained high levels of observed genetic diversity, despite evidence of a population bottleneck in some populations. The lack of correlation between the frequency of resistant plants within populations and the fixation index (F_{IS}) (Spearman's rank coefficient $\rho = -0.02$, $p = .93$), within-population genetic diversity (H_E) (Spearman's rank coefficient $\rho = -0.13$, $p = .67$), and allelic richness (A_R) (Spearman's rank coefficient $\rho = -0.22$, $p = .449$) suggests that weed population size and genetic diversity may have been influenced by other weed control practices or environmental conditions, in addition to treatment with glyphosate. Herbicide rotations and mixtures, tillage, mowing, and pest and environmental pressures present variable selection pressures, which may reduce genetic diversity in all populations, making it difficult to detect reductions in diversity in populations with a high frequency of glyphosate-resistant individuals. Repeated introductions of *L. perenne* seeds from other fields may serve to increase within-population genetic diversity and compensate for alleles lost through selective sweeps and bottlenecks. It is also possible that less-intensively managed *L. perenne* plants growing in roadsides or noncrop areas near the sampled populations may be experiencing less intense selection and subsequently a less intense population bottleneck and that gene flow through pollen exchange with these plants may restore some genetic diversity in populations undergoing selection for resistance.

The genetic diversity observed in weedy populations of *L. perenne* was higher than that observed in other herbicide-resistant agricultural weeds. Populations of another outcrossing grass weed species, *Alopecurus myosuroides*, in France with evolved resistance to ACCase-inhibiting herbicides had lower values (0.246 and 0.240) of expected heterozygosity based on AFLP markers (Delye et al., 2010; Menchari et al., 2007; respectively), and populations of a glyphosate-resistant weed with a mixed mating system, *Ipomoea purpurea*, also had lower

total genetic diversity ($H_E = 0.304$) based on microsatellite markers (Kuester et al., 2015). Populations of species with an outcrossing mating system would be expected to have higher genetic diversity than those with predominantly self-pollinating mating systems. Glyphosate-resistant populations of the two closely related selfing species *Conyza canadensis* and *Conyza bonariensis* in California displayed a wide range of genetic diversities ($H_E = 0.0$–0.45 and $H_E = 0.009$–0.513, respectively) based on microsatellite markers (Okada et al., 2013, 2014).

While genetic diversity observed in weedy *L. perenne* populations was higher than that observed in other species of weeds, the amount of genetic diversity detected in this study was similar to that observed in most other studies of *L. perenne*. Average observed heterozygosity across loci in weedy populations ($H_O = 0.46$) was similar to values observed ($H_O = 0.40$ and $H_O = 0.44$) in some studies of perennial cultivars (Brazauskas et al., 2011 and Wang et al., 2009; respectively) but lower than another study of perennial *L. perenne* cultivars ($H_O = 0.62$) (Kubik et al., 2001) based on microsatellite markers. Expected heterozygosity in wild European *L. perenne* populations ranged from 0.233 to 0.359 based on AFLP markers (McGrath et al., 2007). The high number of alleles detected per locus averaged across populations ($N_A = 10.5$) in California populations was also similar or higher than that seen in studies of *L. perenne* cultivars using some of the same microsatellite markers ($N_A = 19.4$, 13.3, and 9.9; Kubik et al., 2001; Wang et al., 2009; Brazauskas et al., 2011, respectively). Interestingly, the forage cultivars of *L. perenne* do not show lower genetic diversity than their wild and weedy relatives despite many generations of breeding. Rather, *L. perenne* seems to display a very high level of genetic diversity, regardless of origin.

4.2 | Spatial and genetic structuring of populations

Glyphosate-resistant plants were detected in *L. perenne* populations from all sampled areas (Table 2). The Butte County population, located near where glyphosate resistance was first detected in California, contained a high proportion of resistant individuals (Figure 1). In Lake County, the high frequency of resistant plants in populations 12–14 was consistent with grower reports of increased difficulty controlling plants, with a lower frequency further away. In Sonoma County, a very low frequency of resistant individuals was observed in the southern end of the county, with moderate frequency in the northern end. The frequency of individuals surviving glyphosate treatment in Sonoma County was substantially lower than the other areas despite similar reports of weed control failure from growers. Glyphosate has been used for decades as the primary herbicide for weed control in orchards and vineyards, and noncrop areas of California. The appearance of glyphosate resistance is an adaptive response to the widespread and repeated use of glyphosate as a weed management strategy.

The observed microsatellite variation indicates genetic structuring of *L. perenne* populations in northwest California. Principal coordinate analysis of genetic distances indicates that populations located close to each other tend to be genetically similar, as is seen in the grouping of most populations in southern Sonoma County and of populations in northern Sonoma County together with population 13 (Figure 2).

However, geographic proximity does not necessarily indicate genetic similarity. The principal coordinate analysis (Figure 2) and pairwise F_{ST} (Table S3) both reveal larger genetic distance between populations within Lake County than between populations within Sonoma County. This indicates that despite the relatively larger geographic distances, Sonoma County populations are closely related to each other due to either higher genetic exchange between them or less differentiation over time, possibly related to the relative homogeneity of the vineyard cropping system across Sonoma County compared to the mix of perennial crops grown in Lake County. Differences in water availability in primarily drip-irrigated vineyards compared to sprinkler-irrigated orchards may lead to local adaptation or phenological differences due to water stress may contribute to population differentiation in Lake County populations. The lack of correlation between genetic distance and geographic distance found by a Mantel test indicates that genetic distance between populations is not due mainly to factors associated with isolation by distance, but to some other factor such as local adaptation or differences in the strength of selection pressures across the landscape. Extensive long-distance gene flow may also erode the relationship between genetic distance and geographic distance, especially if gene flow is not homogeneous across the entire range. An analysis of molecular variance indicates a significant amount of genetic variation is distributed at the county and population level (Table 5). However, both county and population differences are outweighed by the high genetic variation among individuals, as might be expected in this highly diverse outcrossing species (Brazauskas et al., 2011).

Bayesian clustering STRUCTURE analysis indicates the presence of two distinct gene pools or genetic clusters (Figure 3a). While the individuals assigning to cluster 1 are primarily from Butte County and Sonoma County, individuals assigning to cluster 2 come from all three counties and include almost all individuals from Lake County (Figure 4a). This indicates that there is little admixture between individuals in Lake County and individuals in the other sampled areas or that L. perenne has been introduced into the region too recently for substantial admixture to have occurred. In the subclustering analysis, there is little apparent spatial structure among individuals assigning to subclusters (Figure 4b,c). Many individuals had admixed genomes assigning partially to multiple subclusters, indicating little differentiation or high gene flow between individuals assigning to these subclusters.

4.3 | Evolution of resistance and potential for spread of resistance alleles

Together, data on spatial patterns of population structuring and frequencies of glyphosate-resistant phenotypes allow comparison of hypotheses regarding single or multiple evolutionary origins and subsequent spread of glyphosate resistance in L. perenne populations in northwest California. Populations with moderate frequencies of resistant individuals in Sonoma County (30% > R > 80%, populations 7, 9, 10, 11) contain mostly individuals that assign to cluster 1 with $q > 0.7$, while populations with high frequencies of resistant individuals (R > 80%, populations 12, 13, 14) contain a large proportion of individuals that assign to cluster 2 with $q > 0.7$ (Figure 4a, Table 2).

This suggests that glyphosate resistance has likely evolved independently in individuals that assign to each cluster. Unfortunately, it was not possible to genotype and phenotype the same individuals, which would have allowed stronger inference. In addition, the high genetic diversity observed among individuals within populations and the low percentage of genetic variation observed among populations make it difficult to infer single or multiple origins of resistance using only neutral genetic variation. However, our sequencing of the gene encoding glyphosate's target enzyme in resistant and susceptible plants confirms multiple target-site mutations, which have previously been identified to cause resistance to glyphosate, and thus multiple independent origins of glyphosate resistance (Karn & Jasieniuk, In review), in agreement with the results of STRUCTURE analysis in this study. There is also evidence that some plants are resistant due to a mechanism other than target-site mutations, indicating that multiple mechanisms of resistance are present in the region resulting from at least one additional independent origin of resistance (Karn and Jasieniuk, in review). It is possible that in the future, additional novel mutations may result in yet more independent origins of resistance.

STRUCTURE analysis revealed potential for future spread of resistance alleles through gene flow. Many admixed individuals with genotypes that assigned partially to each cluster were identified (Figure 4). While the majority of individuals assign highly to a single cluster, the considerable number of admixed individuals in many populations indicates that gene flow is common. Localized gene flow between populations located near each other may be pollen-mediated, and pollen movement over distances of 3 km has been documented in L. perenne ssp. rigidum (Busi, Yu, Barrett-Lennard, & Powles, 2008). Short- and long-distance gene flow may also be mediated by seed movement on agricultural machinery and vehicles, or over short distances by wind or animals. The higher levels of admixture detected in this study compared to studies of other herbicide-resistant weeds (Kuester et al., 2015; Okada et al., 2013, 2014) likely relate to the outcrossing nature of L. perenne. Spread of resistance alleles through gene flow could also result in populations and individuals containing multiple mechanisms of glyphosate resistance.

Successful management of glyphosate-resistant L. perenne populations in perennial cropping systems will likely require implementation of integrated pest management programs that include chemical and nonchemical techniques to not only control currently resistant plants, but reduce the intensity of selection pressure for future independent origins of resistance and limit the spread of resistance through gene flow. Increasing tillage where possible, mowing to reduce seed set, applying herbicide alternatives to glyphosate, or applying glyphosate in mixtures with other herbicides may help prevent or delay future origins of resistance by reducing population sizes and reducing the selection pressure of glyphosate in these systems. To limit spread of resistance through gene flow, cleaning weed seed from equipment and shoes moved between infested fields may help reduce long-distance seed transfer. Mowing or tillage may also reduce short-distance gene flow through pollen dispersal by reducing the number of flowers resistant plants produce, although these management techniques may not be able to entirely eliminate pollen production. It is not yet known whether a fitness cost is associated with glyphosate resistance in these *L. perenne* populations, and whether the frequency of resistance would be maintained in the absence of continued glyphosate use, as no fitness studies have yet been conducted in California populations of *L. perenne* and fitness costs associated with herbicide resistance vary depending on mechanism, the measure of fitness used, and genetic background of the population (Giacomini, Westra, & Ward, 2014; Preston et al., 2009; Vila-Aiub, Neve, & Powles, 2009). If any of the mechanisms of resistance present in the sampled area do confer a fitness cost, discontinuing use of glyphosate could result in a gradual decrease in the frequency of resistant individuals in populations. However, glyphosate is still effective on many other weed species, and will likely continue to be used in weed management in perennial crops. If gene flow between populations acts to produce populations containing multiple separate mechanisms of resistance, this could complicate their management if two mechanisms act additively resulting in plants with a very high level of resistance. Separate mechanisms of resistance may also respond differently to management if they confer different fitness penalties in the absence of continued selective pressure by glyphosate.

The results of this study show that glyphosate resistance is common in populations of *L. perenne* in orchards and vineyards of northwest California. In addition, microsatellite marker analyses revealed very high genetic diversity. All populations contained high diversity regardless of the frequency of resistant plants, contrary to what might be expected in populations undergoing strong selection for resistance. Genetic variation among populations and counties was low, in accordance with expectations for widespread and common species with highly outcrossing mating systems. The low genetic differentiation among populations and counties makes it difficult to draw conclusions about the specific origins and routes of spread of glyphosate resistance in northwest California from microsatellite variation. However, we can conclude that there were multiple origins of resistance in the region. In addition, there is potential for future spread of the resistance trait through gene flow, based on the high genetic diversity observed within all populations, the relatively low genetic differentiation among populations and counties, and the considerable admixture among populations detected in this study.

ACKNOWLEDGEMENTS

We are grateful for assistance with field sampling from Carlos Marochio and Vince Harjono; glasshouse screening from Vince Harjono, Aaron Kwong, Domonique Lewis, and Oliver Chen; and with genotyping from Vince Harjono and Aaron Kwong. Funding for this study came from USDA-NIFA-AFRI Award No. 2015-67013-22949, Henry A. Jastro Research Scholarships, and the California Weed Science Society Scholarship.

REFERENCES

Avila-Garcia, W. V., Sanchez-Olguin, E., Hulting, A. G., & Mallory-Smith, C. (2012). Target-site mutation associated with glufosinate resistance in Italian ryegrass (*Lolium perenne* L. ssp. *multiflorum*). *Pest Management Science, 68*, 1248–1254.

Balfourier, F., Charmet, G., & Ravel, C. (1998). Genetic differentiation within and between natural populations of perennial and annual ryegrass (*Lolium perenne* and *L. rigidum*). *Journal of Heredity, 81*, 100–110.

Barrett, S. C. H. (1983). Crop mimicry in weeds. *Economic Botany, 37*, 255–282.

Boutsalis, P. (2001). Syngenta quick-test: A rapid whole-plant test for herbicide resistance. *Weed Technology, 15*, 257–263.

Brazauskas, G., Lenk, I., Pedersen, M., Stender, B., & Lübberstedt, T. (2011). Genetic variation, population structure, and linkage disequilibrium in European elite germplasm of perennial ryegrass. *Plant Science, 181*, 412–420.

Busi, R., Neve, P., & Powles, S. B. (2013). Evolved polygenic herbicide resistance in *Lolium rigidum* by low-dose herbicide selection within standing genetic variation. *Evolutionary Applications, 6*, 231–242.

Busi, R., & Powles, S. B. (2009). Evolution of glyphosate resistance in a *Lolium rigidum* population by glyphosate selection at sublethal doses. *Heredity, 103*, 318–325.

Busi, R., Yu, Q., Barrett-Lennard, R., & Powles, S. (2008). Long-distance pollen-mediated flow of resistance genes in *Lolium rigidum*. *Theoretical and Applied Genetics, 117*, 1281–1290.

Chapuis, M. P., & Estoup, A. (2007). Microsatellite null alleles and estimation of population differentiation. *Molecular Biology and Evolution, 2*, 621–631.

Cornuet, J. M., & Luikart, G. (1997). Description and power analysis of two tests for detecting recent population bottlenecks from allele frequency data. *Genetics, 144*, 2001–2014.

Delye, C., Clement, J. A. J., Pernin, F., Chauvel, B., & Le Corre, V. (2010). High gene flow promotes the genetic homogeneity of arable weed populations at the landscape level. *Basic and Applied Ecology, 11*, 504–512.

Delye, C., Deulvot, C., & Chauvel, B. (2013). DNA analysis of herbarium specimens of the grass weed *Alopecurus myosuroides* reveals herbicide resistance pre-dated herbicides. *PLoS One, 8*(10), e75117. doi:10.1371/journal.pone.0075117

Delye, C., Jasieniuk, M., & Le Corre, V. (2013). Deciphering the evolution of herbicide resistance in weeds. *Trends in Genetics, 29*, 649–658.

Doyle, J. J., & Doyle, J. L. (1987). A rapid DNA isolation procedure for small quantities of fresh leaf tissue. *Phytochemistry Bulletin, 19*, 11–15.

Earl, D. A., & von Holdt, B. M. (2012). STRUCTURE HARVESTER: A website and program for visualizing STRUCTURE output and implementing the Evanno method. *Conservation Genetics Resources, 4*, 359–361.

Evanno, G., Regnaut, S., & Goudet, J. (2005). Detecting the number of clusters of individuals using the software STRUCTURE: A simulation study. *Molecular Ecology, 14*, 2611–2620.

Excoffier, L., Smouse, P., & Quattro, J. M. (1992). Analysis of molecular variance inferred from metric distances among DNA haplotypes: Application to human mitochondrial DNA restriction data. *Genetics, 131*, 479–491.

Fearon, C. H., Hayward, M. D., & Lawrence, M. J. (1983). Self-incompatibility in ryegrass V. Genetic control, linkage, and seed-set in diploid *Lolium multiflorum* Lam. *Heredity, 50*, 35–45.

Gaines, T. A., Lorentz, L., Figge, A., Herrmann, J., Maiwald, F., Ott, M. C., ... Beffa, R. (2014). RNA-Seq transcriptome analysis to identify genes involved in metabolism-based diclofop resistance in *Lolium rigidum*. *The Plant Journal, 78*, 865–876.

Ge, X., d'Avignon, D. A., Ackerman, J. J. H., Collavo, A., Sattin, M., Ostrander, E. L., ... Preston, C. (2012). Vacuolar glyphosate sequestration correlates with glyphosate resistance in ryegrass (Lolium ssp.) from Australia, South America, and Europe: A 31P NMR Investigation. *Journal of Agricultural and Food Chemistry, 60*, 1243–1250.

Giacomini, D., Westra, P., & Ward, S. M. (2014). Impact of genetic background in fitness cost studies: An example from glyphosate-resistant Palmer Amaranth. *Weed Science, 62*, 29–37.

Goudet, J. (1995). FSTAT version 1.2: A computer program to calculate F-statistics. *Journal of Heredity*, *86*, 485–486.

Heap, I. (2016). *International survey of herbicide-resistant weeds*. http:// weedscience.org. (accessed on September 16, 2016).

Jasieniuk, M., Ahmad, R., Sherwood, A. M., Firestone, J. L., Perez-Jones, A., Lanini, W. T., ... Stednick, Z. (2008). Glyphosate-resistant Italian ryegrass (*Lolium multiflorum*) in California: Distribution, response to glyphosate, and molecular evidence for an altered target enzyme. *Weed Science*, *56*, 496–502.

Jasieniuk, M., Brule-Babel, A. L., & Morrison, I. N. (1996). The evolution and genetics of herbicide resistance in weeds. *Weed Science*, *44*, 176–193.

Kalinowski, S. T. (2002). Evolutionary and statistical properties of three genetic distances. *Molecular Ecology*, *11*, 1263–1273.

Karn, E., & Jasieniuk, M. (In review). Nucleotide diversity at site 106 of *EPSPS* in *Lolium perenne* L. ssp. *multiflorum* from California indicates multiple evolutionary origins of glyphosate resistance. In review.

King, J., Thorogood, D., Edwards, K. J., Armstead, I. P., Roberts, L., Skot, K., ... King, I. P. (2008). Development of a genomic microsatellite library in perennial ryegrass (*Lolium perenne*) and its use in trait mapping. *Annals of Botany*, *101*, 845–853.

Kubik, C., Sawkins, M. A., Meyer, W., & Gaut, B. S. (2001). Genetic diversity in seven perennial ryegrass (*Lolium perenne L.*) cultivars based on SSR markers. *Crop Science*, *41*, 1565–1572.

Kuester, A., Chang, S., & Baucom, R. S. (2015). The geographic mosaic of herbicide resistance evolution in the common morning glory, *Ipomoea purpurea*: Evidence for resistance hotspots and low genetic differentiation across the landscape. *Evolutionary Applications*, *8*, 821–833.

Lauvergeat, V., Barre, P., Bonnet, M., & Ghesquiere, M. (2005). Sixty simple sequence repeat markers for use in the *Festuca-Lolium* complex of grasses. *Molecular Ecology Notes*, *5*, 401–405.

Lawton-Rauh, A. (2008). Demographic processes shaping genetic variation. *Current Opinion in Plant Biology*, *11*, 103–109.

Liu, M., Hulting, A. G., & Mallory-Smith, C. (2016). Characterization of multiple herbicide-resistant Italian ryegrass (*Lolium perenne* ssp. *multiflorum*) populations from winter wheat fields in Oregon. *Weed Science*, *64*, 331–338.

Luikart, G., & Cornuet, J. (1998). Empirical evaluation of a test for identifying recently bottlenecked populations from allele frequency data. *Conservation Biology*, *12*, 228–237.

Mahmood, K., Mathiessen, S. K., Kristensen, M., & Kudsk, P. (2016). Multiple herbicide resistance in *Lolium multiflorum* and identification of conserved regulatory elements of herbicide resistance genes. *Frontiers in Plant Science*, *7*, 1160. doi:10.3389/fpls.2016.01160

Mantel, N. (1967). The detection of disease clustering and a generalized regression approach. *Cancer Research*, *27*, 209–220.

Maynard-Smith, J., & Haig, D. (1974). The hitch-hiking effect of a favorable gene. *Genetical Research*, *23*, 23–35.

McGrath, S., Hodkinson, T. R., & Barth, S. (2007). Extremely high cytoplasmic diversity in natural and breeding populations of *Lolium* (Poaceae). *Heredity*, *99*, 531–544.

Menchari, Y., Delye, C., & Le Corre, V. (2007). Genetic variation and population structure in black-grass (*Alopecurus myosuroides* Huds.), a successful, herbicide-resistant, annual grass weed of winter cereal fields. *Molecular Ecology*, *16*, 3161–3172.

Nei, M. (1978). Estimation of average heterozygosity and genetic distance from a number of individuals. *Genetics*, *89*, 583–590.

Neve, P., Vila-Aiub, M., & Roux, F. (2009). Evolutionary thinking in agricultural weed management. *New Phytologist*, *184*, 783–793.

Oerke, E. C. (2006). Crop losses to pests. *Journal of Agricultural Science*, *144*, 31–43.

Okada, M., Hanson, B. D., Hembree, K. J., Peng, Y., Shrestha, A., Stewart, C. N., ... Jasieniuk, M. (2013). Evolution and spread of glyphosate resistance in *Conyza canadensis* in California. *Evolutionary Applications*, *6*, 761–777.

Okada, M., Hanson, B. D., Hembree, K. J., Peng, Y., Shrestha, A., Stewart, C. N., Wright, S. D., & Jasieniuk, M. (2014). Evolution and spread of gly-

phosate resistance in *Conyza bonariensis* in California and a comparison with closely related *Conyza canadensis*. *Weed Research*, *55*, 173–184.

Osuna, M., Okada, M., Ahmad, R., Fischer, A. J., & Jasieniuk, M. (2011). Genetic diversity and spread of thiobencarb resistant early watergrass (*Echinochloa oryzoides*) in California. *Weed Science*, *59*, 195–201.

Owen, M. J., Martinez, N. J., & Powles, S. B. (2014). Multiple herbicide-resistant *Lolium rigidum* (annual ryegrass) now dominates across the Western Australian grain belt. *Weed Research*, *54*, 314–324.

Owen, M. J., Michael, P. J., Renton, M., Steadman, K. J., & Powles, S. B. (2011). Towards large-scale prediction of *Lolium rigidum* emergence. II. Correlation between dormancy and herbicide resistance suggests an impact of cropping systems. *Weed Research*, *51*, 133–141.

Peakall, R., & Smouse, P. E. (2006). GenAlEx 6: genetic analysis in Excel. Population genetic software for teaching and research. *Molecular Ecology Notes*, *6*, 288–295.

Peakall, R., & Smouse, P. E. (2012). GenAlEx 6.5: Genetic analysis in Excel. Population genetic software for teaching and research—An update. *Bioinformatics*, *28*, 2537–2539.

Pimentel, D., Zuniga, R., & Morrison, D. (2005). Update on the environmental and economic costs associated with alien-invasive species in the United States. *Ecological Economics*, *52*, 273–288.

Piry, S., Luikart, G., & Cornuet, J. M. (1999). Bottleneck: A program for detecting recent effective population size reductions from allele data frequencies. *Journal of Heredity*, *90*, 502–503.

Powles, S. B., Lorraine-Colwill, D. F., Dellow, J. J., & Preston, C. (1998). Evolved resistance to glyphosate in rigid ryegrass (*Lolium rigidum*) in Australia. *Weed Science*, *46*, 604–607.

Powles, S. B., & Yu, Q. (2010). Evolution in action: Plants resistant to herbicides. *Annual Reviews in Plant Biology*, *61*, 317–347.

Preston, C., Wakelin, A. M., Dolman, F. C., Bostamam, Y., & Boutsalis, P. (2009). A decade of glyphosate-resistant *Lolium* around the world: Mechanisms, genes, fitness, and agronomic management. *Weed Science*, *57*, 435–441.

Pritchard, J. K., Stephens, M., & Donnelly, P. (2000). Inference of population structure using multilocus genotype data. *Genetics*, *155*, 945–959.

Raymond, M., & Rousset, F. (1995). GENEPOP (version 1.2): Population genetics software for exact tests and ecumenicism. *Journal of Heredity*, *86*, 248–249.

Rice, W. R. (1989). Analyzing tables of statistical tests. *Evolution*, *43*, 223–225.

Rosenberg, N. A. (2004). DISTRUCT: A program for the graphical display of population structure. *Molecular Ecology Notes*, *4*, 137–138.

Rousset, F. (2008). Genepop'007: A complete reimplementation of the Genepop software for Windows and Linux. *Molecular Ecology Resources*, *8*, 103–106.

Sakai, A. K., Allendorf, F. W., Holt, J. S., Lodge, D. M., Molofsky, J., With, K. A., ... Weller, S. G. (2001). The population biology of invasive species. *Annual Review of Ecology and Systematics*, *32*, 305–332.

Simarmata, M., Kaufmann, J. E., & Penner, D. (2003). Potential basis of glyphosate resistance in California rigid ryegrass (*Lolium rigidum*). *Weed Science*, *51*, 678–682.

Van Oosterhout, C., Hutchinson, W. F., Wills, D. P. M., & Shipley, P. (2004). MICRO-CHECKER: Software for identifying and correcting genotyping errors in microsatellite data. *Molecular Ecology Notes*, *4*, 535–538.

Vila-Aiub, M. M., Neve, P., & Powles, S. B. (2009). Fitness costs associated with evolved herbicide resistance alleles in plants. *New Phytologist*, *184*, 751–767.

Wang, J., Dobrowolski, M. P., Cogan, N. O. I., Forster, J. W., & Smith, K. F. (2009). Assignment of individuals genotypes to specific forage cultivars of perennial ryegrass based on SSR markers. *Crop Science*, *49*, 49–58.

Ward, S. M., & Jasieniuk, M. (2009). Sampling weedy and invasive plant populations for genetic diversity analysis. *Weed Science*, *57*, 59–602.

15

Genomic insights into adaptive divergence and speciation among malaria vectors of the *Anopheles nili* group

Caroline Fouet[1] ⓘ | Colince Kamdem[1] | Stephanie Gamez[1] | Bradley J. White[1,2]

[1]Department of Entomology, University of California, Riverside, CA, USA

[2]Center for Disease Vector Research, Institute for Integrative Genome Biology, University of California, Riverside, CA, USA

Correspondence
Caroline Fouet and Bradley J. White, Department of Entomology, University of California, Riverside, CA, USA.
Emails: caroline.fouet@ucr.edu and bwhite@ucr.edu

Funding information
National Institutes of Health, Grant/Award Number: 1R01AI113248 and 1R21AI115271; University of California Riverside

Abstract

Ongoing speciation in the most important African malaria vectors gives rise to cryptic populations, which differ remarkably in their behavior, ecology, and capacity to vector malaria parasites. Understanding the population structure and the drivers of genetic differentiation among mosquitoes is crucial for effective disease control because heterogeneity within vector species contributes to variability in malaria cases and allow fractions of populations to escape control efforts. To examine population structure and the potential impacts of recent large-scale control interventions, we have investigated the genomic patterns of differentiation in mosquitoes belonging to the *Anopheles nili* group—a large taxonomic group that diverged ~3 Myr ago. Using 4,343 single nucleotide polymorphisms (SNPs), we detected strong population structure characterized by high-F_{ST} values between multiple divergent populations adapted to different habitats within the Central African rainforest. Delineating the cryptic species within the *Anopheles nili* group is challenging due to incongruence between morphology, ribosomal DNA, and SNP markers consistent with incomplete lineage sorting and/or interspecific gene flow. A very high proportion of loci are fixed ($F_{ST} = 1$) within the genome of putative species, which suggests that ecological and/or reproductive barriers are maintained by strong selection on a substantial number of genes.

KEYWORDS
Anopheles nili, divergent selection, high-F_{ST} regions, speciation

1 | INTRODUCTION

One of the principal goals of population genetics is to summarize the genetic similarities and differences between populations (Wright, 1984). This task can be relatively straightforward for some taxa, but the genetic relationship among populations can also be difficult to summarize, especially for species whose evolutionary history is complex and reticulate. The best known mosquito species of the genus *Anopheles*—which includes all vectors of human malaria parasites—exhibit very complex rangewide population structure due to the combined effects of cryptic speciation, adaptive flexibility and ongoing gene flow across strong but incomplete reproductive barriers (Harbach, 2013;

Krzywinski & Besansky, 2003). For example, almost all major malaria vectors of the Afrotropical region belong to large taxonomic groups encompassing multiple incipient species relatively isolated reproductively and geographically from one another (reviewed by Sinka et al., 2010; Antonio-Nkondjio & Simard, 2013; Coetzee & Koekemoer, 2013; Dia, Guelbeogo, & Ayala, 2013; Lanzaro & Lee, 2013). These characteristics make them promising model systems to study speciation and the processes which contribute to reproductive barriers (e.g., Turner, Hahn, & Nuzhdin, 2005; Lawniczak et al., 2010; Neafsey et al., 2010; Fontaine et al., 2015; Weng, Yu, Hahn, & Nakhleh, 2016), but can also have far-reaching practical consequences. Both spatial and temporal variabilities in malaria cases and the effectiveness of vector control measures are greatly impacted by heterogeneity within vector species (Molineaux & Gramiccia, 1980; Van Bortel et al., 2001). For these reasons, research on the genetic structure among the major African malaria vector mosquitoes has intensified over the last few decades (Antonio-Nkondjio & Simard, 2013; Coetzee & Koekemoer, 2013; Dia et al., 2013; Lanzaro & Lee, 2013).

The recent scaling up of insecticide-treated nets usage and indoor insecticide spraying to a lesser extent have led to a dramatic reduction in malaria morbidity and mortality across the continent (WHO, 2016). However, other consequences of these large-scale interventions include increased insecticide resistance (reviewed by Hemingway et al., 2016; Ranson & Lissenden, 2016), range shift (e.g., Bøgh, Pedersen, Mukoko, & Ouma, 1998; Derua et al., 2012; Mwangangi et al., 2013) and profound evolutionary changes among vector populations. In contrast to insecticide resistance and range shift, which have been extensively studied, the recent adaptive changes among mosquito populations have yet to be addressed significantly. These changes—which involve local adaptation and genetic differentiation, introgressive hybridization, and selective sweeps across loci conferring resistance to xenobiotics—are particularly evident in the most anthropophilic species (Barnes et al., 2017; Clarkson et al., 2014; Kamdem, Fouet, Gamez, & White, 2017; Norris et al., 2015).

The ecology, taxonomic complexity, geographic distribution, role in transmission, and evolutionary potential of each vector species are unique. Consequently, further research is needed to specifically resolve population structure and the genomic targets of natural selection at a fine scale in all of the important taxa including currently understudied species. The present work focused on a group of malaria vector species representing a large taxonomic unit named *Anopheles nili* group. Despite the significant role some of its species play in sustaining high malaria transmission, this group has received little attention. To date, four species that occur in forested areas of Central and West Africa and are distinguishable by slight morphological variations are known within the *An. nili* group: *An. nili sensu stricto* (hereafter *An. nili*), *An. ovengensis*, *An. carnevalei*, and *An. somalicus* (Awono-Ambene, Kengne, Simard, Antonio-Nkondjio, & Fontenille, 2004; Gillies & Coetzee, 1987; Gillies & De Meillon, 1968). These species are characterized by reticulate evolution and complex phylogenies that have been challenging to resolve so far (Awono-Ambene et al., 2004, 2006; Kengne, Awono-Ambene, Antonio-Nkondjio, Simard, & Fontenille, 2003; Ndo et al., 2010, 2013; Peery et al., 2011; Sharakhova et al., 2013). Populations of *An. nili* and *An. ovengensis* are very anthropophilic and efficient vectors of *Plasmodium* in rural areas where malaria prevalence is particularly high (Antonio-Nkondjio et al., 2006).

To delineate genomic patterns of differentiation, we sampled mosquito populations throughout the range of species of the *An. nili* group in Cameroon and used reduced representation sequencing to develop genomewide SNP markers that we genotyped in 145 individuals. We discovered previously unknown subpopulations characterized by high pairwise differentiation within *An. ovengensis* and *An. nili*. We further explored the genetic differentiation across the genome and revealed the presence of a very high number of outlier loci that are targets of selection among locally adapted subpopulations. These

findings provide significant baseline data on the genetic underpinnings of adaptive divergence and pave the way for further genomic studies in this important group of mosquitoes. Notably, a complete reference genome will enable us to conduct in-depth studies in order to decipher the functional and phenotypic characteristics of the numerous differentiated loci as well as the contribution of recent selective events in ongoing adaptation.

2 | MATERIALS AND METHODS

2.1 | Mosquito species

We surveyed 28 locations within the geographic ranges of species of the *An. nili* group previously described in Cameroon (Figure 1) (Antonio-Nkondjio et al., 2009; Awono-Ambene et al., 2004, 2006; Ndo et al., 2010, 2013). The genetic structure of *Anopheles* species is most often based on macrogeographic or regional subdivisions of gene pools, but can also involve more subtle divergence between larvae and adults, or between adult populations found in or around human dwellings (e.g., Riehle et al., 2011). To effectively estimate

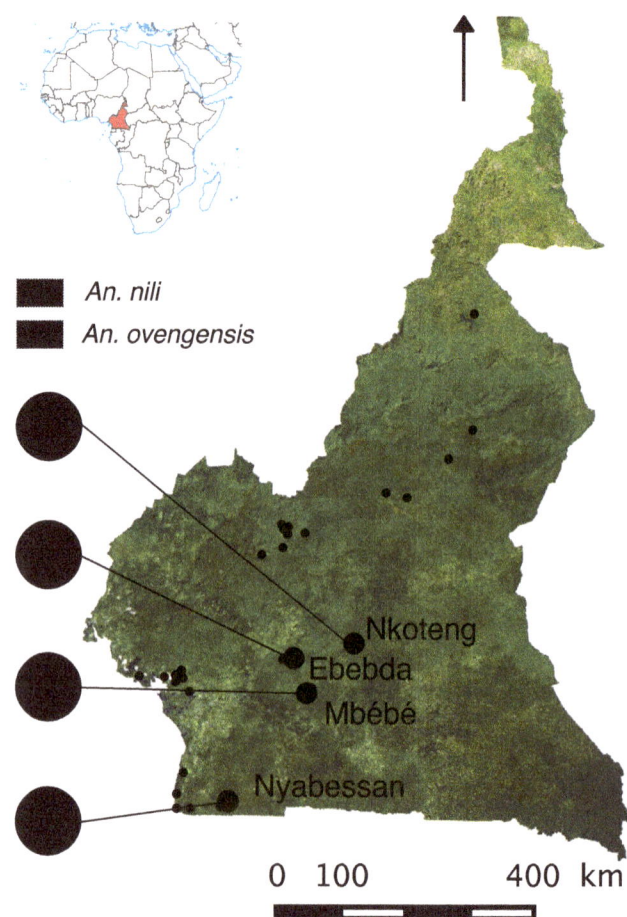

FIGURE 1 Map showing the sampling locations and the relative frequencies of the morphologically defined species *An. nili* and *An. ovengensis* in Cameroon. Small and large black dots indicate, respectively, the 28 locations surveyed and the four sampling sites where mosquitoes were collected

the genetic diversity and identify potential cryptic populations within species, we collected larvae and adult mosquitoes within and around human dwellings using several sampling techniques (Service, 1993) in September–October 2013 (Table S1). To identify the four currently known members of the *An. nili.* group, we used morphological keys and a diagnostic PCR, which discriminates species based on point mutations of the ribosomal DNA (Awono-Ambene et al., 2004; Gillies & Coetzee, 1987; Gillies & De Meillon, 1968; Kengne et al., 2003).

2.2 | Library preparation, sequencing, and SNP discovery

We created double-digest restriction site-associated DNA (ddRAD) libraries as described in Kamdem et al. (2017) using a modified version of the protocol designed by Peterson, Weber, Kay, Fisher, and Hoekstra (2012). Briefly, genomic DNA of mosquitoes was extracted using the DNeasy Blood and Tissue kit (Qiagen) and the Zymo Research MinPrep kit for larvae and adult samples, respectively. Approximately 50 ng (10 μl) of DNA of each mosquito was digested simultaneously with *MluC1* and *NlaIII* restriction enzymes. Digested products were ligated to adapter and barcode sequences enabling identification of individuals. Samples were pooled, purified, and 400-bp fragments selected. The resulting libraries were amplified via PCR and purified, and fragment size distribution was checked using the BioAnalyzer. PCR products were quantified, diluted and single-end sequenced to 100 base reads on Illumina HiSeq2000.

2.3 | SNP discovery and genotyping

The *process_radtags* program of the Stacks v 1.35 pipeline (Catchen, Hohenlohe, Bassham, Amores, & Cresko, 2013; Catchen, Amores, Hohenlohe, Cresko, & Postlethwait, 2011) was used to demultiplex and clean Illumina sequences. Reads that passed quality filters were aligned to the *An. nili* Dinderesso draft genome assembly (Giraldo-Calderon et al., 2015) made up of 51,048 short contigs (~200–30,512 bp long) using Gsnap (Wu & Nacu, 2010). To identify and call SNPs within consensus RAD loci, we utilized the *ref_map.pl* program of Stacks. We set the minimum number of reads required to form a stack to three and allowed two mismatches during catalogue creation. We generated SNP files in different formats for further downstream analyses using the *populations* program of Stacks and Plink v1.09 (Purcell et al., 2007).

2.4 | Population genomics analyses

We analyzed the genetic structure of *An. nili sensu* lato (s.l.) populations using a principal component analysis (PCA) and an unrooted Neighbor-Joining tree (NJ). We also examined ancestry proportions and admixtures between populations in Admixture v1.23 (Alexander, Novembre, & Lange, 2009) and Structure v2.3.4 (Pritchard, Stephens, & Donnelly, 2000). We used the package *adegenet* (Jombart, 2008) to implement the PCA in R (R Development Core Team 2016). The individual-based NJ network was generated from SNP allele frequencies via a matrix

of Euclidian distance using the R package *ape* (Paradis, Claude, & Strimmer, 2004). We ran Admixture with 10-fold cross-validation for values of *k* from 1 through 8. Similarly, we analyzed patterns of ancestry from *k* ancestral populations in Structure, testing five replicates of *k* = 1–8. We used 200,000 iterations and discarded the first 50,000 iterations as burn-in for each Structure run. Clumpp v1.1.2 (Jakobsson & Rosenberg, 2007) was used to summarize assignment results across independent runs. To identify the optimal number of genetic clusters in our sample, we applied simultaneously the lowest cross-validation error in Admixture, the ad hoc statistic deltaK (Earl & VonHoldt, 2012; Evanno, Goudet, & Regnaut, 2005) and the discriminant analysis of principal component (DAPC) method implemented in *adegenet*. To examine the level of genomic divergence among populations, we assessed genetic differentiation (F_{ST}) across SNPs using the *populations* program of the Stacks pipeline. Mean F_{ST} values were also used to quantify pairwise divergence between populations. To infer the demographic history of different populations, we used the diffusion approximation method implemented in the package *∂a∂i* v 1.6.3 (Gutenkunst, Hernandez, Williamson, & Bustamante, 2009). Single-population models were fitted to allele frequency spectra, and the best model was selected using the lowest likelihood and Akaike information criterion as well as visual inspections of residuals.

3 | RESULTS

3.1 | SNP genotyping

We collected mosquitoes from four locations out of 28 sampling sites (Figure 1, Table S1) and sequenced 145 individuals belonging, according to morphological criteria and diagnostic PCRs, to two species (*An. nili* [*n* = 24] and *An. ovengensis* [*n* = 121]). We assembled 197,724 RAD loci that mapped to unique positions throughout the reference genome. After applying stringent filtering rules, 408 loci present in all populations and in at least 50% of individuals in each population were retained. Within these loci, we identified 4,343 high-quality biallelic markers that were used to analyze population structure and genetic differentiation.

3.2 | Morphologically defined species do not correspond to genetic clusters

The PCA and the NJ tree show that the genetic variation across 4,343 SNPs is best explained by more than two clusters, implying subdivisions within *An. nili* and *An. ovengensis* (Figure 2). Three subgroups are apparent within *An. nili* while two distinct clusters segregate in *An. ovengensis* (hereafter referred to as *An. nili* group 1, *An. nili* group 2, *An. nili* group 3, *An. ovengensis* group 1 and *An. ovengensis* group 2). These five subpopulations are strongly correlated with the different sampling sites suggesting local adaptation of divergent populations. Importantly, Structure and Admixture analyses reveal that, at *k* = 2, one population identified by morphology and the diagnostic PCR as *An. nili* has almost the same ancestry pattern as the largest *An. ovengensis* cluster (Figure 3). Such discrepancies between morphology-based and

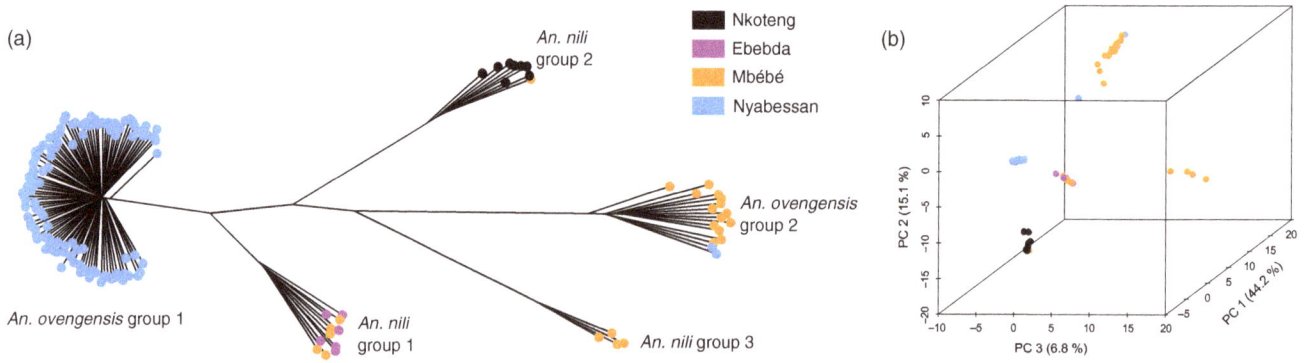

FIGURE 2 Population genetic structure inferred from 4,343 SNPs using a PCA (a) and a neighbor-joining tree (b). The percentage of variance explained is indicated on each PCA axis. Note the strong association between the five genetic clusters and the different sampling locations

molecular taxonomies can be due to a variety of processes including phenotypic plasticity, introgressive hybridization, or incomplete lineage sorting (i.e., when independent loci have different genealogies by chance) (Arnold, 1997; Combosch & Vollmer, 2015; Fontaine et al., 2015; Weng et al., 2016). At $k = 2$ and $k = 3$, some populations also exhibit half ancestry from each morphological species suggestive of gene flow. We found a conflicting number of genetic clusters in our samples likely reflecting the complex history of subdivisions and admixtures among populations (Figure 4). The Evanno et al. (2005) method, which highlights the early stages of divergence between *An. nili* and *An. ovengensis*, indicates two probable ancestors. DAPC and the Admixture cross-validation error, which are more sensitive to recent hierarchical population subdivisions, show five or more distinct clusters as revealed by the PCA and the NJ tree (Figure 4).

As suggested by the long internal branches, which connect subpopulations on the NJ tree, there is strong differentiation between and within morphological species characterized by globally high-F_{ST}

values (Table 1). Relatively lower F_{ST} values observed between certain clusters may be due to greater interpopulation migration and intermixing or more recent divergence. The F_{ST} values do not reflect the morphological delimitation of species. Indeed, the level of genetic differentiation is higher between some subpopulations within the same morphological species. Overall, patterns of genetic structure and differentiation reveal a group of populations whose phylogenies and species status are likely confounded by hybridization and/or incomplete lineage sorting. We argue that the current taxonomy based on morphology and ribosomal DNA does not capture the optimal reproductive units among populations of this group of mosquitoes.

3.3 | Genomic signatures of divergent selection and demographic history

We analyzed patterns of genetic differentiation across SNP loci throughout the genome. Pairwise comparisons are based on filtered variants that satisfy all criteria to be present in both populations, which explains the discrepancy in the number of SNPs observed between specific paired comparisons (Figure 5). The distribution of locus-specific F_{ST} values between the five subpopulations revealed a U-shape characterized by two peaks around 0 and 1. The large majority of SNPs have low-to-moderate divergence, but a substantial number of variants are extremely differentiated between populations. The maximum F_{ST} among SNPs is 1, and the proportion of loci with $F_{ST} = 1$ varies from 6.52% between the populations we termed *An. nili* group 1 and *An. ovengensis* group 1 to 44.74% between the subgroups called *An. nili* group 2 and *An. nili* group 3 (Figure 5). This pattern of genomewide divergence suggests that a very high number of sites with abrupt differentiation—which likely contain genes that contribute to divergent selection and/or reproductive isolation—coexist with regions of weak divergence that can be freely exchanged between species. As is the case with the overall genetic differentiation, morphology is not a reliable predictor of locus-specific divergence. Precisely, the lowest percentage of fixed SNPs is found between *An. ovengensis* from Nyabessan and *An. nili* collected from Mbébé and Ebebda (Figures 1 and 5). In contrast, the largest proportion of fixed loci is observed between locally adapted subgroups within the same morphological species: *An. nili*. The draft reference genome made

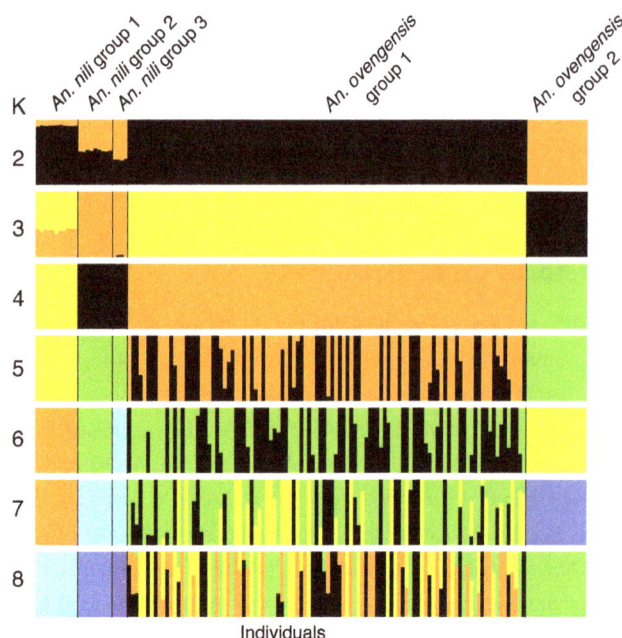

FIGURE 3 Ancestry proportions inferred in Admixture with $k = 2–8$

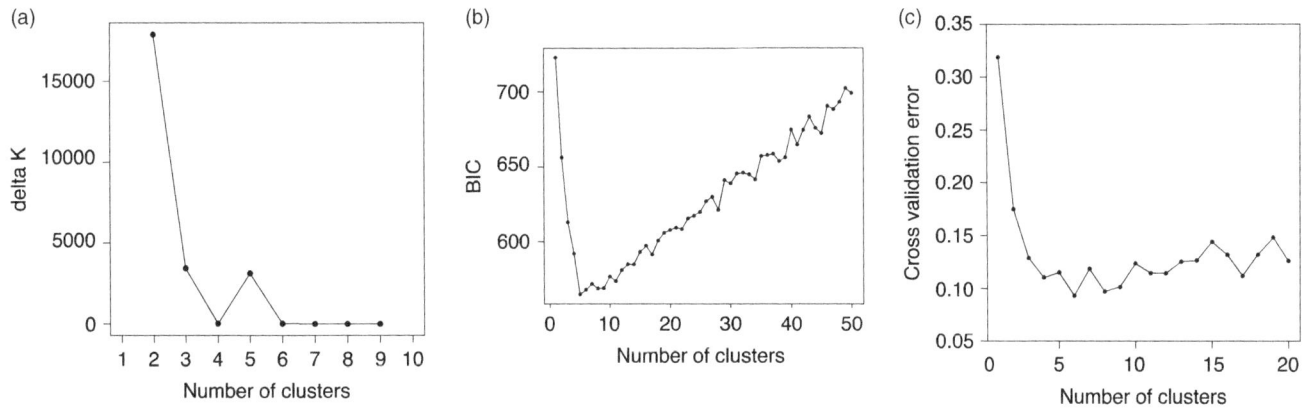

FIGURE 4 Identification of the optimal number of genetic clusters using the delta k method of Evanno et al. (2005) (a), DAPC (b) and 10-fold cross-validation in Admixture (c). The lowest Bayesian information criterion (BIC) and cross-validation error and the highest delta k indicate the most probable number of clusters

TABLE 1 Pairwise F_{ST} between divergent subpopulations of An. nili s.l.

F_{ST}	An. nili group 1	An. nili group 2	An. nili group 3	An. ovengensis group 1	An. ovengensis group 2
An. nili group 1	–				
An. nili group 2	0.374	–			
An. nili group 3	0.506	0.552	–		
An. ovengensis group 1	0.135	0.275	0.364	–	
An. ovengensis group 2	0.432	0.458	0.492	0.349	–

up of short contigs did not enable us to test hypotheses about the genomic distribution of differentiated loci. For example, it remains unknown whether the numerous SNPs that are fixed among populations are spread throughout the entire genome or clustered within genomic regions of low recombination including chromosomal inversions and chromosome centers (Nosil & Feder, 2012; Roesti, Hendry, Salzburger, & Berner, 2012).

Models of population demography indicate that all subgroups have experienced an increase in effective size in a more or less recent past (Table 2). Nevertheless, confidence intervals of population parameters are high in some populations, and our results should be interpreted with the necessary precautions. The population growth is less significant in An. nili group 1.

4 | DISCUSSION

4.1 | Genetic differentiation

Advances in sequencing and analytical approaches have opened new avenues for the study of genomes of disease vectors. We have focused on malaria mosquitoes of the An. nili group, whose taxonomy and population structure have been challenging to resolve with low-resolution markers. We analyzed genetic structure using genomewide SNPs and found strong differentiation and local adaption among populations belonging to the two morphologically defined species An. nili and An.

ovengensis. The exact number of subpopulations remains contentious, with the suggested number of divergent clusters varying from two to five. Significant population structure at eight microsatellite loci has been described among An. nili populations from Cameroon, with F_{ST} values as high as 0.48 between samples from the rainforest area (Ndo et al., 2013). By contrast, An. ovengensis was discovered recently and the genetic structure of this vector remains understudied. This species was initially considered as a sibling of An. nili (Awono-Ambene et al., 2004, 2006; Kengne et al., 2003), but more recent studies have started to challenge the assumed relatedness between the two species due to the high divergence revealed by polytene chromosomes (Sharakhova et al., 2013). Our findings call for a careful review of the current taxonomy within this group of species, which is a necessary first step for accurately delineating the role played by the different subpopulations in malaria transmission.

Our samples were collected from locations characterized by a more or less degraded forest within the rainforest area of Cameroon. In these habitats, larvae of An. nili s.l. exploit relatively similar breeding sites consisting of slow-moving rivers (Antonio-Nkondjio et al., 2009). The ecological drivers of genetic differentiation remain unknown, and will be difficult to infer from our data given the apparent similarity of habitats among the divergent populations we described. Further study is needed to clearly address the environmental variables that may be correlated with ongoing adaptive divergence at adult and larval stages. One of the most expected outcomes of current large-scale malaria

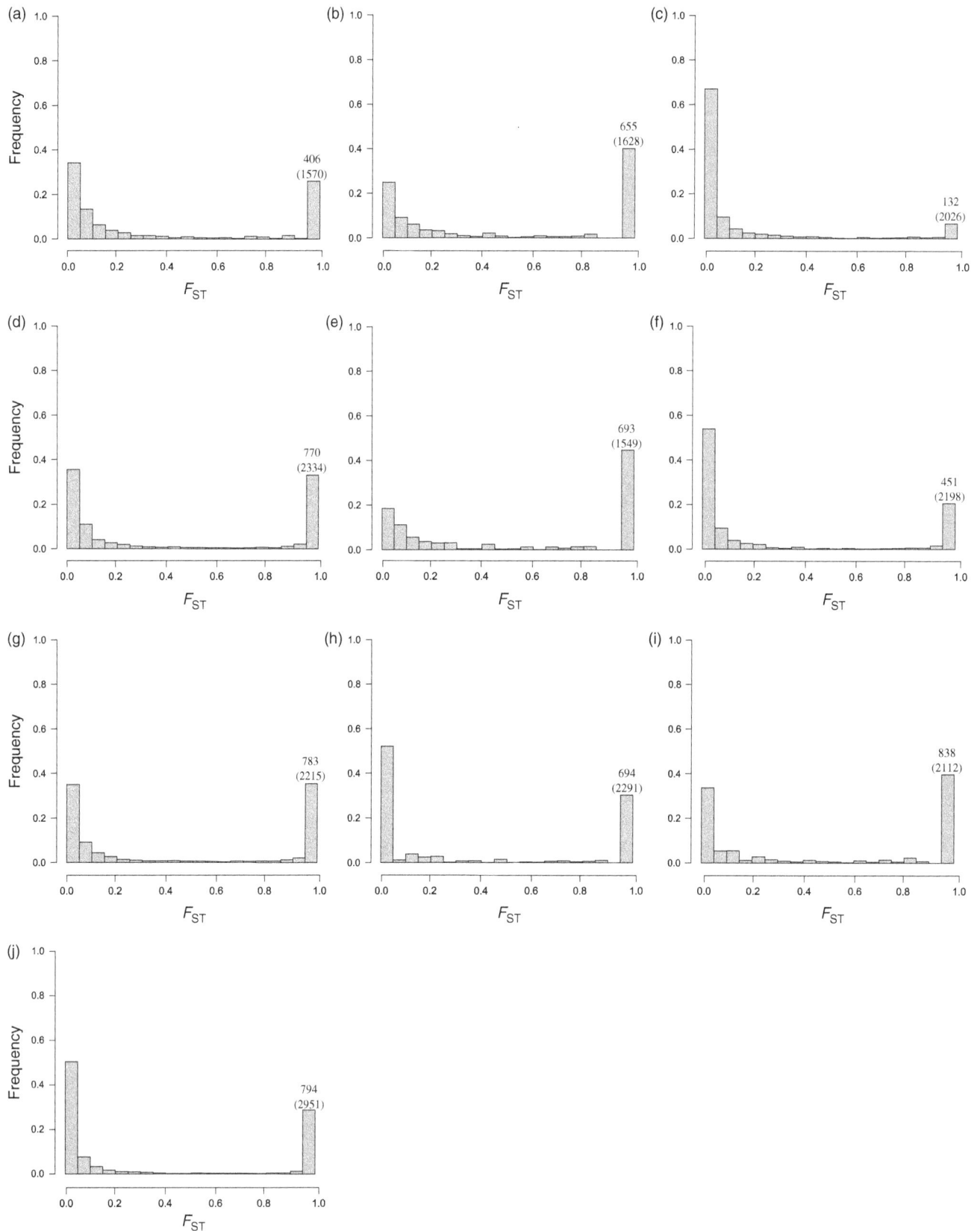

FIGURE 5 Distribution of F_{ST} values throughout the genome between *An. nili* group 1 and *An. nili* group 2 (a); *An. nili* group 1 and *An. nili* group 3 (b); *An. nili* group 1 and *An. ovengensis* group 1 (c); *An. nili* group 1 and *An. ovengensis* group 2 (d); *An. nili* group 2 and *An. nili* group 3 (e); *An. nili* group 2 and *An. ovengensis* group 1 (f); *An. nili* group 2 and *An. ovengensis* group 2 (g); *An. nili* group 3 and *An. ovengensis* group 1 (h); *An. nili* group 3 and *An. ovengensis* group 2 (i); *An. ovengensis* group 1 and *An. ovengensis* group 2 (j). The number of SNPs with F_{ST} = 1 is indicated in each pairwise comparison as well as the total number of SNPs in parenthesis

TABLE 2 Demographic models of different subgroups of *An. nili* s.l.

Population	Best model	Log-likelihood	Final population size[a] (95% CI)	Time[b] (95% CI)
An. nili group 1	*Growth*	−18.42	6.41 (5.326–20.71)	3.70 (1.11–13.31)
An. nili group 2	*Two-epoch*	−19.97	17.87 (9.33–35.50)	11.27 (4.93–19.64)
An. ovengensis group 1	*Growth*	−112.18	13.04 (12.15–17.26)	0.70 (0.58–1.08)
An. Ovengensis group 2	*Growth*	−22.98	19.95 (14.45–45.70)	5.11 (2.33–15.13)

[a]Relative to ancestral population size.
[b]Expressed in units 2Ne generations from start of growth to present.

control measures that are underway in sub-Saharan African countries concerns the effects of increased insecticide exposure on the genetic diversity and population demography of vectors. A substantial population decline that may considerably affect the adaptive potential of vector species has been occasionally reported following a major insecticide-treated bed net distribution campaign and/or indoor residual house spraying (e.g., Athrey et al., 2012). The inferred demographic history of the different subpopulations within the *An. nili* group does not reveal signatures of bottlenecks that can be potentially correlated with increased usage of insecticides and insecticide-treated nets. This result is consistent with the demography of several other important malaria vectors of the Afrotropical region, including *An. gambiae, An. coluzzii, An. funestus* and *An. moucheti*, which reveals a substantial population increase suggesting that intense insecticide exposure has yet to leave deep or detectable impacts on patterns of genetic variation among mosquito populations (Fouet, Kamdem, Gamez, & White, 2017; Kamdem et al., 2017; O'Loughlin et al., 2014).

4.2 | Genomic architecture of geographic and reproductive isolation

Understanding the genomic architecture of reproductive isolation may reveal crucial information on the sequence of events that occur from the initial stages of divergence among populations to the onset of strong reproductive barriers between species (e.g., Turner et al., 2005; Harr, 2006; Nadeau et al., 2012; Ellegren et al., 2012; Carneiro et al., 2014; Burri et al., 2015). One influential concept of speciation coined the "genic view of species" proposes that boundaries between species are properties of individual genes or genome regions and not of whole organisms or lineages (Barton & Hewitt, 1985; Harrison & Larson, 2014; Harrison, 1990; Key, 1968; Nosil & Feder, 2012; Rieseberg, Whitton, & Gardner, 1999; Wu, 2001). We have discovered a substantially high number of SNPs that are strongly differentiated between populations and often fixed within subgroups of *An. nili* s.l. Interpreting this intriguing pattern of genomic differentiation is not straightforward due to the complex interactions between numerous forces—including positive or negative selection, recombination, introgressive hybridization and incomplete lineage sorting—that can affect the level of divergence among SNPs (Begun & Aquadro, 1992; Cutter & Payseur, 2013; Harrison & Larson, 2016; Nachman & Payseur, 2012; Roesti et al., 2012). Some of these variants exhibiting high divergence among populations certainly contain markers of ecological and/or reproductive isolation. However, as far as reproductive

barriers are concerned, recent studies have indicated a complex relationship between the degree of genetic differentiation and gene flow at the genome level (e.g., Gompert et al., 2012; Hamilton, Lexer, & Aitken, 2013a,b; Larson, Andrés, Bogdanowicz, & Harrison, 2013; Larson, White, Ross, & Harrison, 2014; Parchman et al., 2013; Taylor, Curry, White, Ferretti, & Lovette, 2014). Highly divergent genomic regions do not necessarily coincide with regions of reduced gene flow among established or emerging species. Several alternative interpretations exist for the numerous high-F_{ST} regions we detected in all pairwise comparisons (Cruickshank & Hahn, 2014; Delmore et al., 2015; Nachman & Payseur, 2012; Noor & Bennett, 2009). Nevertheless, careful examination of these outliers of differentiation may reveal significant insights into the wide range of genes and traits that contribute to ecological divergence and/or reproductive isolation between subgroups of *An. nili* s.l. A complete genome assembly will be necessary to better delineate specific regions of the genome under natural selection, and therefore clarify the genomic basis of phenotypic fitness differences between divergent populations. This will also help understand the extent to which recent selection associated with human interventions contribute to local adaptation and genetic differentiation as observed in *An. gambiae* and *An. coluzzii* (Kamdem et al., 2017).

Signals consistent with gene flow between *An. nili* and *An. ovengensis* are apparent in our data although it has been proposed that the two morphological species diverged ~3 Myr ago (Ndo et al., 2013). Some individuals display almost half ancestry from each morphological species. The disagreement between morphology/PCR and molecular taxonomies observed in Structure and Admixture analyses also suggests that incongruent genealogies may be widespread along chromosomes due to hybridization. However, hybridization can be difficult to detect because other factors such as incomplete lineage sorting or technical artifacts can leave signatures that are similar to those of interspecific gene flow (Liu et al., 2014; Patterson et al., 2012). A complete reference genome is also needed to analyze the detailed distribution of genealogies across small genomic windows and to disentangle the relative contribution of processes that generate the putative admixtures and species confusion observed among divergent populations (Fontaine et al., 2015; Martin et al., 2013; Weng et al., 2016).

5 | CONCLUSIONS AND IMPLICATIONS

Delineating the fine-scale population structure of mosquito populations is crucial for understanding their epidemiological significance and their potential response to vector control measures. Moreover, recent malaria control efforts affect interspecific gene flow, genetic differentiation, population demography and natural selection in mosquitoes (Athrey et al., 2012; Barnes et al., 2017; Clarkson et al., 2014; Kamdem et al., 2017; Norris et al., 2015). Deciphering the signatures of these processes across mosquito genomes is important to minimize their negative impacts on vector control. Our findings shed some light on the complex evolutionary history and provide a framework for future investigations into the genetic basis of ecological and reproductive barriers among species of the *An. nili* group.

ACKNOWLEDGEMENTS

Funding for this project was provided by the University of California Riverside and NIH grants 1R01AI113248 and 1R21AI115271 to BJW. We thank inhabitants and administrative authorities of the sampling sites included in this study for their collaboration.

AUTHOR CONTRIBUTIONS

CF, CK, and BJW conceived and designed the experiments. CF, CK, SG, and BJW performed the experiments. CF and CK analyzed the data. CF, CK, and BJW wrote the manuscript.

REFERENCES

Alexander, D. H., Novembre, J., & Lange, K. (2009). Fast model-based estimation of ancestry in unrelated individuals. *Genome Research*, 19, 1655–1664.

Antonio-Nkondjio, C., Kerah, C. H., Simard, F., Awono-Ambene, P., Chouaibou, M., Tchuinkam, T., & Fontenille, D. (2006). Complexity of the malaria vectorial system in Cameroon: Contribution of secondary vectors to malaria transmission. *Journal of Medical Entomology*, 43, 1215–1221. https://doi.org/10.1603/0022-2585 (2006)43[1215:COTMVS]2.0.CO;2

Antonio-Nkondjio, C., Ndo, C., Costantini, C., Awono-Ambene, P., Fontenille, D., & Simard, F. (2009). Distribution and larval habitat characterization of Anopheles moucheti, Anopheles nili, and other malaria vectors in river networks of southern Cameroon. *Acta Tropica*, 112(3), 270–276. https://doi.org/10.1016/j.actatropica.2009.08.009

Antonio-Nkondjio, C., & Simard, F. (2013). Highlights on Anopheles nili and Anopheles moucheti, Malaria Vectors in Africa. In S. Manguin (Ed.), *Anopheles mosquitoes - New insights into malaria vectors (INTECH)* (pp. 828). Croatia, European Union: InTech. ISBN 978-953-51-1188-7.

Arnold, M. L. (1997). *Natural hybridization and evolution*. Oxford: Oxford University Press.

Athrey, G., Hodges, T. K., Reddy, M. R., Overgaard, H. J., Matias, A., & Ridl, F. C., ... Slotman, M. A. (2012). The effective population size of malaria mosquitoes: Large impact of vector control. *PLoS Genetics*, 8(12), e1003097. https://doi.org/10.1371/journal.pgen.1003097

Awono-Ambene, H. P., Kengne, P., Simard, F., Antonio-Nkondjio, C., & Fontenille, D. (2004). Description and bionomics of Anopheles (Cellia) ovengensis (Diptera: Culicidae), a new malaria vector species of the Anopheles nili group from south Cameroon. *Journal of Medical Entomology*, 41, 561–568. https://doi.org/10.1603/0022-2585-41.4.561

Awono-Ambene, H. P., Simard, F., Antonio-Nkondjio, C., Cohuet, A., Kengne, P., & Fontenille, D. (2006). Multilocus enzyme electrophoresis supports speciation within the Anopheles nili group of malaria vectors in Cameroon. *The American Journal of Tropical Medicine and Hygiene, 75*,

656–658. https://doi.org/75/4/656 [pii]

Barnes, K. G., Weedall, G. D., Ndula, M., Irving, H., Mzihalowa, T., Hemingway, J., & Wondji, C. S. (2017). Genomic Footprints of Selective Sweeps from Metabolic Resistance to Pyrethroids in African Malaria Vectors Are Driven by Scale up of Insecticide-Based Vector Control. *PLoS Genetics*, 13(2), e10. 1–22. https://doi.org/10.1371/journal.pgen.1006539

Barton, N., & Hewitt, G. (1985). Analysis of hybrid zones. *Annual Review of Ecology and Systematics*, 16, 113–148.

Begun, D., & Aquadro, C. (1992). Levels of naturally occurring DNA polymorphism correlate with recombination rates in *D. melanogaster*. *Nature*, 356, 519–520.

Bøgh, C., Pedersen, E. M., Mukoko, D. A, & Ouma, J. H. (1998). Permethrin-impregnated bednet effects on resting and feeding behaviour of lymphatic filariasis vector mosquitoes in Kenya. *Medical and Veterinary Entomology*, 12(1), 52–59.

Burri, R., Nater, A., Kawakami, T., Mugal, C. F., Olason, P. I., & Smeds, L, ... Ellegren, H. (2015). Linked selection and recombination rate variation drive the evolution of the genomic landscape of differentiation across the speciation continuum of Ficedula flycatchers. *Genome Research*, 25(11), 1656–1665. https://doi.org/10.1101/gr.196485.115

Carneiro, M., Albert, F. W., Afonso, S., Pereira, R. J., Burbano, H., & Campos, R, ... Ferrand, N. (2014). The genomic architecture of population divergence between subspecies of the European Rabbit. *PLoS Genetics*, 10(8), e1003519. https://doi.org/10.1371/journal.pgen.1003519

Catchen, J. M., Amores, A., Hohenlohe, P., Cresko, W., & Postlethwait, J. H. (2011). Stacks: Building and genotyping Loci de novo from short-read sequences. *G3 (Bethesda, Md.)*, 1(3), 171–182. https://doi.org/10.1534/g3.111.000240

Catchen, J., Hohenlohe, P. a., Bassham, S., Amores, A., & Cresko, W. (2013). Stacks: An analysis tool set for population genomics. *Molecular Ecology*, 22, 3124–3140. https://doi.org/10.1111/mec.12354

Clarkson, C. S., Weetman, D., Essandoh, J., Yawson, A. E., Maslen, G., & Manske, M, ... Donnelly, M. J. (2014). Adaptive introgression between Anopheles sibling species eliminates a major genomic island but not reproductive isolation. *Nature Communications*, 5, 4248. https://doi.org/10.1038/ncomms5248

Coetzee, M., & Koekemoer, L. (2013). Molecular Systematics and Insecticide Resistance in the Major African Malaria Vector Anopheles funestus. *Annual Review of Entomology*, 58, 393–412.

Combosch, D. J., & Vollmer, S. V. (2015). Trans-Pacific RAD-Seq population genomics confirms introgressive hybridization in Eastern Pacific Pocillopora corals. *Molecular Phylogenetics and Evolution*, 88, 154–162. https://doi.org/10.1016/j.ympev.2015.03.022

Cruickshank, T. E., & Hahn, M. W. (2014). Reanalysis suggests that genomic islands of speciation are due to reduced diversity, not reduced gene flow. *Molecular Ecology*, 23(13), 3133–3157. https://doi.org/10.1111/mec.12796

Cutter, A. D., & Payseur, B. A. (2013). Genomic signatures of selection at linked sites: Unifying the disparity among species. *Nature Reviews. Genetics*, 14(4), 262–274. https://doi.org/10.1038/nrg3425

Delmore, K. E., Hübner, S., Kane, N. C., Schuster, R., Andrew, R. L., & Câmara, F, ... Irwin, D. E. (2015). Genomic analysis of a migratory divide reveals candidate genes for migration and implicates selective sweeps in generating islands of differentiation. *Molecular Ecology*, 24(8), 1873–1888.

Derua, Y. a, Alifrangis, M., Hosea, K. M., Meyrowitsch, D. W., Magesa, S. M., Pedersen, E. M., & Simonsen, P. E. (2012). Change in composition of the Anopheles gambiae complex and its possible implications for the transmission of malaria and lymphatic filariasis in north-eastern Tanzania. *Malaria Journal*, 11(1), 188. https://doi.org/10.1186/1475-2875-11-188

Dia, I., Guelbeogo, M., & Ayala, D. (2013). Advances and Perspectives in the Study of the Malaria Mosquito Anopheles funestus. In S. Manguin (Ed.), *Anopheles mosquitoes - New insights into malaria vectors (INTECH)* (pp. 828). Croatia, European Union: InTech.

Earl, D. A., & VonHoldt, B. M. (2012). STRUCTURE HARVESTER: A website and program for visualizing STRUCTURE output and implementing the Evanno method. *Conservation Genetics Resources, 4*(359–361).

Ellegren, H., Smeds, L., Burri, R., Olason, P. I., Backström, N., & Kawakami, T., ... Wolf, J. B. W. (2012). The genomic landscape of species divergence in Ficedula flycatchers. *Nature, 491*(7426), 756–760. https://doi.org/10.1038/nature11584

Evanno, G., Goudet, J., & Regnaut, S. (2005). Detecting the number of clusters of individuals using the software structure: A simulation study. *Molecular Ecology, 14*, 2611–2620.

Fontaine, M. C., Pease, J. B., Steele, a., Waterhouse, R. M., Neafsey, D. E., & Sharakhov, I. V, ... Besansky, N. J. (2015). Extensive introgression in a malaria vector species complex revealed by phylogenomics. *Science, 347*(6217), 1258524. https://doi.org/10.1126/science.1258524

Fouet, C., Kamdem, C., Gamez, S., & White, B. J. (2017). Extensive genetic diversity among populations of the malaria mosquito Anopheles moucheti revealed by population genomics. *Infection, Genetics and Evolution, 48*, 27–33. https://doi.org/10.1016/j.meegid.2016.12.006

Gillies, M. T., & Coetzee, M. (1987). *A supplement to the Anophelinae of Africa south of the Sahara.* Johannesburg: The South African Institute for Medical Research.

Gillies, M. T., & De Meillon, B. (1968). *The Anophelinae of Africa South of the Sahara*, 2nd edn. Johannesburg: Publications of the South African Institute for Medical Research.

Giraldo-Calderon, G. I., Emrich, S. J., MacCallum, R. M., Maslen, G., Dialynas, E., & Topalis, P, ... Lawson, D. (2015). VectorBase: An updated bioinformatics resource for invertebrate vectors and other organisms related with human diseases. *Nucleic Acids Research, 43*(D1), D707–D713. https://doi.org/10.1093/nar/gku1117

Gompert, Z., Lucas, L. K., Nice, C. C., Fordyce, J. A., Forister, M. L., & Buerkle, C. A. (2012). Genomic regions with a history of divergent selection affect fitness of hybrids between two butterfly species. *Evolution, 66*(7), 2167–2181. https://doi.org/10.5061/dryad.f0b2f083

Gutenkunst, R. N., Hernandez, R. D., Williamson, S. H., & Bustamante, C. D. (2009). Inferring the joint demographic history of multiple populations from multidimensional SNP frequency data. *PLoS Genetics, 5*(10), e1000695. https://doi.org/10.1371/journal.pgen.1000695

Hamilton, J. a., Lexer, C., & Aitken, S. N. (2013a). Differential introgression reveals candidate genes for selection across a spruce (Picea sitchensis × P. glauca) hybrid zone. *New Phytologist, 197*(3), 927–938. https://doi.org/10.1111/nph.12055

Hamilton, J. A., Lexer, C., & Aitken, S. N. (2013b). Genomic and phenotypic architecture of a spruce hybrid zone (Picea sitchensis × P. glauca). *Molecular Ecology, 22*(3), 827–841. https://doi.org/10.1111/mec.12007

Harbach, R. E. (2013). The Phylogeny and Classification of Anopheles. In S. Manguin (Ed.), *Anopheles mosquitoes - New insights into malaria vectors (INTECH)* (pp. 828) Croatia, European Union: InTech. ISBN 978-953-51-1188-7.

Harr, B. (2006). Genomic islands of differentiation between house mouse subspecies, 730–737. https://doi.org/10.1101/gr.5045006.entiation

Harrison, R. G. (1990). Hybrid zones: Windows on evolutionary process. *Oxford Surveys in Evolutionary Biology, 7*, 69–128.

Harrison, R. G., & Larson, E. L. (2014). Hybridization, introgression, and the nature of species boundaries. *Journal of Heredity, 105*(S1), 795–809. https://doi.org/10.1093/jhered/esu033

Harrison, R. G., & Larson, E. L. (2016). Heterogeneous genome divergence, differential introgression, and the origin and structure of hybrid zones. *Molecular Ecology' 25'* 2454–2466. https://doi.org/10.1111/mec.13582

Hemingway, J., Ranson, H., Magill, A., Kolaczinski, J., Fornadel, C., & Gimnig, J., ... Hamon, N. (2016). Averting a malaria disaster: Will insecticide resistance derail malaria control? *The Lancet, 387*(10029), 1785–1788. https://doi.org/10.1016/s0140-6736(15)00417-1

Jakobsson, M., & Rosenberg, N. (2007). CLUMPP: A cluster matching and permutation program for dealing with multimodality in analysis of population structure. *Bioinformatics, 23*, 1801–1806.

Jombart, T. (2008). adegenet: A R package for the multivariate analysis of genetic markers. *Bioinformatics, 24*, 1403–1405.

Kamdem, C., Fouet, C., Gamez, S., & White, B. J. (2017). Pollutants and insecticides drive local adaptation in African malaria mosquitoes. *Molecular Biology and Evolution, 34*(5), 1261–1275.

Kengne, P., Awono-Ambene, H. P., Antonio-Nkondjio, C., Simard, F., & Fontenille, D. (2003). Molecular identification of the Anopheles nili group African malaria vectors. *Medical and Veterinary Entomology, 17*, 67–74.

Key, K. H. L. (1968). The concept of stasipatric speciation. *Systematic Zoology, 17*, 14–22.

Krzywinski, J., & Besansky, N. J. (2003). Molecular systematics of Anopheles : From subgenera to subpopulations. *Annual Review of Entomology, 48*(1), 111–139. https://doi.org/10.1146/annurev.ento.48.091801.112647

Lanzaro, G. C., & Lee, Y. (2013). Speciation in Anopheles gambiae — The Distribution of Genetic Polymorphism and Patterns of Reproductive Isolation Among Natural Populations. In S. Manguin (Ed.), *Anopheles mosquitoes - New insights into malaria vectors (INTECH).* (pp. 828). Croatia, European Union: InTech.

Larson, E. L., Andrés, J. A., Bogdanowicz, S. M., & Harrison, R. G. (2013). Differential introgression in a mosaic hybrid zone reveals candidate barrier genes. *Evolution, 67*(12), 3653–3661. https://doi.org/10.1111/evo.12205

Larson, E. L., White, T. A., Ross, C. L., & Harrison, R. G. (2014). Gene flow and the maintenance of species boundaries. *Molecular Ecology, 23*(7), 1668–1678. https://doi.org/10.1111/mec.12601

Lawniczak, M. K. N., Emrich, S. J., Holloway, A. K., Regier, A. P., Olson, M., & White, B, ... Besansky, N. J. (2010). Widespread divergence between incipient Anopheles gambiae species revealed by whole genome sequences. *Science (New York, N.Y.), 330*(6003), 512–514. https://doi.org/10.1126/science.1195755

Liu, K. J., Dai, J., Truong, K., Song, Y., Kohn, M. H., & Nakhleh, L. (2014). An HMM-Based Comparative Genomic Framework for Detecting Introgression in Eukaryotes. *PLoS Computational Biology, 10*(6), e1003649. https://doi.org/10.1371/journal.pcbi.1003649

Martin, S. H., Dasmahapatra, K. K., Nadeau, N. J., Salazar, C., Walters, J. R., Simpson, F., ... Jiggins, C. D. (2013). Genome-wide evidence for speciation with gene flow in Heliconius butterflies. *Genome Research, 23*(11), 1817–1828. https://doi.org/10.1101/gr.159426.113

Molineaux, L., & Gramiccia, G. (1980). *The Garki Project. Research on the epidemiology and control of malaria in the sudan savanna of West Africa.* Geneva: World Health Organization, 311 pp. ISBN: 9241560614, 9789241560610.

Mwangangi, J. M., Mbogo, C. M., Orindi, B. O., Muturi, E. J., Midega, J. T., Nzovu, J., ... Beier, J. C. (2013). Shifts in malaria vector species composition and transmission dynamics along the Kenyan coast over the past 20 years. *Malaria Journal, 12*, 13. https://doi.org/10.1186/1475-2875-12-13

Nachman, M. W., & Payseur, B. A. (2012). Recombination rate variation and speciation : Theoretical predictions and empirical results from rabbits and mice. *Philosophical Transactions of the Royal Society B: Biological Sciences, 367*, 409–421. https://doi.org/10.1098/rstb.2011.0249

Nadeau, N. J., Whibley, A., Jones, R. T., Davey, J. W., Dasmahapatra, K. K., & Baxter, S. W, ... Jiggins, C. D.. (2012). Genomic islands of divergence in hybridizing Heliconius butterflies identified by large-scale targeted sequencing, *Philosophical Transactions of the Royal Society of London. Series B, Biological Sciences, 367*(1587), 343–353.

Ndo, C., Antonio-Nkondjio, C., Cohuet, A., Ayala, D., Kengne, P., Morlais, I., ... Simard, F. (2010). Population genetic structure of the malaria vector Anopheles nili in sub-Saharan Africa. *Malaria Journal, 9*, 161. https://doi.org/10.1186/1475-2875-9-161

Ndo, C., Simard, F., Kengne, P., Awono-Ambene, P., Morlais, I., Sharakhov, I., ... Antonio-Nkondjio, C. (2013). Cryptic Genetic Diversity within the Anopheles nili group of Malaria Vectors in the Equatorial Forest Area of Cameroon (Central Africa). *PLoS One, 8*(3), 1–12. https://doi.org/10.1371/journal.pone.0058862

Neafsey, D. E., Lawniczak, M. K. N., Park, D. J., Redmond, S. N., Coulibaly, M. B., Traoré, S. F., ... Muskavitch, M. A. T. (2010). SNP genotyping defines complex gene-flow boundaries among African malaria vector mosquitoes. *Science (New York, N.Y.), 330*(6003), 514–517. https://doi.org/10.1126/science.1193036

Noor, M. A. F., & Bennett, S. M. (2009). Islands of speciation or mirages in the desert? Examining the role of restricted recombination in maintaining species *Heredity, 103*(6), 439–444. https://doi.org/10.1038/hdy.2010.13

Norris, L. C., Main, B. J., Lee, Y., Collier, T. C., Fofana, A., Cornel, A. J., & Lanzaro, G. C. (2015). Adaptive introgression in an African malaria mosquito coincident with the increased usage of insecticide-treated bed nets. *Proceedings of the National Academy of Sciences, 2014*, 18892. https://doi.org/10.1073/pnas.1418892112

Nosil, P., & Feder, J. L. (2012). Widespread yet heterogeneous genomic divergence. *Molecular Ecology, 21*(12), 2829–2832. https://doi.org/10.1111/j.1365-294x.2012.05580.x

O'Loughlin, S. M., Magesa, S., Mbogo, C., Mosha, F., Midega, J., Lomas, S., & Burt, A. (2014). Genomic analyses of three malaria vectors reveals extensive shared polymorphism but contrasting population histories. *Molecular Biology and Evolution, 1*, 14. https://doi.org/10.1093/molbev/msu040

Paradis, E., Claude, J., & Strimmer, K. (2004). Analyses of Phylogenetics and Evolution in R language. *Bioinformatics, 20*(2), 289–290.

Parchman, T. L., Gompert, Z., Braun, M. J., Brumfield, R. T., McDonald, D. B., Uy, J. A. C., ... Buerkle, C. A. (2013). The genomic consequences of adaptive divergence and reproductive isolation between species of manakins. *Molecular Ecology, 22*(12), 3304–3317. https://doi.org/10.1111/mec.12201

Patterson, N., Moorjani, P., Luo, Y., Mallick, S., Rohland, N., Zhan, Y., ... Reich, D. (2012). Ancient admixture in human history. *Genetics, 192*(3), 1065–1093. https://doi.org/10.1534/genetics.112.145037

Peery, A., Sharakhova, M. V, Antonio-Nkondjio, C., Ndo, C., Weill, M., Simard, F., & Sharakhov, I. V. (2011). Improving the population genetics toolbox for the study of the African malaria vector Anopheles nili: Microsatellite mapping to chromosomes. *Parasites & Vectors, 4*(1), 202. https://doi.org/10.1186/1756-3305-4-202

Peterson, B. K., Weber, J. N., Kay, E. H., Fisher, H. S., & Hoekstra, H. E. (2012). Double Digest RADseq: An Inexpensive Method for De Novo SNP Discovery and Genotyping in Model and Non-Model Species. *PLoS One, 7*(5), e37135.https://doi.org/10.1371/journal.pone.0037135

Pritchard, J. K., Stephens, M., & Donnelly, P. (2000). Inference of population structure using multilocus genotype data. *Genetics, 155*, 945–959.

Purcell, S., Neale, B., Todd-Brown, K., Thomas, L., Ferreira, M., Bender, D., ... Sham, P. (2007). PLINK: A toolset for whole-genome association and population-based linkage analysis. *American Journal of Human Genetics, 81*(3), 559–575.

R Development Core Team (2016). *R: A language and environment for statistical computing*. Vienna, Austria: R Foundation for Statistical Computing.

Ranson, H., & Lissenden, N. (2016). Insecticide Resistance in African Anopheles Mosquitoes: A Worsening Situation that Needs Urgent Action to Maintain Malaria Control. *Trends in Parasitology, 32*(3), 187–196. https://doi.org/10.1016/j.pt.2015.11.010

Riehle, M. M., Guelbeogo, W. M., Gneme, A., Eiglmeier, K., Holm, I., Bischoff, E., ... Vernick, K. D. (2011). A cryptic subgroup of Anopheles gambiae is highly susceptible to human malaria parasites. *Science (New York, N.Y.), 331*(6017), 596–598. https://doi.org/10.1126/science.1196759

Rieseberg, L. H., Whitton, J., & Gardner, K. (1999). Hybrid Zones and the Genetic Architecture of a Barrier to Gene Flow Between Two Sunflower Species.

Roesti, M., Hendry, A. P., Salzburger, W., & Berner, D. (2012). Genome divergence during evolutionary diversification as revealed in replicate lake-stream stickleback population pairs. *Molecular Ecology, 21*(12), 2852–2862. https://doi.org/10.1111/j.1365-294x.2012.05509.x

Service, M. W. (1993). *Mosquito ecology: Field sampling methods.* London: UK Elsevier Applied Science, Ed.

Sharakhova, M. V., Peery, A., Antonio-Nkondjio, C., Xia, A., Ndo, C., Awono-Ambene, P., ... Sharakhov, I. V. (2013). Cytogenetic analysis of Anopheles ovengensis revealed high structural divergence of chromosomes in the Anopheles nili group. *Infection, Genetics and Evolution, 16*, 341–348. https://doi.org/10.1016/j.meegid.2013.03.010

Sinka, M. E., Bangs, M. J., Manguin, S., Coetzee, M., Mbogo, C. M., Hemingway, J., ... Hay, S. I. (2010). The dominant Anopheles vectors of human malaria in Africa, Europe and the Middle East: Occurrence data, distribution maps and bionomic précis. *Parasites & Vectors, 3*(1), 117. https://doi.org/10.1186/1756-3305-3-117

Taylor, S. A., Curry, R. L., White, T. A., Ferretti, V., & Lovette, I. (2014). Spatiotemporally consistent genomic signatures of reproductive isolation in a moving hybrid zone. *Evolution, 68*(11), 3066–3081. https://doi.org/10.1111/evo.12510

Turner, T. L., Hahn, M. W., & Nuzhdin, S. V. (2005). Genomic islands of speciation in Anopheles gambiae. *PLoS Biology, 3*(9), 1572–1578. https://doi.org/10.1371/journal.pbio.0030285

Van Bortel, W., Harbach, R. E., Trung, H. D., Roelants, P., Backeljau, T., & Coosemans, M. (2001). Confirmation of Anopheles varuna in vietnam, previously misidentified and mistargeted as the malaria vector Anopheles minimus, *The American Journal of Tropical Medicine and Hygiene, 65*(6), 729–732.

Weng, D., Yu, Y., Hahn, M. W., & Nakhleh, L. (2016). Reticulate evolutionary history and extensive introgression in mosquito species revealed by phylogenetic network analysis. *Molecular Ecology, 25*(11), 2361–2372.

WHO. (2016). *World malaria report 2016.* Geneva: WHO

Wright, S. (1984). *Evolution and the genetics of populations: Genetics and biometric foundations v. 1.* Chicago, IL: University of Chicago. 480 pp. ISBN-13: 978-0226910383.

Wu, C. (2001). The genic view of the process of speciation. *Journal of Evolutionary Biology, 14*, 851–865.

Wu, T. D., & Nacu, S. (2010). Fast and SNP-tolerant detection of complex variants and splicing in short reads. *Bioinformatics, 26*(7), 873–881.

Permissions

List of Contributors

Lindsay Chaney
Plant and Wildlife Sciences, Brigham Young University, Provo, UT, USA

Bryce A. Richardson
USDA Forest Service, Rocky Mountain Research Station, Provo, UT, USA

Matthew J. Germino
U.S. Geological Survey, Forest and Rangeland Ecosystem Science Center, Boise, ID, USA

Christopher T. Frye
Natural Heritage Program, Maryland Department of Natural Resources, Wildlife and Heritage Service, Wye Mills, MD, USA
Department of Plant Science and Landscape Architecture, University of Maryland, College Park, MD, USA

Maile C. Neel
Department of Plant Science and Landscape Architecture and Department of Entomology, University of Maryland, College Park, MD, USA

Min A. Hahn
Department of Botany and Biodiversity Research Centre, University of British Columbia, Vancouver, BC, Canada

Loren H. Rieseberg
Department of Botany and Biodiversity Research Centre, University of British Columbia, Vancouver, BC, Canada
Department of Biology, Indiana University, Bloomington, IN, USA

Ulf Dieckmann
Department of Forest Genetics, Federal Research and Training Centre for Forests, Natural Hazards and Landscape, Vienna, Austria

Stefan Kapeller
Department of Forest Genetics, Federal Research and Training Centre for Forests, Natural Hazards and Landscape, Vienna, Austria
Evolution and Ecology Program, International Institute for Applied Systems Analysis, Laxenburg, Austria

Silvio Schueler
Evolution and Ecology Program, International Institute for Applied Systems Analysis, Laxenburg, Austria

Michael E. Donaldson
Environmental and Life Sciences Graduate Program, Trent University, Peterborough, ON, Canada

Christina M. Davy
Environmental and Life Sciences Graduate Program, Trent University, Peterborough, ON, Canada
Wildlife Research and Monitoring Section, Ontario Ministry of Natural Resources and Forestry, Peterborough, ON, Canada

Craig K. R. Willis
Department of Biology and Centre for Forest Interdisciplinary Research (C-FIR), University of Winnipeg, Winnipeg, MB, Canada

Scott McBurney and Allysia Park
Canadian Wildlife Health Cooperative, Atlantic Region, Atlantic Veterinary College, University of Prince Edward Island, Charlottetown, PEI, Canada

Christopher J. Kyle
Forensic Science Department, Trent University, Peterborough, ON, Canada

Jakob Thyrring and Mikael Kristian Sejr
Department of Bioscience, Arctic Research Centre, Aarhus University, Aarhus C, Denmark

Sofie Smedegaard Mathiesen
Department of Bioscience, Arctic Research Centre, Aarhus University, Aarhus C, Denmark
Section for Marine Living Resources, National Institute of Aquatic Resources, Technical University of Denmark, Silkeborg, Denmark

Einar Eg Nielsen and Jakob Hemmer-Hansen
Section for Marine Living Resources, National Institute of Aquatic Resources, Technical University of Denmark, Silkeborg, Denmark

Peter Leopold
Faculty of Biosciences, Fisheries and Economics, UiT The Arctic University of Norway, Tromsø, Norway

Jørgen Berge
Faculty of Biosciences, Fisheries and Economics, UiT The Arctic University of Norway, Tromsø, Norway
The University Centre in Svalbard, Longyearbyen, Norway

Alexey Sukhotin
White Sea Biological Station, Zoological Institute of Russian Academy of Sciences, St.Petersburg, Russia Invertebrate Zoology Department, St. Petersburg State University, St. Petersburg, Russia

Michaël Bekaert
Institute of Aquaculture, University of Stirling, Stirling, UK

Alexandra Pavlova, Annika M. Lamb and Paul Sunnucks
School of Biological Sciences, Clayton Campus, Monash University, Clayton, VIC, Australia

Katherine A. Harrisson
School of Biological Sciences, Clayton Campus, Monash University, Clayton, VIC, Australia
Department of Environment, Land Water and Planning, Arthur Rylah Institute, Land, Fire and Environment, Heidelberg, VIC, Australia
Department of Ecology Environment and Evolution, School of Life Sciences, La Trobe University, Bundoora, Victoria, 3083, Australia

Luciano B. Beheregaray and Minami Sasaki
School of Biological Sciences, Flinders University, Adelaide, SA, Australia

Rhys Coleman
Applied Research, Melbourne Water, Docklands, VIC, Australia

Dean Gilligan
Freshwater Ecosystems Research, NSW Department of Primary Industries – Fisheries, Batemans Bay, NSW, Australia

Joanne Kearns, Jarod Lyon and Zeb Tonkin
Department of Environment, Land Water and Planning, Arthur Rylah Institute, Land, Fire and Environment, Heidelberg, VIC, Australia

Brett A. Ingram
Department of Economic Development, Jobs, Transport and Resources, Fisheries Victoria, Alexandra, VIC, Australia

Mark Lintermans
Institute for Applied Ecology, University of Canberra, Canberra, ACT, Australia

Thuy T. T. Nguyen
Agriculture Victoria, AgriBio, Centre for AgriBioscience, Bundoora, VIC, Australia

Jian D. L. Yen
School of Physics and Astronomy, Clayton Campus, Monash University, Clayton, VIC, Australia

Jonathan L. Richardson and Mary K. Burak
Department of Biology, Providence College, Providence, RI, USA

Christian Hernandez, James M. Shirvell, Carol Mariani and Adalgisa Caccone
Department of Ecology and Evolutionary Biology, Yale University, New Haven, CT, USA

Ticiana S. A. Carvalho-Pereira, Arsinoê C. Pertile, Jesus A. Panti-May, Gabriel G. Pedra, Soledad Serrano, Josh Taylor and Mayara Carvalho
Centro de Pesquisas Gonçalo Moniz, Fundação Oswaldo Cruz, Ministério da Saúde, Salvador, Brazil

Albert I. Ko
Centro de Pesquisas Gonçalo Moniz, Fundação Oswaldo Cruz, Ministério da Saúde, Salvador, Brazil
Department of Epidemiology of Microbial Disease, Yale School of Public Health, New Haven, CT, USA

Gorete Rodrigues
Centro de Controle de Zoonoses, Secretaria Municipal de Saúde, Ministério da Saúde, Salvador, Brazil

Federico Costa
Instituto de Saúde Coletiva, Universidade Federal da Bahia, UFBA, Salvador, Brazil

James E. Childs
Department of Epidemiology of Microbial Disease, Yale School of Public Health, New Haven, CT, USA

Marta Robertson and Christina Richards
Department of Integrative Biology, University of South Florida, Tampa, FL, USA

Aaron Schrey
Department of Biology, Armstrong State University, Savannah, GA, USA

Ashley Shayter
Rehabilitation Institute, Southern Illinois University, Carbondale, IL, USA

Christina J Moss
Department of Cell Biology, Microbiology and Molecular Biology, University of South Florida, Tampa, FL, USA

Pierre Saumitou-Laprade, Philippe Vernet, Xavier Vekemans, Sylvain Billiard and Sophie Gallina
CNRS, UMR 8198 Evo-Eco-Paleo, Université de Lille - Sciences et Technologies, Villeneuve d'Ascq, France

Laila Essalouh
Montpellier SupAgro, UMR 1334 AGAP, Montpellier, France

Ali Mhaïs
Montpellier SupAgro, UMR 1334 AGAP, Montpellier, France
INRA, UR Amélioration des Plantes, Marrakech, Morocco
Laboratoire AgroBiotechL02B005, Faculté des Sciences et Techniques Guéliz, University Cadi Ayyad, Marrakech, Morocco

Bouchaïb Khadari
Montpellier SupAgro, UMR 1334 AGAP, Montpellier, France
INRA/CBNMed, UMR 1334 Amélioration Génétique et Adaptation des Plantes (AGAP), Montpellier, France

Abdelmajid Moukhli
INRA, UR Amélioration des Plantes, Marrakech, Morocco

Ahmed El Bakkali
INRA, UR Amélioration des Plantes et Conservation des Ressources Phytogénétiques, Meknès, Morocco

Gianni Barcaccia
Laboratory of Genomics and Plant Breeding, DAFNAE - University of Padova, Legnaro, PD, Italy

Fiammetta Alagna
Research Unit for Table Grapes and Wine Growing in Mediterranean Environment, CREA, Turi, BA, Italy
CNR, Institute of Biosciences and Bioresources, Perugia, Italy

Roberto Mariotti, Nicolò G. M. Cultrera, Saverio Pandolfi, Martina Rossi and Luciana Baldoni
CNR, Institute of Biosciences and Bioresources, Perugia, Italy

Kyle W. Wellband
Great Lakes Institute for Environmental Research, University of Windsor, Windsor, ON, Canada

Daniel D. Heath
Great Lakes Institute for Environmental Research, University of Windsor, Windsor, ON, Canada
Department of Biological Sciences, University of Windsor, Windsor, ON, Canada

K. Mathias Wegner
Wadden Sea Station Sylt, Alfred Wegener Institute, Helmholtz Centre for Polar and Marine Research, List, Germany

Carolin C. Wendling
Wadden Sea Station Sylt, Alfred Wegener Institute, Helmholtz Centre for Polar and Marine Research, List, Germany
GEOMAR, Helmholtz Centre for Ocean Research, Kiel, Germany

Armin G. Fabritzek
Department of Ecology, Institute of Zoology, Johannes Gutenberg-University of Mainz, Mainz, Germany

Cong Zeng
College of Animal Science and Technology, Hunan Agricultural University, Changsha, China
School of Biological Sciences, Victoria University of Wellington, Wellington, New Zealand
National Institute for Water and Atmospheric Research, Kilbirnie, Wellington, New Zealand

Jonathan P. A. Gardner
School of Biological Sciences, Victoria University of Wellington, Wellington, New Zealand

Ashley A. Rowden and Malcolm R. Clark
National Institute for Water and Atmospheric Research, Kilbirnie, Wellington, New Zealand

Atal Saha, Torild Johansen and Jon-Ivar Westgaard
Tromsø Department, Institute of Marine Research, Tromsø, Norway

Benjamin Planque
Tromsø Department, Institute of Marine Research, Tromsø, Norway
Hjort Centre for Marine Ecosystem Dynamics, Bergen, Norway

Rasmus Hedeholm
Greenland Institute of Natural Resources, Nuuk, Greenland

Jesper Boje
Greenland Institute of Natural Resources, Nuuk, Greenland
DTU Aqua – National Institute of Aquatic Resources, Charlottenlund, Denmark

Einar E. Nielsen
DTU Aqua – National Institute of Aquatic Resources, Charlottenlund, Denmark

Lorenz Hauser
School of Aquatic and Fishery Sciences, University of Washington, Seattle, WA, USA

Steven X. Cadrin
School for Marine Science and Technology, University of Massachusetts Darmouth, Fairhaven, MA, USA

Elizabeth Karn and Marie Jasieniuk
University of California Davis, Department of Plant Sciences, Davis, CA, USA

Caroline Fouet, Colince Kamdem and Stephanie Gamez
Department of Entomology, University of California, Riverside, CA, USA

Bradley J. White
Department of Entomology, University of California, Riverside, CA, USA
Center for Disease Vector Research, Institute for Integrative Genome Biology, University of California, Riverside, CA, USA

Index

www.ingramcontent.com/pod-product-compliance
Lightning Source LLC
Chambersburg PA
CBHW080653200326
41458CB00013B/4844